Remote Sensing and Digi

AF213047

Volume 22

More information about this series at http://www.springer.com/series/6477

Claudia Kuenzer • Stefan Dech
Wolfgang Wagner
Editors

Remote Sensing Time Series

Revealing Land Surface Dynamics

Springer

Editors
Claudia Kuenzer
German Remote Sensing Data Center, DFD
German Aerospace Center, DLR
Wessling, Germany

Stefan Dech
German Remote Sensing Data Center, DFD
German Aerospace Center, DLR
Wessling, Germany

Institute for Geography and Geology
University of Wuerzburg
Wuerzburg, Germany

Wolfgang Wagner
Department of Geodesy and Geoinformation
Vienna University of Technology
Vienna, Austria

Responsible Series Editor: A. Marçal

ISSN 1567-3200 ISSN 2215-1842 (electronic)
Remote Sensing and Digital Image Processing
ISBN 978-3-319-35243-5 ISBN 978-3-319-15967-6 (eBook)
DOI 10.1007/978-3-319-15967-6

Springer Cham Heidelberg New York Dordrecht London

Softcover reprint of the hardcover 1st edition 2015

Printed on acid-free paper

Springer International Publishing AG Switzerland is part of Springer Science+Business Media (www.springer.com)

Foreword

Our fragile planet experiences global change at unprecedented speed. Much of this change is man-made. We harness rivers and lakes, clear-cut forests, transfer natural ecosystems into agricultural land, and extract underground resources. Through urbanization as well as socio-economic transformation more and more pristine habitats come under pressure. Changes in climate variability impact the dynamics of snow cover and water bodies, soil moisture and vegetation phenology.

Satellite-based earth observation technology allows us to monitor and quantify these changes. Satellite remote sensing – and here especially the analysis of long-term time series – enables us to reveal land surface dynamics that otherwise might remain hidden to the human eye. The book *Remote Sensing Time Series Revealing Land Surface Dynamics* focuses on exactly this potential of space-borne earth observation.

What can earth observation contribute to the understanding of global change? Which satellite sensors exist? Which data really allow for long-term monitoring and time-series analysis? When is a time series long enough to shed light on climate variability? Which challenges face scientists who use remote sensing satellite data to further knowledge about our planet? How do different ecosystems change over time?

This book, which has been initiated by scientists of DLR's German Remote Sensing Data Center (DFD), addresses all these questions. Experts from all over Europe, the USA, and China have contributed to this comprehensive volume.

In recent years many satellite data archives have been made available to the public. The USA made nearly 40 years of Landsat data accessible free of charge. Data archives like this one, which allow us to look into the past for several decades, are of immense value. Medium resolution data collected by the US AVHRR sensor have also been available for several decades. Additionally, since the year 2000, MODIS data also allow us to monitor our entire planet at daily intervals. Europe has followed this lead to facilitate satellite data access. Free optical, thermal infrared, and radar data from sensors on board the ESA research satellite ENVISAT, enabled dense, multifaceted analysis of the land surface between 2002 and 2012. Furthermore, data of the novel and upcoming European Sentinel missions operated by the European Space Agency on behalf of the European Union are currently being made available free of charge both to scientific and commercial users.

Amongst other institutions, our German Remote Sensing Data Center will provide and operate the processing and archiving facilities for the data from Sentinel-1 C-band SAR sensors that grants continuity for ENVISAT-ASAR, as well as for Sentinel-3 OLCI data, continuing the ENVISAT MERIS instrument. Sentinel-5 precursor data will also be processed and archived at DFD.

However, the large amount of earth observation data contained in satellite data archives globally also poses great challenges for the science community. The analysis of time series of data is much more complex than just comparing a few multitemporal satellite scenes. Time-series analysis requires the processing of hundreds, thousands, or even a hundreds of thousands of data sets. This "big data" needs to be calibrated, preprocessed, harmonized, interpolated, and statistically analyzed. At frequent intervals – monthly or annually – time series have to be re-processed to derive updated mean, minima, maxima, variability and anomalies. This is an extremely demanding task. Additionally, the challenge of big data and the challenge to fully exploit all the wealth of information that is contained in data archives are getting bigger every day. More and more space nations launch satellites into orbit. The life span of sensors increases. And data access is eased further. The computer and programming literacy of scientists and young people in general is rapidly increasing. Envisioning this trend, one of our technical answers is that algorithms should rather come to the data instead of routing mass-data archives to the analysts. So, hopefully, as the challenges grow, so will our means to address them. I am confident that we will be able to manage the upcoming challenges, and that the European Copernicus program will help to blaze the trail.

I hope that this book will trigger or deepen your interest in remote sensing time-series analysis as a valuable means to assess the state of our Planet Earth.

I wish you stimulating reading.

Prof. Dr.-Ing. Johann-Dietrich Wörner
Chairman of the Executive Board
German Aerospace Center DLR
Cologne, Germany

Acknowledgements

The chapters presented in this book were contributed by renowned scientists in the field of remote sensing time series analysis from around the world. This work would not have been possible without each author's voluntary and enthusiastic commitment to this book project. We thank all authors for the very smooth process of chapter compilation.

We are extremely grateful to the following experts who have volunteered to peer-review the chapters of this book. Investing their time and knowledge, the reviewers significantly helped to shape the final chapters with their critical comments and suggestions.

Alphabetically, we express our sincere thanks to:

Terry Arvidson (Lockheed Martin, Greenbelt, MD, USA); Kirsten de Beurs (Department of Geography, University of Oklahoma, Norman, OK, USA); Christopher Conrad (Institute of Geography and Geology, University of Würzburg, Würzburg, Germany); Stefan Dech (German Aerospace Center, DLR, Earth Observation Center, EOC, German Remote Sensing Data Center, DFD, Oberpfaffenhofen, Germany); Andreas Dietz (German Aerospace Center, DLR, Earth Observation Center, EOC, German Remote Sensing Data Center, DFD, Oberpfaffenhofen, Germany); Jeffery Eidenshink (Earth Resources Observation and Science, EROS, U.S. Geological Survey, Sioux Falls, SD, USA); Christina Eisfelder (German Aerospace Center, DLR, Earth Observation Center, EOC, German Remote Sensing Data Center, DFD, Oberpfaffenhofen, Germany); Ursula Gessner (German Aerospace Center, DLR, Earth Observation Center, EOC, German Remote Sensing Data Center, DFD, Oberpfaffenhofen, Germany); Huadong Guo (Key Laboratory of Digital Earth Sciences, Institute of Remote Sensing and Digital Earth, RADI, Chinese Academy of Sciences, CAS, Beijing, China); Xulin Guo (Department of Geography and Planning, University of Saskatchewan, Canada); Geoffrey M. Henebry (Geospatial Sciences Center of Excellence, GSCE, South Dakota State University, Brooking, SD, USA); Martin Herold (Wageningen University and Research Centre, WUR, Laboratory of Geo-information Science and Remote Sensing, Wageningen, The Netherlands); Koen Hufkens (Department of Organismic & Evolutionary Biology, Harvard University, Cambridge, MA, USA); Juliane Huth (German Aerospace Center, DLR, Earth Observation Center, EOC, German Remote Sensing Data Center, DFD, Oberpfaffenhofen, Germany); Igor Klein (German Aerospace Center, DLR, Earth Observation Center, EOC, German Remote Sensing Data Center, DFD, Oberpfaffenhofen, Germany); Kim Knauer (German Aerospace Center, DLR, Earth Observation Center, EOC, German Remote Sensing Data Center, DFD, Oberpfaffenhofen, Germany); Patrick Leinenkugel (German Aerospace Center, DLR, Earth Observation Center, EOC, German Remote Sensing Data Center, DFD, Oberpfaffenhofen, Germany); Sebastian van der Linden (Geographisches Institut, Humbold-Universität Berlin, Berlin, Germany); V. E. Nethaji Mariappan (Centre for Remote Sensing and Geoinformatics, Sathyabama University, Chennai, India); Markus Niklaus (German Aerospace Center, DLR, Earth Observation Center, EOC, German Remote Sensing Data Center, DFD, Oberpfaffenhofen, Germany); Marco Ottinger (German Aerospace Center, DLR, Earth Observation Center, EOC, German Remote Sensing Data Center, DFD, Oberpfaffenhofen, Germany); Stuard Phinn (School of Geography, Planning and Environmental Management, University of Queensland, Brisbane, Australia); Andreas Rienow (Department of Geography, University Bonn, Bonn, Germany); Achim Röder (Umweltfernerkundung und Geoinformatik, Universität Trier, Germany); David Roy (Geospatial Science Center of Excellence, South Dakota State University, Brookings, SD, USA); Mauritio Santoro (GAMMA Remote Sensing Research and Consulting AG, Gümligen, Switzerland); Joseph O. Sexton (Global Land Cover Facility, University of Maryland, MD, USA); Pascal Sirguey (School of Surveying, University of Otago, Dunedin, New Zealand); Ruth Sonnenschein (Institute for Applied Remote sensing, European Academy of Bolzano/Bozen, EURAC, Bolzano, Italy); Wolfgang Wagner (Department of Geodesy and Geoinformation, Vienna University of Technology, Vienna, Austria); Jun Wang (Beijing University, China).

Special thanks go to Susan Giegerich for her support in proofreading and language editing selected chapters, as well as to Nils Sparwasser (Science Visualization and Geoinformation, DLR, EOC, DFD) for designing the title page.

Furthermore, we are especially grateful to Philipp Koch for his excellent cooperation and editing work during the preparation of this volume.

Our gratitude also goes to the European Association of Remote Sensing Laboratories (EARSeL) for supporting this book project.

Contents

Contributors

Martin Bachmann German Remote Sensing Data Center, DFD, Earth Observation Center, EOC, German Aerospace Center, DLR, Oberpfaffenhofen, Germany

Stefan Dech German Remote Sensing Data Center, DFD, German Aerospace Center, DLR, Wessling, Germany

Institute for Geography and Geology, University of Wuerzburg, Wuerzburg, Germany

Andreas J. Dietz German Remote Sensing Data Center, DFD, Earth Observation Center, EOC, German Aerospace Center, DLR, Oberpfaffenhofen, Germany

Christina Eisfelder German Remote Sensing Data Center, DFD, Earth Observation Center, EOC, German Aerospace Center, DLR, Oberpfaffenhofen, Germany

Andrea Ehammer Section of Geography, Department of Geosciences and Natural Resource Management, Faculty of Science, University of Copenhagen, Copenhagen K, Denmark

Lars Eklundh Department of Physical Geography and Ecosystem Science, Lund University, Lund, Sweden

Rasmus Fensholt Section of Geography, Department of Geosciences and Natural Resource Management, Faculty of Science, University of Copenhagen, Copenhagen K, Denmark

Corinne Myrtha Frey German Remote Sensing Data Center, DFD, Earth Observation Center, EOC, German Aerospace Center, DLR, Oberpfaffenhofen, Germany

Ursula Gessner German Remote Sensing Data Center, DFD, Earth Observation Center, EOC, German Aerospace Center, DLR, Oberpfaffenhofen, Germany

Felix Greifeneder European Academy of Bozen/Bolzano, EURAC, Bolzano-Bozen, Italy

Patrick Griffiths Geography Department, Humboldt-Universität zu Berlin, Berlin, Germany

Kenneth Grogan Section of Geography, Department of Geosciences and Natural Resource Management, Faculty of Science, University of Copenhagen, Copenhagen K, Denmark

Huadong Guo Key Laboratory of Digital Earth Sciences, Institute of Remote Sensing and Digital Earth, RADI, Chinese Academy of Sciences, CAS, Beijing, China

Patrick Hostert Geography Department, Humboldt-Universität zu Berlin, Berlin, Germany

IRI THESys, Humboldt-Universität zu Berlin, Berlin, Germany

Stephanie Horion Section of Geography, Department of Geosciences and Natural Resource Management, Faculty of Science, University of Copenhagen, Copenhagen K, Denmark

Silvia Huber DHI GRAS, Copenhagen K, Denmark

Juliane Huth German Remote Sensing Data Center, DFD, Earth Observation Center, EOC, German Aerospace Center, DLR, Oberpfaffenhofen, Germany

Per Jönsson Group for Materials Science and Applied Mathematics, Malmö University, Malmö, Sweden

Igor Klein German Remote Sensing Data Center, DFD, Earth Observation Center, EOC, German Aerospace Center, DLR, Oberpfaffenhofen, Germany

Kim Knauer German Remote Sensing Data Center, DFD, Earth Observation Center, EOC, German Aerospace Center, DLR, Oberpfaffenhofen, Germany

Remote Sensing, Institute of Geology and Geography, University of Wuerzburg, Wuerzburg, Germany

Indrani Kommareddy Geospatial Science Center of Excellence, South Dakota State University, Brookings, SD, USA

Valeriy Kovalskyy Geospatial Science Center of Excellence, South Dakota State University, Brookings, SD, USA

Claudia Kuenzer German Remote Sensing Data Center, DFD, Earth Observation Center, EOC, German Aerospace Center, DLR, Oberpfaffenhofen, Germany

Xinwu Li Key Laboratory of Digital Earth Sciences, Institute of Remote Sensing and Digital Earth, RADI, Chinese Academy of Sciences, CAS, Beijing, China

Lei Liang Key Laboratory of Digital Earth Sciences, Institute of Remote Sensing and Digital Earth, RADI, Chinese Academy of Sciences, CAS, Beijing, China

Linlin Lu Key Laboratory of Digital Earth Sciences, Institute of Remote Sensing and Digital Earth, RADI, Chinese Academy of Sciences, CAS, Beijing, China

Sandro Martinis German Remote Sensing Data Center, DFD, Earth Observation Center, EOC, German Aerospace Center, DLR, Oberpfaffenhofen, Germany

Vahid Naeimi Department of Geodesy and Geoinformation, Vienna University of Technology, Vienna, Austria

Markus Niklaus German Remote Sensing Data Center, DFD, Earth Observation Center, EOC, German Aerospace Center, DLR, Oberpfaffenhofen, Germany

Carsten Pathe Department for Earth Observation, Institute of Geography, Faculty for Chemistry and Geosciences, Friedrich-Schiller University, Jena, Germany

Dirk Pflugmacher Geography Department, Humboldt-Universität zu Berlin, Berlin, Germany

Stephen D. Prince Department of Geographical Sciences, University of Maryland, College Park, MD, USA

Kjeld Rasmussen Section of Geography, Department of Geosciences and Natural Resource Management, Faculty of Science, University of Copenhagen, Copenhagen K, Denmark

Achim Röder Environmental Remote Sensing and Geoinformatics, Faculty of Regional and Environmental Sciences, University of Trier, Trier, Germany

David P. Roy Geospatial Science Center of Excellence, South Dakota State University, Brookings, SD, USA

Thomas Ruppert German Remote Sensing Data Center, DFD, Earth Observation Center, EOC, German Aerospace Center, DLR, Oberpfaffenhofen, Germany

Daniel Sabel Department of Geodesy and Geoinformation, Vienna University of Technology, Vienna, Austria

Maurizio Santoro GAMMA Remote Sensing Research and Consulting AG, Gümligen, Switzerland

Christiane Schmullius Department for Earth Observation, Institute of Geography, Faculty for Chemistry and Geosciences, Friedrich-Schiller University, Jena, Germany

Guozhuang Shen Key Laboratory of Digital Earth Sciences, Institute of Remote Sensing and Digital Earth, RADI, Chinese Academy of Sciences, CAS, Beijing, China

Marion Stellmes Environmental Remote Sensing and Geoinformatics, Faculty of Regional and Environmental Sciences, University of Trier, Trier, Germany

Zhongchang Sun Key Laboratory of Digital Earth Sciences, Institute of Remote Sensing and Digital Earth, RADI, Chinese Academy of Sciences, CAS, Beijing, China

Torbern Tagesson Section of Geography, Department of Geosciences and Natural Resource Management, Faculty of Science, University of Copenhagen, Copenhagen K, Denmark

Christian Thiel Department for Earth Observation, Institute of Geography, Faculty for Chemistry and Geosciences, Friedrich-Schiller University, Jena, Germany

Feng Tian Section of Geography, Department of Geosciences and Natural Resource Management, Faculty of Science, University of Copenhagen, Copenhagen K, Denmark

Compton J. Tucker NASA Goddard Space Flight Center, Greenbelt, MD, USA

Padsuren Tungalagsaikhan German Remote Sensing Data Center, DFD, Earth Observation Center, EOC, German Aerospace Center, DLR, Oberpfaffenhofen, Germany

Thomas Udelhoven Environmental Remote Sensing and Geoinformatics, Faculty of Regional and Environmental Sciences, University of Trier, Trier, Germany

Sebastian van der Linden Geography Department, Humboldt-Universität zu Berlin, Berlin, Germany

IRI THESys, Humboldt-Universität zu Berlin, Berlin, Germany

Jan Verbesselt Laboratory of Geo-Information Science and Remote Sensing, Wageningen University, Wageningen, The Netherlands

Wolfgang Wagner Department of Geodesy and Geoinformation, Vienna University of Technology, Vienna, Austria

Cuizhen Wang Department of Geography, University of South Carolina, Columbia, SC, USA

Meng Wang Key Laboratory of Digital Earth Sciences, Institute of Remote Sensing and Digital Earth, RADI, Chinese Academy of Sciences, CAS, Beijing, China

Wenjin Wu Key Laboratory of Digital Earth Sciences, Institute of Remote Sensing and Digital Earth, RADI, Chinese Academy of Sciences, CAS, Beijing, China

Lin Yan Geospatial Science Center of Excellence, South Dakota State University, Brookings, SD, USA

Shiyong Yan National Earthquake Response Support Service, Shijingshan District, Beijing, China

Huaining Yang National Earthquake Response Support Service, Shijingshan District, Beijing, China

Hankui Zhang Geospatial Science Center of Excellence, South Dakota State University, Brookings, SD, USA

Lu Zhang Key Laboratory of Digital Earth Sciences, Institute of Remote Sensing and Digital Earth, RADI, Chinese Academy of Sciences, CAS, Beijing, China

Abbreviations

a.s.l.	Above sea level
AATSR	Advanced Along-Track Scanning Radiometer
aET	Actual evaporation
AI	Aridity Index
ALOS	Advanced Land Observing Satellite
AMOC	Acoustic Monitoring of the Ocean Climate
ANPP	Aboveground net primary productivity
APOLLO	*A*VHRR *P*rocessing scheme *O*ver c*L*ouds *L*and and *O*cean
ARMA	Autoregressive moving average
ASAR	Advanced Synthetic Aperture Radar
ASCAT	Advanced scatterometer
ASI	Italian Space Agency
AVHRR	Advanced Very High Resolution Radiometer
BETHY/DLR	Biosphere Energy Transfer Hydrology Model
BFAST	Breaks For Additive Seasonal and Trend
BISE	Best index slope extraction
BMBF	German Federal Ministry of Education and Research
BMWI	Federal Ministry for Economic Affairs and Energy
BOA	Bottom of atmosphere
BRDF	Bidirectional reflectance distribution function
BWI	Basin Water Index
CAO	Carnegie Airborne Observatory
CAS	Chinese Academy of Sciences
CBERS	China-Brazil Earth Resources Satellite
CCI	Climate Change Initiative
CEODE	The Center for Earth Observation and Digital Earth
CFV	Closest Feature Vector
CGIAR-CSI	Consultative Group on International Agricultural Research - Consortium for Spatial Information
CLASS	Comprehensive Large Array-data Stewardship System

CMAP	CPC Merged Analysis of Precipitation
CNES	French Space Agency
CONUS	Conterminous United States
Cosmo-SkyMed	Constellation of Small Satellites for Mediterranean Basin Observation
CRU	Climatic Research Unit
CSA	Canadian Space Agency
dB	Decibel
DEM	Digital elevation model
DFD	German Remote Sensing Data Centre
DFF	Danish Council for Independent Research
DFG	Alliance of Science Organisations in Germany
DGVMs	Dynamic Global Vegetation Models
DI	Disturbance Index
DIMS	Data and Information Management System
DInSAR	Differential Interferometric SAR
DLM	Distributed lag-model
DLR	German Aerospace Centre
DN	Digital number
DORIS	Delft object-oriented radar interferometric software
DOY	Day-of-year
DVT	Dynamic visible threshold
EBVs	Essential biodiversity variables
EC	European Commission
ECMWF	European Centre for Medium-Range Weather Forecasts
ECV	Essential climate variable
ENSO	El Nino Southern Oscillation
ENVISAT	Environmental Satellite
ENVISAT/ASAR	Environmental Satellite/Advanced Synthetic Aperture Radar
EO	Earth observation
EOM	Earth observation monitor
EOS	End of season
EOST	End of season time
EROS	Earth resources observation and science
ERS	Earth Resources Satellite
ERS	European remote sensing
ESA	European Space Agency
ESA GMES	ESA Global Monitoring for Environment and Security
ESDB	European Soil Database
ETM	Enhanced Thematic Mapper
ETM+	Enhanced Thematic Mapper Plus
EVI	Enhanced Vegetation Index
FAO	Food and Agriculture Organization

FAPAR	Fraction of Absorbed Photosynthetically Active Radiation
FOMO	Remote sensing of the forest transition and its ecosystem impacts in mountain environments
FOV	Field of view
FPAR	Fraction of photosynthetically active radiation
FVC	Fraction of vegetation cover
FT	Functional types
GCOS	Global Climate Observing System
GDP	Gross domestic product
GEO BON	Earth Observations Biodiversity Observation Network
GIMMS	Global Inventory Modeling and Mapping Studies
GLASS	Global LAnd Surface Satellite
GLC2000	Global Land Cover
GLS	Generalized least square
GLS	Global Land Survey
GLWD	Global Lakes and Wetlands Dataset
GMES	Global Monitoring for Environment and Security
GPCC	Global Precipitation Climatology Centre
GPCP	Global Precipitation Climatology Project
GPP	Gross primary productivity
GSE	ESA GMES Service Element
GSV	Growing stock volume
HANTS	Harmonic Analyses of NDVI Time Series
HDF	Hierarchical data format
HDF-EOS	Hierarchical Data Format–Earth Observing System
HRPT	High resolution picture transmission
HWSD	Harmonized World Soil Database
IFOV	Instantaneous field of view
IGBP	International Geosphere-Biosphere Programme
IIASA	International Institute for Applied Systems Analysis
InSAR	Interferometric Synthetic Aperture Radar
INPE	Brazilian Space Agency
IOD	Indian Ocean Dipole
IPCC	Intergovernmental Panel on Climate Change
JAXA	Japan Aerospace Exploration Agency
JERS	Japanese Earth Resources Satellite
JPL	Jet Propulsion Laboratory
K.T.	Kaboré-Tambi
KML	Keyhole Markup Language
LAI	Leaf Area Index
LandTrendr	Landsat-based Detection of Trends in Disturbance and Recovery
LCC	Lambert conic conformal
LCU	Land capability unit

LDCM	Landsat Data Continuity Mission
LDI	Land Degradation Index
LiDAR	Light detection and ranging
LMB	Lower Mekong Basin
LNS	Local net primary productivity scaling
LSP	Land surface phenology
LST	Land surface temperature
LST_13year	Land surface temperature 13-year average
LTDR	Land Long Term Data Record
LUE	Light use efficiency
LULC	Land Use Land Cover
LUT	Look-up table
MACs	Multi-sensor Airborne Campaigns
MEA	Millennium Ecosystem Assessment
MEI	Multivariate ENSO Index
MB	Mekong Basin
MetOp	Meteorological Operational Satellites
MERIS	Medium resolution imaging spectrometer
MGET	Marine Geospatial Ecology Tools
MODIS	Moderate-resolution imaging spectroradiometer
MODIS NBAR	MODIS Nadir bidirectional reflectance distribution function adjusted reflectance
MODIS QA	MODIS Quality assessment
MODSCAGvMODIS	Snow-covered area and grain size
MSS	Multispectral Scanner System
MTCI	MERIS Terrestrial Chlorophyll Index
MTInSAR	Multi-Temporal Interferometric Synthetic Aperture Radar
MVA	Mean annual variation
MVIs	Microwave Vis
NAO	North Atlantic Oscillation
NASA	National Aeronautics and Space Administration
NBR	Normalized Burn Ratio
NDSI	Normalized Difference Snow Index
NDVI	Normalized Difference Vegetation Index
NDWI	Normalised Difference Water Index
NEODASS	Earth Observation Data Acquisition and Analysis Service
NEODC	NERC Earth Observation Data Centre
NERC	Natural Environment Research Council
NESDIS	National Environmental Satellite, Data, and Information Service
NEST	Next ESA SAR Toolbox
NEX	NASA Earth Exchange
NIR	Near-infrared
NIR	Surface reflectances in the near infrared

NLCD	National Land Cover Database
NOAA	National Oceanic and Atmospheric Administration
NORAD	North American Aerospace Defense Command
NPP	Net primary production
NPP	Suomi National Polar-orbiting Partnership
NRSC	Normalized radar cross-section
NSCAT	NASA scatterometer
NSIDC	National Snow and Ice Data Center
OA	Overall accuracy
OA	Overall agreement
OK	Over kappa
OLI	Operational Land Imager
OLS	Ordinary least-square
OSO	Office of Satellite Operations
OSPO	Satellite and Product Operations
PALSAR	Phased Array type L-band Synthetic Aperture Radar
PAR	Photosynthetically active radiation
PATMOS-x	AVHRR Pathfinder Atmospheres Extended
PDO	Pacific Decadal Oscillation
PEM	Production efficiency modelling
POES	Polar Orbiting Environmental Satellites
PolInSAR	Polarimetric Synthetic Aperture Radar Interferometry
PPI	Plant Phenology Index
PPS	Precipitation Processing System
PS	Permanent scatterers
PSI	Persistent scatterer interferometry
QuickSCAT	Quick Scatterometer
RADI	Institute of Remote Sensing and Digital Earth
RBSI	Radar Backscatter Index
RCM	Radar Satellite Constellation
RED	Surface reflectances in the red
REDD	Reducing Emissions from Deforestation and Forest Degradation in Developing Countries
RESTREND	Residual Trend Analysis
RF	Random Forest
RLOS	Radar line of sight
RMSE	Root mean square error
ROIs	Regions of interest
RUE	Rain-use efficiency
SAFs	Satellite application facilities
SAR	Synthetic Aperture Radar
SAR-EDU	Radar Remote Sensing Education Initiative
SARs	Synthetic Aperture Radars
SBInSAR	Small Baseline Subset InSAR

SCAT	Scatterometer
SCD	Snow cover duration
SCD_{ES}	Early season snow cover duration
SCD_{LS}	Late season snow cover duration
SCF	Snow cover fraction
SeaWiFS	The Sea-viewing Wide Field-of-view Sensor
SGRT	SAR Geophysical Retrieval Toolbox
SIR	Scatterometer image reconstruction
SIR	Shuttle Imaging Radar
SOS	Start of season
SOST	Start of season time
SPARC	Separation of Pixels Using Aggregated Rating over Canada
SPOT	Satellite Pour l'Observation de la Terre
SPOT-VGT	Satellite Pour l'Observation de la Terre-Vegetation
SRM	Snowmelt Runoff Model
SRTM	Shuttle Radar Topography Mission
SSM	Surface soil moisture
SSM/I	Special sensor microwave/imager
SST	Sea surface temperature
StaMPS	Stanford Method for Permanent Scatterers
STARFM	Spatial and Temporal Adaptive Reflectance Fusion Model
STL	Seasonal trend decomposition by Loess
SVAT	Soil vegetation atmosphere transfer
SVD	Singular value decomposition
SVM	Support vector machines
SWBD	Shuttle Radar Topography Mission Water Body Data
SWI	Soil Water Index
SWIR	Short-wave infrared
TanDEM-L	TerraSAR-L add-on for digital elevation measurement
TC	Tasseled cap
TCW	Tasseled cap wetness
T-D	Threshold delay
TM	Thematic Mapper
TOA	Top-of-atmosphere
TRMM	Tropical Rainfall Measuring Mission
TS	Theil-Sen
UNCCD	United Nations Convention to Combat Desertification
UNDP	United Nations Development Programme
UNEP	United Nations Environment Program
UNFCCC	United Nations Framework Convention on Climate Change
USGS	United States Geological Survey
UTM	Universal Transverse Mercator

VCF	MODIS Vegetation Continuous Fields
VI	Vegetation Index
VIIRS	Visible/Infrared Imager Radiometer Suite
VIP	Vegetation Index and Phenology Earth Science Data Record
VIS	Visible
WCD	Water cover duration
WDC-RSAT	World Data Center for Remote Sensing of the Atmosphere
WELD	Web Enabled Landsat Data
WGS84	World Geodetic System 1984
WSM	Wide Swath Mode
WUE	Water use efficiency
$WUE_{instantaneous}$	Water use efficiency leaf level
WUE_{yield}	Water use efficiency crop level
WYSIWYG	What You See Is What You Get
WS	Wide Swath

Chapter 1
Remote Sensing Time Series Revealing Land Surface Dynamics: Status Quo and the Pathway Ahead

Claudia Kuenzer, Stefan Dech, and Wolfgang Wagner

Abstract The face of our planet is changing at an unprecedented rate. Forest ecosystems diminish at alarming speed, urban and agricultural areas expand into the surrounding natural space, aquaculture is spreading, sea level rise leads to changes in coastal ecosystems, and even without obvious land cover change, land use intensity may change and complex ecosystems may undergo transient changes in composition. Satellite based earth observation is a powerful means to monitor these changes, and especially time series analysis holds the potential to reveal long term land surface dynamics. Whereas in past decades time series analysis was an elaborate undertaking mostly performed by a limited number of experts using coarse resolution data, attention shifts nowadays to open source tools and novel techniques for analyzing time series and the utilization of the same for numerous environmental applications. The reasons are the pressing call for climate-relevant, long term data analyses and value added products revealing past land surface dynamics and trends, the growing demand for global data sets, and the opening up of multidecadal remote sensing data archives, all at a time of considerably-improved hardware power, computer literacy, and a general trend towards cloud solutions and available open source algorithms and programming languages. This chapter presents a comprehensive overview of time series analysis. We introduce currently orbiting optical, radar, and thermal infrared sensors and elucidate which of them are suitable for long term monitoring tasks based on remote sensing time

C. Kuenzer (✉)
German Remote Sensing Data Center, DFD, Earth Observation Center, EOC, German Aerospace Center, DLR, Oberpfaffenhofen, Germany
e-mail: claudia.kuenzer@dlr.de

S. Dech
German Remote Sensing Data Center, DFD, German Aerospace Center, DLR, Wessling, Germany

Institute for Geography and Geology, University of Wuerzburg, Wuerzburg, Germany

W. Wagner
Department of Geodesy and Geoinformation, Vienna University of Technology, Vienna, Austria

C. Kuenzer et al. (eds.), *Remote Sensing Time Series*, Remote Sensing and Digital Image Processing, DOI 10.1007/978-3-319-15967-6_1

series analysis. We briefly summarize the theoretical concept of time series components and important seasonal statistical features and list the types of variables usually analyzed as time series. Furthermore, we address data related, sensor related, location related, and processing related challenges of time series analysis. Lastly, we assess current developments and upcoming opportunities.

1.1 Introduction

Socio-economic transformation, urbanization, climate variability, and natural hazards all shape the face of our planet. Socio-economic transformation leads – among other consequences – to the intensification of agriculture and forestry, the cultivation of formerly non-arable land, and the expansion of infrastructure, all leading to a decline of natural ecosystems. Deforestation of the planet's last primary and secondary forests aims at the provision of space for monoculture plantations or pastures (Leinenkugel et al. 2014; Broich et al. 2011; Kuemmerle et al. 2009), wetlands are drained to create space for cultivating livestock and crops (Rebelo et al. 2009), rivers and lakes are harnessed (Grumbine and Pandit 2013), and irrigation schemes extend the boundaries of cultivatable land at the cost of groundwater depletion and surface subsidence (Fielding et al. 1998; Amelung et al. 1999; Higgins et al. 2013). Urbanization leads to increasing pressure on natural ecosystems (Taubenböck et al. 2012; Haas and Ban 2014) and brings with it soil, water and air pollution (Duh et al. 2006; Ren et al. 2003). These consequences are amplified by such climate related effects as global sea level rise, shifts in local climate patterns, including changes in the frequency of storms, floods, droughts, or pests, as well as habitat shifts (Lambin et al. 2003; Kuenzer et al. 2014). Natural hazards such as earthquakes, tsunamis, volcanic eruptions, or forest fires are singular events which also lead to changes on the land surface (Arnett et al. 2015; Liu et al. 2012). To understand past land surface changes and to be able to predict possible future developments, constant monitoring of our planet is critical. Numerous bodies and initiatives such as the Intergovernmental Panel on Climate Change (IPCC); the United Nations Framework Convention on Climate Change (UNFCCC); the United Nations Environmental Program (UNEP); as well as many other policy makers and stakeholders including national governments, nongovernmental conservation and aid organizations call for reliable information about global environmental change.

In this context, remote sensing based time series analysis is a powerful tool to reveal land surface dynamics and to analyze the magnitude of these dynamics within a defined monitoring time span (Lasaponara and Lanorte 2012). Time series are usually understood as a real-value, continuous, or discrete series of data, where the values refer to equidistant points in time and where the temporal variation exhibits periodic, cyclic, transient, or random behavior – or a combination of these

(Schönwiese 2008). Time series are commonly generated and analyzed in the context of a variety of disciplines such as finance mathematics and econometrics (e.g., stock values), demography (e.g., population development), oceanography (e.g., tide tables), meteorology and weather forecasting (e.g., temperature and precipitation analyses), astronomy and geophysics (e.g., sun spot activity analysis, seismic analyses for earthquake characterization and prediction), and further geosciences (e.g., time series analysis of earth observation data in geography and geology).

This chapter focuses on the time series analysis of earth observation data from spaceborne instruments. In land remote sensing science, time series analysis is usually understood as the temporally dense monitoring of land surface dynamics over a defined period of time. Earth observation based time series analysis has for a long time been associated with studies employing freely available, coarse resolution data from spaceborne instruments offering near global daily coverage. However, in recent years several space agencies and data providers have adopted open access policies leading to unprecedented access to – also higher resolution – data and simultaneously triggering an increased interest in long term monitoring and the development of creative, novel techniques to create densified time series.

In this chapter we assess currently orbiting sensors, discuss past and present sensor preferences for time series analyses, shed light on the typical variables analyzed, present time series analysis approaches, classify time series processing challenges, and take a look at future opportunities and the pathway ahead.

1.2 Earth Observation Sensors for Time Series Analysis

In the past few decades, numerous multitemporal remote sensing studies have been published, many of them focusing on the land cover classification of a few selected time steps and consecutive post classification comparison (Yuan et al. 2005; Ban 2003; Bruzzone and Prieto 2002; Rogan et al. 2002). Such studies typically employ a small number of data sets, and time series techniques for data analysis are not applied. However, the opening up of many satellite data archives has led to free access to several thousand up to several million space images (Woodcock et al. 2008; Wulder et al. 2012; Lasaponara and Lanorte 2012). A milestone in this respect was the opening up of the Landsat archives of the United States Geological Survey, USGS, in 2008, enabling access to over four million Landsat scenes in precision and terrain corrected format. This eased access, combined with the ever growing programing literacy of scientists and increasingly powerful hardware and software, has led to substantial growth of the remote sensing community exploiting space imagery with time series approaches (Müller et al. 2015; Bontemps et al. 2012; Bonano et al. 2012; Gutman and Masek 2012; Hermosilla et al. 2015; Fensholt et al. 2009).

While there are literally hundreds of earth observation satellites in orbit that monitor our planet with optical, thermal infrared, or radar sensors, the fleet of sensors suitable for time series analysis is actually quite limited. Figure 1.1 depicts optical, including multispectral, satellite sensors launched from 1972 until today. Green bars indicate the monitoring time span covered, and national flags indicate the launching and/or operating country. It can be seen that only very few sensors allow for time series analysis in an optical/multispectral domain covering several decades. They include firstly the coarse resolution fleet of Advanced Very High Resolution Radiometer (AVHRR) sensors operated by the National Oceanic and Atmospheric Administration (NOAA) offering daily coverage, and secondly the Landsat fleet with its sensor lines MSS (MultiSpectral Scanner), TM (Thematic Mapper), ETM+ (Enhanced Thematic Mapper Plus), and OLI (Operational Land Imager; also referred to as LDCM = Landsat Data Continuity Mission). Within the optical/multispectral domain only these two sensor lines allow us to take a look back into our past for over three decades. Therefore, it is only based on these two lines that time series analyses focusing on the derivation of climate relevant long term trends can be undertaken. Now that NASA and USGS have both opened their data archives to allow access to this data free of charge, these sensor lines are currently being heavily exploited (Müller et al. 2015; Hermosilla et al. 2015; Fensholt et al. 2009). Sensors such as the U.S. Moderate Resolution Imaging Spectroradiometer, MODIS, on board the Terra and Aqua platforms, the European Advanced Along-Track Scanning Radiometer (AATSR) and the Medium Resolution Imaging Spectrometer (MERIS) both on board the European Environmental Satellite (ENVISAT) are further instruments for which there is easy access to coarse and medium resolution data. However, in contrast to AVHRR and Landsat data these data are available only for the past decade. MODIS was launched on Terra in 1999 and on Aqua in 2002, and the sensors on board ENVISAT collected data only from 2002 to April 2012. From 1998 onwards selected French SPOT (Satellite Pour l'Observation de la Terre) daily coverage VEGETATION data at 1 km are also available, and higher resolution SPOT data are currently also being made available step by step (first for data older than 5 years) in the context of the SPOT World Heritage Programme. The French Space Agency (CNES) announced that in the last 28 years SPOT satellites have acquired over 30 million images. However, as SPOT data had evoked little interest in the past due to the high cost of the data (Lasaponara and Lanorte 2012), there is less knowledge about SPOT data analysis and intercalibration than, for example, for the Landsat fleet. Since data from sensors such as the Advanced Spaceborne Thermal Emission and Reflection Radiometer, ASTER, (also on board the Terra platform) are only acquired when tasked, time series can only be built up for a few selected places on Earth. Other existing data and data now being acquired by many other sensors (Fig. 1.1) are not available to the public. Therefore, next to the already mentioned workhorses AVHRR and Landsat, it is mainly freely available data from MODIS, MERIS, AATSR, VEGETATION and a few other sensors that are used for multispectral time series analysis to reveal land surface dynamics.

Figure 1.2 depicts timelines of radar based earth observing satellites. Also here, easy access to the data from the European Environmental Satellites ERS-1 and ERS-2 with their Scatterometer (SCAT) as well as from the follow on European

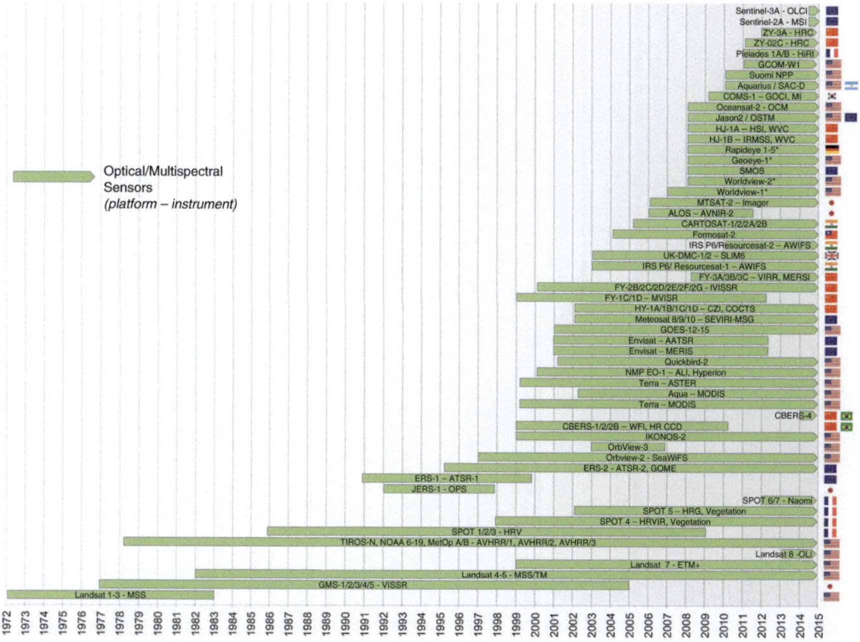

Fig. 1.1 Timelines of optical, including multispectral, earth observation satellites (Modified and adapted from Kuenzer et al. 2014)

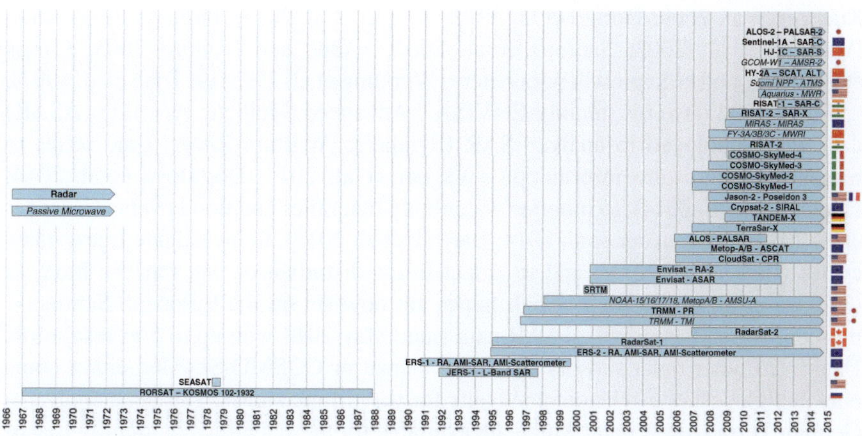

Fig. 1.2 Timelines of radar earth observation satellites (Modified and adapted from Kuenzer et al. 2014)

Meteorological Operational Satellites (MetOp) with their Advanced Scatterometer (ASCAT) has led to their becoming the main workhorses for radar data based time series analyses dating back to the early 1990s. ERS-1/2 and MetOp data have been exploited to derive soil moisture time series (Wagner et al. 1999; Zhao et al. 2008;

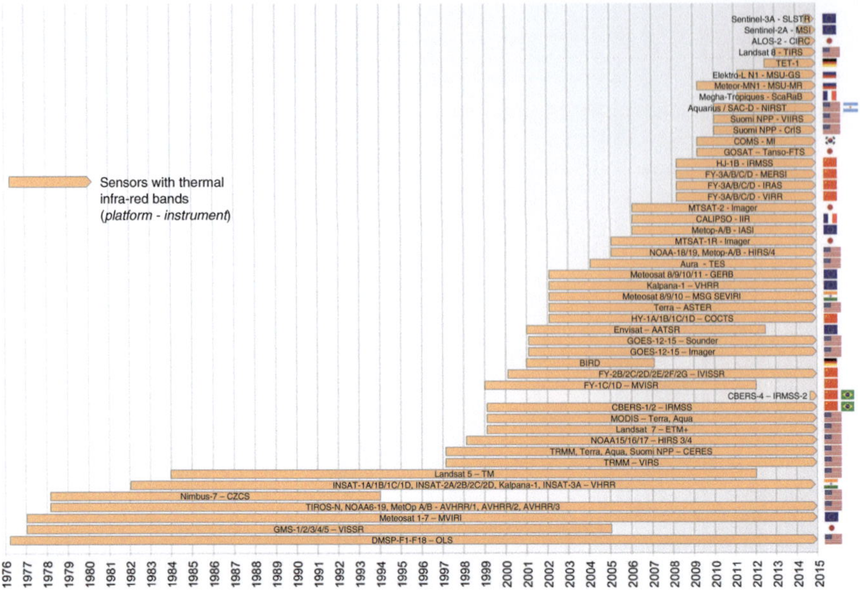

Fig. 1.3 Timelines of thermal infrared earth observation satellites (Modified and adapted from Kuenzer et al. 2014)

Kuenzer et al. 2009; Dorigo et al. 2012; Naeimi et al. 2013), biomass information (Wagner et al. 2003; Santoro et al. 2002; Askne and Santoro 2005; Cartus et al. 2011), and crop growth parameters (Wooding et al. 1995; Le Toan et al. 1997).

From 2002 onwards, mainly ENVISAT Advanced Synthetic Aperture (ASAR) data were employed to analyze SAR (Synthetic Aperture Radar) time series to retrieve soil moisture information (Wagner et al. 2013; Dostálová et al. 2014; Doubková et al. 2014), to monitor water body dynamics and flood events (Kuenzer et al. 2013; Greifeneder et al. 2014; Bartsch et al. 2012), to investigate freeze-thaw cycles (Park et al. 2011), and to detect forest disturbances and retrieve biomass (Santoro et al. 2011, 2015). High hopes are now set on the European Sentinel-1 mission, which has provided SAR data since May 2014 with up to 5 m resolution. Data from other radar sensors such as Radarsat or COSMO-SkyMed either come only at high cost or can be obtained only for very limited areas. The same applies to German TerraSAR-X data, which are available for the scientific community free of charge, but only for small case-study areas, which hampers the generation of national or global products. Furthermore, data acquisition for small study sites can potentially conflict with commercial data requests, so extensive time series can rarely be assembled.

Sensors with thermal infrared bands allowing the analysis of land surface temperature (LST) patterns are listed in Fig. 1.3. In the context of LST it is mainly the AVHRR, MODIS (Frey and Kuenzer 2014; Frey et al. 2012), AATSR, and

Landsat sensors (Song et al. 2015) which have been employed for time series analysis. However, in general, the science community analyzing multiannual up to multidecadal LST patterns and extracting hot spots and anomalies in the context of urban heart islands, long-burning underground coal fires, and other phenomena requiring long term monitoring is relatively small. Thus, the analysis of time series of convincing length is rare in these contexts.

1.3 Remote Sensing Time Series Variables

Time series of remote sensing data from spaceborne sensors can be based on series of raw digital numbers (DN) reflectance values (in %), or on variables commonly derived from the original data prior to analysis (see Fig. 1.4). These variables can be geophysical variables, index variables, thematic variables, topographic variables, and texture variables (Kuenzer et al. 2014). Geophysical variables are defined by a physical unit, a few examples being top-of-the-atmosphere reflectance (TOA), surface reflectance, photosynthetic active radiation, PAR, fraction of absorbed photosynthetic active radiation (FPAR), land surface temperature (LST), leaf area index (LAI), or sediment content. By contrast, index variables are dimensionless, such as the Normalized Difference Vegetation Index (NDVI), the enhanced vegetation index (EVI), the Leaf Area Index (LAI), the soil water index (SWI) and many other indices or feature space components, such as Landsat Tasseled Cap components.

Both geophysical variables and index variables can be used to create long term time series, which can then be analyzed with respect to their daily, weekly,

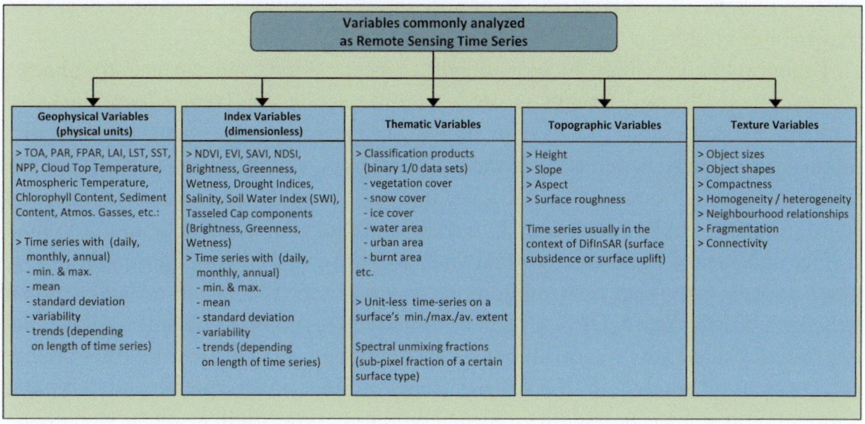

Fig. 1.4 Variables commonly derived in the context of remote sensing based time series analysis. We differentiate Geophysical Variables, Index Variables, Thematic Variables, Topographic Variables, and Texture Variables (Modified and adapted from Kuenzer et al. 2014)

bi-weekly, monthly, annual, or decadal mean value, minima and maxima, standard deviation, variability, anomalies, and trends.

A very commonly used index data set for time series analysis is the GIMMS data set derived from AVHRR data provided by NOAA. GIMMS stands for "Global Inventory Monitoring and Modelling Studies." The spatial resolution of this global area coverage data set is 4 km, and the data have been available since 1981. A comparable vegetation index time series is the global S-10- product, a 10 day synthesis of SPOT-VEGETATION images (VGT-S10) providing NDVI data and all spectral bands in 1 km resolution from 1998 onwards. This data is freely available 3 months after ingestion in the VEGETATION archives. Also from MODIS a synthesized vegetation dataset (MODIS13) exists within the current collection 5, providing NDVI and EVI data at 250 m resolution in 16-day intervals since 1999 (Terra) and 2002 (Aqua).

Thematic variables are variables that are also derived prior to time series analysis based on classification or regression approaches. Typically, data sets with thematic variables are binary data sets with two classes, such as "urban/non-urban," "water / no water," or "snow / no snow," to give only three examples. Stacked as a long term time series these data rows then allow derivations of further statistical products such as mean, maximum, and minimum extent per year (e.g., mean water body extent, minimum water body extent, maximum water body extent), deviations and variability (inundation frequency), as well as long term trends. Several products exist that allow time series analysis based on such thematic layers, such as the Global Snowpack data set (Dietz et al. 2012, 2013, 2014) and the Global WaterPack data set (Klein et al. 2015), both developed by the Land Surface Dynamics Group at the German Remote Sensing Data Center (DFD) of the German Aerospace Center (DLR). As another thematic variable, surface type fractions resulting from spectral unmixing analyses range from 0 to 100 % and indicate how much of a pixel is covered with the respective fraction, and time series of fraction images allow visualization of how sub-pixel components change over time (Kuenzer et al. 2008).

Topographic variables include height, aspect, slope, and surface roughness. However, usually only height is used for time series analyses, where stacks of synthetic aperture radar (SAR) data are used to generate maps of surface deformation based on phase differences of the SAR return pulses (Bonano et al. 2012). The technique applied to analyze SAR data time series in this way, called Differential Interferometric SAR (DInSAR). DInSAR is frequently applied to monitor land surface subsidence in areas of groundwater withdrawal, oil extraction, coal mining, other mining activities, or in regions of ground compaction due to heavy urban structures. Furthermore, DInSAR can be employed to monitor volcanic and tectonic uplifts and glacial movement, amongst other applications. Texture variables are not yet commonly applied in time series analysis, but novel spatial and temporal segmentation approaches and increasing computing power will also lead to increasing exploitation of the related variables.

1.4 Remote Sensing Time Series Components and Features

Each long term remote sensing time series usually consists of three components (see Fig. 1.5): Firstly, a long term directional trend, secondly, seasonal, systematic movements, and thirdly, additional irregular, unsystematic, short term fluctuations. Remote sensing time series analysis aiming at the revelation of land surface dynamics can exploit each of these three components. For example, if we consider a long term time series acquired over a savannah region, the time series will definitely contain a seasonal component (greening and vegetation vigor decline depending on whether it is the rainy or dry season), can contain short term fluctuations based on short term occurrences on land (fire, grazing, etc.), in the atmosphere (e.g. clouds, haze, aerosols, or due to other effects (e.g. bidirectional reflectance distribution function effects (BRDF effects), surface anisotropy, illumination differences, sensor defects etc.), and will exhibit a long term trend over many years (stable conditions, vegetation increase or vegetation degradation). Depending on the research focus, it can be either the seasonal, short term, or trend component which is of special interest – or all three simultaneously (Verbesselt et al. 2010). Whereas climate scientists might be especially interested in long term trends (land surface temperature, snow cover duration, sea level rise, etc.), short term fluctuations are often relevant for an immediate response in managing natural resources (plant disease, fires, natural hazards, etc.). Such short term fluctuations – sometimes also called residuals – are the remaining component if the trend and seasonal components have been removed from a time series. It should be mentioned that this short term residual component of course also contains any noise in the data. An often applied tool to extract time series component is the BFAST tool (Breaks For Additive Season and Trend), which has been developed by Verbesselt et al. (2010). BFAST allows for the

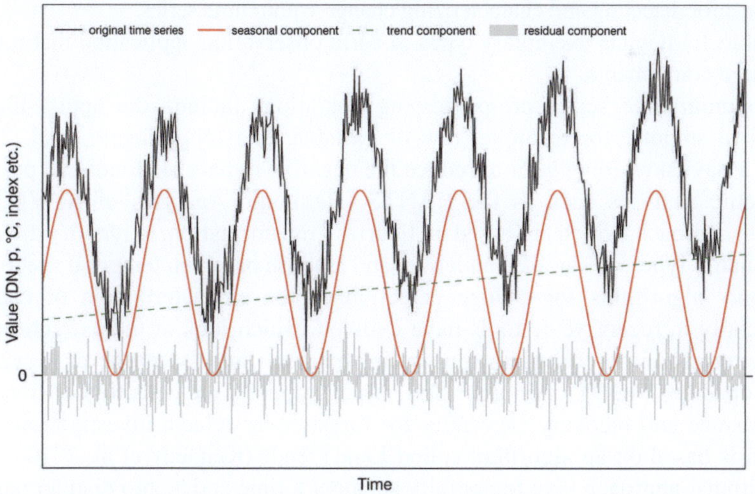

Fig. 1.5 Conceptual sketch of the three components of a time series: trend component, seasonal component, and residual component, all jointly forming the original time series. (Sometimes also referred to as long term, seasonal, and short-term components)

Table 1.1 Examples of earth observation applications relying on residual, seasonal, or trend components of time series

Derivation of	Residual component	Seasonal component	Trend component
Requires	Temporally dense observations with respect to the process to be observed (daily data) Thorough removal of the trend component and the seasonal component to understand the characteristics of the residual signal	An annual or multi-annual time series with dense temporal observations to enable the coverage of all months/seasons	Long term time series with a sufficiently long observation period to derive a significant trend. In the context of geosciences a monitoring period of several decades is necessary
Application examples	Identification of short-term, intra-annual anomalies such as local weather events (strong rain, flash floods, hail), or other short term disturbances, such as pests, lightning, small fires	Annual land use classification based on temporal spectra (e.g., to differentiate different vegetation types or different crop types, e.g., single-, double-, and triple-harvest irrigated rice crops). Identification of annual climate variations	Derivation of vegetation trends (accelerated greening, degradation, shifts in the vegetation period); derivation of trends in inundation dynamics (shift of inundation onset and retreat), derivation of trends in snow cover dynamics (start/duration/melt); trends in land surface temperature, LST

decomposition of time series into trend, season, and remainder components with methods for detecting and characterizing change within time series.

Table 1.1 depicts exemplary types of earth observation application that rely on the three components.

Common time series pre-processing operations include the application of filters to smooth, for example, raw or variable data (Nightingale et al. 2009) (e.g., a Savitsky-Golay filter to reduce the negative impact of clouds), to perform harmonic analyses such as the HANTS (Harmonic Analyses of NDVI Time Series) approach, which are based on Fourier Transformation, to identify different seasonality types, to classify landcover and landuse based on temporal spectra, or to apply algorithms for change detection or the characterization of change. Especially in recent years tools have evolved, which look at the extraction and characterization of change in time series of data. Sulla-Menashe et al. (2014) developed the MODTrendr algorithm, which supports the detection of trends in disturbance and recovery especially for forest cover related investigations, and which is based on an algorithm called LandTrendr (Kennedy et al. 2010). Both algorithms' approach is to temporally segment a time series into distinct periods representing stable conditions, disturbance periods, and recovery. LandTrendr works at a much higher spatial resolution, and was also developed for forest disturbance assessment. LandTrendr (Landsat-based Detection of Trends in Disturbance and Recovery) attempts to capture, label, and map change (Kennedy et al. 2010) exploiting the Landsat archives to the fullest.

For multidecadal time series especially statistical analyses with respect to climate related variations and trends if of interest. The most common statistical parameters which are derived from time series include per pixel:

- mean
- minimum/maximum
- standard deviation
- variability
- anomalies
- turning points
- trends.

Allowing for daily coverage, these parameters can be derived from monthly, annual, multiannual, decadal, or multidecadal time series especially from data from instruments such as AVHRR and MODIS, or ERS and MetOp. Of interest is not only the data value of the minimum, maximum or turning point, etc., but also the timing, usually defined by the day of the year (DOY). Amongst others, DeBeurs and Henerby (2005a) presented a statistical framework for the analysis of long image time series.

In the context of vegetation analysis, one focus of time series analysis is on the extraction of vegetation features, also called phenometrics (see Fig. 1.6), which are metrics defining a plant's vegetative cycle (Lieth 1974; Jönsson and Eklundh 2004). A tool specifically tailored to such applications is TIMESAT, which is a software package for analyzing time series of satellite data. TIMESAT supports the analyses of the seasonality of time series with respect to the dynamic properties of vegetation (Jönsson and Ekludh 2004). Amongst others, TIMESAT allows to extract the following phenometrics from a time series:

- beginning of season (Fig. 1.6a)
- end of season (Fig. 1.6b)
- length of season (Fig. 1.6c)
- base value (Fig. 1.6d)
- time of middle of season (Fig. 1.6e)
- maximum value (Fig. 1.6f)
- amplitude (Fig. 1.6g)
- small integrated value (Fig. 1.6h)
- large integrated value. (Fig. 1.6h, i)

The beginning of season is defined here as the day of the year (and the respective vegetation index value) identified as marking the inception of a consistent upward trend in the vegetation index (e.g., NDVI) time series, depicted as the beginning of measurable photosynthesis in the vegetation canopy. The end of

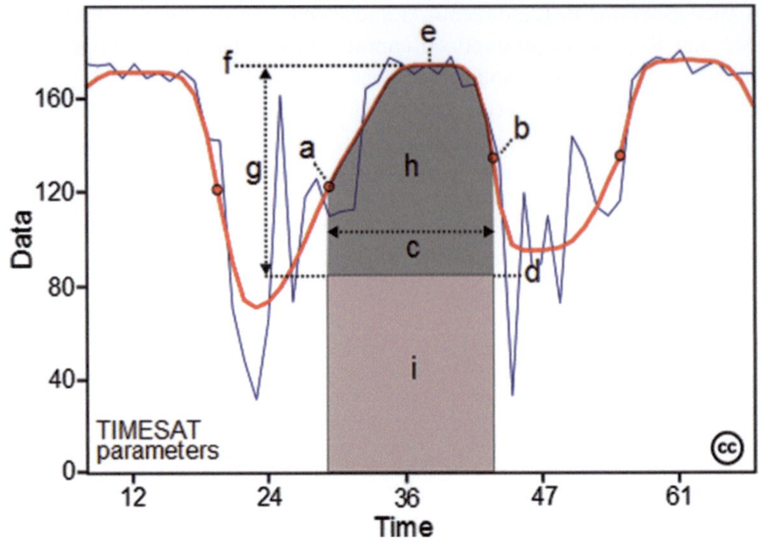

Fig. 1.6 Phenometrics that can be derived from a time series of vegetation index data. Figure supplied as a Creative Commons license figure by the developers of TIMESAT

season is the day of the year (with the corresponding index value) which marks the end of a consistent downward trend in the time series, considered as the end of measurable photosynthesis in the vegetation canopy. The maximum value is the day of the year (and the respective index value) corresponding to the maximum vegetation index value in an annual time series, which is considered to be the time of maximum photosynthesis in the canopy. The length of the season is defined as the number of days from the beginning to the end of the season and defines the duration of photosynthetic activity, whereas the amplitude is defined by the difference between the maximum value and the start of the season values, characterizing the maximum increase in canopy photosynthetic activity above the baseline. The integrated value is not a point in time, but an integral describing the area below the smoothed index time series between the beginning and the end of the season.

The output of TIMESAT is a compilation of files containing phenometric parameters. Nowadays, Timesat can handle different types of remote sensing time series (including MODIS) at different resolutions and can deal with up to two growing seasons per year. Challenges for phenologic analysis are posed by regions which depict a very low vegetation index amplitude over the course of a season or long snow coverage during the winter months or in between the vegetation period, as well as by vegetated areas with several growth cycles per year, such as double, triple, and quadruple cropping cycles (Schwartz 2013; Jönsson and Eklundh 2004). Furthermore, challenges arise from the fact that definitions of phenometrics can vary. Especially Start of Season Time (SOST) and End of Season Time (EOST) can be defined very differently (absolute threshold, maximum curvature, certain fraction of amplitude. Further challenges arise from data availability, irregular observation intervals, or too long compositing periods; but challenges of time series analyses will be addressed in depth at a later stage.

Numerous authors have analyzed phenological metrics over time for selected case study areas, such as Fensholt et al. (2009), White et al. (2009), Ahl et al. (2006), DeBeurs and Henebry (2005b) and Schwartz et al. (2002). Menzel et al. (2006), for example, found that the response of metrics to climate change matches the current global warming pattern. Time series of vegetation indices have – independent of the derivation of phenological metrics – also often been employed to extract an annual land cover classification where the differentiation of individual classes is not based on spectral reflectance, but on annual temporal signatures. Temporal signatures allow for the differentiation of deciduous versus evergreen trees or shrubs as well as the differentiation of harvest cycles and crops types, and therefore enable much more detailed differentiation especially of vegetation classes compared to pixel based approaches (Friedl et al. 2002, 2010; Leinenkugel et al. 2013; Lu et al. 2014).

1.5 Example Applications

An example for a typical display of statistical information derived from a time series of a geophysical variable is displayed in Fig. 1.7. This figure presents the mean monthly Net Primary Productivity (NPP) for China, covering the time span from 1999 to 2012. NPP is the amount of carbon vegetation sequesters in a certain time span (day, week, month, etc.) per unit area minus losses by respiration. NPP was modelled with BETHY/DLR (Eisfelder et al. 2014). In this exemplary case we can observe the following patterns with respect to NPP in China: the country's arid northwest and west exhibit very low year-round NPP values, and visualizes how many days, over the course of 1 year, areas in the Ganges Delta (also known as the Ganges-Brahmaputra Delta) have been inundated. The frequently flooded areas in the Sundarban mangrove forests as well as along the floodplains of northeastern Bangladesh can clearly be noted (Fig. 1.8).

Annual products such as this one informing about inundation length or snow cover duration over the course of 1 year (Dietz et al. 2013) can, if derived for several years, also be jointly analyzed to derive the mean, minimum or maximum inundation or snow cover extent over the course of several years (decades), as well as to derive variability and anomalies hinting at outlier months/years induced by uncommon weather and climate.

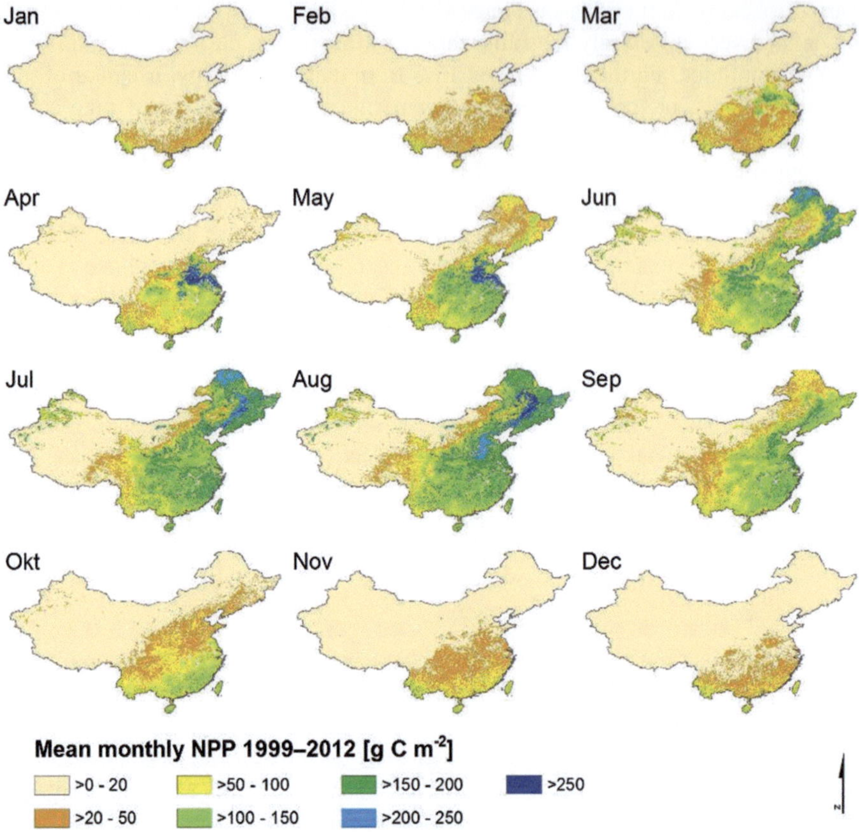

Fig. 1.7 Mean monthly NPP for China for the time period 1999–2012

1.6 Challenges for Remote Sensing Based Time Series Analysis

Challenges for remote sensing based time series analysis can be categorized according to the cause of the challenge. We classify data availability or data policy related challenges, sensor related challenges, location related challenges, and processing related challenges (see Table 1.2). Time series analysis is not possible if the number of available data sets for the area of interest is too small. This is the case for many sensors, which do not have a global or long term acquisition plan. Many sensors require special tasking, which generates high costs. For noncommercial, scientific users it is, for example, unlikely that dense intra-annual or even inter-annual time series can be built up based on high resolution optical WorldView or QuickBird data.

Another data-inherent challenge is the relationship between temporal and spatial resolution. Some satellite sensors delivering coarse-resolution data such as AVHRR

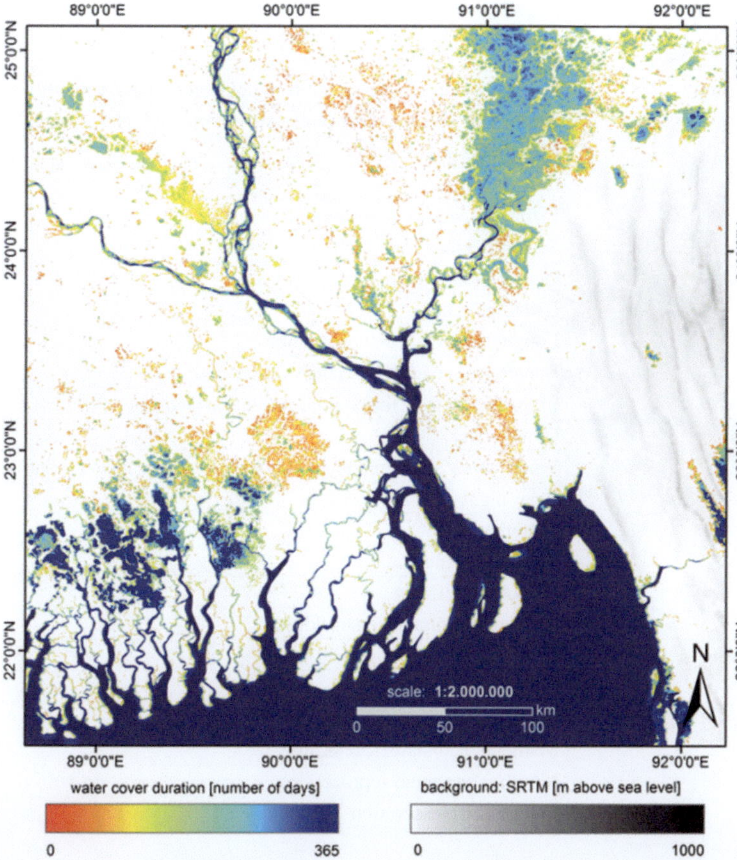

Fig. 1.8 Example of a value-added product derived from a time series of daily, binary thematic products (water/no water) for the year 2013. Inundation (water cover duration) for the Ganges delta (between 0 and 365 times)

or MODIS allow daily coverage of our planet. This allows 365 observations per year under cloud-free conditions. Even in areas with frequent cloud cover a large number of useful observations will still be possible. However, these coarse- or medium-resolution sensors only have spatial resolutions of 1 km to 250 m. On the other hand, sensors offering higher spatial resolution can only cover a smaller part of our planet in 1 day. Figure 1.6 illustrates this dilemma. For a single sensor such as Landsat-7 ETM+, with a revisit rate of 16 days, only about 23 observations of a certain area can be collected, given completely cloud-free conditions. However, under realistic conditions (cloud coverage and past acquisition plans not covering all continents equally) it might only be possible to obtain a handful of observations per year (Fig. 1.9).

Table 1.2 Challenges for remote sensing based time series analysis

Challenge related to	Possible manifestations
Data policy	Limited number of acquisitions (acquisition plan / policy)
	Restricted access to data (via costs or temporal and/or spatial limits)
	Composite products
Sensor	Orbital drift
	Erroneous sensor calibration
	Sensor degradation
	Geolocation errors
	BRDF effects
	Long term sensor line calibration
	Long term sensor line changes in bandwidths
	Long term sensor line spatial resolution changes
Location	Cloud cover
	Water vapor and aerosols in atmosphere (lower NDVI values)
	Polar night
	Extreme terrain
	Sunglint effects
	Anisotropy effects
	Sudden, disconnected subsidence
Processing	Download and storage
	Computing power
	Technical/programming skills
	Mathematical/statistical skills
	Monthly or annual reprocessing
	Long term preservation

Last but not least also the way that datasets are generated and provided can pose a challenge for analyses. Several products are provided as composited datasets – e.g. MODIS 16 day composited NDVI or EVI products. Composite products have the advantage that cloud gaps have been eliminated and that a homogeneous composite can be supplied. However, it should be noted here that composited data sets might not be able to resolve slight climate change related seasonal shifts of only a few days over the course of a few decades. If, for example, we want to detect for past decade's changes which are in the order of a few days (e.g., an on-average 5 day earlier onset of snowmelt, or an on-average 1 week earlier spring greening, etc.), then composites comprising only 8, 10, or 16 days might not be sufficient to resolve such climate related shifts.

Sensor related challenges such as orbital drift, erroneous sensor calibration, sensor degradation, geolocation errors, long term inter-sensor calibration difficulties, and changes in bandwidth or spatial resolution and geolocation accuracy of successive sensors, as well as BRDF effects in data from sensors with a large field of view (FOV), may cause numerous problems, but these can usually be overcome if the related expertise and the time needed for corrective operations are available.

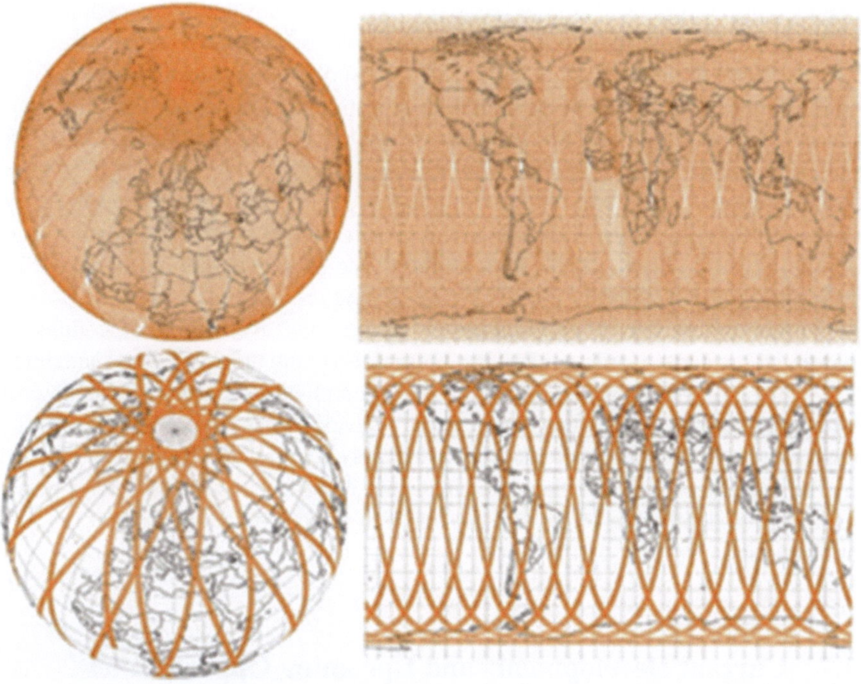

Fig. 1.9 The challenge of temporal coverage versus spatial resolution. The *upper figure* presents the daily coverage of the MODIS sensor (250 m resolution in the VIS, nIR, 500 m resolution in the mIR, 1 km resolution in the TIR), while the *lower figure* presents daily Landsat-7 (ETM+) coverage (30 m spatial resolution in the VIS, nIR, mIR, 60 m spatial resolution in the TIR) (Source: IPF, TU Vienna)

However, this is more challenging for some sensors than for others. Already Brest et al. (1997) noted with respect to AVHRR that "Real decadal scale changes of the earth are much smaller in magnitude than uncertainties in calibration changes and cannot be reliably detected without significant improvements in instrument calibration." In this context Gutman and Masek (2012) reported that such improvements have been achieved for MODIS and also Landsat data.

Location dependent challenges for time series analysis are frequent cloud cover, the influence of water vapor and aerosols, and areas affected by polar night (no sunlight for up to a full half year). Regions with extreme terrain (deeply incised and shadowed valleys) also pose obstacles to time series analysis. Much more rare location related challenges can be sunglint effects; compensation for the resulting spikes or outliers in a time series is usually achieved with simple filter operations. Furthermore, in some regions, very regular patterns (e.g. terraced rice field all aligned in the same direction, sand dunes all aligned in the same direction) can lead to effects of azimuthal anisotropy especially in coarse resolution radar data, and thus can impede time series analyses in certain fields (Bartalis et al. 2006). Another, very rare challenge affecting DInSAR based analysis is sudden land subsidence (e.g., in karst landscapes). Sudden subsidence leads to a loss of connectivity of fringes in an interferogram, so that the subsidence can no longer be quantified (Bonano et al. 2012). For some selected variables snow cover or frequent flooding hamper the generation of suitable time series, such as for the derivation of soil moisture, which cannot be undertaken in near polar latitudes with lengthy snow cover (Wagner et al. 1999). However, snow cover and flooding can also be parameters, which shall be observed, so we do not include them in the list of location related challenges. All these location related challenges can hardly be influenced by the analyst. Prevailing cloud cover and the above mentioned effects cannot be fully compensated, but a gapless time series can be approximated by employing certain restoration and filtering techniques.

Further processing related challenges such as the availability of computing power, storage space, archives for data preservation, and the mandatory mathematical/statistical, and programming skills can – theoretically speaking – be overcome with sufficient financial means and capacity development. However, in reality, a combination of only a few of the time series processing related challenges grouped in Table 1.2 usually leads many scientists and remote sensing laboratories to not engage in in-depth time series analysis and rather focus on multitemporal types of applications.

1.7 Current Developments and Upcoming Opportunities

Whereas in past years time series based investigations of ecosystem properties was mainly performed on coarse (medium) resolution satellite imagery from optical sensors such as AVHRR, MODIS, VEGETATION or MERIS, or radar sensors such as ERS-1/2 SCAT, MetOp ASCAT, and ENVISAT ASAR, more and more studies exploit Landsat resolution type sensors. Recently, tools such as LandTrendr (Landsat-based Detection of Trends in Disturbance and Recovery), which attempt to capture, label, and map change (Kennedy et al. 2010) are made freely available to explore the Landsat archives to the fullest. At the same time, the ESA Sentinel missions will offer advantages and are expected to develop their capacity as future workhorses. ESA Sentinel-1 data available since April 2014 enable the extension of the ENVISAT ASAR archives, and Sentinel-3 will allow the extension of past ENVISAT MERIS time series. Sentinel-2 supplies additional high resolution data at the Landsat scale (Drusch et al. 2012), enabling monitoring approaches that can be compared with temporally dense coarse resolution approaches.

Presuming the sound inter-calibration of different sensors, it is likely that more and more multi-sensor time series will be generated utilizing data from two or more different sensors to overcome data and location related challenges (data availability and access, cloud cover, etc.). For optical instruments it is already common to combine two or several data sets retrieved from Landsat TM, ETM+, and OLI. Additionally, gaps in such data series can be filled with SPOT/Pléiades or future Sentinel-2 data to create denser temporal coverage (time series densification). The creation of time series using multiple sensors of completely different sensor families is a very demanding task which requires utmost adaptation and calibration efforts to ensure that the extracted changes really refer to land surface dynamics and not only to intercalibration related errors. Among others, multisensor time series fusion approaches were already presented by Gao et al. (2006), Zhu et al. (2010), and Hilker et al. (2009), who for example blended Landsat and MODIS surface reflectance data.

Next to the densification of time series with data from sensors of comparable or different resolutions (Landsat and MODIS, etc.), scientists are also starting to exploit opportunities to build up time series with data acquired by passive as well as active sensors (radar and optical data within one time series). For example, the Japanese Advanced Land Observing Satellite (ALOS) which was launched in January 2006 and operated until April 2011, delivers optical data at 10 m spatial resolution with its AVNIR-2 radiometers, as well as L-band radar data at 10 and 100 m spatial resolution. Although ALOS communication failed in April 2011, ALOS-2 was launched just recently in May 2014 and currently several groups globally experiment with a combination of ALOS and Landsat data, including the Web Enabled Landsat Data, WELD, data sets. WELD products are generated from every available Landsat 7 ETM+ scene over the USA with cloud cover below 80 %. The resulting weekly, monthly, seasonal, and annual products allow monitoring tasks at dense temporal intervals. NASA currently also funds the generation of global products centered on six 3-year epochs centered on 1985, 1990, 1995, 2000, 2005, and 2010, respectively. But other data are also suitable for time series densification. Since 2004 the Brazilian Space Agency (INPE) has enabled free access to China-Brazil Earth Resources Satellite (CBERS) data; the ASTER archives have not yet been fully exploited; and missions such as SPOT and the French Pléiades satellites might ease access to their data in the future. Within the next decade the European Sentinels will deliver an unprecedented amount of data that will also be accessible to the scientific community free of charge. Furthermore, data access to imagery acquired by emerging space nations such as China, Thailand, India, Vietnam, and others might be eased. Amongst other bodies, the Group on Earth Observation, GEO, is advancing this goal.

It is the sheer amount of data that need to be automatically received, processed, and archived that poses the greatest challenge over the next decade. The data volume that space agencies and data providers have to safeguard and preserve is enormous, and the amount of data that individual researchers process in the context of their applied studies is also increasing year by year. In these challenging times of "Big Data," cloud solutions for data storage and processing, openly available (open source) processing algorithms and code sharing platforms, and open source programming languages have evolved rapidly. Additionally, novel schemes for re-processing monthly or annual time series are in great demand. Nowadays we can already observe the tendency that it is rather "the algorithms coming to the data" than vice versa. This development holds great potential for international cooperation, including product comparison and validation, and a minimization of redundancies.

Last but not least, the goal is to understand land surface dynamics and the ongoing changes on our planet, and to generate value added products of relevance for informed decision making by stakeholders. Lifting time series analysis from the coarse and medium resolution scale to the Landsat scale is one important task. Another is to fit models to long term time series so that past and present monitoring activities can evolve to innovative forecasting. Such analytical forecasts are commonly undertaken in many disciplines, including economics and demographics. In the geosciences community forecasts are mainly issued by climate scientists, who release temperature, precipitation, or sea level rise projections. A large future challenge for the land remote sensing community will be to realize projections of future land cover and land use developments. This field is still in its infancy, but is expected to develop rapidly over the next few years. Ultimately, any time series analysis is a precursor activity to predicting the future based on trends derived from past observations.

Acknowledgements The authors thank P. Leinenkugel and U. Gessner for discussion of the manuscript. Furthermore, we are grateful to two anonymous reviewers for their valuable comments. Further thanks go to P. Koch for support with the editing of references.

References

Ahl DE, Gower ST, Burrows SN, Shabanov NV, Myneni RB, Knyazikhin Y, Douglas EA (2006) Monitoring spring canopy phenology of a deciduous broadleaf forest using MODIS. Remote Sens Environ 104:88–95

Amelung F, Galloway DL, Bell JW, Zebker HA, Laczniak RJ (1999) Sensing the ups and downs of Las Vegas – InSAR reveals structural control of land subsidence and aquifer-system deformation. Geology 27:483–486

Arnett JTTR, Coops NC, Daniels LD, Falls RW (2015) Detecting forest damage after a low-severity fire using remote sensing at multiple scales. Int J Appl Earth Obs Geoinf 35(Part B):239–246

Askne J, Santoro M (2005) Multitemporal repeat pass SAR interferometry of boreal forests. IEEE Trans Geosci Remote Sens 43(6):1219–1228

Ban Y (2003) Multitemporal ERS-1 SAR and Landsat TM data for agricultural crop classification: comparison and synergy. Can J Remote Sens 29(4):518–526

Bartalis Z, Scipal K, Wagner W (2006) Azimuthal anisotropy of scatterometer measurements over land. IEEE Trans Geosci Remote Sens 44(8):2083–2092

Bartsch A, Trofaier AM, Hayman G, Sabel D, Schlaffer S, Clark DB, Blyth E (2012) Detection of open water dynamics with ENVISAT ASAR in support of land surface modelling at high latitudes. Biogeosciences 9:703–714. doi:10.5194/bg-9-703-2012

Bonano M, Manunta M, Marsella M, Lanari R (2012) Long-term ERS/ENVISAT deformation time-series generation at full spatial resolution via the extended SBAS technique. Int J Remote Sens 33:4756–4783

Bontemps S, Langner A, Defourny P (2012) Monitoring forest changes in Borneo on a yearly basis by an object-based change detection algorithm using SPOTVEGETATION time series. Int J Remote Sens 33:4673–4699

Brest CL, Rossow WB, Roiter MD (1997) Update of radiance calibrations for ISCCP. J Atmos Ocean Technol 14:1091–1109

Broich M, Hansen MC, Potapov P, Adusei B, Lindquist E, Stehman SV (2011) Time-series analysis of multi-resolution optical imagery for quantifying forest cover loss in Sumatra and Kalimantan, Indonesia. Int J Appl Earth Obs Geoinf 13:277–291

Bruzzone L, Prieto DF (2002) A partially unsupervised cascade classifier for the analysis of multitemporal remote-sensing images. Pattern Recogn Lett 23(9):1063–1071

Cartus O, Santoro M, Schmullius C, Li Z (2011) Large area forest stem volume mapping in the boreal zone using synergy of ERS-1/2 tandem coherence and MODIS vegetation continuous fields. Remote Sens Environ 115:931–943

DeBeurs KM, Henebry GM (2005a) A statistical framework for the analysis of long image time series. Int J Remote Sens 26(8):1551–1573. doi:10.1080/01431160512331326657

DeBeurs KM, Henebry GM (2005b) Land surface phenology and temperature variation in the international geosphere-biosphere program high-latitude transects. Glob Chang Biol 11:779–790

Dietz A, Wohner C, Kuenzer C (2012) European snow cover characteristics between 2000 and 2011 derived from improved MODIS daily snow cover products. Remote Sens 4:2432–2454. doi:10.3390/rs4082432

Dietz A, Kuenzer C, Conrad C (2013) Snow cover variability in Central Asia between 2000 and 1 2011 derived from improved MODIS daily snow cover products. Int J Remote Sens 34 (11):3879–3902

Dietz A, Conrad C, Kuenzer C, Gesell G, Dech S (2014) Identifying changing snow cover characteristics in Central Asian between 1986 and 2014 from remote sensing data. Remote Sens 6:12752–12775

Dorigo W, deJeu R, Chung D, Parinussa R, Liu Y, Wagner W, Fernández-Prieto D (2012) Evaluating global trends (1988–2010) in harmonized multi-satellite surface soil moisture. Geophys Res Lett 39(L18405):1–7

Dostálová A, Doubková M, Sabel D, Bauer-Marschallinger B, Wagner W (2014) Seven years of Advanced Synthetic Aperture Radar (ASAR) Global Monitoring (GM) of surface soil moisture over Africa. Remote Sens 6(8):7683–7707

Doubková M, Dostálová A (née Hegyiová), van Dijk AIJM, Blöschl G, Wagner W, Fernández-Prieto D (2014) How do spatial scale, noise, and reference data affect empirical estimates of error in ASAR-derived 1 km resolution soil moisture? IEEE J Sel Top Appl Earth Obs Remote Sens 7(9):3880–3891. doi:10.1109/JSTARS.2014.2324657

Drusch M, Del Bello U, Carlier S, Colin O, Fernandez V, Gascon F, Hoersch B, Isola C, Laberinti P, Martimort P, Meygret A, Spoto F, Sy O, Marchese F, Bargellini P (2012) Sentinel-2: ESA's optical high-resolution mission for GMES operational services. Remote Sens Environ 120:25–36

Duh JD, Shandas V, Chang H, George LA (2006) Rates of urbanisation and the resiliency of air and water quality. Sci Total Environ 400(1–3):238–256

Eisfelder C, Klein I, Niklaus M, Kuenzer C (2014) Net primary productivity in Kazakhstan, its spatio-temporal patterns and relation to meteorological variables. J Arid Environ 103:17–30

Fensholt R, Rasmussen K, Nielsen TT, Mbow C (2009) Evaluation of earth observation based long term vegetation trends – intercomparing NDVI time series trend analysis consistency of Sahel from AVHRR GIMMS, Terra MODIS and SPOT VGT data. Remote Sens Environ 113 (9):1886–1898

Fielding EJ, Blom RG, Goldstein RM (1998) Rapid subsidence over oil fields measured by SAR interferometry. Geophys Res Lett 27:3215–3218

Frey C, Kuenzer C (2014) Land surface temperature dynamics in the Upper Mekong Basin derived from MODIS time series. Int J Remote Sens 35(8):2780–2798

Frey C, Kuenzer C, Dech S (2012) Quantitative comparison of the operational NOAA AVHRR LST product of DLR and the MODIS LST product V005. Int J Remote Sens 33(22):7165–7183

Friedl MA, McIver DK, Hodges JCF, Zhang XY, Muchoney D, Strahler AH, Woodcock CE, Gopal S, Schneider A, Cooper A, Baccini A, Gao F, Schaaf C (2002) Global land cover mapping from MODIS: algorithms and early results. Remote Sens Environ 83:287–302

Friedl MA, Sulla-Menashe D, Tan B, Schneider A, Ramankutty N, Sibley A, Huang X (2010) MODIS collection 5 global land cover: algorithm refinements and characterization of new datasets. Remote Sens Environ 114(1):168–182

Gao F, Masek J, Schwaller M, Hall F (2006) On the blending of the Landsat and MODIS surface reflectance: predicting daily Landsat surface reflectance. IEEE Trans Geosci Remote Sens 44:2207–2218

Greifeneder F, Wagner W, Sabel D, Naeimi V (2014) Suitability of SAR imagery for automatic flood mapping in the Lower Mekong Basin. Int J Remote Sens 35(8):2857–2874

Grumbine E, Pandit M (2013) Threats from India's Himalaya dams. Science 339(6115):36–37. doi:10.1126/science.1227211

Gutman G, Masek JG (2012) Long-term time series of the Earth's land-surface observations from space. Int J Remote Sens 33:4700–4719

Haas J, Ban Y (2014) Urban growth and environmental impacts in Jing-Jin-Ji, the Yangtze River Delta and the Pearl River Delta. Int J Appl Earth Obs Geoinf 30:42–55

Hermosilla T, Wulder MA, White JC, Coops NC, Hobart GW (2015) An integrated Landsat time series protocol for change detection and generation of annual gap-free surface reflectance composites. Remote Sens Environ 158:220–234

Higgins S, Overeem I, Tanaka A, Syvitski J (2013) Land subsidence at aquaculture facilities in the Yellow River Delta, China. Geophys Res Lett 40(15):3898–3902. doi:10.1002/grl.50758

Hilker T, Wulder M, Coops N, Seitz N, White J, Gao F, Masek J, Stenhouse G (2009) Generation of dense time series synthetic Landsat data through data blending with MODIS using a spatial and temporal adaptive reflectance fusion model. Remote Sens Environ 113:1988–1999

Jönsson P, Eklundh L (2004) TIMESAT – a program for analyzing time-series of satellite sensor data. Comput Geosci 30(8):833–845

Kennedy RE, Yang ZQ, Cohen WB (2010) Detecting trends in forest disturbance and recovery using yearly Landsat time series: 1. LandTrendr – temporal segmentation algorithms. Remote Sens Environ 114(12):2897–2910

Klein I, Dietz A, Gessner U, Dech S, Kuenzer C (2015) Preliminary results of the global WaterPack: a novel product to assess inland water body dynamics on a daily basis. Int J Remote Sens 6(1):78–87

Kuemmerle T, Chaskovskyy O, Knorn J, Radeloff VC, Kruhlov I, Keeton WS, Hostert P (2009) Forest cover change and illegal logging in the Ukrainian Carpathians in the transition period from 1988 to 2007. Remote Sens Environ 113(6):1194–1207

Kuenzer C, Bachmann M, Mueller A, Lieckfeld L, Wagner W (2008) Partial unmixing as a tool for single surface class detection and time series analysis. Int J Remote Sens 29(11):1–23. doi:10.1080/01431160701469107

Kuenzer C, Zhao D, Scipal K, Sabel D, Naeimi V, Bartalis Z, Hasenauer S, Mehl H, Dech S, Wagner W (2009) El Niño influences represented in ERS scatterometer derived soil moisture data. Appl Geogr 29(4):463–477

Kuenzer C, Guo H, Leinenkugel L, Huth J, Li X, Dech S (2013) Flood mapping and flood dynamics of the Mekong Delta: an ENVISAT-ASAR-WSM based time series analyses. Remote Sens 5:687–715. doi:10.3390/rs5020687

Kuenzer C, Ottinger M, Wegmann M, Guo H, Wang C, Zhang J, Dech S, Wikelski M (2014) Earth observation satellite sensors for biodiversity monitoring: potentials and bottlenecks. Int J Remote Sens 35(18):6599–6647. doi:10.1080/01431161.2014.964349

Lambin EF, Geist HJ, Lepers E (2003) Dynamics of land-use and land-cover change in tropical regions. Annu Rev Environ Resour 28:205–241

Lasaponara R, Lanorte A (2012) Satellite time-series analysis. Int J Remote Sens 33 (15):4649–4652. doi:10.1080/01431161.2011.638342

Le Toan T, Ribbes F, Floury N, Wang L, Kong JA, Kurosu T, Fujita M (1997) Rice crop mapping and monitoring using ERS-1 data based on experiment and modeling results. IEEE Trans Geosci Remote Sens 35(1):41–56

Leinenkugel P, Kuenzer C, Oppelt N, Dech S (2013) Characterisation of land surface phenology and land cover based on moderate resolution satellite data in cloud prone areas – a novel product for the Mekong Basin. Remote Sens Environ 136:180–198. doi:10.1016/j.rse.2013.05.004

Leinenkugel P, Wolters ML, Oppelt N, Kuenzer C (2014) Tree cover and forest cover dynamics in the Mekong Basin from 2001 to 2011. Remote Sens Environ. doi:10.1016/j.rse.2014.10.021

Lieth H (ed) (1974) Phenology and seasonal modeling. Springer, New York

Liu JG, Mason PJ, Yu E, Wu MC, Tang C, Huang R, Liu H (2012) GIS modelling of earthquake damage zones using satellite remote sensing and DEM data. Geomorphology 139–140:518–535

Lu L, Kuenzer C, Guo H, Li Q, Long T, Li X (2014) A novel land cover classification map based on MODIS time-series in Xinjiang, China. Remote Sens 6:3387–3408. doi:10.3390/rs6043387

Menzel A, Sparks TH, Estrella N, Koch E, Aasa A, Ahas R, Alm-Kubler K, Bissolli P, Braslavska OG, Briede A et al (2006) European phenological response to climate change matches the warming pattern. Glob Chang Biol 12:1969–1976

Müller H, Rufin P, Griffiths P, Siqueira AJB, Hostert P (2015) Mining dense Landsat time series for separating cropland and pasture in a heterogeneous Brazilian savanna landscape. Remote Sens Environ 156:490–499

Naeimi V, Leinenkugel P, Sabel D, Wagner W, Apel H, Kuenzer C (2013) Evaluation of soil moisture retrieval from the ERS and metop scatterometers in the lower Mekong Basin. Remote Sens 5(4):1603–1623

Nightingale J, Morisette J, Wolfe R, Tan B, Gao F, Ederer G, Collatz G, Turner D (2009) Temporally smoothed and gap-filled MODIS land products for carbon modelling: application of the f PAR product. Int J Remote Sens 30:1083–1090

Park SE, Bartsch A, Sabel D, Wagner W, Naeimi V, Yamaguchi Y (2011) Monitoring freeze/thaw cycles using ENVISAT ASAR Global Mode. Remote Sens Environ 115(12):3457–3467

Rebelo LM, Finlayson CM, Nagabhatla N (2009) Remote sensing and GIS for wetland inventory, mapping and change analysis. J Environ Manag 90(7):2144–2153

Ren W, Zhong Y, Meligrana J, Anderson B, Watt WE, Chen J, Leung HL (2003) Urbanization, land use, and water quality in Shanghai: 1947–1996. Environ Int 29(5):649–659

Rogan J, Franklin J, Roberts DA (2002) A comparison of methods for monitoring multitemporal vegetation change using Thematic Mapper imagery. Remote Sens Environ 80(1):143–156

Santoro M, Askne J, Smith G, Fransson JES (2002) Stem volume retrieval in boreal forests from ERS-1/2 interferometry. Remote Sens Environ 81:19–35

Santoro M, Beaudoin A, Beer C, Cartus O, Schmullius C, Shvidenko A, McCallum I, Wegmuller U, Wiesmann A (2011) Retrieval of growing stock volume in boreal forest using hyper-temporal series of Envisat ASAR ScanSAR backscatter measurements. Remote Sens Environ 115(2):490–507

Santoro M, Beaudoin A, Beer C, Cartus O, Fransson J, Hall R, Pathe C, Schmullius C, Shvidenko A, Thurner M, Wegmueller U (2015) Forest growing stock volume of the northern hemisphere: spatially explicit estimates for 2010 derived from Envisat ASAR data. Remote Sens Environ (final review phase)

Schönwiese CD (2008) Klimatologie, 3rd edn. Ulmer (UTB), Stuttgart, 472 pp

Schwartz MD (2013) Phenology: an integrative environmental science. Springer, New York

Schwartz MD, Reed BC, White MA (2002) Assessing satellite-derived start-of-season measures in the conterminous USA. Int J Climatol 22:1793–1805

Song S, Kuenzer C, Zhang Z, Jia Y, Sun Y, Zhang J (2015) Analysis of coal fire dynamics in the Wuda syncline impacted by fire-fighting activities based on in-situ observations and Landsat-8 remote sensing data. Int J Coal Geol (accepted for publication)

Sulla-Menashe D, Kennedy RE, Yang Z, Braaten J, Krankina ON, Friedl MA (2014) Detecting forest disturbance in the Pacific Northwest from MODIS time series using temporal segmentation. Remote Sens Environ 151:114–123

Taubenböck H, Esch T, Felbier A, Wiesner M, Roth A, Dech S (2012) Monitoring urbanization in mega cities from space. Remote Sens Environ 117:162–176

Verbesselt J, Hyndman R, Newnham G, Culvenor D (2010) Detecting trend and seasonal changes in satellite image time series. Remote Sens Environ 114:106–115

Wagner W, Lemoine G, Rott H (1999) A method for estimating soil moisture from ERS scatterometer and soil data. Remote Sens Environ 70:191–207

Wagner W, Luckman A, Vietmeier J, Tansey K, Balzter H, Schmullius C, Davidson M, Gaveau D, Gluck M, Le Toan T, Quegan S, Shvidenko A, Wiesmann A, Jiong Yu JJ (2003) Large-scale mapping of boreal forest in SIBERIA using ERS tandem coherence and JERS backscatter data. Remote Sens Environ 85(2):125–144

Wagner W, Hahn S, Kidd R, Melzer T, Bartalis Z, Hasenauer S, Figa-Saldanã J, de Rosnay P, Jann A, Schneider S, Komma J, Kubu G, Brugger K, Aubrecht C, Züger C, Gangkofer U, Kienberger S, Brocca L, Wang Y, Blöschl G, Eitzinger J, Steinnocher K, Zeil P, Rubel F (2013) The ASCAT soil moisture product: a review of its specifications, validation results, and emerging applications. Meteorol Z 22(1):5–33

White MA, deBeurs KM, Didan K, Inouye DW, Richardson AD, Jensen OP, O'Keefe J, Zhang G, Nemani RR, van Leeuwen WJD et al (2009) Intercomparison, interpretation, and assessment of spring phenology in north America estimated from remote sensing for 1982–2006. Glob Chang Biol 15:2335–2359

Woodcock CE, Allen R, Anderson M, Belward A, Bindschadler R, Cohen W, Gao F, Goward SN, Helder D, Helmer E, Nemani R, Oreopoulos L, Schott J, Thenkabail PS, Vermote EF, Vogelmann J, Wulder MA, Wynne R (2008) Free access to Landsat imagery. Science 320:1011

Wooding M, Attema E, Aschbacher J, Borgeaud M, Cordey RA, De Groot H, Harms J, Lichtenegger J, Nieuwenhuis G, Schmullius C, Zmuda A (1995) Satellite radar in agriculture. Experience with ERS-1. ESA Scientific Publications, Noordwijk, SP-1185. ISBN 92-9092-339-3

Wulder MA, Masek JG, Cohen WB, Loveland TR, Woodcock CE (2012) Opening the archive: how free data has enabled the science and monitoring promise of Landsat. Remote Sens Environ 122:2–10

Yuan F, Sawaya KE, Loeffelholz BC, Bauer ME (2005) Land cover classification and change analysis of the Twin Cities (Minnesota) Metropolitan Area by multitemporal Landsat remote sensing. Remote Sens Environ 98(2–3):317–328

Zhao D, Kuenzer C, Fu C, Wagner W (2008) Evaluation of the ERS scatterometer derived soil water index to monitor water availability and precipitation distribution at three different scales in China. J Hydrometeorol 9:549–562

Zhu XL, Chen J, Gao F, Chen XH, Masek JG (2010) An enhanced spatial and temporal adaptive reflectance fusion model for complex heterogeneous regions. Remote Sens Environ 114 (11):2610–2623

Chapter 2
Time Series Analyses in a New Era of Optical Satellite Data

Patrick Hostert, Patrick Griffiths, Sebastian van der Linden, and Dirk Pflugmacher

Abstract Dense time series of optical remote sensing data have long been the domain of broad-scale sensors with daily near-global coverage, such as the Advanced Very High Resolution Radiometer (AVHRR), the Medium Resolution Imaging Spectrometer (MERIS), the Moderate Resolution Imaging Spectrometer (MODIS) or the Satellite Pour l'Observation de la Terre (SPOT) VEGETATION. More recently, satellite data suitable for fine-scale analyses are becoming attractive for time series approaches. The major reasons for this development are the opening of the United States Geological Survey (USGS) Landsat archive along with a standardized geometric pre-processing including terrain correction. Based on such standardized products, tools for automated atmospheric correction and cloud/cloud shadow masking advanced the capabilities to handle cloud-contamination effectively. Finally, advances in information technology for mass data processing today allow analysing thousands of satellite images with comparatively little effort. Based on these major advancements, time series analyses have become feasible for solving questions across different research domains, while the focus here is on land systems. While early studies focused on better characterising forested ecosystems, now more complex ecosystem regimes, such as shrubland or agricultural system dynamics, come into focus. Despite the evolution of a wealth of novel time series-based applications, coherent analysis schemes and good practice guidelines are scarce. This chapter accordingly strives to structure the different approaches with a focus on potential applications or user needs. We end with an outlook on forthcoming sensor constellations that will greatly advance our opportunities concerning time series analyses.

P. Hostert (✉) • S. van der Linden
Geography Department, Humboldt-Universität zu Berlin, Berlin, Germany

IRI THESys, Humboldt-Universität zu Berlin, Berlin, Germany
e-mail: patrick.hostert@geo.hu-berlin.de

P. Griffiths • D. Pflugmacher
Geography Department, Humboldt-Universität zu Berlin, Berlin, Germany
e-mail: patrick.griffiths@geo.hu-berlin.de

C. Kuenzer et al. (eds.), *Remote Sensing Time Series*, Remote Sensing and Digital Image Processing, DOI 10.1007/978-3-319-15967-6_2

2.1 Introduction

Over the past 50 years, anthropogenic ecosystem changes were more rapid and extensive than in any comparable period in history (MEA 2005) and nowadays no ecosystem is free of pervasive human influence (Vitousek 1997). Changes in land surface characteristics mirror a multitude of processes induced by human alteration of the Earth system as a whole. Earth observation (EO) allows for repeated, synoptic and consistent measurement of the Earth surface and has long been used for environmental assessments (Lambin and Strahler 1994). The process of monitoring changes and modifications of land surface characteristics by means of a series of EO data is often referred to as time series analysis.

Approaches for time series analysis in remote sensing have long been restricted to the domain of wide swath, coarse spatial resolution sensors. Such systems commonly achieve complete global coverage on a (near-) daily basis. Image products can be generated based on a defined temporal interval, such as the 8- or 16-day Normalized Difference Vegetation Index (NDVI) composites derived from AVHRR data. This allows the direct utilization of time series methods, many of which require equidistant observations and were more commonly used in econometric or meteorological sciences. The temporal repeat frequency of higher spatial resolution sensors, however, does not readily allow applying these types of methods. On the one hand, orbital and engineering characteristics drive repeat acquisitions on the order of 7–20 days. On the other hand, acquisition strategies and cloud coverage govern the actual availability of scenes acquired with low or no cloud cover contamination.

Recently, great advances have been made in the use of remote sensing data that allow analyses at landscape to regional scales, such as Landsat's Thematic Mapper (TM), Enhanced Thematic Mapper Plus (ETM+), and Operational Land Imager (OLI). Generally, the availability of optical spaceborne imagery has tremendously increased. This was achieved by numerous new sensors in space, but the sea change is due to open access policies of governmental agencies. While coarse spatial resolution data have traditionally been available free of charge, this was not the case for higher resolution imagery. During 2004, the Brazilian Space Agency (INPE) pioneered free and open access to medium resolution satellite imagery by first providing China-Brazil Earth Resources Satellite (CBERS) data for free and subsequently making the Brazilian Landsat data holdings available at no cost. Most importantly, however, in 2008 the USGS adopted a free data policy for the U.S. Landsat holdings, the largest archive of Landsat data (Woodcock et al. 2008). The global Landsat archive constitutes a unique record of the way humans modify the land surfaces (Roy et al. 2014), and Landsat's spatial resolution enables chronicling of anthropogenic and natural change at all scales (e.g. Gutman et al. 2008). The USGS Landsat archive alone currently contains more than four million scenes and Landsat 8 is currently contributing an unprecedented 650 scenes daily. Additionally, the ongoing Landsat Global Archive Consolidation effort is repatriating unique scenes from data holdings of the network of international receiving stations (Wulder et al. 2012). Next to data availability, providing ready-to-use data is the second prerequisite to promote new time series analysis applications. Landsat data are distributed in precision terrain corrected (L1T) format (Landsat Project Science Office 2010; USGS 2012), providing the radiometric quality and spatial precision needed to directly employ the imagery in end-user workflows. The new L1T processing of Landsat data has set a 'gold standard' for the quality of absolute and relative geometric accuracy. Careful radiometric calibration over the series of Landsat sensors has made consistent conversion of digital counts to radiance and reflectance possible (Chander et al. 2009). Substantial user input and interaction requirements have until recently prevented the correction of atmospheric effects and the subsequent conversion to surface reflectance for large amounts of Landsat data. Automated atmospheric correction algorithms and reliable cloud screening algorithms are available today that allow bulk processing for creating time series of reflectance data (Ju et al. 2012; Masek et al. 2006; Zhu and Woodcock 2012). This increasing availability of Landsat data as well as improved data quality and newly emergent pre-processing algorithms has spurred considerable methodological innovation for time series analyses with Landsat data.

The open data model that was spearheaded with Landsat has since been adapted for several sensor systems, including coarse resolution sensors such as MODIS or MERIS, and higher resolution sensors such as SPOT (i.e. SPOT World Heritage programme, CNES 2014) or the Sentinel programme (ESA 2014). Similarly, the development of important pre-processing (Zhu and Woodcock 2012) and analysis (e.g. Huang et al. 2010; Kennedy et al. 2010; Verbesselt et al. 2010) tools has embarked on a trend towards open source and open access, fostered for example by code sharing platforms and open source programming languages. Distributed and hosted processing services have appeared recently by commercial or scientific institutes (e.g. NASA's Giovanni system, Berrick et al. 2009), several more have been announced in order to address the big data challenges in remote sensing. Simultaneously. Concern regarding observational continuity of long running programs (Wulder et al. 2011) has spurred the discussion regarding suitable imaging systems as a replacement or for different systems to function in a virtual constellation. Time series analytic capabilities can be considerably enhanced in terms of temporal extent or observational density by taking advantage of tools for data fusion (e.g. for merging Landsat and MODIS reflectances, Gao et al. 2006; Zhu et al. 2010) or long term time series models (Jonsson and Eklundh 2004). While these recent developments hold true for a wider range of sensor systems, in the following we focus on Landsat data and time series analyses in the context of land systems.

2.2 Process Dynamics and Time Series Analysis Requirements

Before implementing a time series analysis, it is important to understand which land surface processes are better tackled with time series analyses rather than traditional analyses such as multi-temporal image classification. Generally, analysing phenology-driven, either highly dynamic or gradual and long-term, change processes may profit greatly from the use of time series. Highly dynamic processes require using time series with a sufficient observation density so that 'hot moments' can be captured, while a long-term gradual change signal can only be separated from image or phenological noise when time series are long enough.

Typical processes inducing changes in or modifications of land surface characteristics are manifold. They are often directly driven by human activity (e.g. urbanisation, extension of agricultural areas, open pit mining, deforestation), result partly or indirectly from human action (e.g. climate change-induced vegetation change or glacial melting) or arise from natural processes (e.g. forest windfall, pests, landslides or geological processes, El Ninõ). A detailed monitoring of such processes—for example, quantifying rates of gradual change or detailed identification of dates of abrupt change—requires different analysis schemes. Time series analyses employ the temporal signal at pixel level to derive metrics for mapping or monitoring. Such metrics may include linear or nonlinear trends, amplitude, phase or break points.

Examples from two major domains of remote sensing-based monitoring illustrate the different needs for analyzing different processes: Monitoring agricultural intensification and monitoring deforestation. If we, for example, wish to gain a better understanding of processes related to agricultural intensification in central Europe, we need to get a remote sensing-based characterization of intra- and inter-annual changes in land surface phenology. We need season-specific multi-date measurements across 1 year to describe the diverse agricultural growth, yellowing and harvesting trajectories and we need the same specific multi-date observations between years to describe crop rotation cycles or change trajectories related to cropland-grassland conversions. To map forest changes such as wildfires and deforestation, annual anniversary-date observations may be sufficient. However, in many tropical regions, multiple observations a year are required to obtain a complete cloud-free, annual snap-shot for a region of interest. Actual needs may vary though, depending on the type of forest and ecosystem dynamics.

The examples above show some of the process-related aspects in different ecosystems that steer the applicability of time series approaches. Basically, the process to be monitored, the respective ecosystem dynamics (temporal trajectory of static and changing land use/land cover) and ecosystem heterogeneity (spatial-spectral domain) define the framework needed. In many cases, an ideal approach additionally needs to be adapted to actual data availability, especially when processes are to be assessed over large areas with variable observation density (i.e. extending over several footprints).

2.2.1 Variables Used in Time Series Analysis

Time series analyses have been applied to a wide variety of remote sensing based variables ranging from individual spectral bands and derived indices to remote-sensing-based predictions of land surface or biophysical parameters (Main-Knorn et al. 2013). The choice of variable often depends on the system and change process under study, and also the extent to which a priori information is available to translate surface reflectance to a desired land surface parameter. In the simplest case, a single spectral band might be input to further processing. But in many cases, simple metrics derived from spectral bands enhance the representation of land surface properties such as vegetation greenness—for example, vegetation indices (Choudhury et al. 1994; Huete et al. 2002), tasselled cap transformed bands (Kauth and Thomas 1976) or the outcome of problem-specific spectral mixture analysis (Hostert et al. 2003). It has also been shown that for problems related to classified data, classification probabilities also offer great potential for time series analysis (Yin et al. 2014).

Variables used in time series analyses are, for example:

- Single spectral bands, like the 1.6 μm shortwave infrared in Landsat band 5, which are sensitive for various change processes in forest ecosystems.
- Time series of vegetation indices, commonly used for various vegetation-related analyses.
- Tasseled Cap (TC) components, a combination of brightness and wetness for urban environments, or wetness for forests.
- Integrated indices from tasseled cap; for example, the disturbance index (DI), created for tackling the specific dynamics in temperate forest ecosystems.
- Spectral unmixing fractions, offering a high degree of freedom to focus on question-specific land cover components, such as fractions of concrete in urban environments, soil fractions in agricultural ecosystems or photosynthetic active vegetation for any green vegetation related metrics.
- Biophysical parameters such as canopy cover, leaf area index (LAI), fraction of photosynthetic active radiation (fPAR), and aboveground biomass.

Vegetation indices are commonly used to characterize vegetation dynamics across different temporal scales and processes. In phenological applications, weekly and bi-weekly time series of vegetation indices are used to infer the timing of vegetation green-up and senescence (Zhang et al. 2003). At annual intervals, time series of vegetation indices such as the Normalized Burn Ratio, of TC wetness, and the DI have been used to detect forest changes resulting from clear-cut harvest and insect and fire disturbances (Kennedy et al. 2010). Although, vegetation indices are not tailored to a specific scene or ecosystem, they have been very effective for capturing abrupt and long-term vegetation trends across large areas and multiple biomes. However, in most cases ground measurements are still needed to relate changes in vegetation index derived from time series data to a specific land surface change.

Time series methods also provide a means to directly characterize the temporal dynamics of ecosystem properties (e.g. leaf area index and aboveground biomass) and to improve the mapping of these properties at individual points in time. Biophysical variables are often predicted using radiative transfer models (Myneni et al. 2002) or by building upon the empirical relationship between remote sensing data and ground measurements (Cohen et al. 2003). By incorporating trend information from adjacent time periods, residual noise from atmospheric path radiance and pixel geolocation error can be minimized (smoothing), and predictions can be obtained for periods where surface observations are hindered by clouds or acquisition gaps (gap-filling). For example, Powell et al. (2010) and Main-Knorn et al. (2013) applied a trajectory-based segmentation algorithm (Kennedy et al. 2010) to annual Landsat-based predictions of forest aboveground biomass to characterize forest disturbance related-biomass dynamics. Comparing time-series fitted versus field based biomass estimates showed that the time series algorithm also improved predictions for individual years. Similarly, time series smoothing and gap-filling has been used to improve seasonal representations of biophysical parameters such as MODIS-derived fPAR estimates (Nightingale et al. 2009). Although, seasonal time series of ecosystem properties have traditionally been limited to coarse resolution sensor data like MODIS, such applications will likely be possible for many regions at the Landsat-resolution within the next decade.

Basically, any continuous measure with sensitivity to the process to be monitored may be used as input for a time series analysis. In some cases, as in forestry-related research on temperate forest ecosystems, there are established measures (e.g. TC wetness, DI, Healey et al. 2005). However, tackling new research questions often requires sensitivity analyses on the usefulness of different indicators for running a time series analysis prior to actually doing so.

2.2.2 Implementing Time Series Analyses

It is obvious that different process regimes ask for different temporal resolutions and appropriate methods to deal with analysing temporally varying information. The density of suitable data can be an issue in many time series approaches, despite the enormous increase of data available today.

Theoretically—disregarding cloud cover or downlink capacities—with two Landsat platforms operating simultaneously, an 8-day revisit provides about 45 acquisitions per footprint and year. This allows for assessing most process regimes within terrestrial ecosystem dynamics. However, global availability of Landsat data within the USGS archive shows pronounced regional differences (Kovalskyy and Roy 2013). This is due to cloud cover but also to the fact that a truly global acquisition strategy has only been established during the later years of the Landsat 7 mission. For areas outside of the U.S., less data has been acquired and much of the archived imagery resides in data holdings of the cooperating international receiving stations. Therefore in many areas around the world the ideal time series approach cannot be pursued due to data constraints, and trade-offs need to be considered regarding the established time series and the choice of analysis methods. Scale considerations might also require trade-offs as certain time series characteristics might be ensured for a smaller area, but providing the same characteristics for larger regions such as entire continents is often hardly feasible.

Conceptually, we may consider different types of time series for optical remote sensing data:

- Original image acquisitions: For some applications, such as those with a focus on regions with low cloudiness like deserts and semi-deserts, data availability from Landsat overpasses every 16 days may suffice when using original imagery or mosaicked data ('native data'). For many applications, though, data on a scene-by-scene basis does not suffice and image compositing techniques are required to increase temporal data density.
- Data spacing: Some analyses require equally spaced data, such as annual peak phenology, while others can deal with the entire acquisition record of usually unequally spaced time series.
- Intra- or inter-annual analyses: Depending on the process under research, intra- versus inter-annual time series or combinations of the two might be appropriate.

Summarizing, time series analysis is based on one or several indicators from which appropriate temporal metrics are extracted, such as early maxima of a vegetation index (either allowing or not allowing gaps in the time series), the slope of a linear trend along time series values, a sinusoidal function fitted to a phenological trajectory over several years, or simply a breakpoint in the time series indicating change. Recent research indicates that this selection process can be performed most efficiently by machine learning algorithms, e.g. support vector machines or random forests (Senf et al. 2013). This holds specifically true if tens or hundreds of metrics are calculated. Mapping results from the entire time series analysis process may then again be categorical after classifying the created metrics (classes derived from a time series, such as deforestation, cropping systems or cycles) or continuous (gradients of shrub encroachment on abandoned farmland or soil fraction changes during erosion processes) when using metrics as input to a regression analysis.

2.3 Time Series Analysis Examples

In the following we provide two examples to illustrate the conceptualization of time series analyses. Both examples demonstrate the opportunities and challenges of current time series analysis using freely available Landsat data.

2.3.1 Monitoring Tropical Deforestation

Mapping and monitoring tropical deforestation and forest degradation has been a focus of remote sensing research for several decades (Mayaux and Lambin 1995; Skole and Tucker 1993; Woodwell et al. 1987). Estimating the rate of change in tropical forest cover is a key requirement for understanding the impact of human activities on the Earth's climate, the functioning of ecosystems, and biodiversity (Lambin et al. 2003). While tremendous progress has been made in developing remote sensing-based methods and protocols for estimating tropical deforestation rates from local to pan-tropical scales (GOFC-GOLD 2013), in many regions accurate forest monitoring has only recently become feasible with dense optical time series of high spatial resolution (<50 m). The necessity of high spatial resolution sensor data for monitoring tropical forest changes arises from the complexity of forest change patterns, which range globally from large clearings for mechanized agriculture to fragmented clearings associated with small-holder agriculture. The forest definition, as agreed in the Marrakech accords (UNFCC 2002), establishes a 0.05–1 ha minimum area, a minimum of 10–30 % tree canopy cover, and a potential of 2–5 m tree height for forest. Such small-scale changes cannot be resolved sufficiently with coarse resolution sensor data.

Prior to the opening of the USGS Landsat archive, the Global Land Survey (GLS) epochal dataset (Gutman et al. 2008) provided the first time series of orthorectified, high-resolution images that covered nearly the entire tropical region. However, the epochal resolution has proven to be insufficient for detecting forest change in some regions or for detecting transient forest change. In Indonesia, where the majority of land-use change comes from conversions of tropical forest to tree and oil palm plantations, Broich et al. (2011) showed that time-series approaches based on all good land observations were more accurate in mapping forest cover change than change maps based on epochal image composites. Similarly, shifting cultivation leads to a temporary removal of forest cover followed by a short period of cultivation and subsequent fallow regrowth. In regions of Southeast Asia, where shifting cultivation is the dominant land use, forest changes occur within complex mosaics of primary and secondary forest vegetation (Ziegler et al. 2012). The detection of clearings from shifting cultivation is complicated by the pronounced vegetation phenology in tropical dry and seasonal forest, where the clearing and burning usually occurs towards the end of the dry season, and by the fast recovery of the spectral signal when the land is left fallow (Fig. 2.1).

To detect forest clearings in northern Laos, Pflugmacher et al. (2014) constructed annual time series of the Normalized Burn Ratio (NBR) based on all available Landsat imagery with a cloud cover of less than 75 %. The NBR index contrasts Landsat's 2.2 μm short-wave infrared (SWIR2) band with the near-infrared (NIR) band (NBR = (Band 4 − Band 7) / (Band 4 + Band 7); Key and Benson 2006). NBR can theoretically range from −1 to 1, where low and negative values are associated with sparsely- or un-vegetated areas and high values are associated with dense vegetation. The difference in the NBR between two images

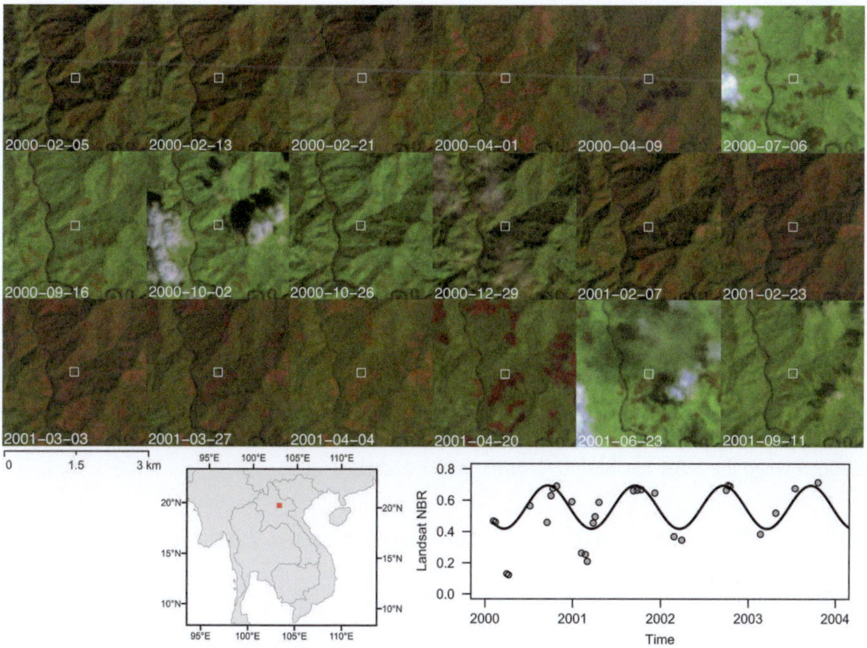

Fig. 2.1 Landsat image chips (bands 5-4-3) showing the spatial and temporal patterns of slash and burn agriculture of a small area (3,000 × 3,000 m) in Houaphan Province, Lao PDR. A single crop of rice is followed by fallow periods of 2–10 years; 5–10 % of the area is used for successive crops of rice for 2–5 years. The time series plot in the lower right shows the Normalized Burn Ratio (NBR) from all cloud-free Landsat imagery between 2000 and 2004. To illustrate the seasonal variations in NBR a simple harmonic function fitted to the entire pixel time series is overlaid

(ΔNBR) has been most prominently used to map burned area and burn severity of vegetation fires (Roy et al. 2008). However, due to its sensitivity to forest structure and vegetation moisture, ΔNBR has also been used to map forest disturbance caused by wind-throw, logging, and insects in temperate forests (Kennedy et al. 2012). Further, NBR is less sensitive to atmospheric water vapour and increased aerosols during the burning season, compared to indices that use Landsat's visible bands. In order to minimize the effects of vegetation phenology on ΔNBR between years, Pflugmacher et al. (2014) used annual, minimum-value composites of NBR for the dry season (Fig. 2.2). The approach requires images that are atmospherically corrected (Masek et al. 2006) and cloud-masked (Zhu and Woodcock 2012).

Time series approaches that incorporate explicit seasonal models (Verbesselt et al. 2010) or more complex temporal segmentation algorithms (Kennedy et al. 2010) have recently become available, but these have not been rigorously tested in the tropics, yet. However, it has already become apparent that the opening of the Landsat archive and the release of the Landsat Climate Data Record in November 2013 have greatly enhanced the monitoring of tropical forest changes.

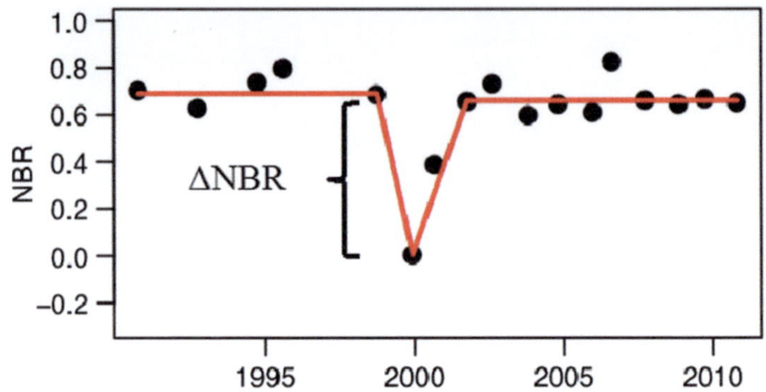

Fig. 2.2 Annual time series of the minimum Normalized Burn Ratio (NBR) for the dry season (October–April) from a single pixel cleared in 2000 (Houaphan, Lao PDR). The red line connects the undisturbed, disturbed and recovered periods via linear segments. The timing of recovery is defined when the NBR reaches 90 % of the pre-disturbance value

2.3.2 Mapping Pan-Carpathian Agricultural Land Use Change

The second example is based in a temperate European setting and illustrates how changes in agriculture production systems can be assessed for larger regions using historical Landsat data. The study focused on the extended Carpathian Mountain range in Eastern Europe, a region comprising of approximately 375,000 km^2 and currently extending over seven individual countries. Until 1989, almost the entire region was under the political influence of the Soviet Union, which included collectivized agricultural production. Being heavily subsidized and connected to large output markets, basically all cultivatable land was under agricultural production. With the collapse of Eastern European Socialism, agricultural production experienced a drastic shock and most production ceased, leading to extensification through conversion of cropland to grassland, or the abandonment of cropland with subsequent shrub encroachment. Since 2004, most countries have been accessioned to the European Union, which again affected agricultural land use considerably, including re-cultivation efforts or intensification (Griffiths et al. 2013a).

The open Landsat archive provides a unique opportunity for assessing the spatio-temporal patterns of these crop regime changes over the entire region since 1984. For this, at least 32 Landsat footprints need to be included into the analysis scheme. The region comprises an extensive altitudinal gradient, from plains over foothills to high mountain valleys. Cloud coverage is generally profound. The USGS Landsat holdings for the period of 1984 (the first year of Landsat 5 TM data) to 2012 are heterogeneous, with relative data scarcity for the 1980s, basically no data available between 1995 and 1999, but better data availability after 2000 including Landsat 7 imagery that suffered from the defect of the scan line corrector after May 2003. From a time series methods perspective, two spectral-temporal features need to be captured when attempting to assess these changes in cropping regimes:

(1) observations from the spring, summer and fall seasons, which are required to reliably separate cropping systems from grassland land use, and (2) a series of such season-specific observations over the different time periods—socialism, transition and post-EU accession—to be able to detect changes among these different agricultural land use regimes. Image compositing algorithms were used to approximate the desired series of season-specific observations over the entire region (Griffiths et al. 2013a, b). Overall, the dataset included cloud-free multi-season composites for 1984, 2000, and 2010 to represent the socialist, transition and post-EU accession periods respectively. In general, multi-year imagery is required to achieve a complete cloud-free coverage for the region.

First, all available Landsat L1T imagery with a metadata-based cloud cover estimate of no more than 70 % was downloaded (approximately 5,000 images). All images were atmospherically corrected and converted to surface reflectance (Masek et al. 2006), and subsequently cloud/shadow masks were developed (Zhu and Woodcock 2012). To ensure the ability to process imagery on a per-pixel basis over several Universal Transverse Mercator (UTM) zones, all data was then transferred into a common projected continental coordinate system. The applied compositing algorithm provides three types of outputs (Fig. 2.3): best observation composites, spectral-temporal variability metrics and metadata layers (Griffiths et al. 2013b). For the first, all available pixel observations for a given time window were evaluated based on the image metadata (e.g. acquisition year and day-of-year (DOY)) and image characteristics (e.g. pixels distance to clouds). A score was produced for each observation based on these different characteristics and the pixel observation with the highest score was transferred into the best observation composite for a certain year and season (for details on the generation of scores, refer to Griffiths et al. (2013b)). Spectral-temporal variability metrics were calculated for defined seasonal windows based on all cloud free observations available for a given yearly range. These metrics were based on the NDVI and included the seasonal mean, range and standard deviation. Metadata layers provided not only the origin of each pixel, but also the evaluated score and number of cloud free observations on a per-pixel level.

Once the data were composited and all metrics were derived, supervised Random Forest (RF) models were used to capture the changes of interest (Griffiths et al. 2013a). Training data were collected for a total of 15 stable or transition classes, including cropland-grassland-forest transitions as well as stable cropland or forest classes. The RF model was then applied to a combined dataset containing the best observation composites (3×6 spectral bands for three reference periods, 54 bands in total), the variability metrics (3×3 metrics for the three reference periods, equalling 27 bands) and selected metadata layers (e.g. day-of-year, year, compositing score), resulting in a total of 100 bands. The overall accuracy of the resulting change map was over 90 %. Individual transition classes were subsequently recoded to assess agricultural regime changes. For example, agricultural abandonment included cropland-grassland conversions but also cropland-forest conversions. Details on the processing, mapping and result interpretation are provided in Griffiths et al. (2013a).

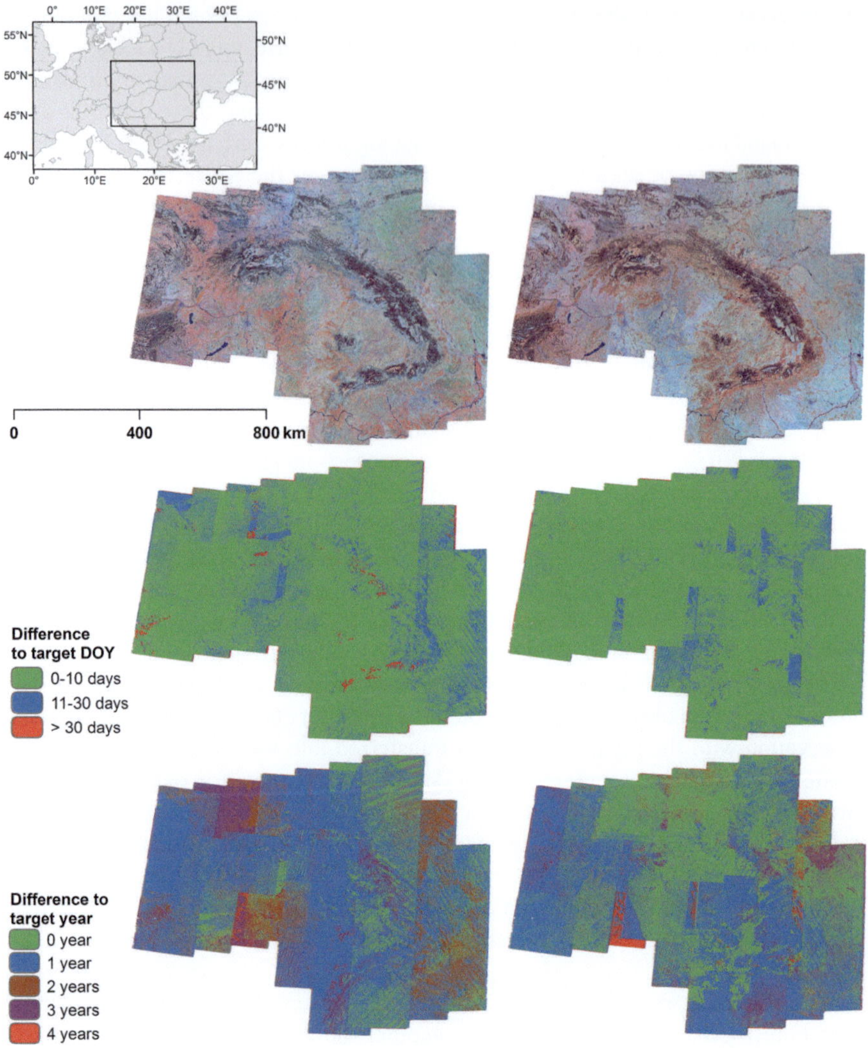

Fig. 2.3 Results of the seasonal compositing of Landsat data for the target year 2010. The top row shows the spring and fall best observation composite on the left and right, respectively (bands 4-5-3). The seasonal and annual characteristics of these composites are then summarized, providing the offset to the respective target day-of-year (*middle*) and the respective acquisition years (*bottom*) for each pixel. The overview map (*top left*) shows the approximate location

2.4 Challenges and Opportunities

The development of time series analysis approaches has grown swiftly during the last years, as the opening of the Landsat archive and streamlined pre-processing and data delivery created a surge of new methods and applications. The ongoing "repatriation" of Landsat imagery from distributed global archives into the USGS archive will create further opportunities; for example, the European archive alone holds 1.5 million unique Landsat scenes not available elsewhere. Yet, the remote sensing community is just at the verge of a new era, where accelerating global change meets increasing capabilities for remote sensing data analysis. Forthcoming challenges and opportunities relate to three intertwined domains.

From an application-focused perspective, the need for new information products is evident. Classic land use / land cover change analyses need to be improved; for example, maps of heterogeneous (e.g. savannah landscapes) or dynamic ecosystems (e.g. shifting cultivation systems in the tropics) are still deficient. Beyond this evident need for enhancements, there is also the need to create more problem-focused products, such as Essential Climate Variables from remote sensing data. Without such seamless global databases, climate impact research will fail to create vital information for climate policy decision makers. Furthermore, information that supports deciphering land use intensification or extensification patterns where land use itself has not changed is urgently needed to understand limitations in our use of natural resources globally. Both will only be possible with spatial high-resolution dense time series analysis capable of capturing subtle and long-term change patterns.

From a methods perspective, time series densification depends on improved image compositing techniques. In a perfect remote sensing setup, data availability would not be an issue. However, useful optical data are not available with every satellite overpass because of varying cloud coverage across the globe, daily variation in illumination and atmospheric conditions, as well as data storage and downlink-related acquisition constraints. Gaining the maximum amount of usable pixels from partially cloud covered datasets is hence key to further increased data density, which in turn is mandatory to successfully monitor processes that are at the same time driven by phenology and land use.

From a sensor-focused perspective, forthcoming improvements in time series analyses will depend on our ability to create time series from multiple sensors, often referred to as "virtual sensor constellations". Few studies have used multiple sensors to improve the temporal resolution of time series, mostly employing different sensors of the Landsat family (Hostert et al. 2003; Main-Knorn et al. 2013; Pflugmacher et al. 2012). Creating sensor constellations is a non-trivial task, though, and the remote sensing community is just starting to develop operational constellation setups. Future virtual constellations will also focus on the integration of optical and radar data or multi-scale optical data. The latter has been pioneered based on multi-sensor fusion approaches such as spatio-temporal adaptive reflective fusion model (STARFM, Hilker et al. 2009). The most promising approach, however, will certainly be the integration of new and next-generation optical land imaging systems with similar characteristics. Specifically, the USGS Landsat family and ESA-based Sentinel-2 missions will create unprecedented virtual constellation synergies (Drusch et al. 2012), coming close to the temporal density of coarse-resolution monitoring systems.

Acknowledgements We greatly acknowledge support by the German Federal Ministry for Economic Affairs and Energy (BMWi) in the frame of the Sense Carbon project (Project no. 50EE1254). We are also grateful to contributions by the EU-FP7-funded research project I-REDD+ (Grant Agreement No 265286). This chapter is part of research framed by the USGS-NASA Landsat Science Team 2012–2016.

References

Berrick SW, Leptoukh G, Farley JD, Rui H (2009) Giovanni: a web service workflow-based data visualization and analysis system. IEEE Trans Geosci Remote Sens 47:106–113

Broich M, Hansen MC, Potapov P, Adusei B, Lindquist E, Stehman SV (2011) Time-series analysis of multi-resolution optical imagery for quantifying forest cover loss in Sumatra and Kalimantan, Indonesia. Int J Appl Earth Obs Geoinf 13:277–291

Chander G, Markham BL, Helder DL (2009) Summary of current radiometric calibration coefficients for Landsat MSS, TM, ETM+, and EO-1 ALI sensors. Remote Sens Environ 113:893–903

Choudhury BJ, Digirolamo NE, Dorman TJ (1994) A comparison of reflectances and vegetation indices from three methods of compositing the AVHRR-GAC data over Northern Africa. Remote Sens Rev 10:245–263

CNES (2014) CNES's SPOT World Heritage programme to provide free SPOT satellite archive imagery over five years old for non-commercial uses [online]. Available from: http://www.cnes.fr/web/CNES-en/7134-press-releases.php?item=8193. Accessed Oct 2014

Cohen WB, Maiersperger TK, Gower ST, Turner DP (2003) An improved strategy for regression of biophysical variables and Landsat ETM+ data. Remote Sens Environ 84:561–571

Drusch M, Del Bello U, Carlier S, Colin O, Fernandez V, Gascon F, Hoersch B, Isola C, Laberinti P, Martimort P, Meygret A, Spoto F, Sy O, Marchese F, Bargellini P (2012) Sentinel-2: ESA's optical high-resolution mission for GMES operational services. Remote Sens Environ 120:25–36

ESA (2014) Free access to COPERNICUS Sentinel satellite data [online]. Available from: http://www.esa.int/Our_Activities/Observing_the_Earth/Copernicus/Free_access_to_Copernicus_Sentinel_satellite_data. Accessed Oct 2014

Gao F, Masek J, Schwaller M, Hall F (2006) On the blending of the Landsat and MODIS surface reflectance: predicting daily Landsat surface reflectance. IEEE Trans Geosci Remote Sens 44:2207–2218

GOFC-GOLD (2013) A sourcebook of methods and procedures for monitoring and reporting anthropogenic greenhouse gas emissions and removals associated with deforestation, gains and losses of carbon stocks in forests remaining forests, and forestation. GOFC-GOLD Land Cover Project Office, Wageningen University, The Netherlands

Griffiths P, Müller D, Kuemmerle T, Hostert P (2013a) Agricultural land change in the Carpathian ecoregion after the breakdown of socialism and expansion of the European Union. Environ Res Lett 8:045024

Griffiths P, van der Linden S, Kuemmerle T, Hostert P (2013b) A pixel-based Landsat compositing algorithm for large area land cover mapping. IEEE J Sel Top Appl Earth Obs Remote Sens 6:2088–2101

Gutman G, Byrnes R, Masek J, Covington S, Justice C, Franks S, Headley R (2008) Towards monitoring land-cover and land-use changes at a global scale: the Global Land Survey 2005. Photogramm Eng Remote Sens 74:6–10

Healey SP, Cohen WB, Yang ZQ, Krankina ON (2005) Comparison of Tasseled Cap-based Landsat data structures for use in forest disturbance detection. Remote Sens Environ 97:301–310

Hilker T, Wulder M, Coops N, Seitz N, White J, Gao F, Masek J, Stenhouse G (2009) Generation of dense time series synthetic Landsat data through data blending with MODIS using a spatial and temporal adaptive reflectance fusion model. Remote Sens Environ 113:1988–1999

Hostert P, Roder A, Hill J (2003) Coupling spectral unmixing and trend analysis for monitoring of long-term vegetation dynamics in Mediterranean rangelands. Remote Sens Environ 87:183–197

Huang C, Goward SN, Masek JG, Thomas N, Zhu Z, Vogelmann JE (2010) An automated approach for reconstructing recent forest disturbance history using dense Landsat time series stacks. Remote Sens Environ 114:183–198

Huete A, Didan K, Miura T, Rodriguez EP, Gao X, Ferreira LG (2002) Overview of the radiometric and biophysical performance of the MODIS vegetation indices. Remote Sens Environ 83:195–213

Jonsson P, Eklundh L (2004) TIMESAT – a program for analyzing time-series of satellite sensor data. Comput Geosci 30:833–845

Ju J, Roy DP, Vermote E, Masek J, Kovalskyy V (2012) Continental-scale validation of MODIS-based and LEDAPS Landsat ETM+ atmospheric correction methods. Remote Sens Environ 122:175–184

Kauth RJ, Thomas GS (1976) The tasselled cap – a graphic description of the spectral-temporal development of agricultural crops as seen by Landsat. Symposium on machine processing of remotely sensed data. The Laboratory for Applications of Remote Sensing, Purdue University, West Lafayette, IN, USA

Kennedy RE, Yang Z, Cohen WB (2010) Detecting trends in forest disturbance and recovery using yearly Landsat time series: 1. LandTrendr – temporal segmentation algorithms. Remote Sens Environ 114:2897–2910

Kennedy RE, Yang Z, Cohen WB, Pfaff E, Braaten J, Nelson P (2012) Spatial and temporal patterns of forest disturbance and growth within the area of the Northwest Forest Plan. Remote Sens Environ 122:117–133

Key CH, Benson NC (2006) Landscape Assessment (LA). FIREMON: Fire effects monitoring and inventory systems. In: Lutes DC, Keane RE, Carati JF, Key CH, Benson NC (eds) General technical report RMRS-GTR-164-CD. USDA Forest Service, Rocky Mountains Research Station, Fort Collins, CO

Kovalskyy V, Roy DP (2013) The global availability of Landsat 5 TM and Landsat 7 ETM+ land surface observations and implications for global 30 m Landsat data product generation. Remote Sens Environ 130:280–293

Lambin EF, Strahler AH (1994) Change-vector analysis in multitemporal space – a tool to detect and categorize land-cover change processes using high temporal-resolution satellite data. Remote Sens Environ 48:231–244

Lambin EF, Geist HJ, Lepers E (2003) Dynamics of land-use and land-cover change in tropical regions. Annu Rev Environ Resour 28:205–241

Landsat Project Science Office (2010) Landsat 7 science data users handbook. Goddard Space Flight Center in Greenbelt, NASA, Maryland

Main-Knorn M, Cohen WB, Kennedy RE, Grodzki W, Pflugmacher D, Griffiths P, Hostert P (2013) Monitoring coniferous forest biomass change using a Landsat trajectory-based approach. Remote Sens Environ 139:277–290

Masek JG, Vermote EF, Saleous NE, Wolfe R, Hall FG, Huemmrich KF, Gao F, Kutler J, Lim TK (2006) A Landsat surface reflectance dataset for North America, 1990–2000. IEEE Geosci Remote Sens Lett 3:68–72

Mayaux P, Lambin EF (1995) Estimation of tropical forest area from coarse spatial-resolution data – a 2-step correction function for proportional errors due to spatial aggregation. Remote Sens Environ 53:1–15

MEA (2005) Ecosystems and human well-being: synthesis. In: World Resources Institute (ed) Millennium ecosystem assessment. Island Press and World Resources Institute, Washington, DC

Myneni RB, Hoffman S, Knyazikhin Y, Privette JL, Glassy J, Tian Y, Wang Y, Song X, Zhang Y, Smith GR, Lotsch A, Friedl M, Morisette JT, Votava P, Nemani RR, Running SW (2002) Global products of vegetation leaf area and fraction absorbed PAR from year one of MODIS data. Remote Sens Environ 83:214–231

Nightingale J, Morisette J, Wolfe R, Tan B, Gao F, Ederer G, Collatz G, Turner D (2009) Temporally smoothed and gap-filled MODIS land products for carbon modelling: application of the f PAR product. Int J Remote Sens 30:1083–1090

Pflugmacher D, Cohen WB, Kennedy RE (2012) Using Landsat-derived disturbance history (1972–2010) to predict current forest structure. Remote Sens Environ 122:146–165

Pflugmacher D, Grogan K, Thongmanivong S, Hett C, Fensholt R, Hostert P (2014) Historic trends of deforestation and forest degradation in Houaphan province, Laos. Impacts of reducing emissions from deforestation and forest degradation and enhancing forest carbon stocks (I-REDD+), Deliverable 3.4

Powell SL, Cohen WB, Healey SP, Kennedy RE, Moisen GG, Pierce KB, Ohmann JL (2010) Quantification of live aboveground forest biomass dynamics with Landsat time-series and field inventory data: a comparison of empirical modeling approaches. Remote Sens Environ 114:1053–1068

Roy DP, Boschetti L, Justice CO, Ju J (2008) The collection 5 MODIS burned area product – global evaluation by comparison with the MODIS active fire product. Remote Sens Environ 112:3690–3707

Roy DP, Wulder MA, Loveland TR, Woodcock CE, Allen RG, Anderson MC, Helder D, Irons JR, Johnson DM, Kennedy R, Scambos TA, Schaaf CB, Schott JR, Sheng Y, Vermote EF, Belward AS, Bindschadler R, Cohen WB, Gao F, Hipple JD, Hostert P, Huntington J, Justice CO, Kilic A, Kovalskyy V, Lee ZP, Lymburner L, Masek JG, McCorkel J, Shuai Y, Trezza R, Vogelmann J, Wynne RH, Zhu Z (2014) Landsat-8: science and product vision for terrestrial global change research. Remote Sens Environ 145:154–172

Senf C, Pflugmacher D, van der Linden S, Hostert P (2013) Mapping rubber plantations and natural forests in Xishuangbanna (Southwest China) using multi-spectral phenological metrics from MODIS time series. Remote Sens 5:2795–2812

Skole D, Tucker C (1993) Tropical deforestation and habitat fragmentation in the Amazon – satellite data from 1978 to 1988. Science 260:1905–1910

UNFCC (2002) Report of the conference of the parties on its seventh session, held in Marrakech from Oct–Nov 2001. In: UNFCCC (eds) Addendum Part Two: action taken by the conference of the parties, FCCC/CP/2001/13/Add. 1, vol I. United Nations Framework Convention on Climate Change Secretariat, Bonn

USGS (2012) Landsat data continuity mission (LDCM) Level 1 (L1) Data format control book (DFCB). USGS, Sioux Falls

Verbesselt J, Hyndman R, Newnham G, Culvenor D (2010) Detecting trend and seasonal changes in satellite image time series. Remote Sens Environ 114:106–115

Vitousek PM (1997) Human domination of Earth's ecosystems. Science 277:494–499

Woodcock CE, Allen R, Anderson M, Belward A, Bindschadler R, Cohen W, Gao F, Goward SN, Helder D, Helmer E, Nemani R, Oreopoulos L, Schott J, Thenkabail PS, Vermote EF, Vogelmann J, Wulder MA, Wynne R (2008) Free access to Landsat imagery. Science 320:1011

Woodwell GM, Houghton RA, Stone TA, Nelson RF, Kovalick W (1987) Deforestation in the tropics – new measurements in the Amazon Basin using Landsat and NOAA advanced very high-resolution radiometer imagery. J Geophys Res-Atmos 92:2157–2163

Wulder MA, White JC, Masek JG, Dwyer J, Roy DP (2011) Continuity of Landsat observations: short term considerations. Remote Sens Environ 115:747–751

Wulder MA, Masek JG, Cohen WB, Loveland TR, Woodcock CE (2012) Opening the archive: how free data has enabled the science and monitoring promise of Landsat. Remote Sens Environ 122:2–10

Yin H, Pflugmacher D, Kennedy RE, Sulla-Menashe D, Hostert P (2014) Mapping annual land use and land cover changes using MODIS time series. IEEE J Sel Top Appl Earth Obs Remote Sens 7:3421–3427

Zhang X, Friedl MA, Schaaf CB, Strahler AH, Hodges JCF, Gao F, Reed BC, Huete A (2003) Monitoring vegetation phenology using MODIS. Remote Sens Environ 84:471–475

Zhu Z, Woodcock CE (2012) Object-based cloud and cloud shadow detection in Landsat imagery. Remote Sens Environ 118:83–94

Zhu XL, Chen J, Gao F, Chen XH, Masek JG (2010) An enhanced spatial and temporal adaptive reflectance fusion model for complex heterogeneous regions. Remote Sens Environ 114:2610–2623

Ziegler AD, Phelps J, Yuen JQ, Webb EL, Lawrence D, Fox JM, Bruun TB, Leisz SJ, Ryan CM, Dressler W, Mertz O, Pascual U, Padoch C, Koh LP (2012) Carbon outcomes of major land-cover transitions in SE Asia: great uncertainties and REDD plus policy implications. Glob Chang Biol 18:3087–3099

Chapter 3
Calibration and Pre-processing of a Multi-decadal AVHRR Time Series

Martin Bachmann, Padsuren Tungalagsaikhan, Thomas Ruppert, and Stefan Dech

Abstract Since the early 1980s, the German Remote Sensing Data Centre (DFD) of the German Aerospace Centre (DLR) has received archived and processed Advanced Very High Resolution Radiometer (AVHRR) data from the Polar Orbiting Environmental Satellites (POES) of the National Oceanic and Atmospheric Administration (NOAA). By December 2013, over 237,000 paths over Europe have since been archived at DLR. Based on these High Resolution Picture Transmission (HRPT) raw datasets, an operational pre-processing and value-adding chain has been developed (Dech et al., Aerosp Sci Technol 2(5):335–346, 1998; Tungalagsaikhan et al., Proc. 23th DGPF (12), 2003). In this chapter, the series of AVHRR sensors is introduced, and information on calibration and system correction procedures is given. Next, the pre-processing part of DLR's processing chain is described, where focus is set on the calibration aspects. Time series examples are provided to show the influence of changes in calibration over time, and to illustrate the need for consistent pre-processing and data harmonization. According to these requirements DLR's multi-decadal archive of AVHRR data will be re-processed in the frame of the TIMELINE project, providing consistent and well-calibrated time series data.

M. Bachmann (✉) • P. Tungalagsaikhan • T. Ruppert
German Remote Sensing Data Center, DFD, Earth Observation Center, EOC,
German Aerospace Center, DLR, Oberpfaffenhofen, Germany
e-mail: Martin.Bachmann@dlr.de

S. Dech
German Remote Sensing Data Center, DFD, German Aerospace Center, DLR, Wessling, Germany

Institute for Geography and Geology, University of Wuerzburg, Wuerzburg, Germany

C. Kuenzer et al. (eds.), *Remote Sensing Time Series*, Remote Sensing and Digital Image Processing, DOI 10.1007/978-3-319-15967-6_3

3.1 Introduction

Within a changing global environment, it is of high importance to monitor the state of the Earth's surface and atmosphere. Therefore high-level initiatives, such as the Global Climate Observing System (GCOS) supported by the United Nations Framework Convention on Climate Change (UNFCCC) and the Intergovernmental Panel on Climate Change (IPCC), have set up requirements for Earth Observation (EO) data products in order to meet the needs of the climate modelling community. The resulting set of Essential Climate Variables (ECVs) includes key atmospheric variables (e.g. aerosol and cloud properties), oceanic variables (e.g. sea surface temperature and ocean color) and terrestrial variables (e.g. albedo and snow cover) (GCOS/WMO 2009). Similar initiatives to harmonize monitoring schemes using Earth observation techniques are in development, such as the Essential Biodiversity Variables (EBVs) from the Group on Earth Observations Biodiversity Observation Network (GEO BON, see Pereira et al. 2013).

As a response, initiatives such as the ESA Climate Change Initiative (CCI) are answering to these needs for EO data, and are thus setting requirements for consistent global time series data using multiple sensors such as MERIS (Medium Resolution Imaging Spectrometer), SPOT (Satellite Pour l'Observation de la Terre) Vegetation 1 & 2, and (A)ATSR ((Advanced) Along Track Scanning Radiometer). These are wide swath sensors that provide data at coarse spatial resolutions (300 m for MODIS to over 1 km for (A)ATSR and SPOT Vegetation), (quasi-) global coverage, and high temporal resolution revisit time of 1–3 days. As data from these sensors has only been available since 1991, in the case of ATSR, and from 2002, in the case of MERIS (or from 1999 onward for the comparable MODIS on Terra), there is a significant demand for EO data dating back even further.

A unique source of EO data with long term continuity dating back as far as 1978 is available from the AVHRR (Advanced Very High Resolution Radiometer) series of sensors. By using AVHRR, regional and global monitoring applications can therefore achieve long-term coverage (e.g. Gutman and Masek 2012; Swinnen and Veroustraete 2008) and consequently are able to provide the required input for global models (e.g. Sellers et al. 1997). Using AVHRR, a variety of thematic products can be derived on an operational basis, including land and sea surface temperature products (LST and SST, e.g. Frey et al. 2012; EUMETSAT 2011a; Casey et al. 2010), vegetation indices such as the Normalized Difference Vegetation Index (NDVI, e.g. Dech et al. 1998; Fontana et al. 2012), and a number of advanced cloud properties, such as cloud top temperatures or the liquid/ice water path (e.g. Kriebel et al. 2003; EUMETSAT 2010).

However, without careful and consistent pre-processing, such applications of AVHRR data are not possible as uncorrected sensor-related effects may be mistaken as trends in the observed thematic time series (e.g. Beck et al. 2011). Thus, the following presents important details on the series of AVHRR sensors, the processing requirements and examples for time series data. Focus is on pre-processing aspects which have a large impact on the consistency of the generated time series data, such as the radiometric calibration procedures of reflective and thermal bands, changes in the spectral response functions, and correction approaches improving the geolocation of the imagery.

3.2 The Advanced Very High Resolution Radiometer (AVHRR) Instrument Series

The NOAA AVHRR fleet is a series of cross-track whisk-broom instruments starting with the launch of the first AVHRR sensor in 1978 on board the NOAA TIROS-N satellite. Since then the AVHRR series of instruments was successfully operated on board of 14 of the NOAA polar orbiting environmental satellites (POES) and on EUMETSAT MetOp-A and B. In addition, AVHRR sensors were also on board of the failed missions NOAA-B and NOAA-13. Up-to-date and historical information on the status of NOAA and MetOp platforms and including the corresponding instrumentations is made available through the Spacecraft Status Summaries hosted at the NOAA Office of Satellite Operations (OSO) (see http://www.oso.noaa.gov/poesstatus/). Regarding data processing aspects of the AVHRR instruments, frequently updated status information is hosted by the NOAA Satellite and Product Operations (OSPO) (see http://www.ospo.noaa.gov/Products/ppp/notices.html).

3.2.1 Overview of the NOAA AVHRR Series

Within the long history of the AVHRR series, three versions of the instrument exist, denoted AVHRR, AVHRR/2 and AHVHRR/3 (Kramer 2002; Kidwell 1998; Robel 2009). Common to all AVHRR sensors is the wide Field of View (FOV) of +/–55.4° resulting in a swath width of ~2,400–3,000 km (depending on the spacecraft orbit), and having an Instantaneous Field of View (IFOV) of approximately 1.4 mrad for each channel. This results in a symmetrical ground-projected IFOV at nadir of approximately 1.1 km, which increases towards the edges of the swath to approximately 6.2 km (cross-track) by 2.3 km (along-track). For all instruments, data is digitized to 10 bit precision.

An important change between the instruments is an increase in the number of bands from four to five and finally to six (Tables 3.1 and 3.2, Fig. 3.1), which have subsequently improved the mapping capabilities in the thermal and short-wave infrared domain. However, as can be seen from Figs. 3.2 and 3.3, the spectral response functions are slightly different for each individual AVHRR instrument, both between and within an instrument series. Note that the underlying response function measurements are derived pre-launch and as changes in orbit have occurred, these response functions are reported to have shifted (Trishchenko 2009).

Table 3.1 Overview of the NOAA AVHRR series of instruments

Instrument series	Satellite	Channels
AVHRR	TIROS-N, NOAA-6, 8 and 10	1, 2, 3, 4
AVHRR/2	NOAA-7, 9, 11, 12 and 14	1, 2, 3, 4, 5
AVHRR/3	NOAA-15 onward to NOAA-19, MetOP-A and -B	1, 2, 3A, 3B, 4, 5

Table 3.2 Channels and spectral sampling range of the NOAA AVHRR series of instruments

Channel	Approximate spectral sampling range
1	0.58–0.68 μm
2	0.72–1.10 μm
3	3.55–3.93 μm
3A	1.58–1.64 μm
3B	3.55–3.93 μm
4	10.5–11.5 μm (first AVHRR series)
	10.3–11.3 μm (AVHRR/2 and/3)
5	11.5–12.5 μm

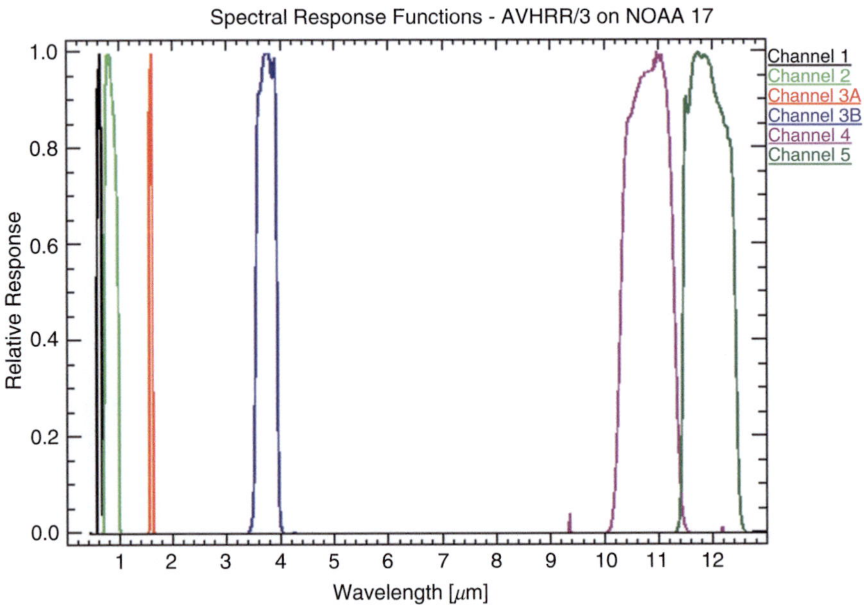

Fig. 3.1 Overview of the spectral response functions of AVHRR/3 on NOAA 17

In order to provide a rough estimation of the influence of different spectral response functions of the various AVHRR sensors, a typical vegetation reflectance signature with a spectral resolution of 5 nm was resampled to the response of channels 1 and 2 of NOAA 18 and 19 (Fig. 3.4). In this example, the relative difference in channel 1 between NOAA 18 and 19 is 0.5 %, for channel 2 the

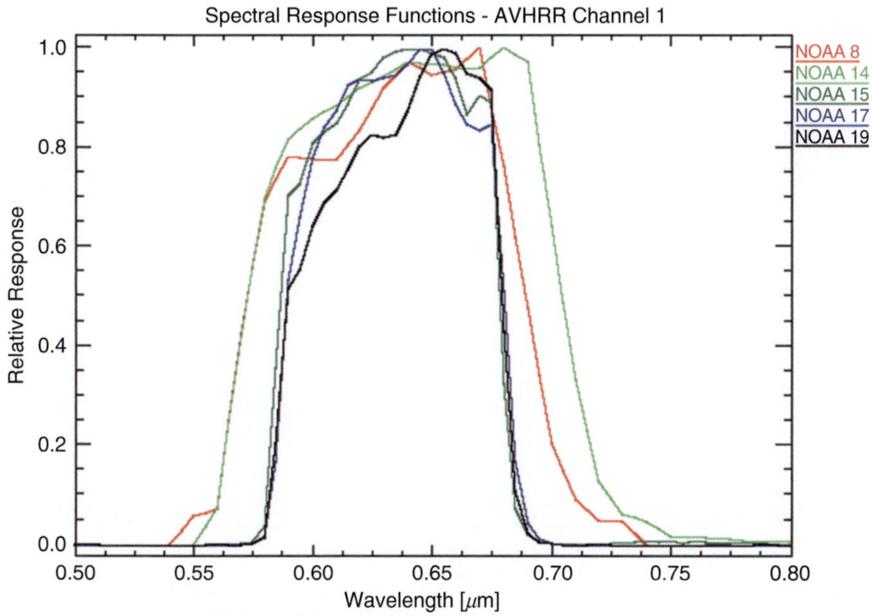

Fig. 3.2 Spectral response functions of channel 1 of various AVHRR instruments

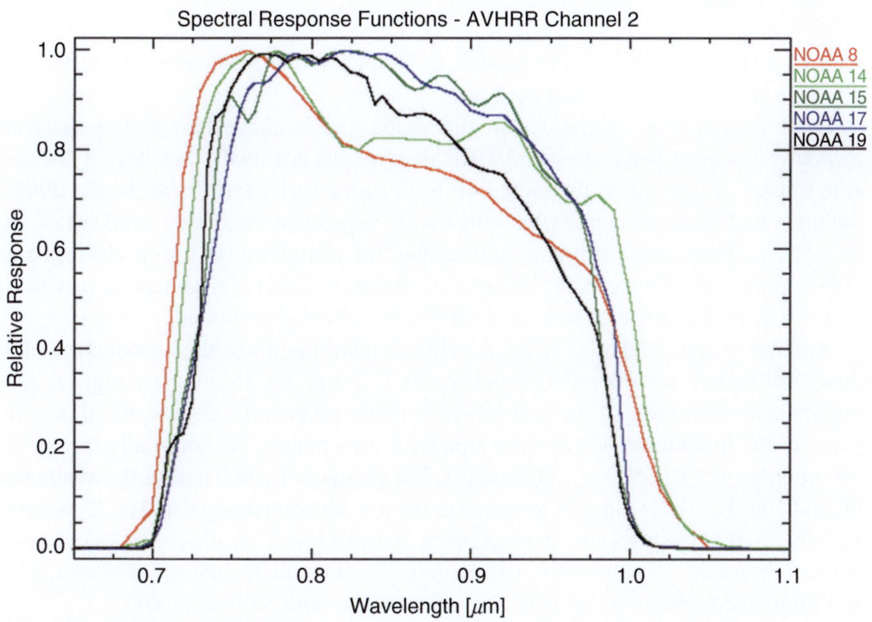

Fig. 3.3 Spectral response functions of channel 2 of various AVHRR instruments

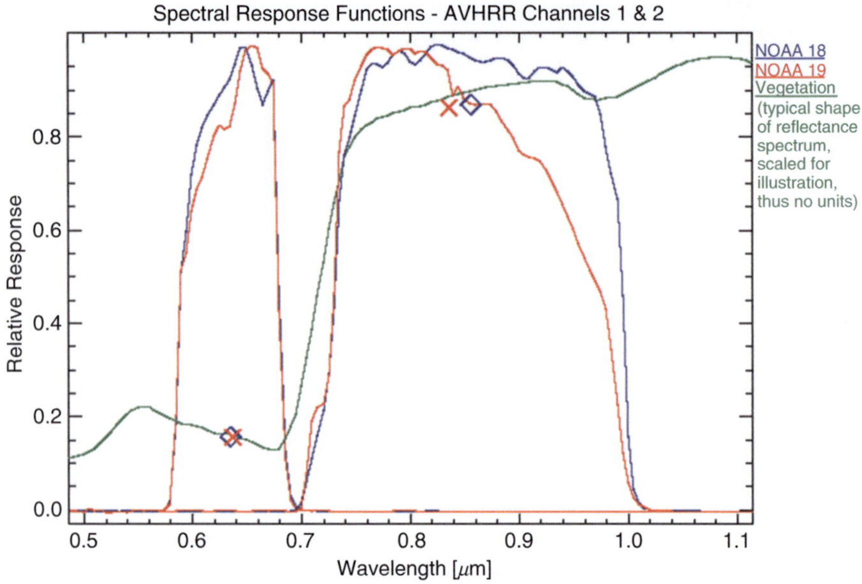

Fig. 3.4 Influence of the spectral response functions of AVHRR on NOAA 18 (*blue curve*) and NOAA 19 (*red curve*) on a typical vegetation spectrum (*green curve*, arbitrary scaling). The *blue diamond* symbols represent the data value of the resampled vegetation spectrum for NOAA 18 channels 1 and 2, the *red crosses* for NOAA 19

relative difference is ~1 %. For band 2, the effective center wavelength shifted by 20 nm, while band 1 is shifted by 2 nm.

Even though these variations in the shape and in the extent of the spectral response functions between the AVHRR instruments are small, they have a noticeable impact on generated products like NDVI time series (see Trishchenko 2009; Swinnen and Veroustraete 2008), with sensor-dependent variations in NDVI of up to ±15 %. Therefore correction approaches for channel-wise top-of-atmosphere (TOA) radiances (Trishchenko 2009) and correction factors for generated products (e.g. Trishchenko 2009; Steven et al. 2003) have been determined.

Another change in design of the AVHRR/3 instrument was the introduction of a dual-gain feature for the reflective channels 1, 2 and 3A. In order to improve the radiometric resolution of the instrument for low reflectance targets, the dynamic range of the instrument was divided equally in two ranges, i.e. nominally from 0 to 500 counts and from 500 to 1,000 counts. For channels 1 and 2 half of the available Digital Number (DN) range is assigned to the low albedo range from 0 to 25 % with the other half to the high albedo range from 26 to 100 %. This allows for an increase in the radiometric resolution for dark targets. For channel 3A, the split between low and high albedo range is set at 12.5 % albedo (Rao and Sullivan 2001).

3.2.2 Calibration and System Correction Procedures

An excellent introduction to the calibration of NOAA AVHRR can be found in Rao and Sullivan (2001). With NOAA-16 still in the commissioning phase in 2001, and three AVHRR instruments yet to be launched, the following text therefore aims to add information on the system correction of the instruments post 2001, and only to briefly summarize the information contained in Rao and Sullivan (2001).

3.2.2.1 Calibration and System Correction of the Reflective Channels

For the whole AVHRR series, no on-board calibration sources exist for the reflective channels 1, 2 and 3A. Consequently the calibration of the system is based on the pre-launch calibration conducted under laboratory conditions. Unfortunately, all instruments did show a different response behavior after launch and have changed significantly over time (see Fig. 3.5). The reasons for this well-known degradation are contamination by launch-vehicle exhaust followed by factors when in space that include the outgassing of the various components, extreme temperature shifts, exposure to ultraviolet radiation and other factors (see Rao and Sullivan 2001). Therefore one of the main objectives of the NOAA National Environmental Satellite, Data, and Information Service (NESDIS, see http://www.nesdis.noaa.gov and NOAA OSPO services as well as the AVHRR Pathfinder programme is to establish a post-launch calibration for the reflective channels (e.g. Rao and Chen 1994, 1996, 1999). To account for the degradation of the sensor, one can make use of vicarious approaches such as the calibration over (pseudo-) invariant desert sites and the intercalibration with airborne and space borne sensors (see Rao and Sullivan (2001) for details). The prime source for the resulting post-launch calibration factors are the Level1B Notices provided by the NOAA OSPO. These reports are available for all operational AVHRR instruments, and date back to the year 2000. Note that they also contain information on instruments and missions before 2000. Since 2006, these previously irregular reports are issued monthly and provide updates of the calibration values as direct input into the system correction.

Having pre- or post-launch calibration factors available, the system correction itself is straightforward for the reflective channels 1, 2 and 3A. The calibration from raw DN count to technical albedo value A (in percent) for channel i is based on the following simple equation:

$$A_i = S_i \ C_{10,\mathrm{i}} + I_i \tag{3.1}$$

where $C_{10,i}$ is the raw DN as 10bit counts for channel i
S_i is the slope (or gain factor) for channel i
I_i is the intersect (or offset factor) for channel i

The pre-launch values of slope and intercept of the AVHRR series until NOAA-14 are documented in Kidwell (1998). Since the beginning of November 1996,

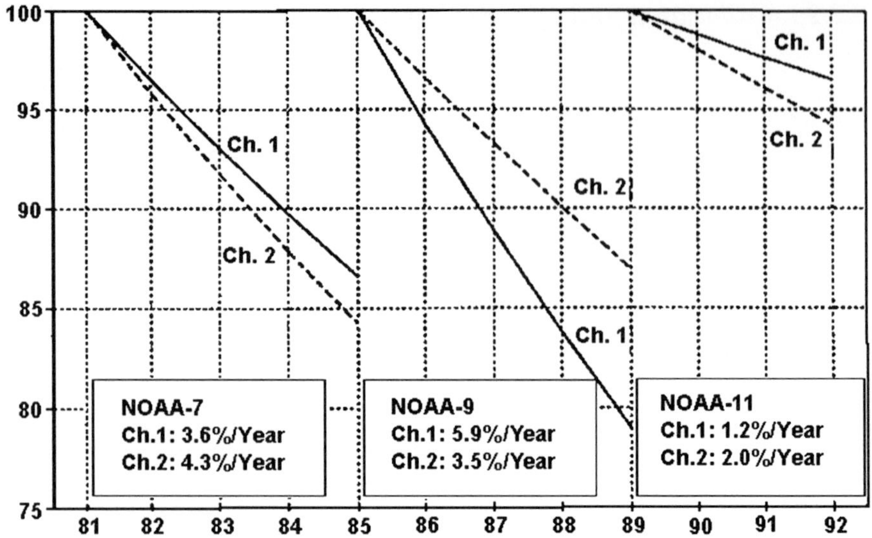

Fig. 3.5 Relative degradation rates of AVHRR channels 1 and 2 on NOAA-7, 9 and 11 (Modified from Rao and Chen 1994)

post-launch calibration updates were carried out at monthly intervals (Rao and Chen 1996), and the resulting calibration factors were made available at various NOAA sites (for an overview, see: http://www.noaasis.noaa.gov/NOAASIS/ml/calibration.html). For observations in-between the dates of the monthly updates, slope and intersect values can be derived by linear interpolation.

During the revision of the post-launch calibration for the NOAA-14 instrument, an additional factor ρ^2 is introduced in the calibration equation which accounts for seasonal variations of TOA solar irradiation (Rao and Chen 1996).

Consequently,

$$A_i^* = \rho^2 [S_i \ C_{10,\mathrm{i}} + I_i] \tag{3.2}$$

where ρ is the Earth-Sun distance expressed in astronomical units.
A_i* is the technical albedo in percent adjusted for solar irradiance changes for channel i

For the reflective channels of the AVHRR/3 series, this equation has to be applied separately for the low and high albedo ranges taking into account the valid DN count separating these ranges. For example, for the AVHRR/3 instrument on NOAA-19, the pre-launch calibration coefficients are defined as listed in Table 3.3 (Robel 2009) with separate slope and intercept values per albedo range.

Note that the exact count where the low and high albedo range intersect is slightly variable between sensors (e.g. for NOAA-19 at 497, 501, and 497 counts for channels 1, 2, and 3A, respectively). This is documented in Robel (2009) and the frequently updated online resource of NOAA OSPO.

Table 3.3 NOAA-19 AVHRR/3 pre-launch calibration coefficients (technical albedo representation)

Channel #	Contents	Slope	Intercept
1	Low albedo range (0–25 %)	0.055091	−2.1415
	High albedo range (26–100 %)	0.16253	−55.863
2	Low albedo range (0–25 %)	0.054892	−2.1288
	High albedo range (26–100 %)	0.16325	−56.445
3A	Low albedo range (0–12.5 %)	0.027174	−1.0881
	High albedo range (12.6–100 %)	0.18798	−81.491

Source: Robel (2009)

In order to derive the corresponding at-sensor radiance L in units of [W m^{-2} μm^{-1} sr^{-1}] from the technical albedo values in [%], the following relationship according to Rao and Chen (1996) is used:

$$A_i^* = (100\ \pi\ L_i\ \omega_i)/(F_{0,i}\ \cos\theta_0) \tag{3.3}$$

with

ω_i: effective width (in μm) of channel i
$F_{0,i}$: extra-terrestrial solar irradiance (W m^{-2}) within the pass of channel i
θ_0: solar zenith angle

In Robel (2009), the corresponding values for the various AVHRR instruments are documented. It is worth noting that this data uses the extra-terrestrial solar irradiance data based on Neckel and Labs (1984). Thus, small discrepancies to at-sensor radiance measurements using the CEOS (Committee on Earth Observation Satellites) endorsed irradiance spectrum by Thuillier exist (Thuillier et al. 2003).

3.2.2.2 Calibration and System Correction of the Thermal Channels

In the following, the calibration of the recent series of AVHRR/3 instruments is outlined; for the older AVHRR and AVHRR/2 instruments, calibration procedures are described in Kidwell (1998). For the AVHRR/3 instruments, the on-board calibration of the thermal channels 3, 3B, 4 and 5 is based on an on-board black body (also denoted as Internal Calibration Target, ICT) and measurements of deep-space (Robel 2009; Trishchenko et al. 2002). The ICT temperature itself is not stable, and is therefore, monitored by multiple Platinum Resistance Thermistors (PRTs). Thus, the temperature is known within an instrument-specific uncertainty (see Trishchenko et al. 2002 for details). For each scan line, the whiskbroom instrument first records 10 readings of deep-space, then the 2,048 readings of the Earth view (i.e., spatial pixel), followed by 10 readings of the ICT. Therefore, a "hot" and a "cold" target with known temperatures are included in every line, allowing for a two-point calibration.

Within the processing, the internal blackbody temperature is calculated by the alternating read-out of the four PRTs (see Robel 2009):

$$T_{PRT1} = d_0 + d_1 C_{PRT1} + d_2 C^2{}_{PRT1} + d_3 C^3{}_{PRT1} + d_4 C^4{}_{PRT1} \quad (3.4)$$

$$T_{PRT2} = d_0 + d_1 C_{PRT2} + d_2 C^2{}_{PRT2} + d_3 C^3{}_{PRT2} + d_4 C^4{}_{PRT2} \quad (3.5)$$

$$T_{PRT3} = d_0 + d_1 C_{PRT3} + d_2 C^2{}_{PRT3} + d_3 C^3{}_{PRT3} + d_4 C^4{}_{PRT3} \quad (3.6)$$

$$T_{PRT4} = d_0 + d_1 C_{PRT4} + d_2 C^2{}_{PRT4} + d_3 C^3{}_{PRT4} + d_4 C^4{}_{PRT4} \quad (3.7)$$

where C_{PRT} is the count value and T_{PRT} the corresponding temperature in Kelvin. The instrument-specific coefficients d_0 to d_4 are listed in Robel (2009). Finally the average temperature reading T_{BB} of all four read-outs is used:

$$T_{BB} = 0.25*(T_{PRT1} + T_{PRT2} + T_{PRT3} + T_{PRT4}) \quad (3.8)$$

Due to effects such as thermal forcing and solar radiation contamination of the ICT (straylight), and owing to errors in the signal processing and transmission, short- and long-term fluctuations of the PRT measurements within an orbit can be observed (Trishchenko et al. 2002; Trishchenko 2002). Therefore, statistical and Fourier filtering approaches can be used to remove unwanted spikes (Trishchenko 2002), and physical models of the ICT can improve the calibration accuracy (Trishchenko et al. 2002).

The spectral radiance N_{BB} in units of [mW m^{-2} cm sr^{-1}] of each thermal channel for the internal blackbody is computed based on the Planck function over the spectral response functions of the channels. This is realized by using a look-up-table (LUT) with a resolution of 0.1 K and for values between 180 and 340 K, which can be approximated by

$$T_{BB}{}^{*} = A + B\, T_{BB} \quad (3.9)$$

$$N_{BB} = \frac{c_1 v_c^3}{e^{\frac{c_2 v_c}{T^*_{BB}}} - 1} \quad (3.10)$$

with $T_{BB}{}^{*}$ denoting the approximated temperature in [K] using the LUT. As before, the values for the centroid wavenumber v_c and the coefficients c_1, c_2, A and B can be found in Robel (2009).

A two-point calibration approach is then used to calculate the Earth view spectral radiance N_E using the output of the blackbody reading C_{BB} and the corresponding blackbody radiance N_{BB} as well as the deep-space reading C_S and deep-space radiance N_S. Note that the calibration read-outs are smoothed to eliminate noise (see Robel 2009 for details).

In a first step, linear behavior of the instrument is assumed, thus:

$$N_{LIN} = N_S + (N_{BB} - N_S)\frac{(C_S - C_E)}{(C_S - C_{BB})} \tag{3.11}$$

where C_E is the count of the Earth view.

However, as known from pre-launch calibrations, the response is non-linear. Therefore, the next step includes the quadratic nonlinearity correction according to:

$$N_{COR} = b_0 + b_1 N_{LIN} + b_2 N_{LIN}^2 \tag{3.12}$$

with the coefficients as documented in Robel (2009). It is worth noting that establishing this non-linearity correction for the thermal channels was one achievement of the NOAA Pathfinder programme (Rao and Sullivan 2001).

Finally, the incoming radiance N_E for each Earth reading is calculated by:

$$N_E = N_{LIN} + N_{COR} \tag{3.13}$$

3.2.2.3 Corrections of Satellite Orbit Changes and Related Effects

As the orbit elements for the NOAA series provided by the North American Aerospace Defense Command (NORAD) are predictions, the resulting uncertainty in platform position and attitude leads to an increased uncertainty in geolocation of the generated products, with errors in position up to 10 km (see Fig. 3.6 for an illustrative example). An approach reducing this effect will be presented in the next section.

In addition, the series of AVHRR sensors on board various NOAA satellites is prone to a satellite orbit drift, where the local time of observation, and thus, the solar zenith angle is slowly changing over the lifetime of the satellite. For reflective bands, the reduction of this effect requires two processing steps: first, an accurate atmospheric correction taking into account the increasing downwelling path length; second, a correction for surface anisotropy with changing sun-target-sensor geometry (i.e., correction of the bi-directional reflectance distribution function (BRDF), see Gutman and Masek 2012). However, as a changing observation time also affects the observed temperatures, a correction of the orbit drift effect for longer time series can hardly be achieved (Gutman 1999).

3.3 DLR's AVHRR Pre-processing Chain

Due to the importance of AVHRR, multiple receiving, archiving and processing facilities exist world-wide. For Europe, this includes EUMETSAT (EUMETSAT 2011a, 2013) with various Satellite Application Facilities (SAFs) hosted at different institutions, the Natural Environment Research Council (NERC) Earth Observation Data Acquisition and Analysis Service (NEODASS) as well as the NERC Earth

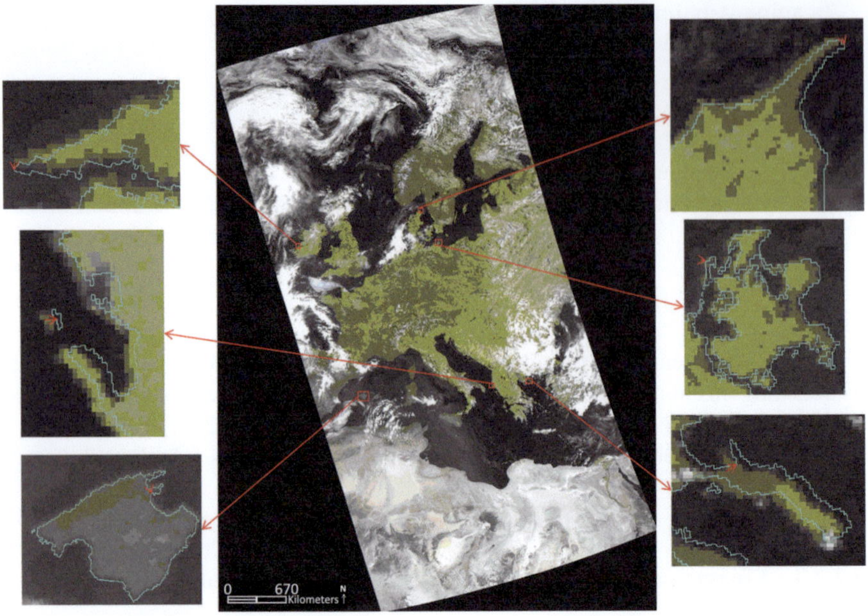

Fig. 3.6 Illustrative example for the discrepancy in geolocation between an AVHRR pass and the coastlines (in *cyan*), with *red arrows* indicating the direction and magnitude of local displacement (See Frey et al. 2014)

Observation Data Centre (NEODC, see Groom et al. 2006 and EUMETSAT 2011b), DLR'S World Data Center for Remote Sensing of the Atmosphere, WDC-RSAT and the AVHRR processing within the Data and Information Management System DIMS (see WDC-RSAT (http://wdc.dlr.de/about/), Dech et al. (1998), and http://www.timeline.dlr.de), and the ESA archive, hosted at ESA Earthnet Online (https://earth.esa.int/web/guest/missions/3rd-party-missions/current-missions/noaa-avhrr). In addition, universities have receiving and archiving facilities, such as the Freie Universität Berlin, Germany (http://www.geo.fu-berlin.de/met/ag/sat/satdaten/index.html) or the University of Bern, Switzerland (http://www.geography.unibe.ch/content/forschungsgruppen/fernerkundung/forschung/noaa_avhrr/avhrr_data_archive/index_eng.html).

In the following, the processing scheme for NOAA AVHRR High Resolution Picture Transmission (HRPT) telemetry data at DLR is outlined (see Dech et al. 1998 and Tungalagsaikhan et al. 2003). Within this processing and data management environment, 30 plus years of NOAA-AVHRR raw data were processed to L1b, L2 and L3 data products, so that by Dec. 2013, over 237,000 scenes over Europe are available at DLR (Fig. 3.7).

The overall workflow of this operational processing scheme is depicted in Fig. 3.8. After reception at DLR's ground segment, the HRPT data is imported,

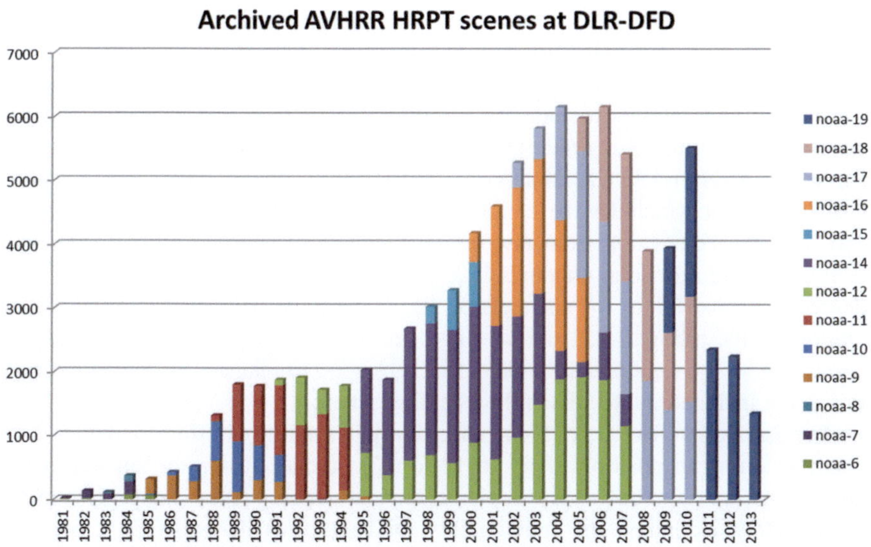

Fig. 3.7 Archived NOAA AVHRR scenes at DLR-DFD (receiving stations DLR Oberpfaffenhofen and FU Berlin)

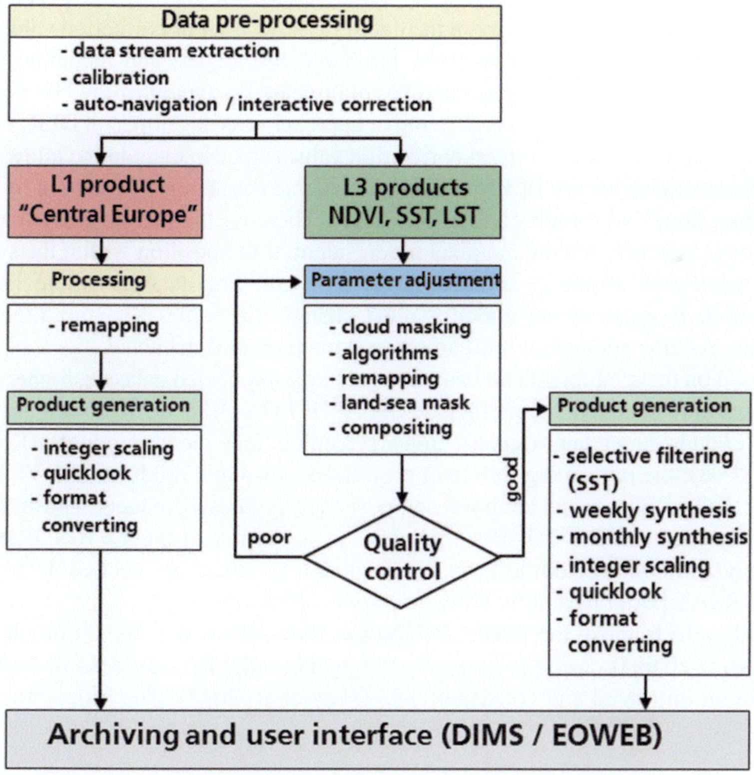

Fig. 3.8 Workflow of DLR's AVHRR processing chain

and archived within DLR's multi-mission infrastructure DIMS (see Heinen et al. 2009). Next, the extraction and generation of metadata is conducted, which is based on functionalities of the TeraScan software by SEASPACE. Extracted parameters and datasets include the satellite ID, the time and date of path, the generation of latitude/longitude grids, of satellite zenith angle grid, of sun zenith angle grid, and of relative azimuth grids. Subsequently the raw HRPT image data is processed, which includes the calibration of the reflective and thermal channels, and thus, the conversion from raw digital numbers to physical units (to percent technical albedo for solar channels 1, 2, and 3A and to brightness temperatures in degrees Celsius/Kelvin for channels 3, 3B, 4 and 5) as described in Sect. 3.2.2. Optionally the filling of bad and missing data by duplicating nearby lines, and the correction for the seasonal variation of the TOA solar irradiance can be applied, but this is not included in the standard processing scheme.

As described in Sect. 3.2.2.3, the uncertainty in satellite exterior orientation can result in errors in position of up to 10 km. To reduce this error, an improved autonavigation approach is implemented at DLR (see Tungalagsaikhan et al. 2003), extending the functionality of TeraScan (SEASPACE). Using an external coastline dataset, first a set of subsets ("boxes") with clearly identifiable features are generated and extracted. Next, a land/water mask is generated using the AVHRR imagery. Then the correlation of the land/water mask and the static coastline is calculated for each box, applying various offsets in x and y directions in order to find the best fit between the datasets. Next, a set of correction values for the satellite exterior orientation (yaw, pitch and roll angles, and including tilt if necessary) and start time is generated based on least-squares fit of the box coordinates. To improve this process, an iterative approach was developed at DLR, where in a first step boxes with medium correlation values are also included to allow for a coarse navigation. In two subsequent iterations, fine navigation is achieved by only including boxes with high correlation values. Thereby, the resulting accuracy in position is typically within one pixel independent of the position within the swath, with worst-case errors in geolocation of 5–6 km. For datasets where human interaction is required, an operator must change the values for start time and satellite exterior orientation until an optimal navigation is achieved.

Based on these calibrated and autonavigated pre-processed datasets, higher level products are generated. These include the APOLLO (*A*VHRR *P*rocessing scheme *O*ver c*L*ouds *L*and and *O*cean) cloud screening and cloud products (Kriebel et al. 2003), the remapping into map projections, thematic products (NDVI, LST, SST, cloud products), and temporal composites of synthesis products. The archived and processed L1 and L3 data products are accessible using DLR'S EOWEB gateway (http://eoweb.dlr.de/), and atmospheric products are accessible via the WDC-RSAT portal hosted by DLR (http://wdc.dlr.de/).

Within the frame of the TIMELINE project (http://www.timeline.dlr.de), DLR's processing chain is currently being re-designed in order to re-process all archived data in an improved and consistent way (Frey et al. 2014). The processing will

Table 3.4 Foreseen TIMELINE products (Frey et al. 2014)

Type	Variable
Calibration	Radiance at sensor
Surface radiative variables	TOA and BOA/Reflectance and brightness temperature
	Land and sea surface temperature (LST/SST)
	Sea surface temperature (SST)
Land surface	Vegetation variables (NDVI, LAI, FAPAR, FVC)
	Burnt area and fire detection ("Hot Spots")
	Water masks
Cryosphere	Snow and ice over land, Sea ice coverage
Atmosphere	Cloud masks
	Degree of cloud cover, thermodynamic cloud phase, cloud top temperature, cloud height, cloud optical thickness, precipitation potential

BOA bottom-of- atmosphere, *LAI* leaf area index, *FAPAR* fraction of absorbed photosynthetically active radiation, *FVC* fraction of vegetation cover

now include an atmospheric correction, the improved APOLLO13 cloud processor, and an increased number of higher-level products (Table 3.4). The current design also includes an improved chip-matching and a subsequent ortho-rectification process to improve existing shortcomings within the geometric correction. For the instrument-dependent spectral response functions, an adjustment similar to the approach by Trishchenko (2009) will be applied. Regarding the radiometric calibration, additional harmonization factors are provided within the metadata. These additional gain and offset values for each dataset are derived from time series analysis over pseudo-invariant calibration sites, so that the consistency of the resulting multi-sensor time series is improved. The importance of such a re-calibration process will be further discussed in the following sections.

3.4 Influence of Calibration and Re-calibration

As described in Sect. 3.2.2.1, the calibration of the reflective channels of the AVHRR series is critical, as the slow degradation of the visible channels can reach a magnitude of a few per cent per year. Therefore, the resulting post-launch values for gain and offset change significantly over time. In order to understand the changes in sensor calibration, a number of experiments were carried out by the authors in the context of the TIMELINE project. These will be described in this section. Approaches towards an improved calibration are then discussed in Sect. 3.5.

Table 3.5 Changes in calibration of AVHRR, NOAA-18, channel 1

Year	Day of year	Low albedo, slope	Low albedo, intercept	Cut (in DN) between low and high gain	High albedo, slope	High albedo, intercept
2005	Pre-launch	0.05400	−2.130	501	0.16100	−55.37
2005	256	0.05045	−1.990	501	0.15040	−51.73
				...		
2012	160	0.05581	−2.200	501	0.16640	−57.22

Table 3.6 Changes in calibration of AVHRR, NOAA-18, channel 2

Year	Day of year	Low albedo, slope	Low albedo, intercept	Cut (in DN) between low and high gain	High albedo, slope	High albedo, intercept
2005	Pre-launch	0.05290	−2.084	501	0.15873	−55.31
2005	256	0.05485	−2.161	501	0.16460	−57.35
				...		
2012	160	0.06444	−2.539	501	0.19340	−67.37

3.4.1 Calibration Over Time

In the following, examples of the variability in the calibration are given. For this purpose, AVHRR/3 on board of NOAA-18 (launched in May 2005), is chosen as this sensor has a long history of data acquisitions in DLR's archive. All values are based on the values included in SEASPACE, and the monthly updates by NOAA OSPO, as used in DLR's processing system. At DLR, NOAA-18 was mainly used until 2012, so the ~7 years between the beginning of the operational phase and 2012 are shown.

When comparing the pre- to the first post-launch calibrations of channel 1 and 2 (Tables 3.5 and 3.6), a strong decrease in the gain settings of channel 1 can be observed, whereas for channel 2 gain has strongly increased. When plotting the gain and offset factors over time (Figs. 3.9, 3.10, 3.11 and 3.12) a larger variability is observed from launch until the beginning of 2007, with the gain in channel 1 reduced by 8 %. Since then, the overall trend of the gain in channels 1 and 2 is a steady increase. By 2012 the gain of channel 1 is close to the pre-launch estimate, whereas channel 2 has seen a relative increase of ~20 % (Figs. 3.13 and 3.14). Within the overall trend of an increasing gain and decreasing offset, shorter phases of opposing trends also occur which cannot be described by a function, but must be provided as individual factors. In Fig. 3.15, these differences are depicted based on the relative changes compared to pre-launch values. The relative changes in gain and offset over time are similar, but also here small differences are obvious (Fig. 3.16).

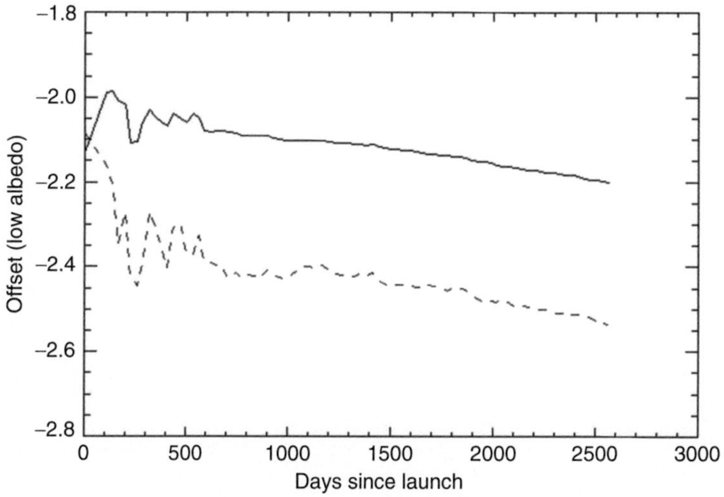

Fig. 3.9 Change of post-launch calibration – offsets (low albedo), NOAA 18, channel 1 (*solid line*), channel 2 (*dotted*)

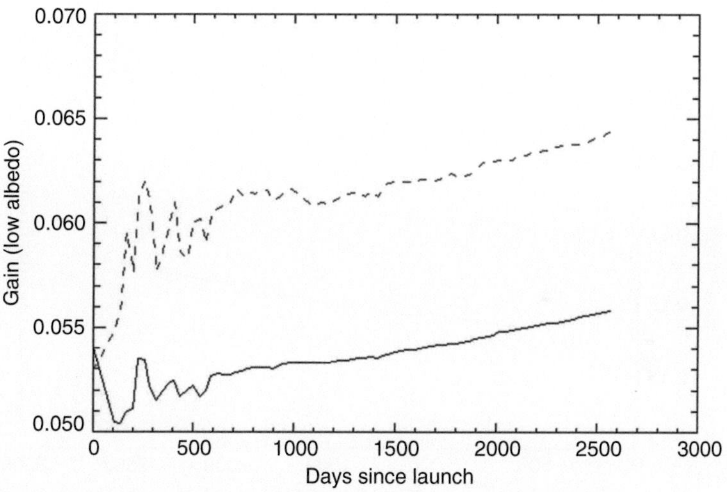

Fig. 3.10 Change of post-launch calibration – gain (low albedo), NOAA 18, channel 1 (*solid line*), channel 2 (*dotted*)

3.4.2 Impact on Generated L1b Products

In order to study the effects of sensor degradation over time, and to assess the influence of a re-calibration in the generated time series, similar approaches as for the post-launch calibration are frequently applied. A recent overview of such approaches including an extensive list of references can be found in Chander et al. (2013).

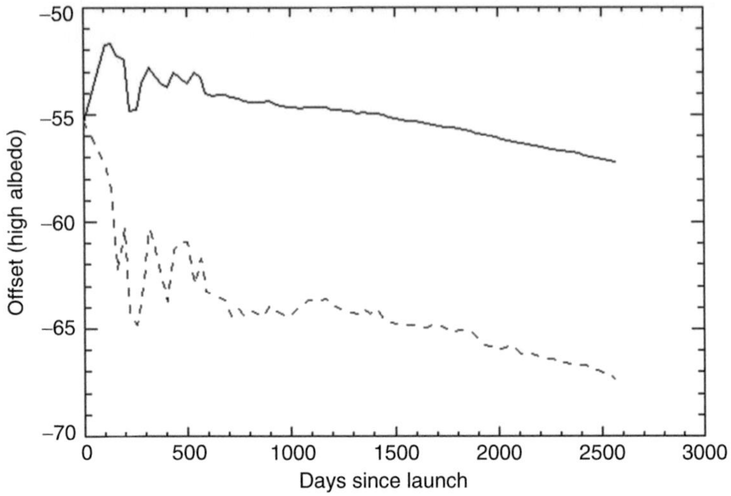

Fig. 3.11 Change of post-launch calibration – offset (high albedo), NOAA 18, channel 1 (*solid line*), channel 2 (*dotted*)

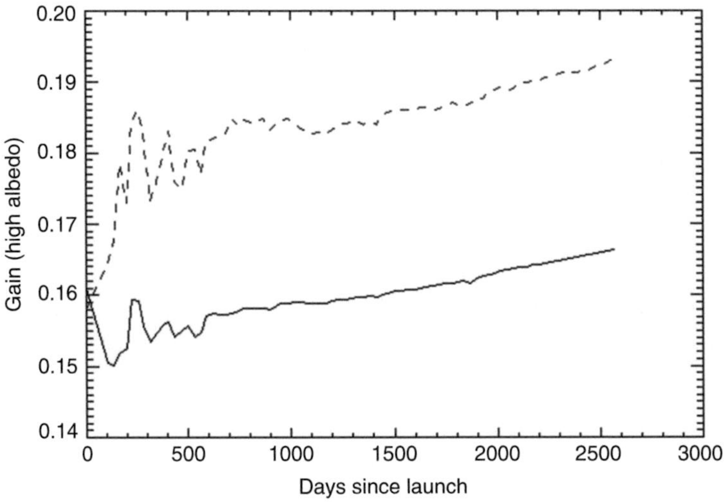

Fig. 3.12 Change of post-launch calibration – gain (high albedo), NOAA 18, channel 1 (*solid line*), channel 2 (*dotted*)

In order to clearly discriminate changes from sensor response to changes of the observed area on the Earth's surface, not every location is suitable. Therefore, a number of requirements for vicarious validation and calibration test sites exist. Suitable sites have to be large in relation to the ground-projected IFOV of the sensors, as well as, spatially and spectrally homogeneous in order to reduce the influence of errors in geolocation. In addition, the sites shall be temporally

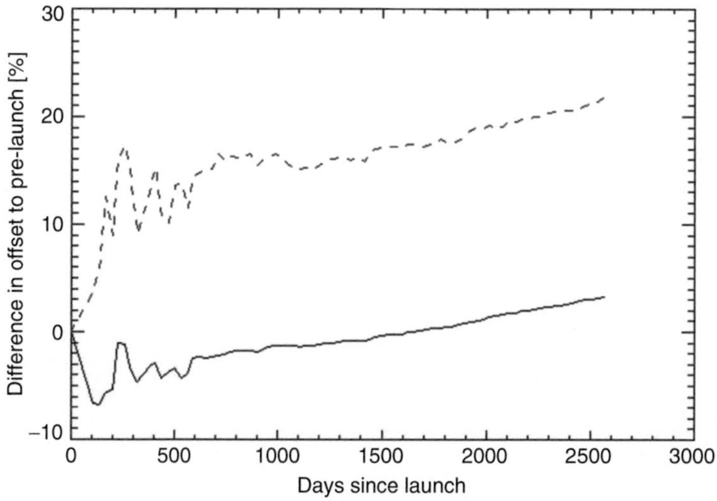

Fig. 3.13 Relative change in low albedo offsets pre- to post-launch; NOAA 18, channel 1 (*solid line*), channel 2 (*dotted*)

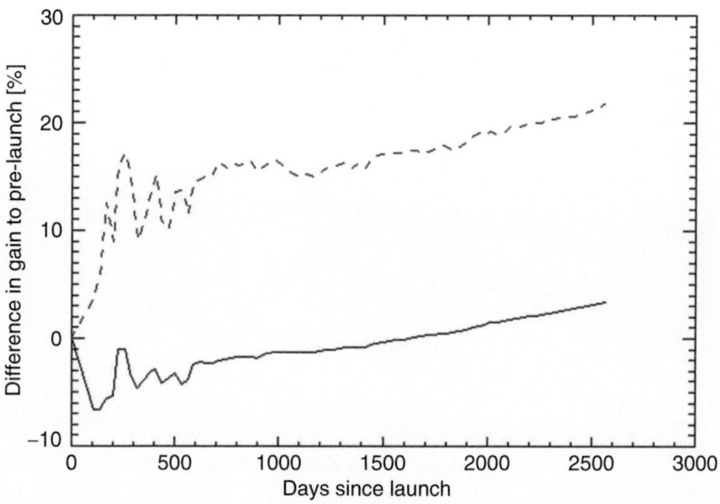

Fig. 3.14 Relative change in low albedo gains pre- to post-launch; NOAA 18, channel 1 (*solid line*), channel 2 (*dotted*)

invariant, both across seasons and over multiple years. Sites with stable climatic conditions and sites located at high altitudes are preferred in order to reduce the influence of the atmosphere. High-albedo surfaces are preferred so that changes in sensor gain can be easily identified. Consequently, non-vegetated deserts and salt lakes are the most suitable calibration targets. A catalogue of well-suited and established calibration sites is given in the Catalogue of CEOS Landnet sites

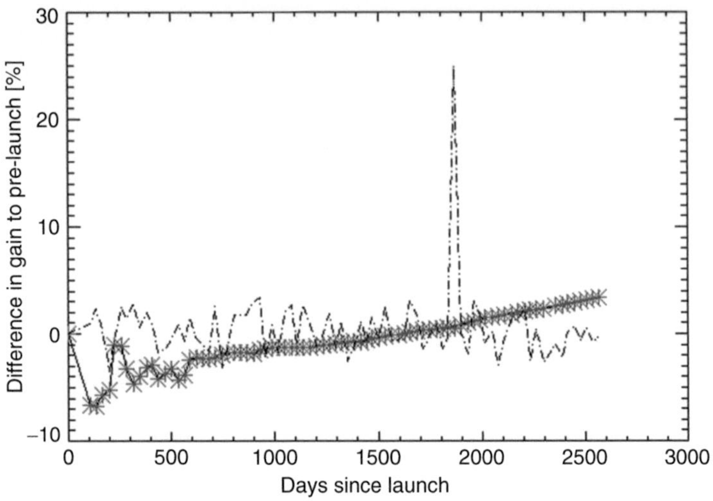

Fig. 3.15 NOAA 18, channel 1. *Solid line*: relative low albedo gain, *green stars*: relative high albedo gain, *dashed line*: difference low – high, magnified by factor 100

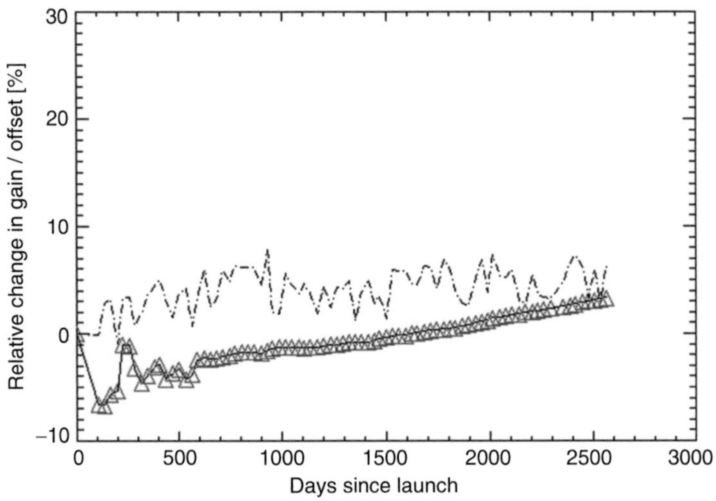

Fig. 3.16 NOAA 18, channel 1, low albedo. *Solid line*: relative low albedo gain to pre-launch, *green triangles*: relative low albedo offsets to pre-launch, *dashed line*: difference relative change in gain – relative change in offset, magnified by factor 100

(CEOS see http://calvalportal.ceos.org/ceos-landnet-sites, USGS see http://calval.cr.usgs.gov/rst-resources/sites_catalog/).

In this study, we present the following examples from the CEOS-endorsed sites Libya 4 and Algeria 3 (see Fig. 3.17 for coordinates and overview imagery), which are bright, spatially and spectrally homogeneous (i.e., dunes at different scales), and

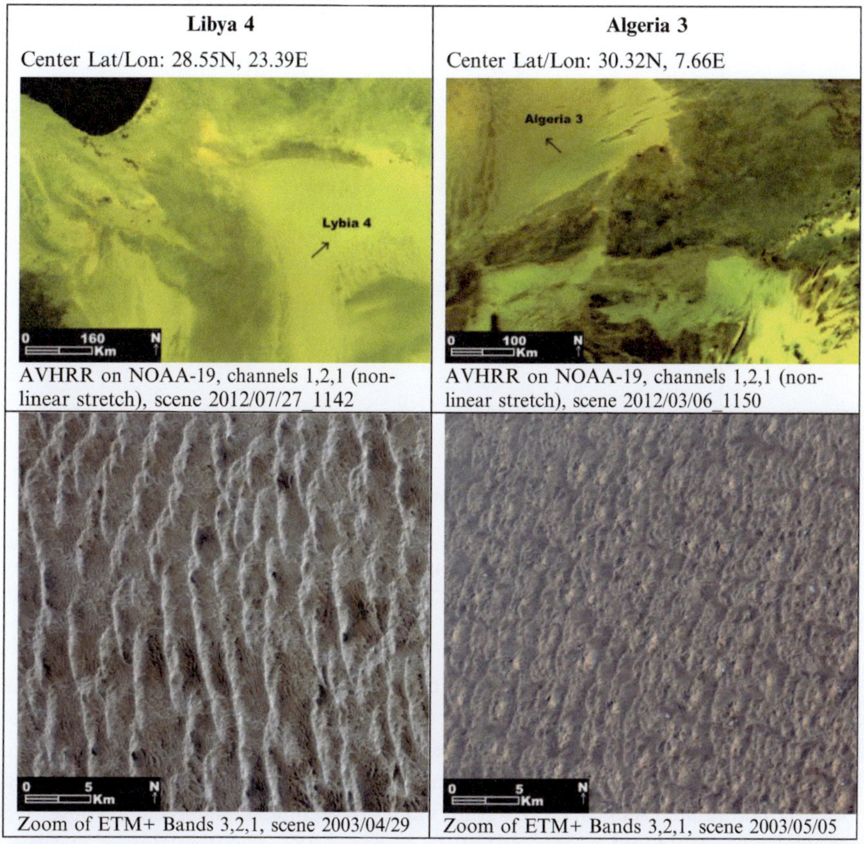

Fig. 3.17 CEOS test sites used in this study (Source of ETM+ imagery: http://calval.cr.usgs.gov/rst-resources/sites_catalog/)

represent large usable areas of 75 by 75 km. These desert targets in Libya have been frequently used for AVHRR re-calibration and inter-satellite standardization (e.g. Rao and Sullivan 2001 for the NOAA/NASA Pathfinder programme, and also Casey et al. 2010 for the Pathfinder SST project).

Additional requirements exist when monitoring a sensor over larger timescales. This is particularly the case when dealing with sensors prone to larger degradation or when inter-calibrating between different sensors. Examples include

- data acquisition at same local time and observation angle, or otherwise modeling of the site BRDF and correction for atmospheric conditions including path length differences
- similarity of spectral response functions, or otherwise modeling approaches accounting for this effect
- similarity of spatial resolution, or otherwise spatial modeling
- exclusion of all datasets with cloud and cloud shadow contamination.

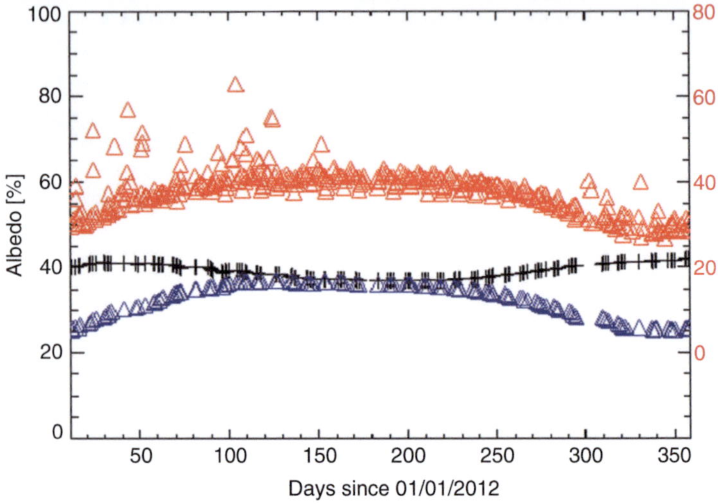

Fig. 3.18 Influence of pre-processing steps. Test site Libya 4, year 2012. *Red triangles*: no solar correction, all observation angles, no filtering, no strict cloud screening, vertical offset for clarity (see *red* scale bar on the *right*); *blue triangles*: no solar correction, close to nadir, Lee-filtered, cloud screening; *black crosses*: solar corrected, close to nadir, Lee-filtered, cloud screening

For AVHRR, the problems of orbit drift, and to a lesser degree, the changes in spectral response functions have to be accounted for. In order to illustrate the effect of local viewing time and differences in the Earth-Sun distance, three different pre-processing steps are depicted in Fig. 3.18. Without the filtering of cloud-contaminated scenes, large outlier values of technical albedo can be observed especially in spring and winter (red curve in Fig. 3.18). The inclusion of observations with larger off-nadir view angles results in a generally large scattering of about an absolute value of 5 % technical albedo, and the seasonally change of the Sun-Earth distance is clearly visible. When including only scenes where the target area is located close to nadir, and with a strict filtering of cloud-contaminated pixels, outliers and scattering from the overall trend are significantly reduced, but still the seasonal effect for this scene has a magnitude of ~10 % absolute in technical albedo (blue curve in Fig. 3.18). When this effect is accounted for, a stable signal without the influence of too many external influences can be observed (black curve). Note that still the technical albedo is not constant over time, but shows a slight seasonal trend with a magnitude of ~3–4 % technical albedo. Nevertheless, with an adequate pre-processing and with an exclusion of unsuitable scenes, it is possible to observe any sensor-related effects. Thus, in the following only scenes with observation times between 11:00 and 13:00 GMT and observation angles for the two test sites close to nadir (within $\pm 10°$) are considered. In addition, the resulting time series were checked for cloud contamination over the sites. When using these cloud and nadir thresholds for the years 2012 and 2013, 308 observations (out of a total of 736) are considered as suitable for Libya4, and 96 observations (out of a total of 524) are suitable for Algeria3. Furthermore, an analysis of

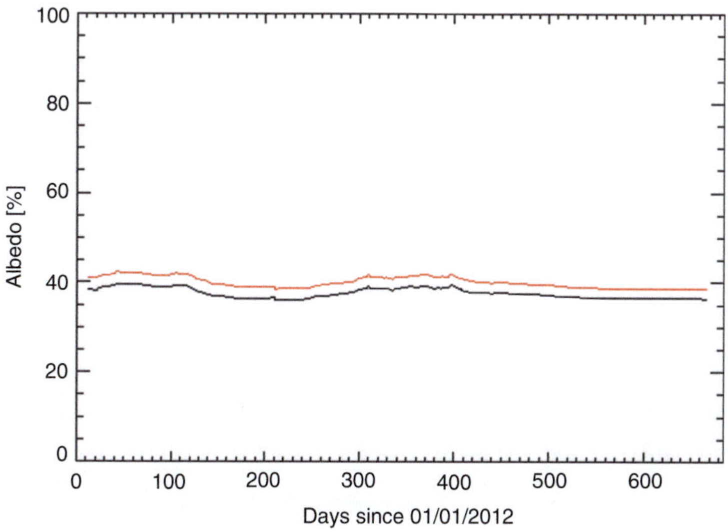

Fig. 3.19 Technical albedo values for AVHRR/3 on NOAA-19: Algeria 3, channel 1; *red*: pre-launch, *black*: post-launch calibration

NDVI time series indicated that vegetation was absent through the year. For a further removal of outliers, a Lee-filter as implemented in IDL (EXELIS VIS) was applied.

When analyzing the resulting time series for both the Algeria 3 (Figs. 3.19 and 3.20) and Libya 4 (Figs. 3.21 and 3.22) test sites, the post-launch calibration resulted in an overall decrease in technical albedo in channel 1, and an increase in technical albedo in channel 2 compared to the pre-launch calibration. The absolute magnitude of this is about −2.5 % technical albedo in channel 1 and +2.7 % technical albedo in channel 2 for Libya4.

When plotting the relative change (Fig. 3.23), this shows a mean difference in technical albedo of −6.1 % in channel 1, and a mean of about +6.6 % in channel 2 (not depicted) for both test areas, with obvious changes over time. As these relative changes are highly consistent for both test sites, there is high confidence in the observed trends. When comparing this change in relative difference in the technical albedo values of the pre-/post-launch with the series of changes in calibration (esp. in slope, see Figs. 3.24 and 3.25), similarities occur. Most prominent, the increase towards the peak around day 70 and the subsequent decrease with its center around day 125 is visible both in the relative technical albedo difference, as well as, in the slope values. In addition, an area of minor changes occurs in both series between days 230 and 300, with an overall increasing trend. Thus, these differences in sensor pre- to post-launch calibration have a direct influence on the technical albedo product of over 6 % in case of AVHRR on NOAA-19 for the years 2012/13.

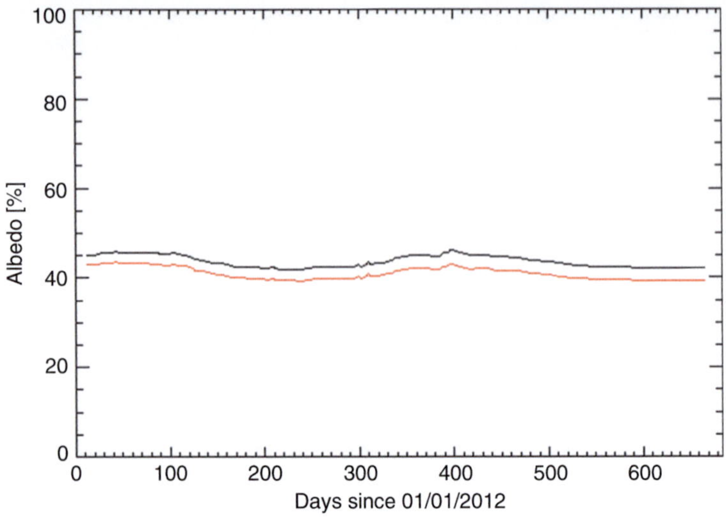

Fig. 3.20 Technical albedo values for AVHRR/3 on NOAA-19: Algeria 3, channel 2 *red*: pre-launch, *black*: post-launch calibration

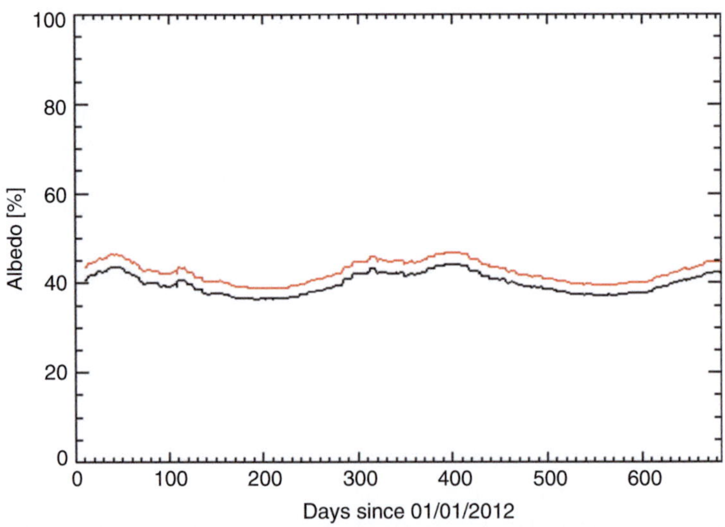

Fig. 3.21 Technical albedo values for AVHRR/3 on NOAA-19: Libya 4, channel 1; *red*: pre-launch, *black*: post-launch calibration

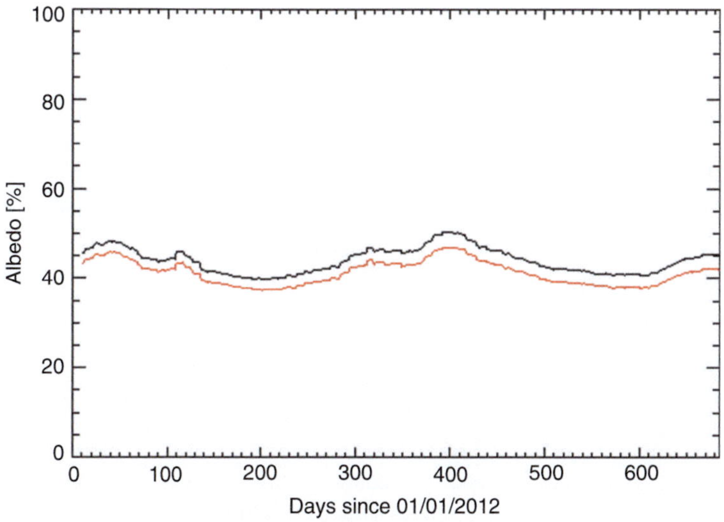

Fig. 3.22 Technical albedo values for AVHRR/3 on NOAA-19: Libya 4, channel 2 *red*: pre-launch, *black*: post-launch calibration

3.5 Towards Data Harmonization and Consensus Calibration

In addition to the post-launch calibration values provided by NOAA OSPO and NOAA NESDIS (i.e., Rao 1987; Rao and Chen 1996, 1999), quite a number of alternative calibration factors exists, including the works by Vermote and Saleous (2006), Che and Price (1992), Teillet and Holben (1994), Swinnen and Veroustraete (2008). In Figs. 3.26 and 3.27, the comparisons of different calibration factors for NOAA-9 (from Rao et al. 1993) and for NOAA-14 (Rao and Sullivan 2001) provide an idea of the spread within different post-launch calibration sets. When creating time series based on data from multiple AVHRR instruments, additional cross-calibration between instruments has to be conducted using approaches based on simultaneous nadir overpass observations and reference targets (Latifovic et al. 2012) or statistical approaches (Schmidt et al. 2008). Still the review paper by Molling et al. (2010) shows that the differences between long-term calibration sets can reach a magnitude of up to 10 %.

As one of the main requirements of long time-series is the consistency in the data, and as these calibration differences lead to non-compatible products between processing facilities, harmonization of the calibration of the AVHRR reflective channels is an important venture. In addition, the presented subtle changes in the spectral response curves and also the usage of different solar irradiance spectra can increase the differences in the processed data of different processing facilities.

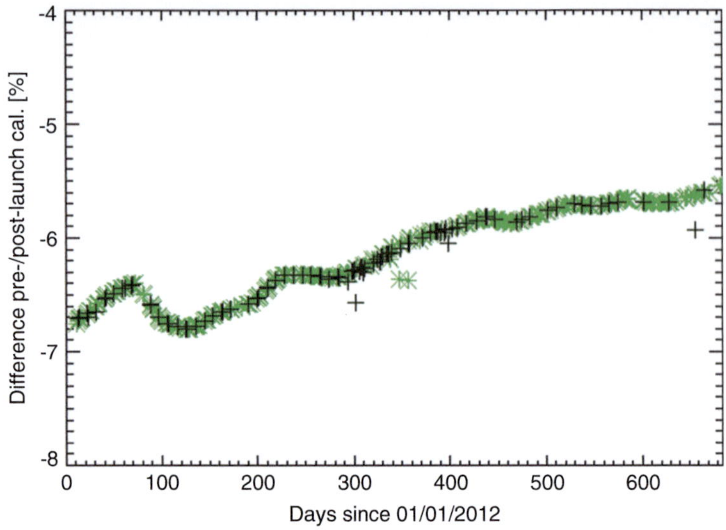

Fig. 3.23 Relative difference in pre- **to** post-launch calibration in channel 1. *Green stars*: Libya; *black crosses*: Algeria. No filtering of data applied

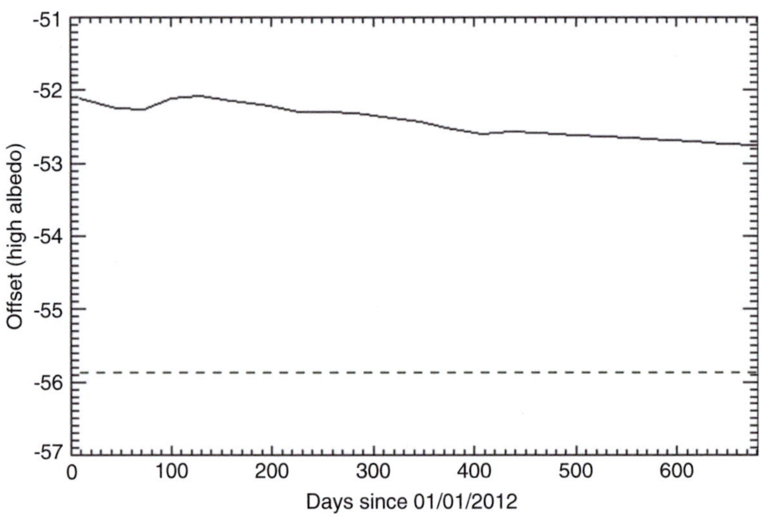

Fig. 3.24 Offset for AVHRR on NOAA 19, channel 1, high albedo, *solid line*: post-launch, *dashed*: pre-flight

In order to improve the consistency in calibration between processing facilities, and to achieve a consensus in the post-launch calibration, the initiative by NOAA/NESDIS (Heidinger et al. 2010 and Molling et al. 2010) is most advanced. One component is the generation of the AVHRR Pathfinder Atmospheres Extended (PATMOS-x) dataset, which is based on the vicarious calibration using MODIS

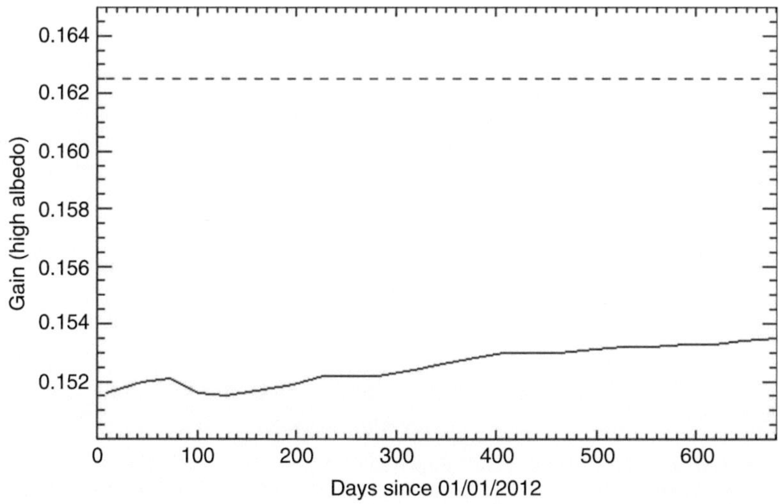

Fig. 3.25 Gain for AVHRR on NOAA 19, channel 1, high albedo, *solid line*: post-launch, *dashed*: pre-flight

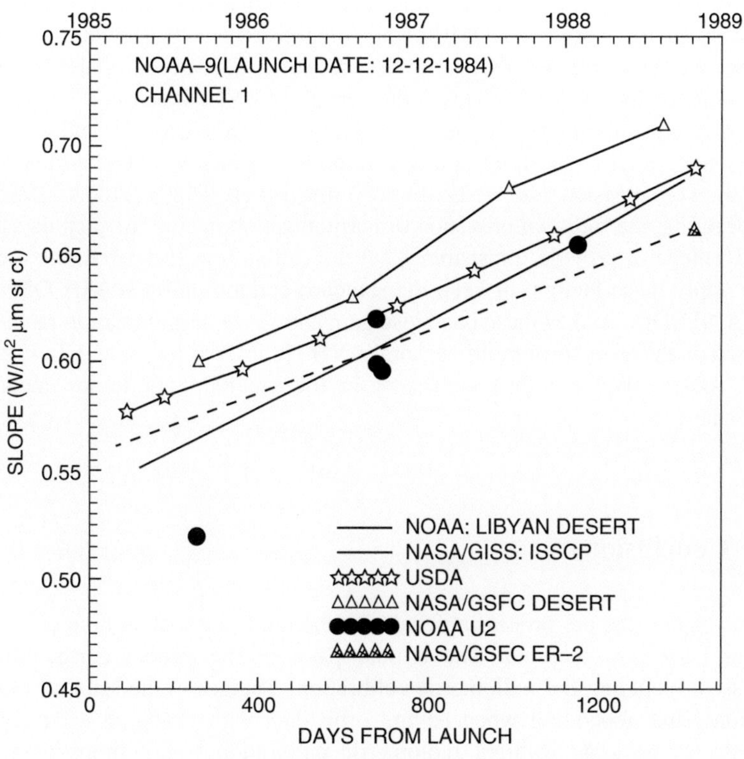

Fig. 3.26 Change in calibration for NOAA-9 according to various studies (From Rao et al. 1993)

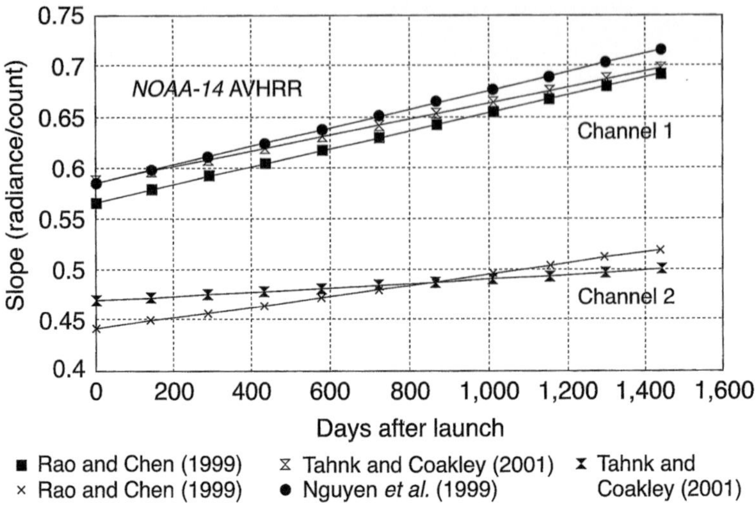

Fig. 3.27 Change in calibration for NOAA-14 according to various studies (From Rao and Sullivan 2001)

as reference for the years from 2000 onwards. Simultaneous nadir observations for a Libyan Desert site and DOME-C in Antarctica is used for inter-calibrating between MODIS and AVHRR, and also between the various AVHRR sensors, thus extending the calibration back to the very early AVHRR sensors. The accuracy of the derived calibration coefficients are within 2 % for channel 1, and within 3 % for channel 2 relative to MODIS, therefore providing an improved baseline for the generation of consistent and well-calibrated time series. Within DLR's TIMELINE re-processing, the foreseen provision of harmonization factors for each dataset has the advantage to provide transparency on the calibration. If desired, the user can easily apply these factors, or keep the standard calibration by NOAA OSPO and NOAA NESDIS. And as these bandwise gain and offset harmonization factors will be based on the analysis of multi-sensoral AVHRR observations over CEOS sites in North Africa, they are thus consistent in the methodology to the mentioned approaches.

3.6 Conclusions

In this chapter, the pre-processing chain for the multi-decadal archive of AVHRR data at DLR was outlined. The critical parts of the system correction were addressed, including the radiometric calibration, changes in the spectral response functions, and geocoding when having orbit data with changing accuracy. The necessity of accurate in-flight radiometric calibration was demonstrated using time series examples for pseudo-invariant CEOS test sites. This is especially crucial for the older AVHRR sensors, with NOAA-9 having a degradation rate of almost 6 % per year. But also the newer sensors require a frequent adjustment of in-flight calibration, with NOAA-18 showing a relative difference in gain and offset of ~20 % in 6 years when compared to laboratory calibration. Next, even though the band widths of the AVHRR channels are comparably broad, changes in the spectral response functions have a significant influence on the sensed signal. As a consequence, derived products such as the NDVI can be inconsistent between different AVHRR sensors, with sensor-dependent variations of over 10 % in NDVI. Regarding the image geometry, both systematic shifts and additional local distortions in geolocation can be observed, resulting in geocoding errors of up to 10 km. Therefore an adjustment of the orbit parameters using coast line data is can significantly improve the geolocation.

As the mentioned factors have a significant influence on the consistency of the generated time series data, alternative correction approaches and advanced methods were briefly presented. Taking these new developments into account, DLR's AVHRR pre-processing chain is currently being re-designed and extended in context of the TIMELINE project (http://www.timeline.dlr.de). This will include an improved geocoding by means of image-to-image matching and subsequent ortho-rectification. The radiometric calibration of the reflective and thermal channels will be improved and harmonized by taking developments towards a consensus calibration into account. Also the effect of different spectral response functions will be corrected using an established approach. Making use of this advanced pre-processing scheme, an improved consistency in thematic data products within the lifetime of one sensor, between sensors of the same series and also between different sensors can be achieved.

Acknowledgements The authors would like to thank the anonymous reviewers for their valuable comments on the manuscript. Dr. Derek Rogge is acknowledged for improving the language and structure of this chapter. Also Dr. Claudia Künzer, Dr. Corinne Frey and Dr. Kurt Günther are acknowledged for helpful comments on the text.

References

Beck HE, McVicar TR, van Dijk A, Schellekens J, de Jeu RAM, Bruijnzeel LA (2011) Global evaluation of four AVHRR-NDVI data sets: intercomparison and assessment against Landsat imagery. Remote Sens Environ 115(10):2547–2563

Casey KS, Brandon TB, Cornillon P, Evans R (2010) The past, present and future of the AVHRR pathfinder SST program. In: Barale V, Gower JFR, Alberotanza L (eds) Oceanography from space: revisited. Springer, Dordrecht

Chander G, Hewison TJ, Fox N, Wu X, Xiong X, Blackwell WJ (2013) Overview of intercalibration of satellite instruments. IEEE Trans Geosci Remote Sens 51(3):1056–1080

Che N, Price JC (1992) Survey of radiometric calibration results and methods for visible and near infrared channels of NOAA-7, -9, and -11 AVHRRs. Remote Sens Environ 41(1):19–27

Dech SW, Tungalagsaikhan P, Preusser C, Meisner RE (1998) Operational value-adding to AVHRR data over Europe: methods, results, and prospects. Aerosp Sci Technol 2(5):335–346
EUMETSAT (2010) Algorithm theoretical basis document cloud physical products AVHRR/SEVIRI, Ref SAF/CM/KNMI/ATBD/CPP, Issue 1.1, Date 17 June 2010. http://www.cmsaf.eu/bvbw/generator/CMSAF/Content/Publication/atbd__pdf/SAF__CM__KNMI__ATBD__CPP__1.1,templateId=raw,property=publicationFile.pdf/SAF_CM_KNMI_ATBD_CPP_1.pdf. Accessed 3 Feb 2014
EUMETSAT (2011a) AVHRR Level 1b product guide Ref.: EUM/OPS-EPS/MAN/04/0029 Issue: v3A Date: 21 Jan 2011 http://oiswww.eumetsat.org/WEBOPS/eps-pg/AVHRR/AVHRR-PG-index.htm. Accessed 3 Feb 2014
EUMETSAT (2011b) The advanced very high resolution radiometer (AVHRR/3). NERC Earth Observation Data Centre, 20 Dec 2013. http://badc.nerc.ac.uk/view/neodc.nerc.ac.uk__ATOM__dataent_12447304699327717. Accessed 3 Feb 2014
EUMETSAT (2013) Operational services specification Ref: EUM/OPS/SPE/09/0810 Issue: v2c Date 29.05.2013 http://www.eumetsat.int/website/wcm/idc/idcplg?IdcService=GET_FILE&dDocName=PDF_TEN_302979_OSS&RevisionSelectionMethod=LatestReleased&Rendition=Web. Accessed 3 Feb 2014
EXELIS VIS: IDL 8.2.3. http://www.exelisvis.com/ProductsServices/IDL.aspx. Accessed 3 Feb 2014
Fontana FMA, Coops NC, Khlopenkov KV, Trishchenko AP, Riffler M, Wulder MA (2012) Generation of a novel 1 km NDVI data set over Canada, the northern United States, and Greenland based on historical AVHRR data. Remote Sens Environ 121:171–185
Frey CM, Kuenzer C, Dech S (2012) Quantitative comparison of the operational NOAA-AVHRR LST product of DLR and the MODIS LST product V005. Int J Remote Sens 33(22):7165–7183
Frey C, Dietz A, Bachmann M, Bernhard EM, Ruppert T, Molch K, Wolfmüller M, Kuenzer C, Mueller A, Dech S (2014) TIMELINE – processing of AVHRR time series over Europe and North Africa. In: Proceedings 34th EARSeL symposium, Warsaw, 2014
GCOS/WMO (2009) Guideline for the generation of satellite-based datasets and products meeting GCOS requirements. GCOS-128, WMO/TD No. 1488, http://www.wmo.int/pages/prog/gcos/Publications/gcos-128.pdf. Accessed 3 Feb 2014
Groom S, Miller P, Shutler J, Lonie N, Parkes S (2006) NERC Earth observation data acquisition and analysis service: a new partnership. RSPSoc Annual Conference, Cambridge, 2006. Available from http://www.neodaas.ac.uk/docs/Groom_RSPSoc2006.pdf. Accessed 3 Feb 2014
Gutman GG (1999) On the monitoring of land surface temperatures with the NOAA/AVHRR: removing the effect of satellite orbit drift. Int J Remote Sens 20(17):3407–3413
Gutman G, Masek JG (2012) Long-term time series of the Earth's land-surface observations from space. Int J Remote Sens 33(15):4700–4719
Heidinger AK, Straka WC, Molling CC, Sullivan JT, Wu X (2010) Deriving an inter-sensor consistent calibration for the AVHRR solar reflectance data record. Int J Remote Sens 31 (24):6493–6517
Heinen T, Kiemle S, Buckl B, Mikusch E, Lyola D (2009) The geospatial service infrastructure for DLR's national remote sensing data library. IEEE J Sel Top Appl Earth Obs Remote Sens 4:260–269
Kidwell KB (ed) (1998) NOAA Polar Orbiter Data (POD) user's guide – satellite and data description for NOAA-14 and earlier, Nov 1998 revision. http://www.ncdc.noaa.gov/oa/pod-guide/ncdc/docs/podug/index.htm. Accessed 3 Feb 2014
Kramer HJ (2002) Observation of the earth and its environment. Survey of missions and sensor, 4th edn. Springer, Berlin
Kriebel KT, Gesell G, Kästner M, Mannstein H (2003) The cloud analysis tool APOLLO: improvements and validations. Int J Remote Sens 24(12):2389–2408
Latifovic R, Pouliot D, Dillabaugh C (2012) Identification and correction of systematic error in NOAA AVHRR long-term satellite data record. Remote Sens Environ 127:84–97

Molling CC, Heidinger AK, Straka WC, Wu X (2010) Calibrations for AVHRR channels 1 and 2: review and path towards consensus. Int J Remote Sens 31(24):6519–6540

Neckel H, Labs D (1984) The solar radiation between 3300 and 12500 Angstroms. Sol Phys 90:2

Pereira HM, Ferrier S, Walters M, Geller GN, Jongman RH, Scholes RJ, Bruford MW, Brummitt N, Butchart SH, Cardoso AC, Coops NC, Dulloo E, Faith DP, Freyhof J, Gregory RD, Heip C, Höft R, Hurtt G, Jetz W, Karp DS, McGeoch MA, Obura D, Onoda Y, Pettorelli N, Reyers B, Sayre R, Scharlemann JP, Stuart SN, Turak E, Walpole M, Wegmann M (2013) Ecology. Essential biodiversity variables. Science 339(6117):277–278

Rao CRN (1987) Pre-launch calibration of channels 1 and 2 of advanced very high resolution radiometer, NOAA Technical Report NESDIS 36. http://www.nesdis.noaa.gov/technical_reports/. Last accessed 3 Feb 2014

Rao CRN, Chen J (1994) Post-launch calibration of the visible and near-infrared channels of the advanced very high resolution radiometer on NOAA-7, -9, and -11 spacecraft, NOAA Technical Report NESDIS 78. http://www.nesdis.noaa.gov/technical_reports/. Accessed 3 Feb 2014

Rao CRN, Chen J (1996) Post-launch calibration of the visible and near-infrared channels of the advanced very high resolution radiometer on NOAA-14 spacecraft. http://noaasis.noaa.gov/NOAASIS/ml/aboutn14vis.html. Accessed 3 Feb 2014

Rao CRN, Chen J (1999) Revised post-launch calibration of the visible and near-infrared channels of the Advanced Very High Resolution Radiometer (AVHRR) on the NOAA-14 spacecraft. Int J Remote Sens 20(18):3485–3491

Rao CRN, Sullivan JT (2001) Calibration of the advanced very high resolution radiometer. In: Cracknell AP (ed) Remote sensing and climate change: the role of earth observation. Springer, Berlin, pp 117–157

Rao CRN, Chen J, Staylor FW, Able P, Kaufman YJ, Vermota E, Rossow WR, Brest C (1993) Degradation of the visible and near-infrared channels of the advanced very high resolution radiometer on the NOAA-9 spacecraft: assessment and recommendations for corrections, NOAA Technical Report NESDIS 70. http://www.nesdis.noaa.gov/technical_reports/. Accessed 3 Feb 2014

Robel J (ed) (2009) NOAA KLM user's guide – satellite and data description of NOAA's polar-orbiting satellites from NOAA-15 and later, Feb 2009 revision. http://www.ncdc.noaa.gov/oa/pod-guide/ncdc/docs/klm/cover.htm. Accessed 3 Feb 2014

Schmidt M, King EA, McVicar TR (2008) A method for operational calibration of AVHRR reflective time series data. Remote Sens Environ 112(3):1117–1129

SEASPACE: Tera Scan Suite http://www.seaspace.com/software.php. Accessed 3 Feb 2014

Sellers PJ, Dickinson RE, Randall DA, Betts AK, Hall FG, Berry JA, Collatz GJ, Denning AS, Mooney HA, Nobre CA, Sato N, Field CB, Henderson-Sellers A (1997) Modeling the exchanges of energy, water, and carbon between continents and the atmosphere. Science 275 (5299):502

Steven MD, Malthus TJ, Baret F, Xu H, Chopping MJ (2003) Intercalibration of vegetation indices from different sensor systems. Remote Sens Environ 88(4):412–422

Swinnen E, Veroustraete F (2008) Extending the SPOT-VEGETATION NDVI time series (1998–2006) back in time with NOAA-AVHRR data (1985–1998) for southern Africa. IEEE Trans Geosci Remote Sens 46(2):558–572

Teillet PM, Holben BN (1994) Towards operational radiometric calibration of NOAA AVHRR imagery in the visible and near-infrared channels. Can J Remote Sens 20:1–10

Thuillier G, Herse M, Labs D, Foujols T, Peetermans W, Gillotay D, Simon PC, Mandel H (2003) The solar spectral irradiance from 200 to 2400 nm as measured by the SOLSPEC spectrometer from the atlas and Eureca missions. Sol Phys 214(1):1–22

Trishchenko AP (2002) Removing unwanted fluctuations in the AVHRR thermal calibration data using robust techniques. J Atmos Ocean Technol 19:1939–1954

Trishchenko AP (2009) Effects of spectral response function on surface reflectance and NDVI measured with moderate resolution satellite sensors: extension to AVHRR NOAA-17, 18 and METOP-A. Remote Sens Environ 113(2):335–341

Trishchenko AP, Fedosejevs G, Li Z, Cihlar J (2002) Trends and uncertainties in thermal calibration of AVHRR radiometers onboard NOAA-9 to NOAA-16. J Geophys Res: Atmos 107:ACL 17-1–ACL 4778–4790

Tungalagsaikhan P, Günther KP, Gesell G, Dech SW, Ruppert T (2003) Operational processing of AVHRR data at DFD. In: Tagungsbände der Wissenschaftlich-Technischen Jahrestagungen der DGPF, Band 12: Auf dem Weg zu operationellen Prozessketten in Photogrammetrie, Fernerkundung und Geoinformation, 23. Jahrestagung der DGPF, Bochum

Vermote EF, Saleous NZ (2006) Calibration of NOAA16 AVHRR over a desert site using MODIS data. Remote Sens Environ 105(3):214–220

Chapter 4
Analysis of Snow Cover Time Series – Opportunities and Techniques

Andreas J. Dietz, Claudia Kuenzer, and Stefan Dech

Abstract Snow cover is one of the most dynamic land cover parameters that can be monitored from space and plays an important role for the Earth's climate system and hydrological circle. While the spatial extent can be limited to narrow mountain ridges during summer, the snow cover percentage on the Northern Hemisphere may exceed 50 % (Lemke et al., Observations: changes in snow, ice and frozen ground. In: Solomon S, Qin D, Manning M, Chen Z, Marquis MC, Averyt K, Tignor M, Miller HL (eds) Climate change 2007: the physical science basis. Contributions of Working Group 1 to the fourth assessment report of the intergovernmental panel on climate change. Cambridge University Press, Cambridge and New York, pp 337–383, 2007) of the total land surface (~45 million km^2) during winter seasons (Barry et al., Global outlook for ice & snow. United Nations Environment Programme, Hertfordshire, 2007). Remote sensing has been used since the early 1970s to map terrestrial snow cover (Brown, J Clim 13:2339–2355, 2000) and both – sensors as well as retrieval algorithms – have undergone a substantial development since that time. This chapter will give a short introduction on how snow cover can be monitored from space. Furthermore, techniques will be outlined that show how time series analyses can be applied to remotely sensed snow cover products to reduce the compromising effect of cloud cover and to investigate the fundamental characteristics of snow. Time series of snow cover data allow for various analyses covering the fields of hydrology, climate research, flood prediction and management, and economy. Short term variations and extreme events can be analysed as well as long term climatological trends, constituting time series of snow cover data a valuable tool for a large bandwidth of applications.

A.J. Dietz (✉) • C. Kuenzer
German Remote Sensing Data Center, DFD, Earth Observation Center, EOC, German Aerospace Center, DLR, Oberpfaffenhofen, Germany
e-mail: andreas.dietz@dlr.de; claudia.kuenzer@dlr.de

S. Dech
German Remote Sensing Data Center, DFD, German Aerospace Center, DLR, Wessling, Germany

Institute for Geography and Geology, University of Wuerzburg, Wuerzburg, Germany

C. Kuenzer et al. (eds.), *Remote Sensing Time Series*, Remote Sensing and Digital Image Processing, DOI 10.1007/978-3-319-15967-6_4

4.1 Introduction

Changes in Global snow cover characteristics have been observed that have several severe impacts on various aspects of the environment, including available fresh water resources, intensity and quantity of occurring floods and droughts, hydropower generation, winter tourism, food production, permafrost, glacier mass balance, and atmospheric circulation patterns. Because of all these implications and the fact that changes are predicted to intensify in the future, snow cover was identified as an Essential Climate Variable (ECV) within the Global Climate Observing System (GCOS) (WMO and GCOS 2011) and addressed as a most critical climate component by the Intergovernmental Panel on Climate Change (IPCC) (Lemke et al. 2007) and the United Nations Environment Program (UNEP) (Barry et al. 2007). The principles of how to map snow cover from space have been summarized in various review articles before (e.g. Dietz et al. 2012a; Frei et al. 2012; Solberg et al. 2006) and will therefore only be referred to briefly. Although it is generally possible to use Synthetic Aperture Radar (SAR) data as well as data from passive microwave (PM) sensors and the reflective part of the spectrum to estimate snow cover parameters, only the latter two sources are common within the context of snow cover time series (König et al. 2001). This chapter will be limited to snow cover analyses based on observations from the reflective part of the spectrum. A special focus will be directed onto processing and analysis of snow cover time series. Only by analysing time series of daily snow cover conditions over multiple years possible long term changes of snow cover can be recognized. Cloud coverage, polar darkness, and additional limitations arising from forested areas and mountainous regions can cause problems during the preparation of gapless time series (Salminen et al. 2009; Vikhamar and Solberg 2002). In the following it will outline how these problems can be dealt with.

4.2 Remote Sensing of Snow

The basis for detecting snow cover from the reflective part of the spectrum relies on its distinct spectral properties. Depending on the available sensor, a combination of these properties renders snow a unique land cover feature and allows discriminating between snow, ice, clouds, and the remaining land cover types. Challenges exist that emerge from the high variability of snow, the surrounding land cover itself, and sensor limitations like resolution and spectral band widths. These challenges will be addressed to in the following section before techniques to classify snow and process time series of snow cover data follow in the subsequent sections.

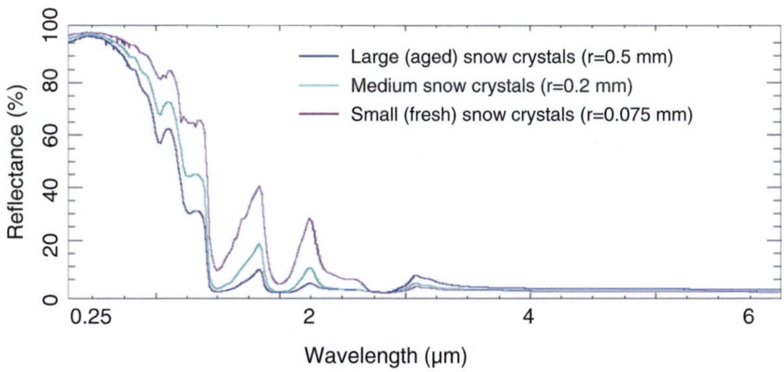

Fig. 4.1 Spectral characteristics of different snow types (Modified from Dietz 2013)

4.2.1 Snow Cover Mapping from Space – Physical Background and Limitations

Freshly fallen snow reflects around 90 % of radiation in visible wavelengths (Hall and Martinec 1985). This value drops near zero for longer wavelengths in the near infrared region, creating an explicit gradient that can easily be monitored and exploited for snow cover mapping from operational satellite sensors (Pepe et al. 2005). This gradient is especially useful to discriminate between clouds and snow, as clouds reflect a higher proportion of incoming solar radiation in the short wave infrared. Figure 4.1 illustrates the spectral signature of different snow types: freshly fallen snow crystals (purple plot), intermediate (cyan plot), and aged snow crystals (dark blue plot).

The different signatures in Fig. 4.1 are the result of multiple modifying effects occurring once the snow crystals have accumulated on the ground: Grain size and shape are changing while aging, leading to larger crystal sizes. This metamorphosis of snow crystals can be caused by the pressure within the snow pack or melt- and refreeze processes (Foster et al. 1999; Rango 1996). Furthermore, impurities like soot or dust may reduce the reflectance by up to 5 % while they also increase the absorption of incoming solar radiation (Aoki et al. 2007; Hadley and Kirchstetter 2012).

Remote sensing sensors observing the reflective and emissive part of the spectrum provide the bandwidths necessary to detect snow cover relying on the characteristics presented in Fig. 4.1. The minimum requirements – bands covering the visible and short wave infrared – are provided by all instruments enlisted in Table 4.1 that includes a selection of sensors commonly used for snow cover mapping along with some additional information about revisit time, resolution, and swath width.

Many different techniques exist to map snow cover based on the instruments listed in Table 4.1. We again refer to (Frei et al. 2012) and (Dietz et al. 2012a) for a detailed overview. The strong gradient between visible and short wave infrared

Table 4.1 Remote sensing instruments used to map snow cover from the reflective part of the spectrum

Satellite(s) – Instrument(s)	Operational since/ until	Revisit time	Spatial resolution	Swath width
Landsat – MSS/TM/ETM+/ OLI	1972/Today	16–18 days	30–100 m	185 km
Terra, Aqua – MODIS	2000/Today	Twice per day	250–1,000 m	2,330 km
TIROS/NOAA/Metop – AVHRR	1978/Today	At least daily	1,100 m	2,400 km
Envisat/AATSR	2002/2012	2–3 days	1,000 m	500 km
Envisat/MERIS	2002/2012	2–3 days	300 m	1,150 km
ERS-2/ATSR-2	1995/2011	2–3 days	1,000 m	512 km
Sentinel 2	To be launched in 2015	3–5 days	10–60 m	290 km
Sentinel 3 – OLCI/SLSTR	To be launched in 2015	1–2 days	300–500 m	1,270 km

Sources: Donlon et al. (2012), Dozier (1989), Drusch et al. (2012), ESA (2013a, b), Hall et al. (1995), Malenovský et al. (2012) and Solberg et al. (2010)

(Fig. 4.1) constitutes the basis for methods aiming to detect snow cover in remotely sensed data from the reflective part of the spectrum. The Normalized Difference Snow Index (NDSI) is often used to automatically classify snow covered surfaces. Introduced by (Crane and Anderson 1984), the NDSI is based on the difference between the visible and shortwave infrared reflectance:

$$NDSI = \frac{(VIS - SWIR)}{(VIS + SWIR)} \tag{4.1}$$

NDSI values greater than 0.4 are usually a good indicator for the presence of snow (Hall et al. 1995), though mixed pixel/fractional snow cover effects, forest cover, or impurities within the snowpack can reduce the NDSI. Land cover induced limitations like snow cover underestimation within forested regions can be mitigated by providing additional input data: A forest-inventory map can help to increase the accuracy of snow classifications as certain thresholds (e.g. for NDSI) can be adjusted to local land cover conditions (Vikhamar and Solberg 2003). The NDSI can be used to detect snow in Landsat scenes (Ault et al. 2006) and it serves as classification basis for the operational MODIS snow cover products (Hall et al. 2002). The *snowmap* algorithm (Hall et al. 1995) that produces the operational MODIS snow products (MOD 10 product line) differentiates between forested and non-forested regions: The NDSI threshold of 0.4 (for non-forested areas) is reduced in regions with forest cover to prevent underestimation of actual snow cover. A second test is added to prevent overestimation of snow: The reflectance of MODIS band 4 must exceed 10 % for each pixel in order to be mapped as snow (Klein et al. 1998). Dark surfaces like water bodies tend to receive NDSI values greater 0.4

and by introducing the 10 % threshold of band 4, misclassifications can be reduced to a minimum. The overall accuracy of the operational MODIS snow products reaches 93 % (Hall and Riggs 2007).

One of the first fully automated snow cover classification algorithms was the AVHRR Processing scheme Over Land, cLoud and Ocean (APOLLO). AVHRR data are suitable for time series analysis because these sensors exist since 1979, observing the Earth on a daily basis and with a resolution that fits the requirements stated by GCOS. Therefore, an automated method to extract snow cover information from AVHRR data and discriminate between clouds and snow is highly desirable. APOLLO consists of various tests for cloud detection based on distinct thresholds. Details about the cloud screening can be found in (Saunders and Kriebel 1988) – they are not in the focus of the presented methods to process snow cover data. The module to detect snow cover however is illustrated in Fig. 4.2. It is part of an extension to map ice and snow from AVHRR (Gesell 1989). As Fig. 4.2 shows, the snow cover detection is based on several tests involving all available AVHRR channels. The extension was added to APOLLO because snow was often mistakenly classified as cloud cover. The Dynamic Visible Threshold (DVT) test, the ratio between channel 2 and channel 1 reflectance (R21T), and the temperature difference between channel 4 and channel 5 (T45T) were especially susceptible to these misclassifications. In the extension, a channel 3 threshold test is added to validate the cloud flagged pixels. In this step a pixel may be recoded to snow, increasing the accuracy of the snow cover classification. APOLLO was developed between 1986 and 1988 and may therefore be presumed to be outdated. The snow and ice extension and additional other improvements (Kriebel et al. 2003) make it an attractive alternative if additional input data (as they are required for the SPARC routine introduced in the next paragraph) are not available.

Another method that has been developed especially for AVHRR data is Separation of Pixels Using Aggregated Rating over Canada (SPARC) (Khlopenkov and Trishchenko 2007). SPARC consists basically of three cloud tests and additional snow and thin cirrus tests, each producing a score value that when summed up from all tests will give the final aggregated rating value. The difference between this approach and many other commonly applied methods (like APOLLO, CLAVR, MODIS scheme) is that SPARC performs all tests independently from each other. The branching approach within most alternative methods will decide step by step (often threshold by threshold, see Fig. 4.2) if a pixel may be covered by clouds or snow but it is usually impossible to go back through these steps and change a prior decision based on a following test result. In SPARC, tests for temperature (difference between ground temperature and satellite-based brightness temperature), reflectance brightness (based on the reflectance in band 1 over land and band 2 over water), and reflectance (involving the reflectance in band 3) are performed and completed by additional snow, uniformity, and thin cirrus tests. The sum of all these test results can then be used to detect the snow. This method was used for Canada (Khlopenkov and Trishchenko 2007) and in a modified version also for Europe (Hüsler et al. 2012). The drawback of SPARC is that additional input

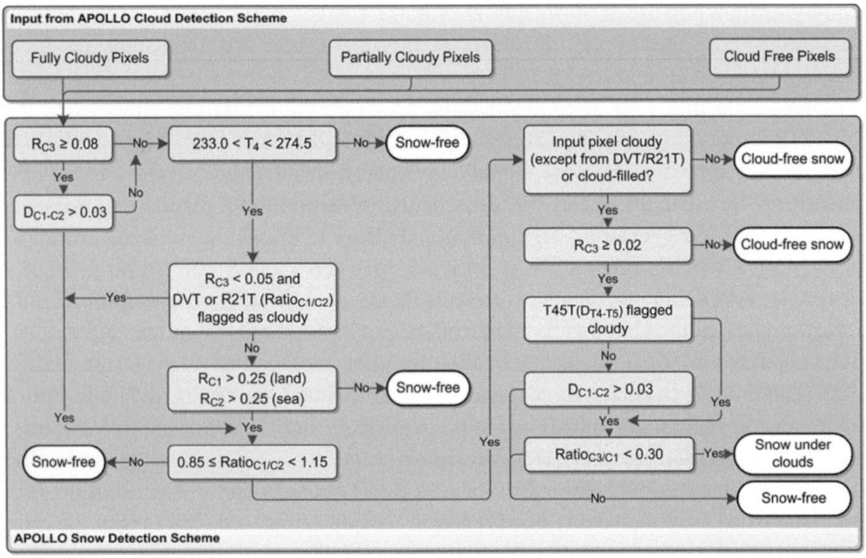

Fig. 4.2 APOLLO scheme to detect snow cover from AVHRR data (*RC3*: Reflectance in channel 3; *DC1-C2*: Difference between channel 1 and channel 2; *T4*: brightness temperature in channel 4; *RatioC1/C2*: ration between channel 1 and channel 2; *DVT*: dynamic visible threshold; *R21T*: ratio between channel 1 and channel 2 – cloud test; *T45T*: difference between brightness temperatures of channels 4 and 5)

parameters (skin temperature) are required. These may not always be available for the respective study region or time span.

Unfortunately, operational snow products based on AVHRR at 1 km resolution are not freely available for download at the moment. Users interested in time series of medium resolution snow cover data are therefore reliant on the MODIS snow cover products. If time series prior to the year 2000 are required, snow cover can be derived from raw AVHRR data relying on e.g. the SPARC or APOLLO routines. The GlobSnow snow extent products are available on a daily basis since 1995, but as the snow cover information is derived from Envisat AATSR and ERS-2 ATSR-2 sensors, a full global coverage is not achieved on a daily (not even weekly) basis (Luojus et al. 2013).

While *snowmap,* SPARC, and APOLLO only discriminate between two classes (snow covered and snow free pixels), fractional snow cover algorithms exist that detect the snow cover percentage within each pixel. Though the binary information is usually enough for large scale planetary or continental studies, information about the snow cover fraction within each pixel may be required for small scale investigations. Hydrological applications can benefit from fractional snow cover data especially when analysing or predicting runoff in small hydrological catchments (Bales et al. 2008; Metsamaki et al. 2005).

Fractional snow cover data is not a necessary input for the presented methods to analyse time series of snow cover data. That is because these methods aim to make statements about temporal aspects like duration (overall, early season, late season) of one or multiple snow seasons. Binary snow cover information is sufficient for studies about such topics. However: systematic errors may occur especially in mountainous regions if fractional snow cover is not considered (Dozier et al. 2008). Therefore, one technique for fractional snow cover mapping is presented here. For additional references, Frei et al. (2012) and Dietz et al. (2012a) should be referred. The MODIS Snow-Covered Area and Grain size (MODSCAG) model relies on spectral unmixing to assess the proportional snow cover fraction within each observed pixel (Painter et al. 2009). A library of different snow endmembers (basically consisting of several snow grain sizes) is used additionally to rock, soil, vegetation, and lake ice endmembers. The grain size influences the spectral characteristics of snow as already illustrated in Fig. 4.1. Therefore, an estimation of fractional snow cover is only feasible when also estimating the grain size of the snow pack. MODSCAG applies linear spectral mixture analysis to each possible combination of endmembers from the spectral libraries. Snow cover fraction and grain size are selected according to the model that fits best for each endmember. The root mean square error (RMSE) for the fractional snow cover classification ranges from 1 % to 13 % while the mean RMSE accounts for 5 %.

The presented methods to classify snow from remotely sensed data should be understood as a brief excerpt of what has been developed since the 1970s. Selecting the right algorithm and data source depends on the individual research problem, the size of the study region, and the time frame. Several reviews about snow cover mapping have already been introduced within this section that may help during the decision making. The following sections will outline how time series of snow cover data can be processed and analysed.

4.2.2 Processing of Snow Cover Time Series

Before snow cover time series can be processed the data basis for the analyses must be selected and prepared. There are many different algorithms and satellite data available to produce snow cover information from remote sensing data as it was briefly indicated in Sect. 4.2.1 and Table 4.1. For the following sections we rely on the MODIS daily snow cover products MOD10A1 and MYD10A1 (Hall et al. 2000; Riggs and Hall 2004), provided by the National Snow and Ice Data Center (NSIDC 2013). This product has been used and validated in many different studies (e.g. Dietz et al. 2013; Huang et al. 2011; Simic et al. 2004; Tekeli et al. 2005; Zhou et al. 2013) and is available on a daily basis for the whole globe. The presented methods may be applied to any other snow cover time series derived from any arbitrary sensor. The only prerequisite is the availability of binary information about snow presence per pixel (which could also be prepared from fractional snow cover data). Once the data has been downloaded from NSIDC, processing of the time series is necessary to extract desirable snow cover parameters. It was already stated in the introduction that changes in climate – both sustainable effects as well as short-term variability – will affect snow cover conditions on the ground. It is important to understand these changes for several reasons: Earths radiation budget, vegetation and irrigation, tourism and hydropower generation may all be influenced by changed Snow Cover Duration (SCD). These changes may be the consequence from longer or shorter SCD or a shifted snow cover season towards earlier or later snow cover melt (expressed by Early Season Snow Cover Duration SCD_{ES} and Late Season Snow Cover Duration SCD_{LS}) SCD, SCD_{ES} and SCD_{LS} may help to predict runoff characteristics during spring, can be used as input parameters for hydrological models, and may serve as an indicator for upcoming natural disasters like floods or droughts (Muntán et al. 2009; Rango 1996; Tekeli et al. 2005; Thurman 2011). Regional variability of snow cover changes may vary significantly (Räisänen and Eklund 2012). Therefore it is important to map SCD, $SCD_{ES,}$ and SCD_{LS} with high spatial and temporal resolution. To derive these parameters, the time series of snow cover data must pass through several processing steps for generally two reasons:

First, a snow cover map produced from a single satellite observation only represents the snow status for a single point in time. This status can change quickly within only very few days. Therefore, only the combination of multiple observations for a given time series can allow further statements about SCD.

Second, the limitation of cloud coverage hinders sophisticated calculations of SCD, SCD_{ES} and SCD_{LS}. The frequency and distribution of cloud coverage varies significantly over the globe and with time. According to (Wylie et al. 2005), the average cloud cover frequency accounts for ~ 60 % in latitudes prone to snow cover. We investigated the cloud cover distribution for the part of the Northern Hemisphere where snow cover is common (north of 30°N): Fig. 4.3 illustrates the number of cloud covered (a) days for the hydrological year 2003/2004 (September 1st 2003 until August 31st 2004) as extracted from the MODIS daily snow cover product MOD10A1. The number of days affected by polar darkness is also depicted (b). Around 34.000 MODIS tiles were processed to generate the cloud cover statistics presented in Fig. 4.3. The average clear sky rate (days without cloud cover or polar darkness) for any pixel shown in Fig. 4.3 accounts for 46.3 %. As a consequence, the status of snow cover on the ground also remains unknown for 53.7 % of the satellite observations. Untreated, this problem would lead to severe uncertainties in the calculation of snow cover parameters. Therefore, various algorithms have been developed to estimate the snow cover status even under cloudy conditions. The following paragraph gives an overview of some of these methods. For a more detailed overview we refer to (Dietz et al. 2012a).

4.2.2.1 Estimation of Snow Cover Status Below Clouds

Figure 4.3 illustrates the spatial distribution and duration of cloud cover and polar darkness for the regions north of 30°N. 53.7 % of the observations are flagged as cloud or darkness in the daily snow cover product MOD10A1 (Hall et al. 2000).

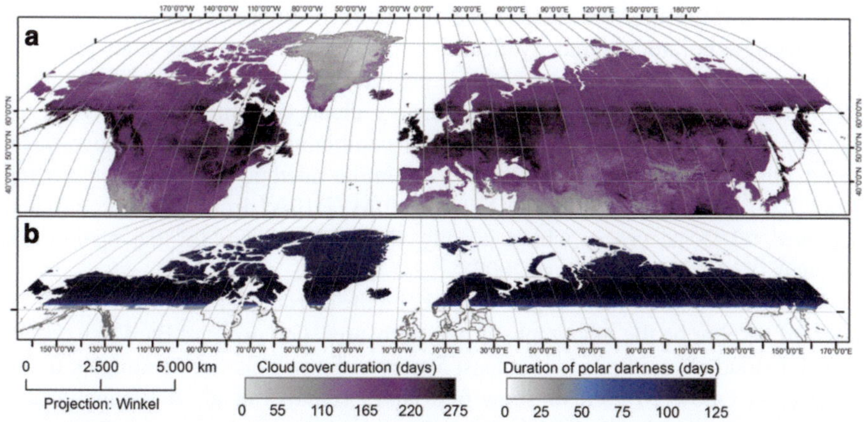

Fig. 4.3 Northern Hemisphere cloud cover duration (**a**) and polar darkness duration (**b**) extracted from daily snow cover product MOD10A1 for the hydrological year 2003/2004

Some regions are affected more frequently than others: The east coast of North America (Quebec, Newfoundland) as well as large parts of Europe (Great Britain, Middle Europe) and eastern Asia (Kamchatka, southern Siberia) are examples for heavily cloud covered areas. The sharp border of cloud cover duration at ~ 65° N in Fig. 4.3a arises from the beginning presence of polar darkness in this region. Together, cloud cover and polar darkness obscure large proportions of snow prone areas especially in the winter season. Therefore, techniques to estimate the snow cover status below clouds are required before a reasonable calculation of SCD, SCD_{ES} and SCD_{LS} becomes possible. Generally, two basic approaches to assess the snow cover status below clouds can be followed when relying on remote sensing data originating from the reflective part of the spectrum: Temporal or spatial interpolation. Both techniques are based on assumptions of how snow cover develops with time and/or local terrain settings. As these general assumptions may not always reflect the natural conditions, applying methods to estimate the snow cover status below clouds always reduces the accuracy of the initial snow cover classification result. A combination of multiple techniques is therefore advisable as some methods are more reliable than others – though they may not be suited to clear the whole scene from clouds. Combinations of multiple cloud removal algorithms have been applied by (Dietz et al. 2012b, 2013; Gafurov and Bárdossy 2009) before.

Temporal interpolation techniques rely on the fact that cloud cover positions are highly variable over time. Combining snow cover products from different observations helps to reduce the overall cloud percentage. These combinations can involve observations from the same day (e.g. originating from different satellite platforms or overpasses) or multiple days from the past (and in case of historical data also from the future). As snow cover may be highly variable especially during autumn und spring season, combining snow cover information from several days potentially affects the accuracy of the final result. The MODIS snow cover products

MOD10A2/MYD10A2 are based on a combination of 8 observation days, therefore minimizing – but not completely clearing – cloud coverage (Riggs et al. 2006).

Figure 4.4 illustrates a temporal interpolation for the 1st and 2nd of January, 2014: Both days are characterized by heavy cloud contamination (45.7 % for January 1st, 61.6 % for January 2nd). By combining the two successive days, cloud cover was reduced to 37.5 %. A similar method was used by (Wang et al. 2009): They combined MOD10A1 and MYD10A1 from the same day and achieved 92 % accuracy, which is only slightly less than for the operational products (93 %, Hall and Riggs 2007).

Spatial interpolation techniques can rely on the direct neighbourhood of a cloud affected pixel or on the topography of the study area. Gafurov and Bárdossy (2009) recoded cloud covered pixels if at least 5 surrounding pixels were belonging to the same class (snow covered or snow-free land). (Parajka et al. 2010) included a Digital Elevation Model (DEM) and extracted upper and lower snowlines from the daily snow cover datasets: The upper snow line represents the elevation where all cloud-free pixels are classified as snow. Above this altitude, all cloud-covered pixels can also be assumed snow-covered -given a certain overall cloud cover percentage is not reached. The lower snowline on the other hand identifies the altitude below which all cloud-free pixels are also snow-free. Once determined, all cloud-covered pixels below the lower snowline can be assumed snow-free. The snowline method exploits the fact that snow cover increases with elevation: In mountainous regions, the mean snow cover duration increases by ~4 days per 100 m altitude (Dietz et al. 2012b, 2013; Parajka et al. 2010). Figure 4.5 illustrates the snowline analysis for January 2nd 2014 – again for the same MODIS tile already used in Fig. 4.4. The lower snowline was identified at 134 m above sea level (a.s.l.) while the upper snowline was found at an altitude of 2,768 m. This information can be used to clear another 6 % of cloud cover from the scene as all cloud covered pixels below 134 m can be recoded to snow free land and all pixels above 2,768 m can be set to snow covered. The cloud covered pixels in between remain unknown. This method should only be used if the overall cloud cover percentage of the scene does not exceed a certain value. (Gafurov and Bárdossy 2009) suggest 30 % as an upper cloud cover threshold.

The accuracy of interpolation techniques varies and therefore, a combination of different methods may be appropriate (Gafurov and Bárdossy 2009). Figure 4.6 illustrates a combination of temporal and spatial interpolation techniques as it was applied by Dietz et al. (2013). The sequence of steps is ordered according to their accuracy. This ensures that the effect on the accuracy stays minimal. However: The overall accuracy of the final product will be lower than of the initial snow cover products. The combination from Fig. 4.6 was used for studies in Central Asia (Dietz et al. 2013) and in a slightly modified version for Europe (Dietz et al. 2012b). The accuracy of the daily snow maps was assessed to be around 85 % – always varying with respect to local land cover conditions, topography, and season.

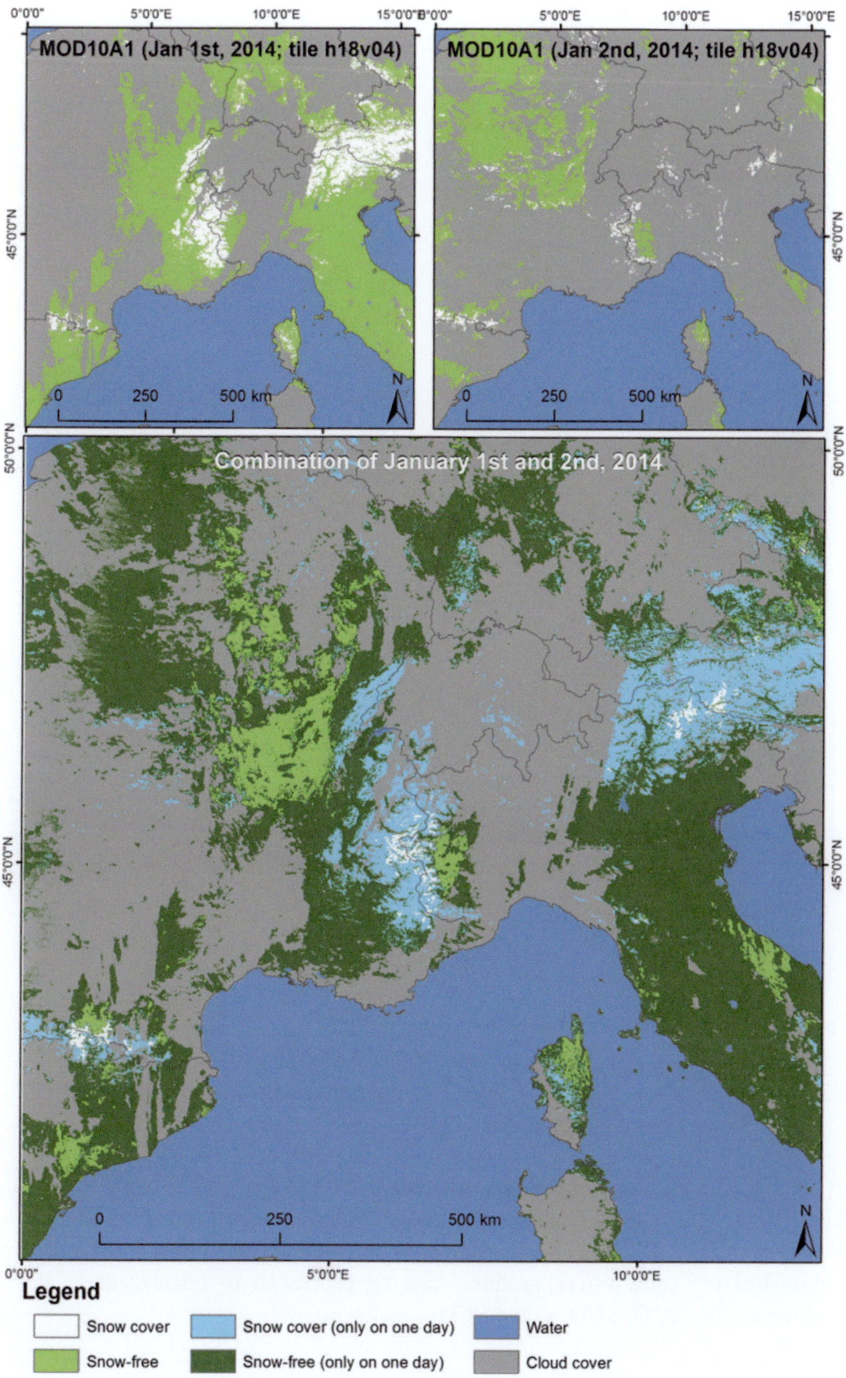

Fig. 4.4 Temporal combination of snow cover maps

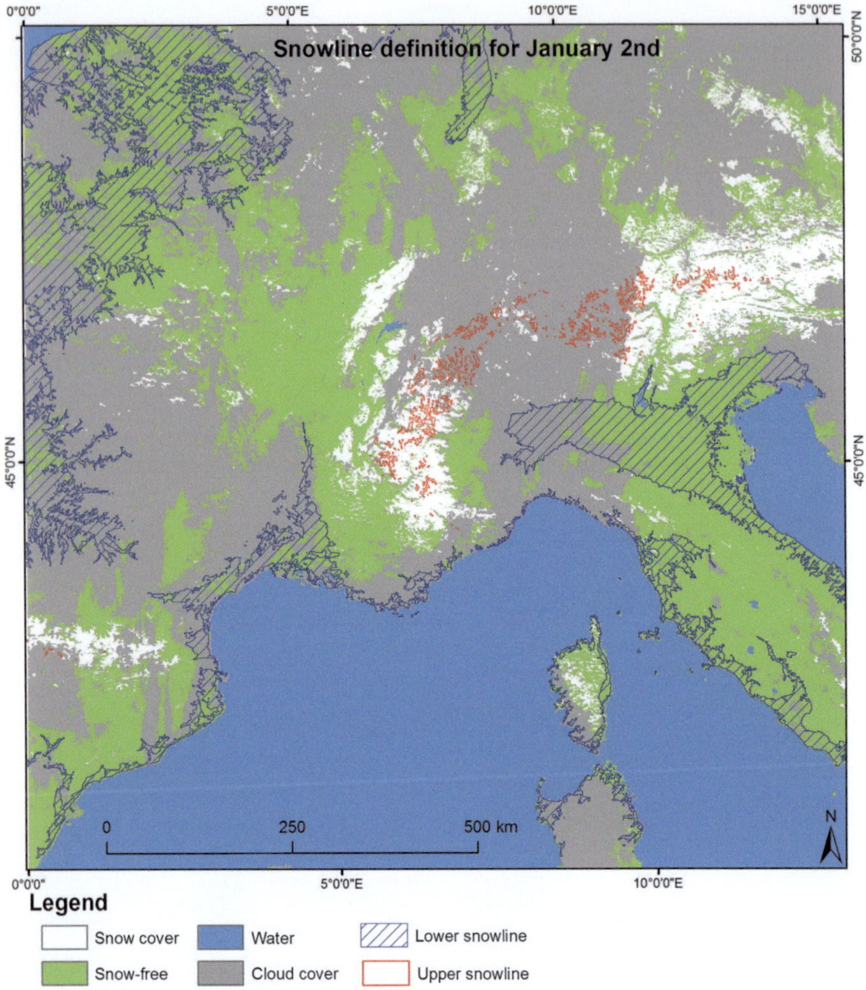

Fig. 4.5 Upper and lower snowline analysis using a DEM

4.2.2.2 Generation of Snow Cover Parameters From Cloud-Cleared Snow Cover Time Series

In Sect. 4.2.2.1 the snow cover data was cleared from any cloud covered or missing data pixels. This step is necessary to allow for the analysis of the snow data as presented in Sect. 4.2.2.2: The time series of daily (or in some cases multi-temporal composites of) snow cover products can be processed to retrieve snow cover parameters like SCD, SCD_{ES}, and SCD_{LS} using Eq. 4.2:

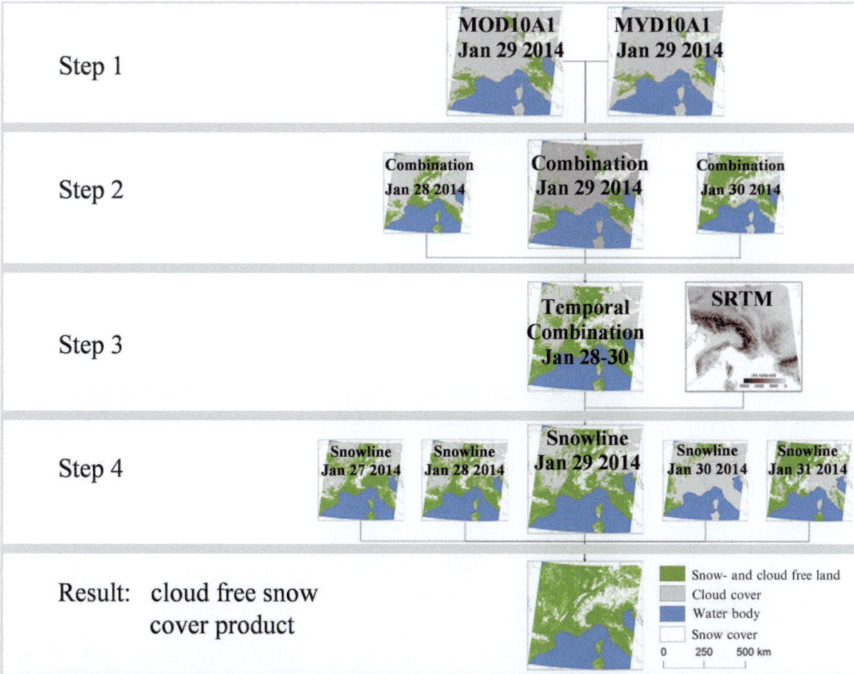

Fig. 4.6 Combination of cloud interpolation techniques (schematic overview): Operational MODIS snow cover products MOD10A1 and MYD10A1 (input) are merged for each calendar day (step 1), over a three-day period (step 2), combined with a DEM to detect snowlines (step 3), and interpolated based on the full time series of a hydrological year (step 4). The output is 100 % cloud free

$$SCD = \sum_{i=1}^{n} (s_i) \tag{4.2}$$

Where n is the number of observations (e.g. beginning on Sept. 1st of a year and ending on Aug. 31st of the following year), and s represents the cloud-free snow cover dataset. SCD_{ES} and SCD_{LS} base on the same equation, but the number of observations differs: For the Northern Hemisphere SCD_{ES} often refers to the SCD between September 1st and January 15th while SCD_{LS} is calculated from the time span between January 16th and August 31st. The exact dates can be modified depending on the hydrology of the study region. A hydrological year should be defined according to the analysis of long term meteorological data from within the study region (Zhou et al. 2013). Therefore, also the date to discriminate between SCD_{ES} and SCD_{LS} may vary. For the Southern Hemisphere a hydrological year may range from March 1st of a given year until February 28th of the next year. The date to separate early season from late season in this case could be September 1st.

Together, SCD, SCD_{ES} and SCD_{LS} describe the snow cover characteristics of a hydrological year. They should always be analysed as a complete set as changes in

SCD_{ES} and SCD_{LS} might not influence overall SCD while longer/shorter SCD may not necessarily be caused by both, changed SCD_{ES} and SCD_{LS}. Variations in any of these parameters may induce multiple effects: A change in overall SCD affects the radiation budget, leading to an increase in absorbed incoming solar radiation if SCD is decreased. This again may contribute to global climate change (Fernandes and Zhao 2008; Klein et al. 2000). Changes in SCD_{ES} and/or SCD_{LS} might influence vegetation growth (Grippa et al. 2005; Scott and Wayne 1995) and runoff behaviour during spring season (Klein and Barnett 2003). Additionally, information about SCD_{LS} can be used as an input parameter in flood prediction (Tekeli et al. 2005; Thirel et al. 2013). Therefore, the calculation of snow cover parameters can constitute a useful tool for various applications. Section 4.3 will outline some of these applications in more detail.

Figure 4.7 illustrates the SCD for the hydrological year 2012/2013 derived from the MODIS daily snow cover products MOD10A1 and MYD10A1 for the Northern

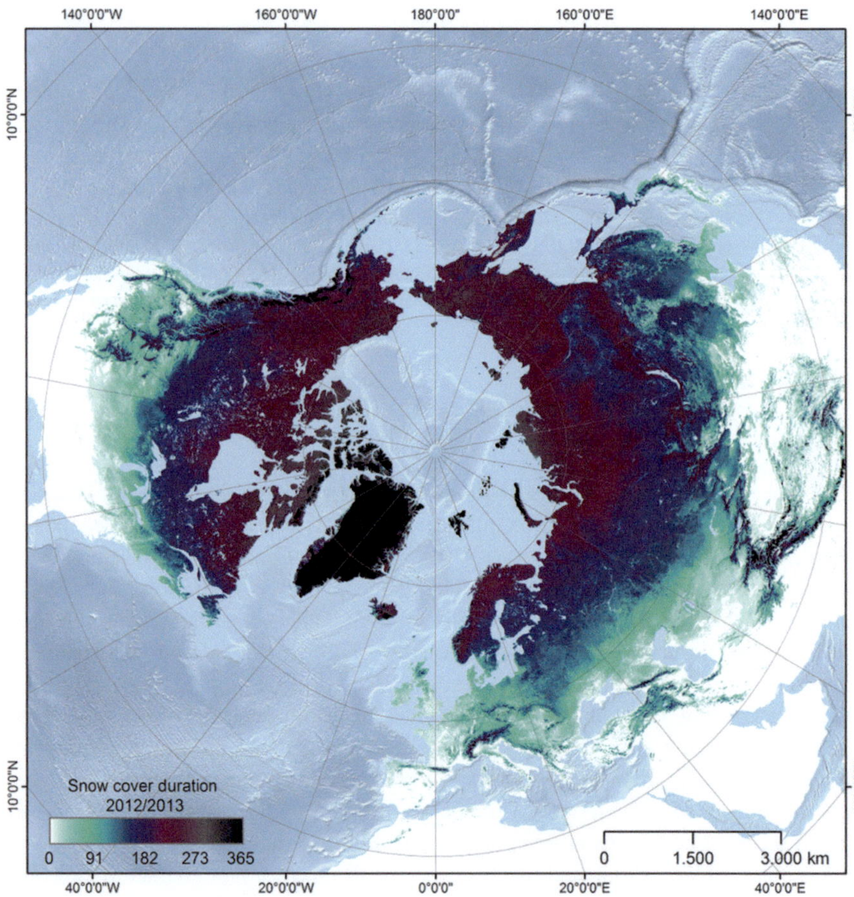

Fig. 4.7 Map of the snow cover duration on the Northern Hemisphere for the hydrological year 2012/2013 (September 2012–August 2013) derived from MODIS time series

Hemisphere. The time series forming the basis of Fig. 4.7 was processed according to the methods described in Sect. 4.2.2 and Figs. 4.4, 4.5 and 4.6: Combination of Aqua and Terra MODIS (MOD and MYD10A1), temporal interpolation in two different steps, and snowline detection. The resolution of 500 m allows for both, regional and large-scale studies of snow cover characteristics. In Fig. 4.7 the typical increase in SCD according to the latitude is clearly visible, which is only interrupted by mountain ranges or terrain depressions. Studies in Central Asia and Europe have revealed that the SCD increase accounts for ~ 5 days per degree latitude while in mountainous terrain, SCD increases by ~ 4 days per 100 m elevation (Dietz et al. 2012b, 2013).

4.3 Analysis of Snow Cover Time Series

Both the snow cover time series as well as the derived snow cover parameters can be analysed in various ways. As already stated in the introduction, snow cover affects a wide range of hydrological, climatological, and anthropological aspects. Figure 4.7 already illustrates a SCD product and indicates some possible ideas about where potential analyses may be directed to. The range of possible applications is wide and therefore, a few examples will be outlined in this section.

Runoff in many of Eurasia's and North America's largest rivers is dominated by snow: More than 50 % of the stream-flow are generated by snowmelt in these catchments, meaning that snow cover variations affect the runoff regime in these rivers considerably (Barnett et al. 2005). Daily snow cover information as prepared by the cloud interpolation scheme (Fig. 4.6) can be analysed to identify the (mean) Snow Cover Fraction (SCF) within a hydrological catchment (Eq. 4.3):

$$SCF = \frac{A_{snow}}{A_{total}} * 100 \tag{4.3}$$

With A_{snow} standing for the snow covered area within a hydrological catchment while A_{total} refers to the total area of the catchment.

Figure 4.8 illustrates the mean SCF (a, b) between 2000 and 2013 for the Volga and Danube river catchments as well as the standard deviation for this time span. Both rivers are situated in the snow-dominated runoff zone (Barnett et al. 2005), but the SCF-graphs differ considerably: Maximum SCF within the Volga catchment (Fig. 4.8a) accounts for more than 90 % in mid-winter with deviations reaching their maximum during autumn and spring seasons. Danube River only accounts for around 50 % SCF and deviations are much higher with a maximum during winter season. The graphs in Fig. 4.5c, d depict the daily SCF deviations during the 13 years of observation. Some extreme events like the negative outlier in 2006/2007 for the Danube catchment are visible from the time series and can be connected to respective low flow events in the river's runoff. Spring 2007 was a season with exceptionally low runoff, which was caused not by less precipitation

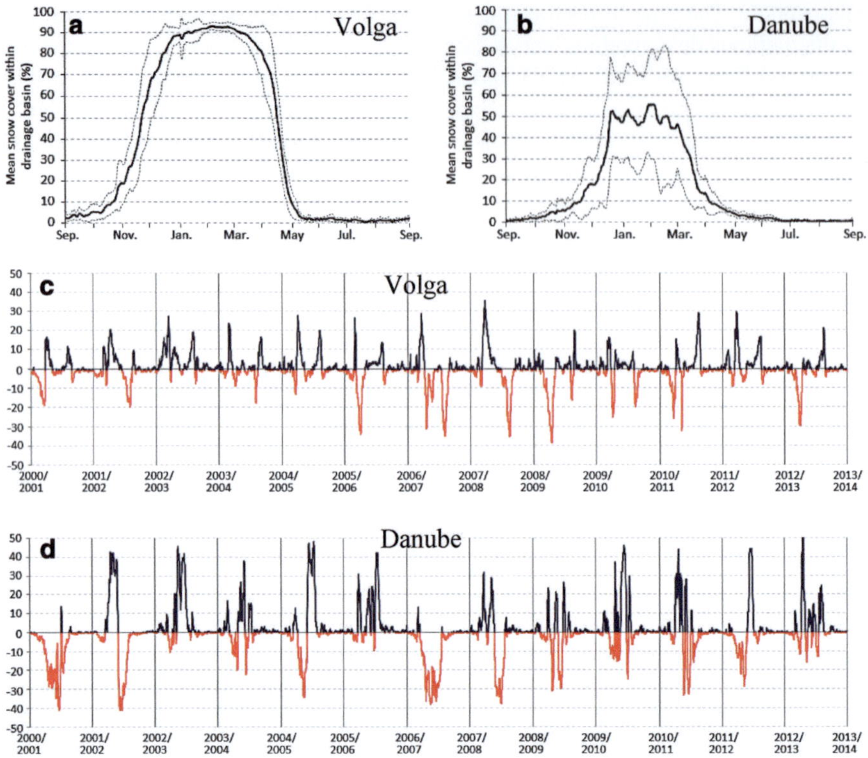

Fig. 4.8 Mean SCF for Volga (**a**) and Danube (**b**) hydrological catchments. The *solid line* represents the mean SCF while the *grey* and *dashed line* illustrates the standard deviation of SCF between 2000 and 2013. Figures (**c**) and (**d**) depict the deviation (in %) from the mean SCF per hydrological year

during the winter months but the fact that the temperature was too high to allow for snow accumulation. Consequently, melt water amounts during spring season turned out to be much less than under normal conditions (Belz and Rademacher 2007). The time series of snow cover data can therefore be analysed with regards to abnormal runoff events, because the negative SCF can be recognized from the time series way before the intrinsic runoff is being released through snow melt. Runoff originating from snowmelt contributes up to 40 % to the discharge volume of some of Danube's subcatchments (Weber et al. 2010). The mountainous upstream region of Danube River is of eminent importance for the downstream water availability especially during spring and early summer. Climate change will most likely alter these conditions, as less snow will be accumulated during winter while snowmelt will occur earlier due to increased temperatures. The effect of these changes may be extensive but is difficult to predict. It is important to identify already the changes in SCD, SCD_{ES} and SCD_{LS} to be able to adapt existing hydrological models (Mauser and Bach 2009; Weber et al. 2010).

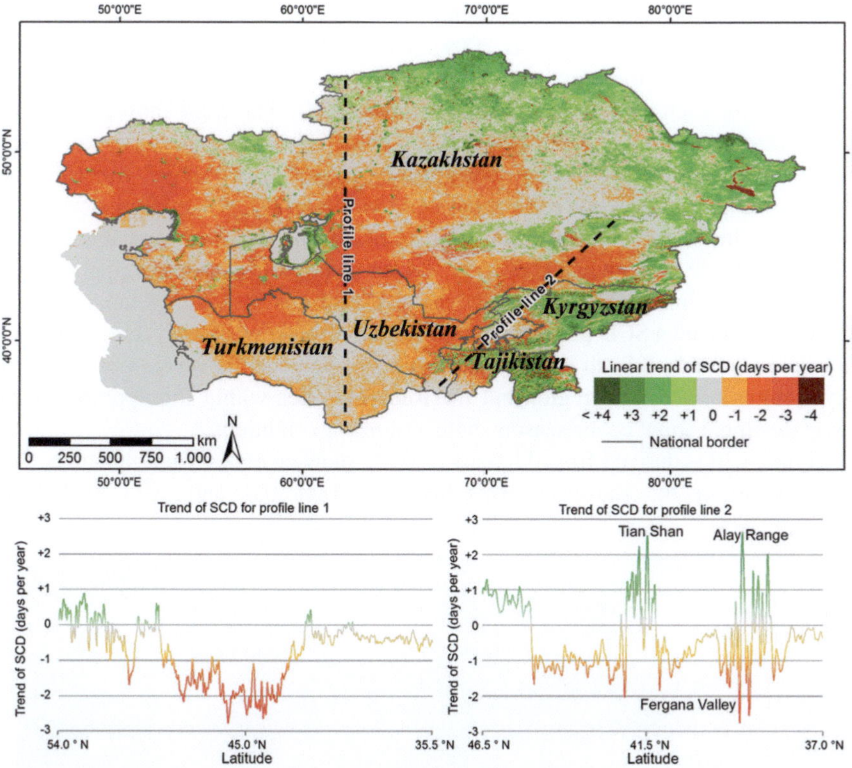

Fig. 4.9 Linear trend of SCD between 1991/1992 and 2011/2012 for Central Asia

Another way to leverage the time series of snow cover parameters is the trend analysis of multiple years of SCD, SCD_{ES} and SCD_{LS}. Climate change leads to changes in SCD, often accompanied by shifts in SCD_{ES} and SCD_{LS} as well (Brown 2000; Lemke et al. 2007; Räisänen 2007). To identify the extent and intensity of these changes, time series of snow cover parameters can be analysed.

For Fig. 4.9, time series of SCD for the years between 1991/1992 and 2011/2012 have been analysed in order to detect possible linear trends. The snow cover data used as a basis for the SCD calculations are originating from the operational MOD10A1 and MYD10A1 data again. As this source only ranges back until 2000, AVHRR was used in addition to the MODIS data. As already stated in Sect. 4.2.2, no operational snow cover product at full AVHRR resolution is available for download. Therefore the raw Level 1B AVHRR data were obtained from NOAA CLASS (National Oceanic and Atmospheric Administration – Comprehensive Large Array-data Stewardship System) (NOAA n.d.) and processed relying on the APOLLO scheme also introduced in Sect. 4.2.1.

At the first glance, the negative trends between ~40 and 50°N depicted in Fig. 4.9 conform to hemispherical studies (Brown 2000; Choi et al. 2010) while also the slight increase in the most northern regions of Central Asia has been reported before (Brown and Mote 2009). The resolution of MODIS and AVHRR is higher compared to coarse resolution data commonly used in climate studies. Examples for such studies are those of (Brown 2000) who used snow cover data with a resolution of 125–200 km and (Choi et al. 2010), who used the same data source. Because of the short time series of only 20 years and the high variability of snow cover within the study region, the trend results in Fig. 4.9 are statistically not significant in all regions. Especially the trend results for the mountainous area remain arguable. However: They serve as an example for possible applications given that the time series of snow cover parameters is sufficiently long Fig. 4.7 already visualized a SCD product derived from a time series of daily, cloud-cleared snow cover products. Such information is useful for a general overview of the snow cover status of a certain region but the true value lies within the opportunity to compare single snow cover seasons with e.g. mean conditions.

Figure 4.10 is derived from 11 years of daily snow cover data and illustrates the mean SCD for the European Alps between 2000/2001 and 2011/2012. The

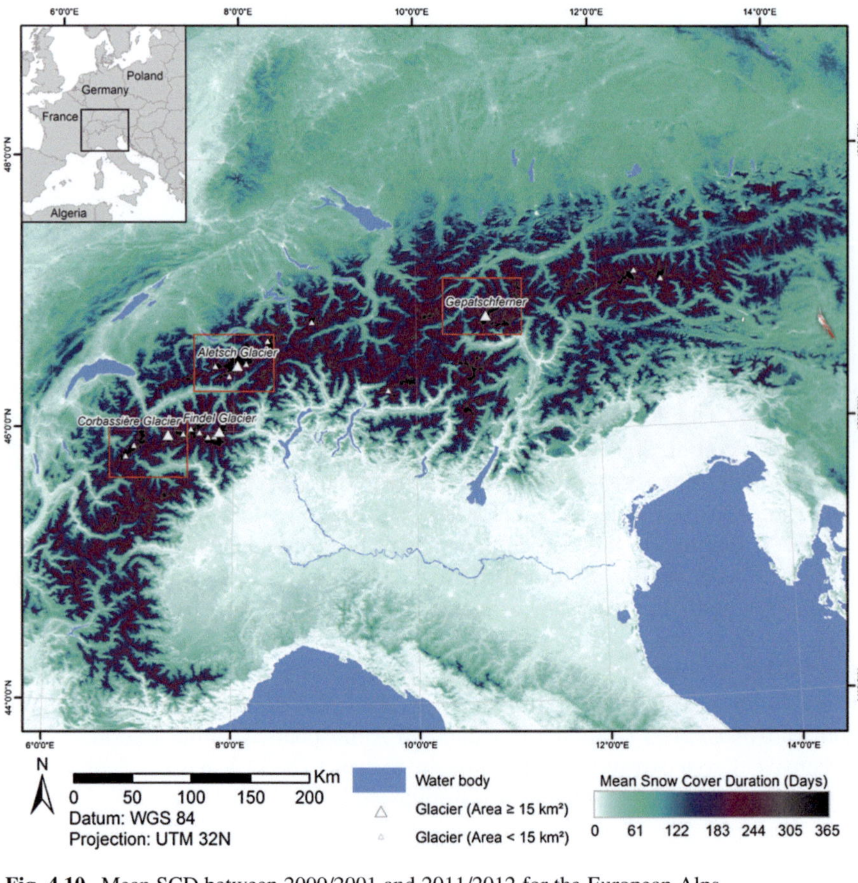

Fig. 4.10 Mean SCD between 2000/2001 and 2011/2012 for the European Alps

resolution of 500 m allows identifying many regional details, like e.g. the SCD gradient (4 days per 100 m elevation) within the mountainous regions. The mean conditions shown in Fig. 4.10 can also serve as a basis for the comparison with single year SCD conditions. This allows analyzing regions with abnormal snow cover characteristics within a single snow cover season. The usefulness of such analyses has been proven in (Dietz et al. 2012b). The impact of weather phenomena such as exceptionally warm winter months or high/low precipitation rates on the snow cover parameters can be detected. This information is useful with regards to the management of water reservoirs, hydropower generation, flood prediction, and runoff simulations (Butt and Bilal 2011; Pepe et al. 2005; Solberg and Andersen 1994; Zhao et al. 2009). As climate is changing towards warmer winter seasons, analyses of SCD, SCD_{ES} and SCD_{LS} will become more important.

The Snowmelt Runoff Model (SRM) is another possible application of the processed snow cover data. The SRM is a tool to forecast the runoff in mountain basins by simulating the snowmelt (Martinec et al. 1998). The model requires several input parameters, including daily precipitation, temperatures, and snow covered area. Once these data are available, the SRM is capable of predicting runoff in mountain basins successfully (Rango et al. 1990). The quality of the results obtained from such models depends on the quality of the input data. If 60 % of the pixels remain unknown due to cloud coverage the results will be unstable. By processing the daily snow cover maps with methods to interpolate below clouds, the overall accuracy of the model prediction can be increased. Today, snow cover products derived from remotely sensed data can contribute significantly to various hydrological research and analysis questions (Wagner et al. 2009). Techniques to improve the snow cover time series as presented in this section are therefore a necessary processing step to provide the required quality.

4.4 Summary and Conclusions

Snow cover constitutes an important land cover parameter, as it influences a wide range of climatological and hydrological aspects. Therefore, analysing snow cover parameters like Snow Cover Duration (SCD), Early Season SCD (SCD_{ES}) and Late Season SCD (SCD_{LS}) can serve as the basis for many different applications, a few of which have been outlined in this chapter. The generation of snow cover time series from medium resolution remote sensing data requires several steps, beginning with the detection of snow cover and the separation between clouds and snow in a first step. The Normalized Difference Snow Index (NDSI) as part of the *snowmap* algorithm is referred to in this context as well as the APOLLO and SPARC routine – both suitable to classify snow from MODIS and AVHRR data. MODSCAG may be consulted whenever fractional snow cover information is required. Cloud contamination is the biggest challenge when it comes to the generation of snow cover parameters from the time series of daily data as up to 60 % of the observations – especially during winter season – can be cloud covered.

Thus, a combination of different methods to estimate the snow cover status below clouds is applied to the time series before SCD, SCD_{ES}, and SCD_{LS} can be calculated.

A few examples for possible applications are included in this chapter but they are far from being complete. The spatial and temporal resolution of the time series and the derived products allow for both, regional studies as well as large scale analyses of snow cover characteristics. Depending on the research question and the available auxiliary data, many additional applications and combinations are possible. The snow cover data can be combined with runoff data, temperature and/or precipitation data, information about permafrost, soil moisture, vegetation, and Digital Elevation Models (DEMs). Flood and drought prediction, potential exposure to avalanches or analyses with regards to winter tourism become possible. The high spatial and temporal variability of snow cover can pose a challenge when dealing with such problems. Therefore, a sophisticated scheme to process the snow cover data and reduce the effect of cloud coverage is required in order to ensure a high quality of results. Finally, trend analyses could be performed based on long term time series of snow cover parameters SCD, SCD_{ES}, or SCD_{LS} that would identify potential effects of climate variability on the snow cover characteristics of a region. Such changes may have severe consequences for the climate system, causing negative feedbacks due to changed planetary albedo. Long time series of snow cover are required to derive sophisticated results, but with the upcoming Sentinel sensors the continuous data availability of medium resolution remote sensing data suited to derive global snow cover parameters is ensured.

References

Aoki T, Motoyoshi H, Kodama Y, Yasunari TJ, Sugiura K (2007) Variations of the snow physical parameters and their effects on albedo in Sapporo. Ann Glaciol 46:375–381

Ault TW, Czajkowski KP, Benko T, Coss J, Struble J, Spongberg A, Templin M, Gross C (2006) Validation of the MODIS snow product and cloud mask using student and NWS cooperative station observations in the Lower Great Lakes Region. Remote Sens Environ 105:341–353. doi:10.1016/j.rse.2006.07.004

Bales RC, Dressler KA, Imam B, Fassnacht SR, Lampkin D (2008) Fractional snow cover in the Colorado and Rio Grande basins, 1995–2002. Water Resour Res 44:1–10. doi:10.1029/2006WR005377

Barnett TP, Adam JC, Lettenmaier DP (2005) Potential impacts of a warming climate on water availability in snow-dominated regions. Nature 438:303–309. doi:10.1038/nature04141

Barry RG, Armstrong R, Callaghan T, Cherry J, Gearheard S, Nolin A, Russel D, Zaeckler C (2007) Global outlook for ice & snow. United Nations Environment Programme, Hertfordshire

Belz JU, Rademacher S (2007) Die Niedrigwasser-Situation in den Bundeswasserstraßen im April 2007 (in German). Koblenz

Brown RD (2000) Northern Hemisphere snow cover variability and change, 1915–97. J Clim 13:2339–2355

Brown RD, Mote PW (2009) The response of Northern Hemisphere snow cover to a changing climate. J Clim 22:2124–2145. doi:10.1175/2008JCLI2665.1

Butt MJ, Bilal M (2011) Application of snowmelt runoff model for water resource management. Hydrol Process 25:3735–3747. doi:10.1002/hyp.8099
Choi G, Robinson DA, Kang S (2010) Changing Northern Hemisphere snow seasons. J Clim 23:5305–5310. doi:10.1175/2010JCLI3644.1
Crane RG, Anderson MR (1984) Satellite discrimination of snow/cloud surfaces. Int J Remote Sens 5:213–223. doi:10.1080/01431168408948799
Dietz AJ (2013) Central Asian snow cover characteristics between 1986 and 2012 derived from time series of medium resolution remote sensing data. Wuerzburg
Dietz AJ, Kuenzer C, Gessner U, Dech S (2012a) Remote sensing of snow – a review of available methods. Int J Remote Sens 13:4094–4134
Dietz AJ, Wohner C, Kuenzer C (2012b) European snow cover characteristics between 2000 and 2011 derived from improved MODIS daily snow cover products. Remote Sens 4:2432–2454. doi:10.3390/rs4082432
Dietz AJ, Kuenzer C, Conrad C (2013) Snow-cover variability in central Asia between 2000 and 2011 derived from improved MODIS daily snow-cover products. Int J Remote Sens 34:3879–3902
Donlon C, Berruti B, Buongiorno A, Ferreira MH, Féménias P, Frerick J, Goryl P, Klein U, Laur H, Mavrocordatos C, Nieke J, Rebhan H, Seitz B, Stroede J, Sciarra R (2012) The Global Monitoring for Environment and Security (GMES) Sentinel-3 mission. Remote Sens Environ 120:37–57. doi:10.1016/j.rse.2011.07.024
Dozier J (1989) Spectral signature of alpine snow cover from the Landsat thematic mapper. Remote Sens Environ 28:9–22
Dozier J, Painter TH, Rittger K, Frew JE (2008) Time–space continuity of daily maps of fractional snow cover and albedo from MODIS. Adv Water Resour 31:1515–1526. doi:10.1016/j.advwatres.2008.08.011
Drusch M, Del Bello U, Carlier S, Colin O, Fernandez V, Gascon F, Hoersch B, Isola C, Laberinti P, Martimort P, Meygret A, Spoto F, Sy O, Marchese F, Bargellini P (2012) Sentinel-2: ESA's optical high-resolution mission for GMES operational services. Remote Sens Environ 120:25–36. doi:10.1016/j.rse.2011.11.026
ESA (2013a) ATSR. https://earth.esa.int/web/guest/missions/esa-operational-eo-missions/ers/instruments/atsr. Accessed 30 Nov 2014
ESA (2013b) AATSR. https://earth.esa.int/web/guest/missions/esa-operational-eo-missions/envisat/instruments/aatsr. Accessed 30 Nov 2014
Fernandes R, Zhao H (2008) Mapping daily snow cover extent over land surfaces using NOAA AVHRR imagery. In: Proceedings of the 5th EARSeL Workshop: remote sensing of land ice and snow. Bern, 11–13 Feb, pp 1–8
Foster JL, Hall DK, Chang ATC, Rango A, Wergin W, Erbe E (1999) Effects of snow crystal shape on the scattering of passive microwave radiation. IEEE Trans Geosci Remote Sens 37:1165–1168
Frei A, Tedesco M, Lee S, Foster J, Hall DK, Kelly R, Robinson DA (2012) A review of global satellite-derived snow products. Adv Space Res 50:1007–1029. doi:10.1016/j.asr.2011.12.021
Gafurov A, Bárdossy A (2009) Cloud removal methodology from MODIS snow cover product. Hydrol Earth Syst Sci 13:1361–1373
Gesell G (1989) An algorithm for snow and ice detection using AVHRR data. An extension to the APOLLO software package. Int J Remote Sens 10:897–905. doi:10.1080/01431168908903929
Grippa M, Kergoat L, Le Toan T, Mognard NM, Delbart N, L'Hermitte J, Vicente-Serrano SM (2005) The impact of snow depth and snowmelt on the vegetation variability over central Siberia. Geophys Res Lett 32:2–5. doi:10.1029/2005GL024286
Hadley OL, Kirchstetter TW (2012) Black-carbon reduction of snow albedo. Nat Clim Chang 2:437–440. doi:10.1038/nclimate1433
Hall DK, Martinec J (1985) Remote sensing of ice and snow. Chapman & Hall, London
Hall D, Riggs G (2007) Accuracy assessment of the MODIS snow products. Hydrol Process 21:1534–1547. doi:10.1002/hyp

Hall DK, Riggs GA, Salomonson VV (1995) Development of methods for mapping global snow cover using moderate resolution imaging spectroradiometer data. Remote Sens Environ 54:127–140

Hall DK, Riggs GA, Salomonson, VV (2000) MODIS/Terra Snow Cover Daily L3 Global 500m Grid V005. National Snow and Ice Data Center. Digital media, Boulder, Colorado USA. http://nsidc.org/data/docs/daac/modis_v5/myd10a2_modis_aqua_snow_8-day_global_500m_grid.gd.html. Accessed 30 Nov 2014

Hall DK, Riggs GA, Salomonson VV, Digirolamo NE, Bayr KJ (2002) MODIS snow-cover products. Remote Sens Environ 83:181–194

Huang X, Liang T, Zhang X, Guo Z (2011) Validation of MODIS snow cover products using Landsat and ground measurements during the 2001–2005 snow seasons over northern Xinjiang, China. Int J Remote Sens 32:133–152. doi:10.1080/01431160903439924

Hüsler F, Jonas T, Wunderle S, Albrecht S (2012) Validation of a modified snow cover retrieval algorithm from historical 1-km AVHRR data over the European Alps. Remote Sens Environ 121:497–515. doi:10.1016/j.rse.2012.02.018

Khlopenkov KV, Trishchenko AP (2007) SPARC: new cloud, snow, and cloud shadow detection scheme for historical 1-km AVHHR data over Canada. J Atmos Ocean Technol 24:322–343. doi:10.1175/JTECH1987.1

Klein AG, Barnett AC (2003) Validation of daily MODIS snow cover maps of the Upper Rio Grande River Basin for the 2000–2001 snow year. Remote Sens Environ 86:162–176. doi:10.1016/S0034-4257(03)00097-X

Klein AG, Hall DK, Riggs GA (1998) Improving snow cover mapping in forests through the use of a canopy reflectance model. Hydrol Process 12:1723–1744. doi:10.1002/(SICI)1099-1085(199808/09)12:10/11<1723::AID-HYP691>3.3.CO;2-U

Klein AG, Hall DK, Nolin AW (2000) Development of a prototype snow albedo algorithm for the NASA MODIS instrument. In: 57th eastern snow conference, Syracuse, New York, USA, pp 143–158

König M, Winther J, Isaksson E (2001) Measuring snow and glacier ice properties from satellite. Rev Geophys 39:1–27

Kriebel KT, Gesell G, Kästner M, Mannstein H (2003) The cloud analysis tool APOLLO: improvements and validations. Int J Remote Sens 24:2389–2408. doi:10.1080/01431160210163065

Lemke P, Ren J, Alley RB, Allison I, Carrasco J, Flato G, Fujii Y, Kaser G, Mote P, Thomas RH, Zhang T (2007) Observations: changes in snow, ice and frozen ground. In: Solomon S, Qin D, Manning M, Chen Z, Marquis MC, Averyt K, Tignor M, Miller HL (eds) Climate change 2007: the physical science basis. Contributions of Working Group 1 to the fourth assessment report of the intergovernmental panel on climate change. Cambridge University Press, Cambridge, UK/-New York, pp 337–383

Luojus KP, Pulliainen JT, Takala M, Kangwa M, Smolander T (2013) GlobSnow-2 product user guide. Finnish Meteorological Institute, Helsinki

Malenovský Z, Rott H, Cihlar J, Schaepman ME, García-Santos G, Fernandes R, Berger M (2012) Sentinels for science: potential of Sentinel-1, -2, and -3 missions for scientific observations of ocean, cryosphere, and land. Remote Sens Environ 120:91–101. doi:10.1016/j.rse.2011.09.026

Martinec J, Rango A, Roberts R (1998) Snowmelt Runoff Model (SRM) User's Manual (Updated Edition 1998), 4.0 ed. USDA Hydrology Laboratory, Beltsville, MD, USA

Mauser W, Bach H (2009) PROMET – large scale distributed hydrological modelling to study the impact of climate change on the water flows of mountain watersheds. J Hydrol 376:362–377. doi:10.1016/j.jhydrol.2009.07.046

Metsamaki S, Anttila S, Markus H, Vepsalainen J (2005) A feasible method for fractional snow cover mapping in boreal zone based on a reflectance model. Remote Sens Environ 95:77–95. doi:10.1016/j.rse.2004.11.013

Muntán E, Garcia C, Oller P, Marti G, Garcia A, Gutiérrez E (2009) Reconstructing snow avalanches in the Southeastern Pyrenees. Nat Hazards Earth Syst Sci 9:1599–1612

NOAA (n.d.) Comprehensive large array-data stewardship system. http://www.class.noaa.gov. Accessed 18 Sept 2012

NSIDC (2013) NSIDC website. http://nsidc.org/. Accessed 28 Feb 2013

Painter TH, Rittger K, Mckenzie C, Slaughter P, Davis RE, Dozier J (2009) Retrieval of subpixel snow covered area, grain size, and albedo from MODIS. Remote Sens Environ 113:868–879. doi:10.1016/j.rse.2009.01.001

Parajka J, Pepe M, Rampini A, Rossi S, Blöschl G (2010) A regional snow-line method for estimating snow cover from MODIS during cloud cover. J Hydrol 381:203–212. doi:10.1016/j.jhydrol.2009.11.042

Pepe M, Brivio PA, Rampini A, Nodari FR, Boschetti M (2005) Snow cover monitoring in Alpine regions using ENVISAT optical data. Int J Remote Sens 26:4661–4667. doi:10.1080/01431160500206635

Räisänen J (2007) Warmer climate: less or more snow? Clim Dyn 30:307–319. doi:10.1007/s00382-007-0289-y

Räisänen J, Eklund J (2012) 21st century changes in snow climate in Northern Europe: a high-resolution view from ENSEMBLES regional climate models. Clim Dyn 38:2575–2591. doi:10.1007/s00382-011-1076-3

Rango A (1996) Spaceborne remote sensing for snow hydrology applications. Hydrol Sci J 41:477–494. doi:10.1080/02626669609491521

Rango A, van Katwijk V, Martinec J (1990) Snowmelt runoff forecasts in Colorado with remote sensing. Hydrology in mountainous regions. I – hydrological measurements; the water cycle. In: Proceedings of two Lausanne symposia, Aug 1990, pp 627–634

Riggs G, Hall DK (2004) Snow mapping with the MODIS aqua instrument. In: 61st eastern snow conference, Portland, Maine, USA, pp 81–84

Riggs GA, Hall DK, Salomonson VV (2006) MODIS snow products user guide to collection 5. Boulder, CO, USA

Salminen M, Pulliainen J, Metsämäki S, Kontu A, Suokanerva H (2009) The behaviour of snow and snow-free surface reflectance in boreal forests: implications to the performance of snow covered area monitoring. Remote Sens Environ 113:907–918. doi:10.1016/j.rse.2008.12.008

Saunders RW, Kriebel KT (1988) An improved method for detecting clear sky and cloudy radiances from AVHRR data. Int J Remote Sens 9:123–150. doi:10.1080/01431168808954841

Scott PA, Wayne R (1995) Impacts of increased winter snow cover on upland tundra vegetation : a case example. Clim Res 5:25–30

Simic A, Fernandes R, Brown R, Romanov P, Park W (2004) Validation of VEGETATION, MODIS, and GOES+ SSM/I snow-cover products over Canada based on surface snow depth observations. Hydrol Process 18:1089–1104. doi:10.1002/hyp.5509

Solberg R, Andersen T (1994) An automatic system for operational snow-cover monitoring in the Norwegian mountain regions. In: Proceedings of the International Geoscience and Remote Sensing Symposium, Pasadena, CA, USA, 8–12 Aug 1994, pp 2084–2086

Solberg R, Koren H, Amlien J (2006) A review of optical snow cover algorithms. Norwegian Computing Center Note, no. SAMBA/40/06

Solberg R, Wangensteen B, Amlien J, Koren H, Metsämäki S, Nagler T, Luojus K, Pulliainen J (2010) A new global snow extent product based on ATSR-2 and AATSR. In: Proceedings of the 2010 I.E. International Geoscience and Remote Sensing Symposium, 25–30 July 2010, Honolulu, Hawaii, USA, p 4

Tekeli AE, Akyürek Z, Şorman AA, Şensoy A, Şorman ÜA (2005) Using MODIS snow cover maps in modeling snowmelt runoff process in the eastern part of Turkey. Remote Sens Environ 97(2):216–230. doi:10.1016/j.rse.2005.03.013

Thirel G, Salamon P, Burek P, Kalas M (2013) Assimilation of MODIS snow cover area data in a distributed hydrological model using the particle filter. Remote Sens 5:5825–5850. doi:10.3390/rs5115825

Thurman M (2011) Natural disaster risks in central Asia : a synthesis table of contents. UNDP Bureau for Crisis Prevention and Recovery

Vikhamar D, Solberg R (2002) Subpixel mapping of snow cover in forests by optical remote sensing. Remote Sens Environ 84:69–82

Vikhamar D, Solberg R (2003) Snow-cover mapping in forests by constrained linear spectral unmixing of MODIS data. Remote Sens Environ 88:309–323. doi:10.1016/j.rse.2003.06.004

Wagner W, Verhoest NEC, Ludwig R, Tedesco M (2009) Remote sensing in hydrological sciences. Hydrol Earth Syst Sci 13:813–817

Wang X, Xie H, Liang T (2009) Development and assessment of combined Terra and Aqua MODIS snow cover products in Colorado. J Appl Remote Sens 3:1–15

Weber M, Braun L, Mauser W, Prasch M (2010) Contribution of rain, snow- and icemelt in the Upper Danube discharge today and in the future. Geografia Fisica e Dinamica Quaternaria 33:221–230

WMO and GCOS (2011) Systematic observation requirements for satellite-based data products for climate – 2011 update, GCOS – 154 ed. WMO and GCOS, Geneva, Switzerland

Wylie D, Jackson DL, Menzel WP, Bates JJ (2005) Trends in global cloud cover in two decades of HIRS observations. J Clim 18:3021–3031

Zhao Q, Liu Z, Ye B, Qin Y, Wei Z, Fang S (2009) A snowmelt runoff forecasting model coupling WRF and DHSVM. Hydrol Earth Syst Sci 13:1897–1906

Zhou H, Aizen E, Aizen V (2013) Deriving long term snow cover extent dataset from AVHRR and MODIS data: Central Asia case study. Remote Sens Environ 136:146–162. doi:10.1016/j.rse.2013.04.015

Chapter 5
Global WaterPack: Intra-annual Assessment of Spatio-Temporal Variability of Inland Water Bodies

Igor Klein, Andreas J. Dietz, Ursula Gessner, and Claudia Kuenzer

Abstract The knowledge and understanding of intra- and inter annual characteristics of inland water bodies, such as natural lakes and artificial reservoirs are crucial for many reasons. Inland water bodies are sensitive to environmental variations and human impact which is reflected in spatial and temporal dynamics of surface extent. A time-series of areal surface extent of lakes and reservoirs might be a helpful dataset to understand the complex system and the spatio-temporal patterns of natural lakes and artificial reservoirs. In this study, we describe an approach to detect water bodies based on dynamical thresholding on daily basis and utilizing high frequency observations. Daily MODIS (Moderate Resolution Imaging Spectrometer) products were used to generate water masks for the year 2013 on global scale. The results indicate that time series of water bodies' extent are important especially for those inland water bodies which are dominated by temporal changes and fluctuation through the year. In combination with ancillary data, our understanding of environmental and human interaction and the reaction of water bodies will be improved. Such information is critical to support sustainable water management, as well as for climate change discussion since many inland water bodies are sensitive to short- and long term environmental alterations.

5.1 Introduction

Human activities, climate change, environmental alterations and meteorological variability are the drivers for changes of inland water bodies throughout time. For example, natural lakes shrink and vanish due to overexploitation of water resources and new water bodies arise due to construction of reservoirs or ice melting caused

I. Klein (✉) • A.J. Dietz • U. Gessner • C. Kuenzer
German Remote Sensing Data Center, DFD, Earth Observation Center, EOC,
German Aerospace Center, DLR, Oberpfaffenhofen, Germany
e-mail: igor.klein@dlr.de; andreas.dietz@dlr.de; ursula.gessner@dlr.de;
claudia.kuenzer@dlr.de

C. Kuenzer et al. (eds.), *Remote Sensing Time Series*, Remote Sensing and Digital Image Processing, DOI 10.1007/978-3-319-15967-6_5

by rising temperatures. Lake levels and volumes have been identified as one of the Terrestrial Essential Climate Variables and thus current and historical information are required for research of United Nations Framework Convention on Climate Change (UNFCCC) and Intergovernmental Panel on Climate Change (IPCC) (GTOS 2008). Remote sensing data and methodology provide unique opportunities to detect water bodies on different spatial and temporal scales. Traditionally, there are manifold studies focussing on water detection from remote sensing data. Most of them are dealing with the accurate delineation between water pixel and land pixel as well as the unmixing of pixels which cover both types (e.g. Gstaiger et al. 2012; Jain et al. 2005; Kuenzer et al. 2013; Martinis et al. 2015; Verpoorter et al. 2012; Weiss and Crabtree 2011). Currently, there exist comprehensive and sophisticated global datasets of inland water bodies such as Global Lakes and Wetlands Dataset (GLWD) (Lehner and Döll 2004), Shuttle Radar Topography Mission (SRTM) water body data (SWBD) (Slater et al. 2006), the recent 250 m MODIS (Moderate Resolution Imaging Spectrometer) water mask (Carroll et al. 2009) and the recent Global Water Bodies product at 300 m resolution based on full ENVISAT-ASAR (Environmental Satellite- Advanced Synthetic Aperture Radar) dataset which was generated within the framework of European Space Agency (ESA) Climate Change Initiative (CCI) (ESA-Land Cover-CCI 2014). These datasets are very valuable and are available on a global scale. However, they only provide one temporal snapshot (Fichtelmann and Borg 2012) and do not account the changes of water bodies over time. Especially the dynamics and variations of inland water bodies are crucial information for many reasons and are required not only by different scientific disciplines but also by political decision makers and stake holders. The understanding of the interaction between different physical parameters, human impact, climate variability and their influence on water bodies is essential. Therefore, it is important to detect inland water bodies with high temporal resolution to mirror their dynamical characteristics and thus enable further investigation about the drivers of diverse variations and possible consequences for the environment and human beings.

Over the last years, the availability of free of charge remote sensing data has increased dramatically and large datasets have been accessible to researchers all over the world. The analyses of time-series has reached a new dimension in terms of data storage space and processing time. This brings up new challenges to acquire accurate information of high spatial and temporal resolution. The demand for temporal information and dynamic characteristics of inland water bodies has been increased since the availability of remote sensing datasets with high temporal resolution. Many authors published interesting results presenting the reconstruction of water bodies over the past decades (Feng et al. 2012; Fichtelmann and Borg 2012; Haas et al. 2009; Klein et al. 2014a; Pekel et al. 2014). These products are based on data composites over several days for specific study regions.

In this chapter, we describe our approach to classify surface waters based on daily temporal resolution remote sensing data products originating from MODIS on board of Terra and Aqua satellites. The concept presented in Klein et al. (2014b) will be described in more details. We describe our strategy to utilize different datasets in order to acquire cloud free water masks with high temporal resolution. The MODIS near-infrared (NIR) measurements, cloud information and derived dynamic thresholds are mainly used to delineate between water and land pixels for the year 2013. The generated time series of water masks is used to remove misclassification originating from cloud shadow and also to interpolate missing values. The time-series of water masks over 2013 reflects the areal evaluation of inland water bodies which is presented based on selected test sites. This chapter gives an overview of the Global WaterPack product by presenting some exemplary cases. Additionally, we discuss further required development to avoid misclassification errors resulting from ambiguous spectral signature of different land surfaces.

5.2 Study Area and Materials

5.2.1 Study Area

The processing was performed based on 200 MODIS tiles which cover main land masses on global scale as highlighted in Fig. 5.1. Polar Regions dominated by polar nights were excluded due to the lack of usable observations of the optical MODIS sensor. Furthermore, tiles which cover remote islands distributed all over the oceans were not processed.

5.2.2 Data

5.2.2.1 Data for Water Detection

The MODIS instrument on board of sun-synchronic satellites Terra and Aqua provides a unique possibility of land cover observation because of its high temporal resolution. Due to different orbits, the Terra and Aqua satellites are viewing one and the same area twice a day which enable two observations with different cloud conditions of the same area. The acquired NIR datasets were downloaded as part of MOD09GQ and MYD09GQ products which represents surface spectral reflectance in 620–670 nm (band 1) and in 841–876 nm (band 2) spectral width as it would be measured at ground level without the atmospheric influence (Land Processes Distributed Active Archive Center 2014). The datasets are distributed as gridded level-2 (L2G) products in sinusoidal projection and feature a spatial resolution of 250 m. Second important input datasets were the MOD10A1 and MYD10A1 snow products. The snow products are available with same temporal resolution as 09GQ products, however with a coarser spatial resolution of 500 m. This L3G product contains thematic information such as cloud cover, lake ice, snow cover and missing data which are generated with a range of different algorithms (Hall and

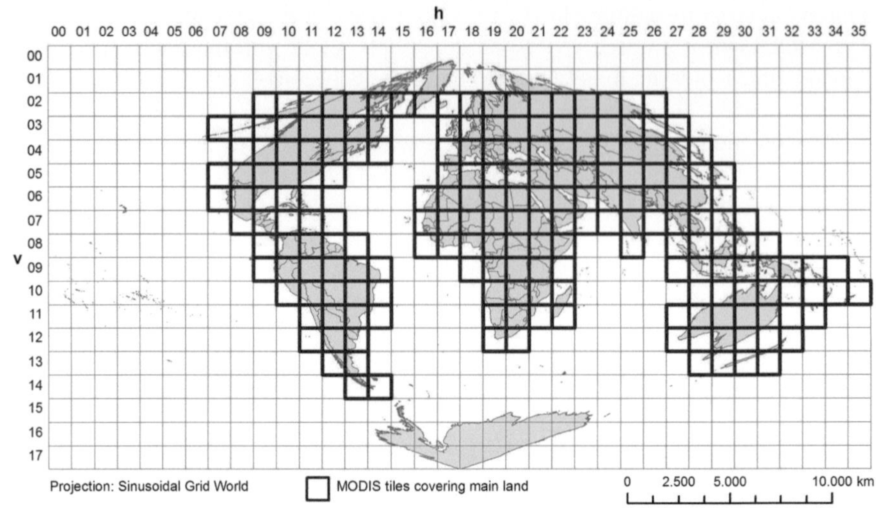

Fig. 5.1 Overview of processed MODIS tiles which cover main land masses (without Polar Regions and remote islands)

Riggs 2007; Riggs et al. 2006). It also contains static information of ocean coverage and inland lakes based on an outdated static water mask. Additionally, we used the MOD44W product which is an improved static water mask based on SRTM Water Body Dataset (SWBD) and on 8+ years of Terra MODIS data; and 6+ years of Aqua MODIS data (Carroll et al. 2009). All mentioned MODIS products were acquired in compressed Hierarchical Data Format – Earth Observing System (HDF-EOS) format. The used datasets and main information are summarized in Table 5.1.

5.2.2.2 Ancillary Data

In addition to MODIS datasets we acquired the digital elevation model (DEM) of SRTM with a spatial resolution of 90 m (CGIAR-CSI 2008). Furthermore, 40 Landsat images for selected locations (in Argentina, Bangladesh, Brazil, China, Kazakhstan, USA) were downloaded to perform a first accuracy assessment of the presented product.

5.3 Methodology

Remote sensing data and especially the detected radiance in NIR spectrum are well suited to monitor water bodies. Figure 5.2 shows spectral characteristics of different water types in comparison with wetlands and vegetation (based on laboratory measurements from Clark et al. 2007). Bukata et al. (1995) stated that the total

Table 5.1 Used datasets for Global WaterPack for the year 2013 (for each MODIS tile)

Dataset	Number of datasets per year	Available time interval/date of release	Spatial resolution	Used information/ spectral band width
MOD09GQ	365	2000-open	250 m	0.841–0.876 μm
MYD09GQ	365	2002-open	250 m	0.841–0.876 μm
MOD10A1	365	2000-open	500 m	Cloud, lake ice
MYD10A1	365	2002-open	500 m	Cloud, lake ice
MOD44W	1	2009	250 m	Inland water
DEM	1	2005	90 m	m above SL

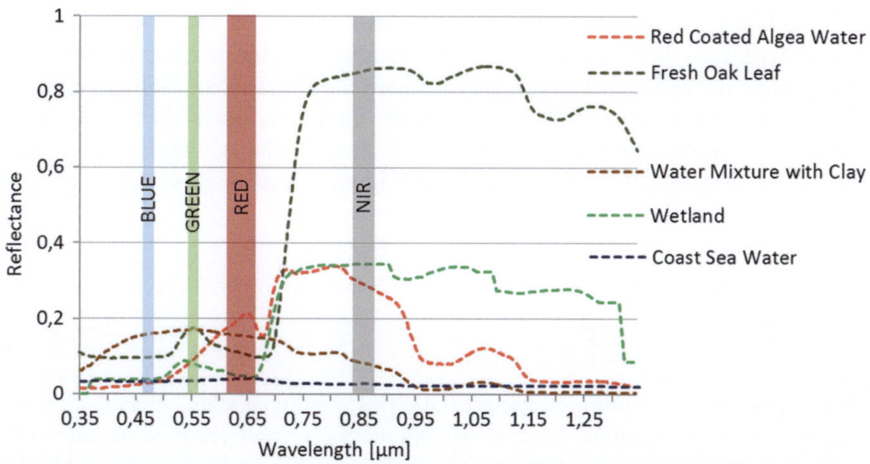

Fig. 5.2 Spectral characteristics of different water types, wetland and fresh oak leaf (measurements from Clark et al. 2007). Visible and near infra-red ranges of MODIS sensor are marked

radiance of a water pixel recorded by remote sensors is a function of water surface radiance, subsurface volumetric radiance and radiance from the bottom of the pixel. Compared to water, most land surfaces feature higher reflection in NIR band. This fact has been utilized for water detection in many studies in a range of different approaches (e.g. Gao 1996; McFeeters 1996; Ouma and Tateishi 2006; Ji et al. 2009).

There exists a variety of different remote sensing models to delineate between water and no water, whereby mostly the NIR was utilized either in a single band slicing model (Ryu et al. 2002) difference model, ratio model (Sheng et al. 2001) or spectral water index model (Gao 1996; McFeeters 1996; Xu 2006). Depending on study objectives, study area, time interval of interest, available datasets, and defined maximum processing time, each of those methods has its advantages and disadvantages. Our goals were to process daily datasets on a global scale, and generate a product with information about how often a pixel was covered by water. The massive amount of data on a global scale and the required processing time were the main reasons to use a single band threshold method instead of band combinations. The utilization of a single band method for water classification has been widely used because of the relatively fast computing approach (Feyisa et al. 2014; Ji et al. 2009; Ryu et al. 2002). However, we implemented a dynamic thresholding allowing calculation of different thresholds for individual scenes. Fichtelmann and Borg (2012) introduced a complex method for accurate water mask extraction from Advanced Along-Track Scanning Radiometer (AATSR). We followed a similar approach using a static water masks as a priori knowledge to derive dynamic thresholds for multiple time steps. There are many reasons for using different thresholds for individual observations especially at global scale with high temporal frequency of observations. The spectral characteristics of one and the same water body can differ significantly e.g. depending on different water conditions such as chlorophyll content, suspended particles, surface roughness and water depth. Various combinations of the named conditions and the fact that the quality of atmospheric correction is not constant (Vermote and Kotchenova 2008) give pixels different spectral values at different time steps. Therefore, a single threshold value derived for one image might not be suitable for another. The individual steps of pre-processing, classification and post-processing are discussed in the following sections.

5.3.1 Pre-processing

Before processing all corresponding MODIS tiles automatically, several requirements had to be fulfilled. First of all, all datasets were required in identical projection, spatial resolution and spatial extent. Because our procedure was mainly based on MODIS datasets, we reprojected and spatially extracted the DEM data into square tiles with sinusoidal projection and 250 m spatial resolution adequate to original MODIS datasets. The 90 m DEM was resampled to 250 m pixel size by using nearest neighbour resampling method. Afterwards, slopes of the terrain were calculated using the maximal changes in elevation values of the DEM considering a 3×3 pixel moving window.

The *MOD10A1* and *MYD10A1* products are available with a spatial resolution of 500 m (2,400*2,400 pixels per tile). Therefore, the datasets were downscaled to 250 m to enable automatic pixel based calculation.

5.3.2 Water Classification Based on Dynamic Thresholding

The applied algorithm operates on two main procedures. Within the first procedure intermediate results are calculated which are being used for the second step (compare Fig. 5.3 orange and blue coloured sub-processes). The first procedure is the actual water detection which is based on dynamic NIR band thresholding. At the beginning, the static water information from *MOD44W* dataset is claimed to be the

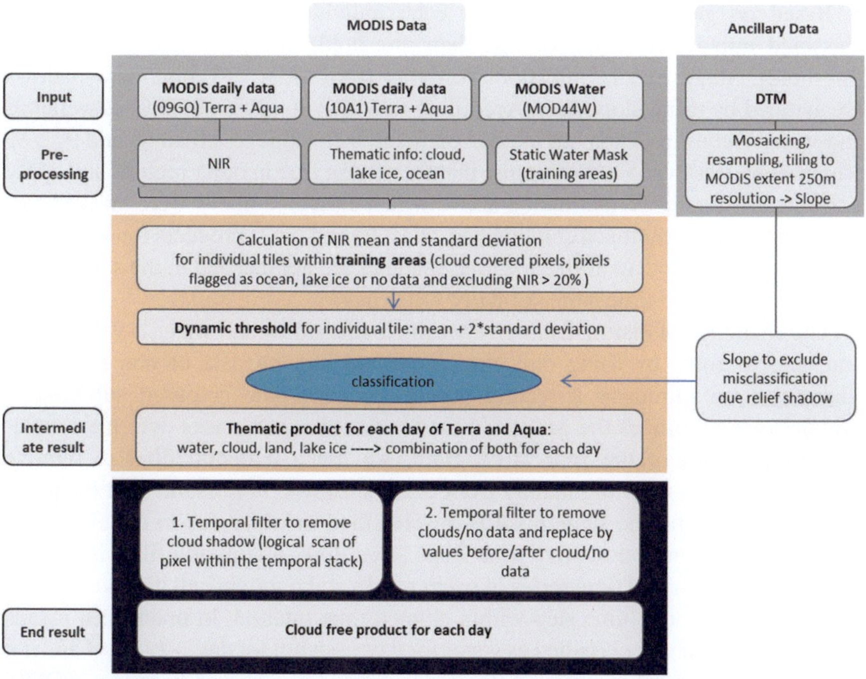

Fig. 5.3 Workflow Global WaterPack 2013 (Modified from Klein et al. 2014b)

initial training areas for dynamic threshold calculation for each individual scene. We assume that many water bodies shrink and increase in their areal extent. Therefore, some assumptions must be taken into account before any definitive decision on training areas is made. We utilize the information of cloud coverage, lake ice and ocean from corresponding *10A1* products. Although some authors stated that the cloud detection used in *10A1* product is improvable (Dietz et al. 2013; Leinenkugel et al. 2013), we use this information because the level of its quality does not influence our detection negatively. Every pixel which is assigned as water in *MOD44W* static mask, and for considered day as ocean, lake ice or cloud coverage in *10A1* product is excluded from the training pool. Furthermore, we defined a maximum threshold value of 20 % in NIR reflectance, above which all pixel are excluded even though none of the former conditions applied. In this manner we receive only those pixels for training which were (i) classified as water in *MOD44W* and at the same time (ii) feature less than 20 % in NIR reflectance on considered day, (iii) are not assigned as cloud, ocean or lake ice in comprehensive *10A1* dataset. Therefore, the large amount of ocean water does not fudge the dynamic threshold for inland water bodies. Same is true for pixels which are assigned as lake ice, cloud coverage and no data (e.g. due to polar night). Based on the remaining pixels within the training area, mean value and standard deviation of NIR are calculated. The dynamic value for each individual image is calculated as the sum of the mean value plus twice the standard deviation. Figure 5.3 visualizes the workflow of described processes.

Based on generated threshold for each individual scene, the NIR band is classified into water (equal or less than threshold) and no water (greater than threshold). Major misclassifications resulting from relief and mountain shadows are avoided by using slope data. According to Niu et al. (2009) 99.2 % of wetlands and water bodies in China are located on areas with slope less than 8° and 97.5 % with a slope less than 5°. We assume that this is also true in other regions and apply a threshold value of 5° inclination above which classified pixels were masked out. The intermediate result is a combination of *MOD* and *MYD* products for 1 day. The error of misclassified areas due to cloud shadows are approached in the second step using the generated time series of daily water masks.

The challenge of distinguishing between cloud shadow and water is well known and hard to solve by using only optical remote sensing data of one time step. Therefore, we introduce a second procedure step (blue coloured sub-process Fig. 5.3) which utilizes the generated time series of water masks over the considered time period of the year 2013. Here, we use a temporal filter to remove misclassifications resulting mainly from cloud shadows. We assume that the probability that one and the same pixel is misclassified as water due to cloud shadow being very low for consecutive time steps. Therefore, a temporal filter is applied scanning the generated time series of water masks and removing all pixels classified as water for only one time step within defined time interval. In practice, it means that a pixel which was classified as water for day x but not for day x-1 *till* x-4 and not for day $x+1$ *till* $x+4$ is removed (temporal interval was chosen based on MODIS 8 day composites as a benchmark). In this manner, most cloud shadows are excluded. Furthermore, the temporal composite of water masks (after the first temporal filter without errors due to cloud shadow) is used to interpolate and replace values which were covered by clouds or no data values. Here, the classification of water or no water from temporally closest clear sky observation is used to fill the gaps resulting from cloud coverage or no data. This is critical in regions with temporal long and permanent cloud cover. However, without including additional remotes sensing data (e.g. Synthetic Aperture Radar (SAR) data) we only assume the situation beneath cloud coverage. As an end result, we generated a layer with information about how often a pixel was classified as water for the year 2013 and name it Water Cover Duration (WCD).

The described classification process and the automatic selection of training pixels are illustrated in Fig. 5.4. It visualizes the generated temporal result of water cover duration compared to the static water product which represents a temporal snapshot. It is recognizable that some lakes are only filled with water seasonally. Furthermore, some temporal flooded areas are not detected in *MOD44W* (highlighted by black circles in Fig. 5.4).

5.3.3 Accuracy Assessment

Ideally, results from remote sensing data should be validated with ground truth measurements. However, there are no known datasets with information about

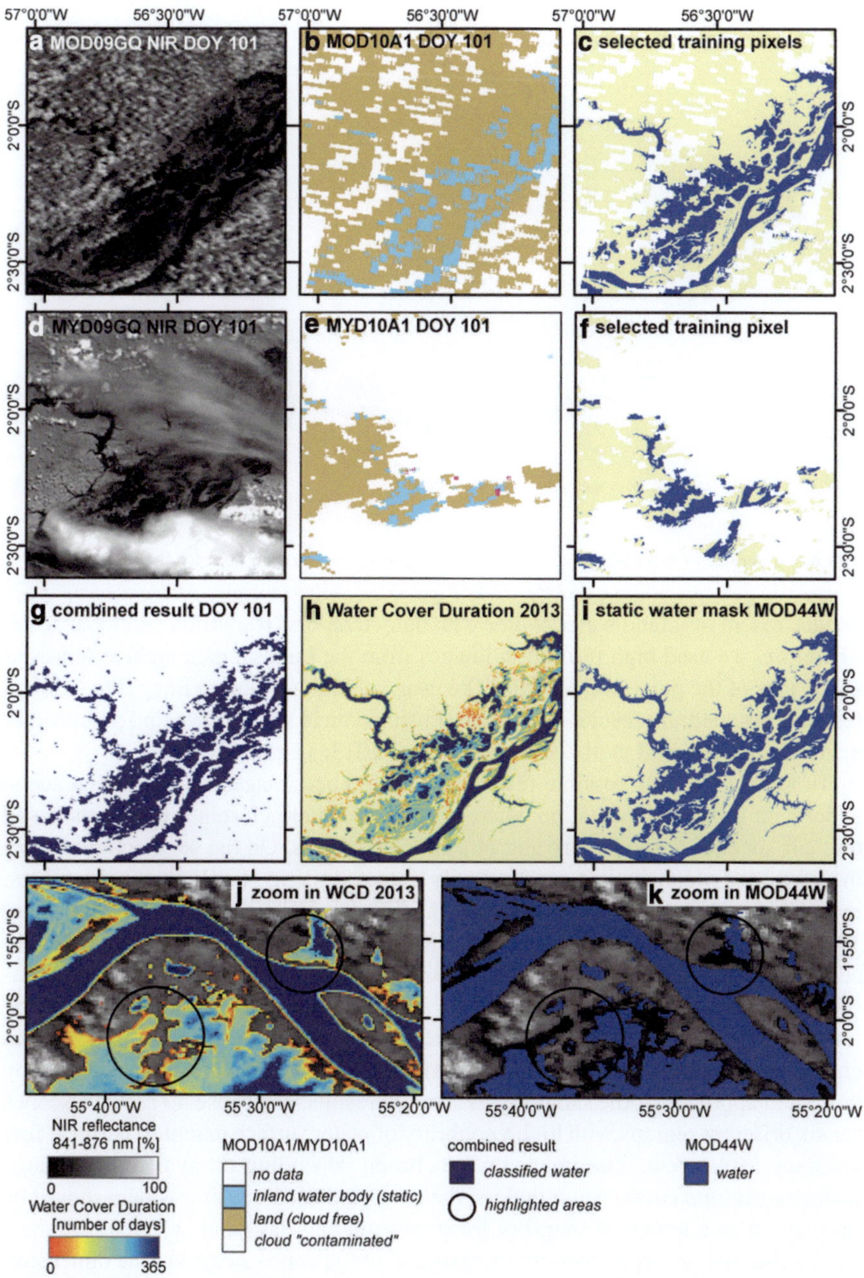

Fig. 5.4 Visualization of training pixel selection on 11 April 2013 (DOY 101) in floodplains along the Amazonas River. (**a**) near-infrared *MOD09GQ*, (**b**) corresponding thematic product *MOD10A1*, (**c**) used pixels for training based on *MOD44W*, (**d**) near-infrared *MYD09GQ*, (**e**) corresponding thematic product *MYD10A1*, (**f**) used pixels for training based on *MOD44W*, (**g**) classified water for day 101 based on both results from *MOD* and *MYD* products, (**h**) water cover duration 2013, (**i**) MOD44W

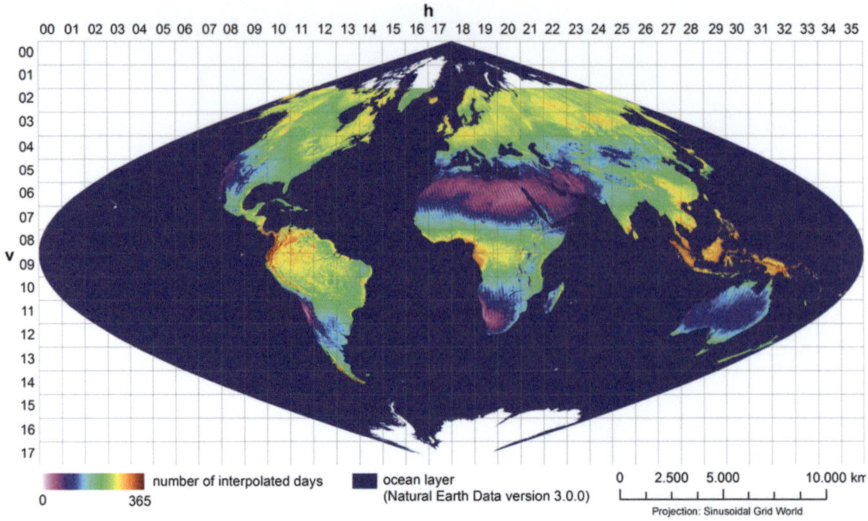

Fig. 5.5 Number of days per pixel which were cover by cloud or no data for the year 2013

surface extent of inland water bodies with daily temporal resolution on global scale. Therefore, we used high resolution images from the Landsat data archive to assess the quality of the presented product. The usage of high resolution images to validate results from medium resolution remote sensing data is well known and practised in many studies (Carroll et al. 2009; Dietz et al. 2013; Li et al. 2013).

To enable the best possible result of the reference images, each Landsat scene underwent image segmentation where several pixels are clustered to objects based on their spectral homogeneity and object compactness. On the segmentation level the lakes and water bodies can be easily detected using simple threshold values, band combination (NDWI (Normalised Difference Water Index) after McFeeters 1996; Gao 1996, NDVI (Normalized Difference Vegetation Index) after Tucker et al. 2005) and additional object features. The results underwent visual interpretation to guarantee best possible result of created reference dataset. The generated reference water masks from Landsat were finally rescaled to MODIS pixel size to enable a pixel-to-pixel comparison with comprehensive results derived by presented approach of the same day as Landsat results. We chose 40 Landsat scenes for six different regions with high variability of water surface extent to perform first accuracy assessment. The images were chosen depending on availability, image quality and cloud cover. Nevertheless, the presented accuracy assessment should be understood as a temporal snapshot for corresponding days and selected test sites. Comprehensive in-depth accuracy assessment of the entire area and time dimension is a very time consuming process which would be beyond the scope of a book chapter.

Additionally, a layer which includes information about the amount of interpolated days was generated (Fig. 5.5). The layer contains the information on how many days each pixel was interpolated by temporal filter and how often the water was detected directly. The assessment layer concludes which areas have the more meaningful results or questionable results due to a lack of clear sky observations.

5.4 Results

5.4.1 Temporal Development of Water Bodies

Figure 5.6 presents the result for all processed tiles for the year 2013 which is based on over 292.000 input datasets. The permanently water-covered areas (oceans, permanent lakes and reservoirs) are recognizable as areas of dark blue colour. Beside these water bodies, there are many lakes and water reservoirs on a smaller scale that show high dynamic behaviour around a core area of permanently existing water bodies (compare Figs. 5.7 and 5.8). Dynamic water bodies are found all over the world. Natural lakes might dry out during dry seasons and be filled during wet seasons. Furthermore, artificial water reservoirs show high variability in their areal extent depending on wet season and water use in dry season. Besides the dynamic increase and decrease of surface area, there are also examples of constantly decreasing water bodies including the popular examples of Aral Sea or Lake Chad which were shrinking during the last century. To illustrate different evolutions, we present some exemplary case studies (marked in Fig. 5.6) which are illustrated in Figs. 5.7 and 5.8.

Figure 5.7 illustrates the results for four highly dynamic water bodies from different regions with an exemplary profile line. Along the profile line we plotted

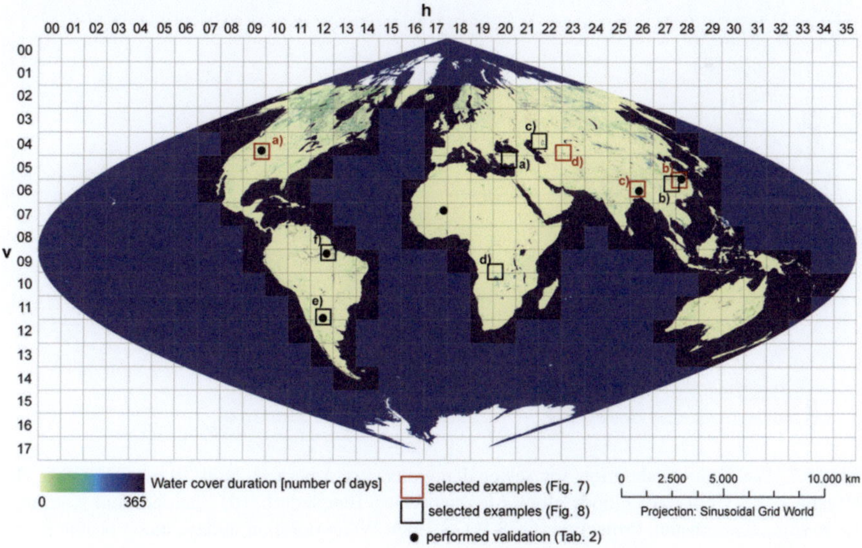

Fig. 5.6 Water Cover Duration 2013. Number of days classified as water

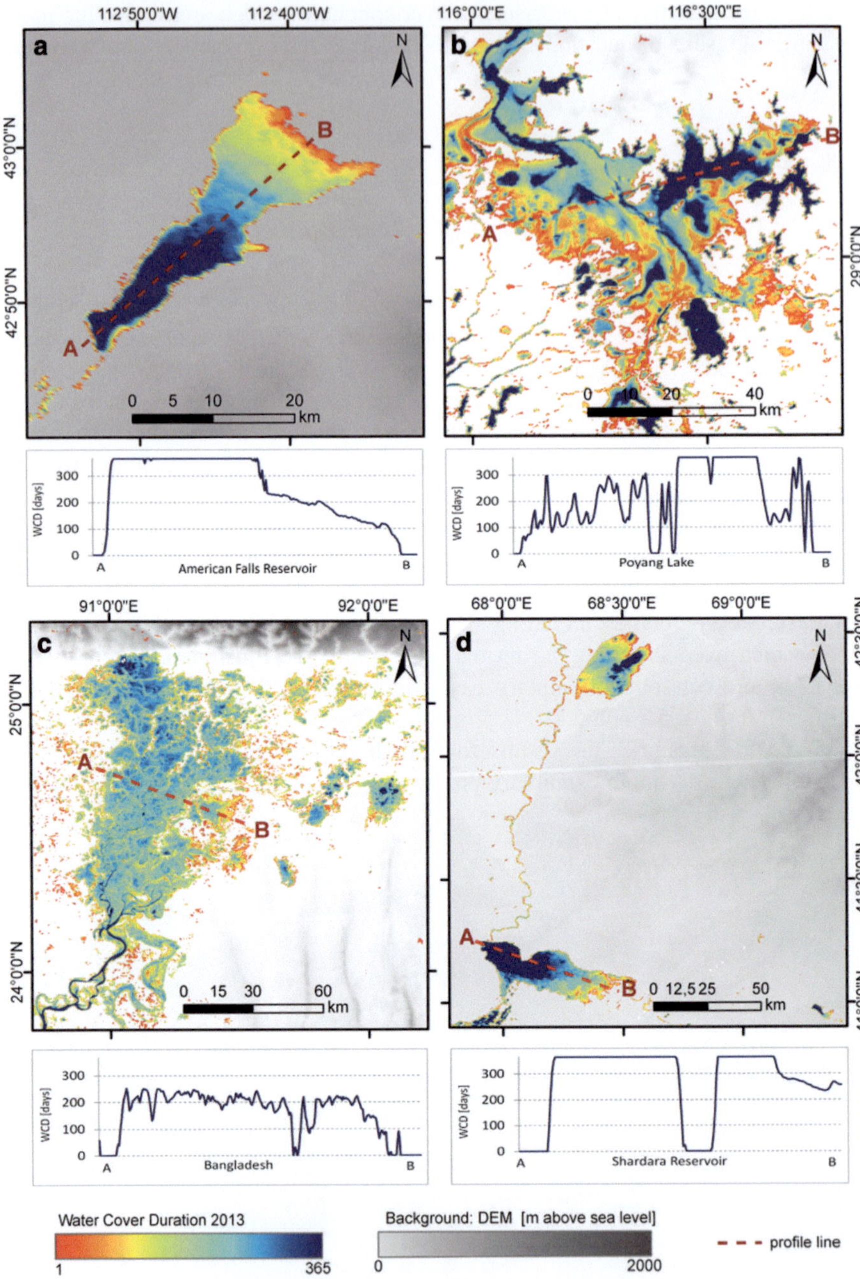

Fig. 5.7 Water cover duration for selected regions: (**a**) American Falls Reservoir, USA, (**b**) Poyang Lake, China, (**c**) Floods along Meghna River, Bangladesh, (**d**) Shardara and Koksaray Reservoirs, Kazakhstan. Projection: GCS WGS 1984. Water duration in days along profile lines

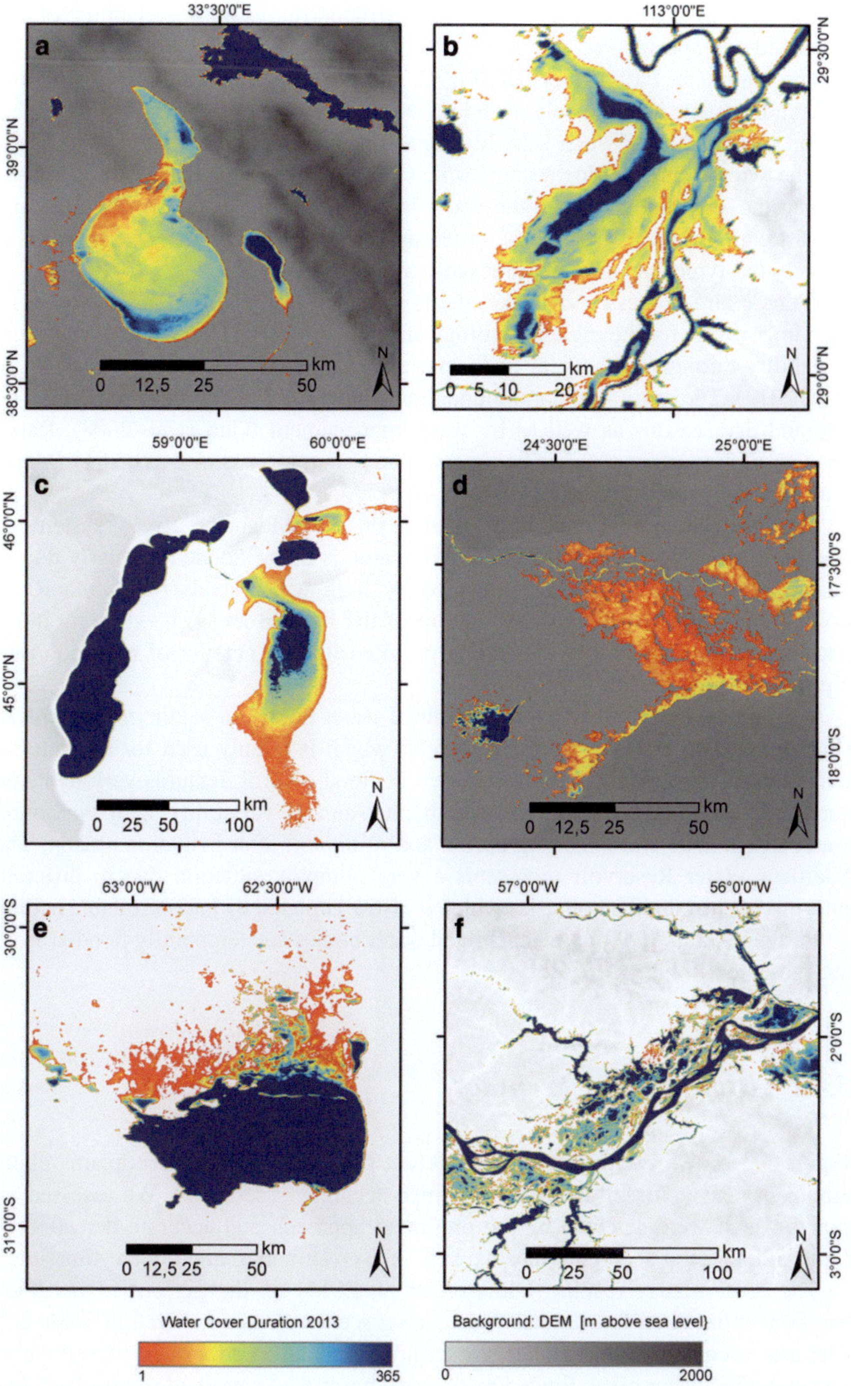

Fig. 5.8 Water cover duration for selected regions: (**a**) Lake Tuz, Turkey, (**b**) Lake Dongting, China, (**c**) Aral Sea, Uzbekistan, (**d**) Chobe floodplain and Lake Liambezi, Namibia, (**e**) Mar Chiquita, Argentina (**f**) Floodplain along Amazonas River, Brazil. Projection: GCS WGS 1984

the number of days covered with water to underline spatial variability of the discussed examples.

The American Falls Reservoir (Fig. 5.7a) is an artificial water body which is used and regulated extensively for a large agricultural irrigation community in this region. The amount of water is highly depending on snow pack conserved throughout the winter and melting waters which feed the reservoir during spring time. During the vegetation season, the water is being withdrawn for surrounding and downstream irrigated agriculture, whereby the water inflow is decreasing. This leads to the typical spatio-temporal pattern of the reservoir.

China's largest fresh water lake, the Poyang Lake (Fig. 5.7b), is known for its high importance for the regional ecology and economy. The lake features very high variability throughout the year and rapid changes in lake inundation areas (Feng et al. 2012). The spatio-temporal inundation patterns of Poyang Lake are driven by wet and dry seasons as well as by water management (Guo et al. 2008). Rapid changes of the surface area can be observed on a short term as well as on a long term scale (Feng et al. 2012).

The spatio-temporal flooding pattern of the Meghana River system in Bangladesh is visualized in Fig. 5.7c. This area is being flooded regularly due to heavy monsoon rains and is controlled by high water levels of Padma River downstream. Magnitude and areal extent of the flooded areas show strong inter- and intra-annual variability depending on intensity and amount of monsoon precipitation during the wet season.

The last example illustrates two artificial water bodies in southern Kazakhstan (Fig. 5.7d). The Shardara Water Reservoir which is mainly used for agricultural irrigation, hydropower generation as well as flood control, features very dynamic surface area. The reservoir shows high intra-annual variability with maximum water extent after snow melting period and minimum after irrigation season. The Shardara Water Reservoir represents a very complex situation due to different interests of surrounding countries which is also completed by uncontrollable nature.

Further examples of dynamical inland water bodies and temporally flooded areas are presented in Fig. 5.8.

5.4.2 *Evaluation of Accuracy*

Figure 5.9 shows exemplary three different time steps for the floodplains along Amazonas River, Brazil and the performed accuracy assessment. We assessed all pixels which were detected by the presented approach and accurate detection of Landsat images at accordant day, as well as overestimation and underestimation.

The error matrix results (Olofsson et al. 2014) of the performed accuracy assessment based on the 40 analysed Landsat scenes are summarized in Table 5.2. The user accuracy (complementary measure commission error) describes the correct classified water pixel divided by the total number of classified water pixel. The producer accuracy (complementary measure omission error) is the deviation of total

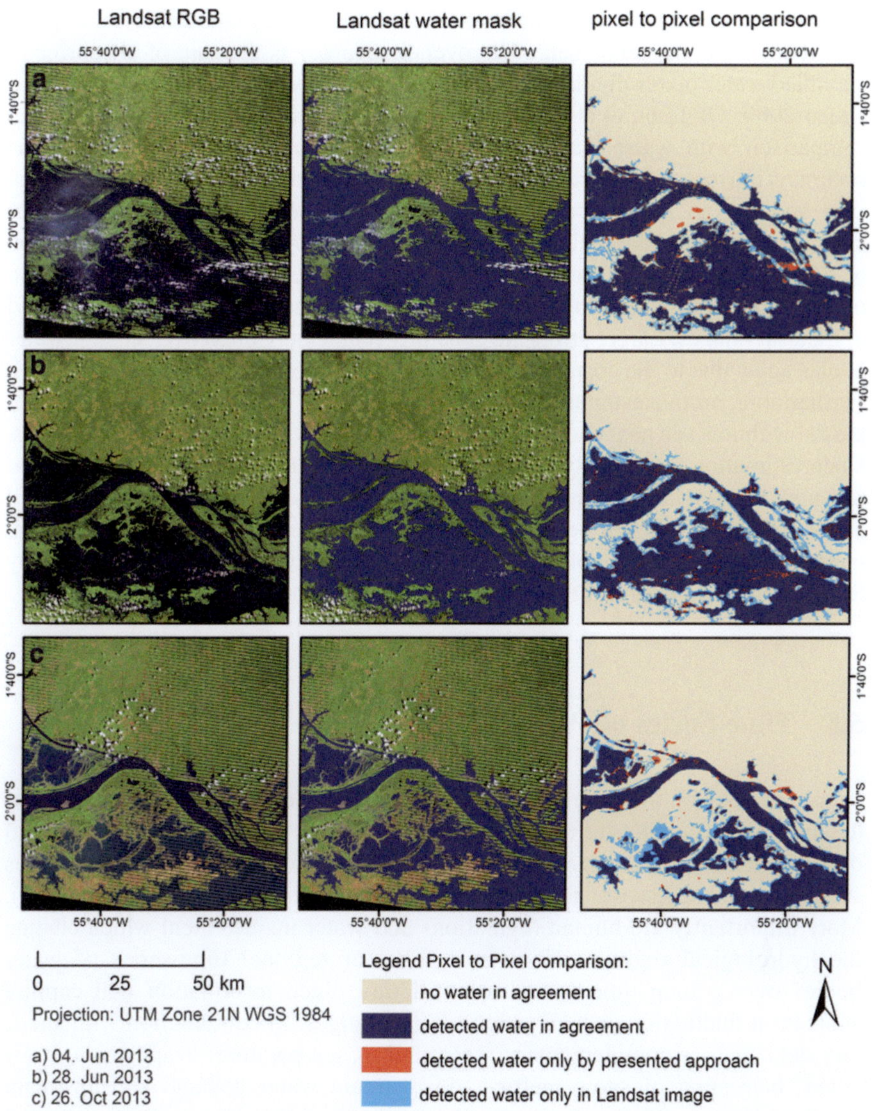

Fig. 5.9 Exemplary validation for three time steps (Floodplain along Amazonas River, Brazil)

Table 5.2 Summarized results of accuracy assessment based on 40 Landsat images

	Mar Chiquita		Meghana River		Poyang Lake	
	User Acc.	*Prod. Acc.*	*User Acc.*	*Prod. Acc.*	*User Acc.*	*Prod. Acc.*
	0.91	0.97	0.87	0.94	0.68	0.90
	Niger Wetland		Floodplain, Amazonas		Salt Lakes	
	User Acc.	*Prod. Acc.*	*User Acc.*	*Prod. Acc.*	*User Acc.*	*Prod. Acc.*
	0.94	0.98	0.78	0.93	0.89	0.94
Overall accuracy			**0.89**			

number of correctly classified water pixel divided by total number of water pixel identified in the reference data. The overall accuracy is the sum of all correctly classified water pixels divided by the total number of water pixels (Congalton and Green 2009; Olofsson et al. 2014). The overall accuracy based on pixel-to-pixel comparison with water masks derived from Landsat images was 89 %. This accuracy mirrors a temporal snapshot for corresponding days and selected test sites and not an in-depth assessment. Generally, the results of MODIS water masks overestimate the real water bodies' size due to the coarse pixel resolution of 250 m compared to the finer Landsat resolution. Carroll et al. (2009) also observed overestimation of their static MODIS water mask product when compared to Landsat images. Larger errors are generally found more often over smaller water bodies again due to the coarse resolution of MODIS also stated by Li et al. (2013). Furthermore, an overestimation was observed in high latitude areas during winter season with low sun angle and resulting unfavourable satellite detection conditions. Underestimation was mainly observed for longer cloud covered periods (tropical regions), for wetlands and swamps with high aquatic vegetation production as well as for waters with higher sediment load due to flooding situation. The underestimation of presented example (Fig. 5.9) illustrates a case where waters are highly loaded with sediments and thus were underestimated by the approach.

5.5 Time-Series of Water Bodies

In the context of this work, the question arises upon why information about dynamic characteristics of inland water bodies is crucial. First of all, the reaction of lakes in endorheic basins is a very good proxy for climatological variables such as annual and seasonal precipitation and snow melting (Mason et al. 1994). More importantly are human regulations and water management which change the hydrological regimes. However, global or regional time-series of water bodies over a long time period are still rare. Such information will capture short-term fluctuation as well as long-term changes. In combination with ancillary data such as complementary precipitation, temperature, evaporation, snow cover, beginning of snow melting, downstream water demand and upstream human activities, a better understanding of the world's lakes and reservoirs can be achieved. Further studies must focus on data fusion combining different meteorological parameters and human impact to enable a comprehensive understanding of the entire system. This practically applies to different fragile ecosystems with low annual precipitation, high temperatures, environmental change and strong human interaction.

5.6 Conclusion

The goal of this study was to highlight the need of time-series information for inland water bodies by presenting our approach and the results for the year 2013. In this chapter, we discuss our classification strategy to detect water bodies on a global scale with high temporal resolution based on medium spatial resolution remote sensing data and a priori knowledge. The approach is a fast process which uses only a single band, digital elevation model, cloud cover information and a static water mask.

Many inland water bodies feature high intra- and inter-annual surface extent variability. Relief and hydrological characteristics of some catchments induce those water bodies to be highly responsive to climate variation and human activities (Friedrich and Oberhaensli 2004). The presented examples underline intra-annual variability in a demonstrative way. Such intra-annual dynamics cannot be captured by a single image or a static water mask (Feng et al. 2012). Furthermore, the examples illustrate that various water bodies progress differently in their areal extent due to natural environment depending on e.g. the size of hydrological basin, relief, precipitation, land cover, geology and anthropogenic impact (e.g. construction of dams, political and economic decisions on water discharge, irrigation). Therefore, it is important to monitor and reconstruct the temporal development of water bodies globally and put it into comprehensive context. The documentation of past conditions of inland water bodies and their variability over time, in combination with any driving forces and observed alterations, might improve our understanding of such complex systems.

In this study, we proved that medium-resolution data can be effectively used to detect information such as areal extent of inland water bodies on a global scale with high temporal resolution. Future work will focus on the implementation of an additional set of MODIS products accounting for different geographical regions and dealing with the challenges mentioned in this chapter.

Acknowledgement We highly appreciate that all MODIS, Landsat and DEM data were provided free of charge by the National Aeronautics and Space Administration(NASA), the National Oceanic and Atmospheric Administration (NOAA), the United States Geological Survey (USGS) and the Consultative Group on International Agricultural Research – Consortium for Spatial Information (CGIAR-CSI). Furthermore we would like to thank the anonymous reviewers for their valuable comments on the manuscript.

References

Bukata RP, Jerome JH, Kondratyev KY, Pozdnyakov DV (1995) Optical properties and remote sensing of inland and coastal waters. CRC Press, New York

Carroll ML, Townshend JR, DiMiceli CM, Noojipady P, Sohlberg RA (2009) A new global raster water mask at 250 m resolution. Int J Digit Earth 2:291–308. doi:10.1080/17538940902951401

CGIAR-CSI (2008) What is CGIAR-CSI? http://csi.cgiar.org/WhtIsCGIAR_CSI.asp. Accessed 19 Feb 2014

Clark RN, Swayze GA, Wise R, Livo E, Hoefen T, Kokaly R, Sutley SJ (2007) USGS digital spectral library splib06a, Digital data series 231. U.S. Geological Survey, Denver

Congalton RG, Green K (2009) Assessing the accuracy of remotely sensed data: principles and practices. Lewis Publishers, Boca Raton/London/New York

Dietz AJ, Kuenzer C, Conrad C (2013) Snow-cover variability in central Asia between 2000 and 2011 derived from improved MODIS daily snow-cover products. Int J Remote Sens 34:3879–3902. doi:10.1080/01431161.2013.767480

ESA-Land Cover-CCI (2014) http://www.esa-landcover-cci.org/?q=node/156. 11 Dec 2014

Feng L, Hu C, Chen X, Cai X, Tian L, Gan W (2012) Assessment of inundation changes of Poyang Lake using MODIS observations between 2000 and 2010. Remote Sens Environ 121:80–92. doi:10.1016/j.rse.2012.01.014

Feyisa GL, Meilby H, Fensholt R, Proud SR (2014) Automated water extraction index: a new technique for surface water mapping using Landsat imagery. Remote Sens Environ 140:23–35. doi:10.1016/j.rse.2013.08.029

Fichtelmann B, Borg E (2012) A new self-learning algorithm for dynamic classification of water bodies. In: Murgante B, Gervasi O, Misra S, Nedjah N, Rocha AMAC, Taniar D, Apduhan BO (eds) Computational science an its applications – ICCSA 2012, Part III. Springer, Salvador de Bahia, pp 457–470

Friedrich J, Oberhaensli H (2004) Hydrochemical properties of the Aral Sea water in summer 2002. J Mar Syst 47:77–88

Gao B (1996) NDWI a normalized difference water index for remote sensing of vegetation liquid water from space. Remote Sens Environ 58:257–266

Gstaiger V, Gbhardt S, Huth J, Wehrmann T, Kuenzer C (2012) Multi-sensoral derivation of inundated areas using TerraSAR-X and ENVISAT ASAR data. Int J Remote Sens 33 (22):7291–7304. doi:10.1080/01431161.2012.700421

GTOS (2008) Terrestrial essential climate variables for climate change assessment, mitigation and adaptation. FAO, Rome

Guo H, Dong Z, Chen X, Ma X, Zhang P (2008) Impact analysis of Poyang lake after three Gorges project's running by using ANFIS. 2008 international conference on multimedia and information technology, pp 741–743

Haas EM, Bartholomé E, Combal B (2009) Time series analysis of optical remote sensing data for the mapping of temporary surface water bodies in sub-Saharan Western Africa. J Hydrol 370:52–63. doi:10.1016/j.jhydrol.2009.02.052

Hall DK, Riggs GA (2007) Accuracy assessment of the MODIS snow products. Hydrol Process 1547:1534–1547. doi:10.1002/hyp

Jain SK, Singh RD, Jain MK, Lohani AK (2005) Delineation of flood-prone areas using remote sensing techniques. Water Resour Manag 19:333–347. doi:10.1007/s11269-005-3281-5

Ji L, Zhang L, Wylie B (2009) Analysis of dynamic thresholds for the normalized difference water index. Photogramm Eng Remote Sens 75:1307–1317

Klein I, Dietz AJ, Gessner U, Galayeva A, Myrzakhmetov A, Kuenzer C (2014a) Evaluation of seasonal water body extents in Central Asia over the past 27 years derived from medium-resolution remote sensing data. Int J Appl Earth Obs Geoinf 26:335–349. doi:10.1016/j.jag.2013.08.004

Klein I, Dietz AJ, Gessner U, Dech S, Kuenzer C (2014b) Preliminary results of Global WaterPack: a novel product to assess inland water bodies dynamics on daily basis. Remote Sens Lett (in print)

Kuenzer C, Guo H, Leinenkugel L, Huth J, Li X, Dech S (2013) Flood mapping and flood dynamics of the Mekong Delta: an ENVISAT-ASAR-WSM based time series analyses. Remote Sens 5:687–715. doi:10.3390/rs5020687

Land Processes Distributed Active Archive Center (2014) https://lpdaac.usgs.gov/products/modis_products_table. Accessed 16 May 2014

Lehner B, Döll P (2004) Development and validation of a global database of lakes, reservoirs and wetlands. J Hydrol 296:1–22. doi:10.1016/j.jhydrol.2004.03.028

Leinenkugel P, Kuenzer C, Oppelt N, Dech S (2013) Characterisation of land surface phenology and land cover based on moderate resolution satellite data in cloud prone areas – a novel product for the Mekong Basin. Remote Sens Environ 136:180–198. doi:10.1016/j.rse.2013.05.004

Li S, Sun D, Goldberg M, Stefanidis A (2013) Derivation of 30-m-resolution water maps from TERRA/MODIS and SRTM. Remote Sens Environ 134:417–430. doi:10.1016/j.rse.2013.03.015

Martinis S, Kuenzer C, Twele A (2015) Flood studies using synthetic aperture radar data. In: Thenkabail P (ed) Remote sensing handbook. In print at Taylor & Francis Inc./CRC Press

Mason IM, Guzkowska MAJ, Rapley CG, Street-Perrott FA (1994) The response of lake levels and areas to climatic changes. Clim Change 27:161–197

McFeeters SK (1996) The use of the normalized difference water index (NDWI) in the delineation of open water features. Int J Remote Sens 17:1425–1432. doi:10.1080/01431169608948714

Niu Z, Gong P, Cheng X, Guo J, Wang L, Huang H, Shen S, Wu J, Wang X, Wamg X, Ying Q, Liang L, Zhang L, Wang L, Yao Q, Yang Z, Guo Z, Dai Y (2009) Geographical characteristics of China's wetlands derived from remotely sensed data. Sci China Ser D Earth Sci 54:723–738

Olofsson P, Foody GM, Herold M, Stehman SV, Woodcock CE, Wulder MA (2014) Good practices for estimating area and assessing accuracy of land change. Remote Sens Environ 148:42–57. doi:10.1016/j.rse.2014.02.015

Ouma YO, Tateishi R (2006) A water index for rapid mapping of shoreline changes of five East African Rift Valley lakes: an empirical analysis using Landsat TM and ETM+ data. Int J Remote Sens 27:3153–3181. doi:10.1080/01431160500309934

Pekel JF, Vancutsem C, Bastin L, Clerici M, Vanbogaert E, Bartholomé E, Defourny P (2014) A near real-time water surface detection method based on HSV transformation of MODIS multi-spectral time series data. Remote Sens Environ 140:704–716. doi:10.1016/j.rse.2013.10.008

Riggs GA, Hall DK, Salomonson VV (2006) MODIS snow products user guide to collection 5. National Snow and Ice Data Center, Boulder

Ryu J, Won J, Min KD (2002) Waterline extraction from Landsat TM data in a tidal flat a case study in Gomso Bay, Korea. Remote Sens Environ 83:442–456

Sheng Y, Gong P, Xiao G (2001) Quantitative dynamic flood monitoring with NOAA AVHRR. Int J Remote Sens 22(9):1709–1724

Slater JA, Garvey G, Johnston C, Haase J, Heady B, Kroenung G, Little J (2006) The SRTM data finishing process and products. Photogramm Eng Remote Sens 72(3):237–247

Tucker C, Pinzon J, Brown M, Slayback D, Pak E, Mahoney R, Vermote E, El Saleous N (2005) An extended AVHRR 8-km NDVI dataset compatible with MODIS and SPOT vegetation NDVI data. Int J Remote Sens 26:4485–4498. doi:10.1080/01431160500168686

Vermote EF, Kotchenova S (2008) Atmospheric correction for the monitoring of land surfaces. J Geophys Res 113:D23S90. doi:10.1029/2007JD009662

Verpoorter C, Kutser T, Tranvik L (2012) Automated mapping of water bodies using Landsat multispectral data. Limnol Oceanogr Methods 10:1037–1050. doi:10.4319/lom.2012.10.1037

Weiss DJ, Crabtree RL (2011) Percent surface water estimation from MODIS BRDF 16-day image composites. Remote Sens Environ 115:2035–2046. doi:10.1016/j.rse.2011.04.005

Xu H (2006) Modification of normalised difference water index (NDWI) to enhance open water features in remotely sensed imagery. Int J Remote Sens 27:3025–3033. doi:10.1080/01431160600589179

Chapter 6
Analysing a 13 Years MODIS Land Surface Temperature Time Series in the Mekong Basin

Corinne Myrtha Frey and Claudia Kuenzer

Abstract Land surface temperature (LST) is an important parameter in the climate system, impacting vegetation development, snow cover, runoff, and human livelihoods. Knowledge of LST dynamics can furthermore be used as an indicator for climate variability and change. LST is regularly measured from satellite sensors on a broad spatial scale and with a high temporal resolution. In this research, MODIS (Moderate Resolution Imaging Spectroradiometer) sensor data are used to assess the spatial and temporal patterns of LST in the Mekong Basin (MB) including its temporal variability. The dataset contains 13 years of MODIS LST data, a unique measurement time series in terms of resolution, accuracy, and homogeneity. In the analysis a temporal granularity of 8-days was used. The MB was divided into six physiographically homogenous regions. The height and magnitude of annual LST curves differ between the regions and prove to be strongly dependent on the topography. Large intra-annual magnitudes and low temperatures (daytime/nighttime annual regional means are 14 °C/−7 °C) are found in the northern areas, mainly in the high-lying Tibetan Plateau. The more southern areas are characterized by low LST seasonality and high temperatures (daytime/nighttime annual regional means of these regions range from 25 °C to 30 °C/19 °C to 25 °C). The year-to-year variability of LST is similar in all regions (regional weekly daytime/nighttime deviations lower than 4 °C/6 °C, except for the Tibetan Plateau, where regional weekly daytime/nighttime deviations reach 6 °C/18 °C). In summer, 42 % of daytime LST could be explained by topographic height. In winter and in nighttime scenes, topography explained 89–97 % of the LST distribution. Land use and land use change further influence the LST pattern, mainly in the daytime. An example of rising LST due to deforestation is given. This study allows for an improved understanding of temperature dynamics in one of the world's largest river basins.

C.M. Frey (✉) • C. Kuenzer
German Remote Sensing Data Center, DFD, Earth Observation Center, EOC, German Aerospace Center, DLR, Oberpfaffenhofen, Germany
e-mail: corinne.frey@dlr.de; claudia.kuenzer@dlr.de

C. Kuenzer et al. (eds.), *Remote Sensing Time Series*, Remote Sensing and Digital Image Processing, DOI 10.1007/978-3-319-15967-6_6

6.1 Introduction

Land surface temperature (LST) is a highly variable quantity of the Earth's surface, in both space and time. Its temporal variability mainly results from the annual and daily cycles of solar irradiation, which are further influenced by cloud cover and general weather situations. Spatial variability is governed by surface characteristics like albedo, emissivity, soil moisture, heat capacity of the surface soil layers, and topography. LST influences the surface energy budget, for example it largely controls the magnitude of the sensible heat flux and thus the air temperature. In this view, LST can be considered as an important indicator for climate variability (Kuenzer and Dech 2013).

The measurement of LST is conducted by radiometers which are sensitive in the thermal infrared domain (usually 8–14 μm). Thus the measurement can be conducted *in situ* with radiometers, or by airborne or space borne sensors. Measurements from satellites have the advantage that LST can be captured regularly with a high temporal resolution on a broad spatial scale. Due to the measurement geometry, remotely sensed LST is also a function of surface roughness and topography –the interpretation and correction of thermal anisotropy has been researched only recently (Vinnikov et al. 2012; Guillevic et al. 2013).

Due to its complexity, LST-related research of the past decades often focussed on the analysis of single scenes. Recently, researchers started shifting their focus toward time series, like long-standing Sea Surface Temperature (Merchant 2013), because the potential for applications is high (Jin and Dickinson 2010; Tian et al. 2012; Sobrino and Julien 2013). This study is an example for a regional application study using LST time series acquired over the Mekong Basin (MB).

The MB is a well suited area for analyses, as it is characterized by heterogeneous natural and anthropogenic strata. Surface heights range from 0 km at the coastal zones up to 6,000 m in the Himalayan mountain range. This complex distribution of surface characteristics must lead to significant differences in LST over the spatial domain as well as in the temporal shape. The overall goal of this study is to describe LST spatial behavior and temporal dynamics in the MB using statistical parameters and figures showing LST or quantitative derivatives of LST. A special focus is given to the influence of land cover and land cover change on LST. Input for the analysis is a 13 years' time series of MODIS (Moderate Resolution Imaging Spectroradiometer) LST data.

6.2 Study Region

Study region is the Mekong Basin, the catchment of the Mekong River. The Mekong River runs from the Tibetan Plateau through six countries: the People's Republic of China, the Union of Myanmar, Lao People's Democratic Republic, Kingdom of Thailand, Kingdom of Cambodia, and the Socialist Republic of

Vietnam (Fig. 6.1). The MB is home to nearly 80 million inhabitants. Rapid development, land cover changes (especially deforestation), and regulatory measures along the river (hydropower) have attracted the attention of many stakeholders, decision makers and researchers (e.g. Kuenzer et al. 2012, 2013; Leinenkugel et al. 2013b; Son et al. 2012; Shrestha et al. 2013). Leinenkugel

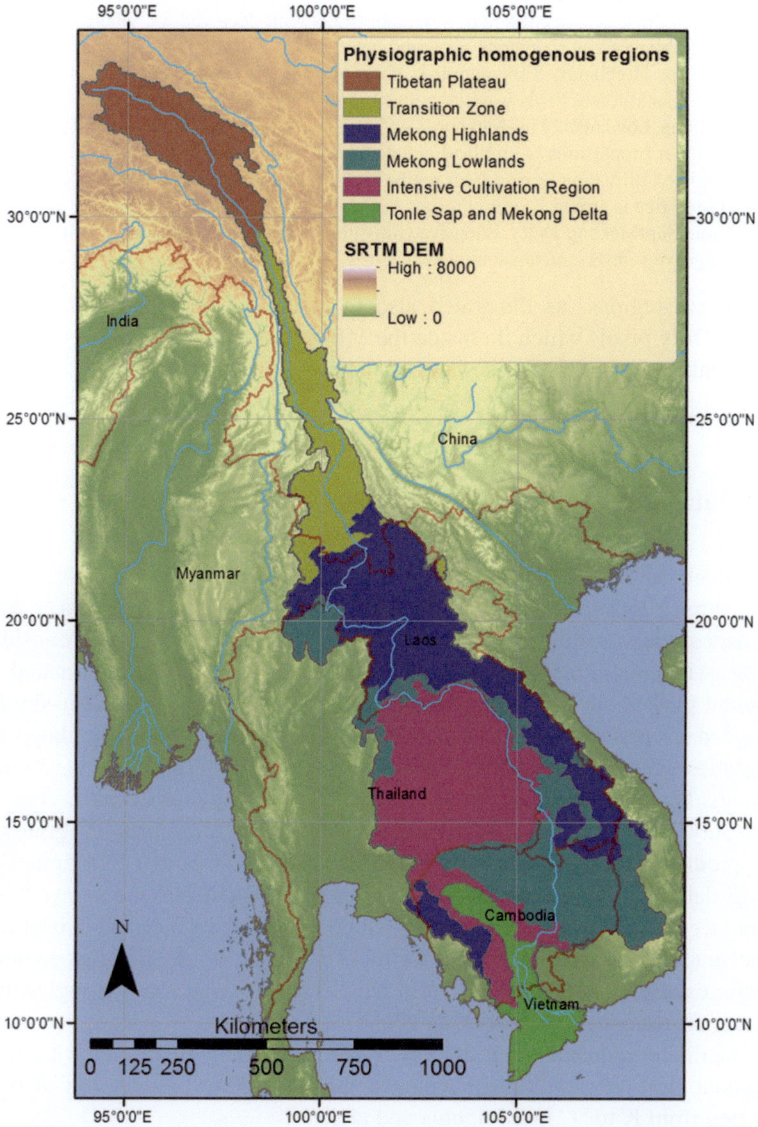

Fig. 6.1 The Mekong Basin study area divided in the six physiographic homogenous regions as published by Leinenkugel et al. 2013b. The background shows the surface heights as given by SRTM (Shuttle Radar Topography Mission) data

et al. (2013b), following Gupta (2009), distinguish six physiographically homogeneous regions in the MB:

(I) Tibetan Plateau (though only parts of the Tibetan Plateau are part of the MB). This region is dominated by a mostly high-altitude, polar tundra climate. Above the treeline the surface is covered with alpine grasslands, below needle-leaved forests and shrubland is found.
(II) Transition Zone (in the article called the Lancangjiang Basin). Here, strong elevation gradients dominate connecting high mountain peaks with deep valleys, where coniferous forests, shrublands, and croplands are found.
(III) Mekong Highlands. This region is characterised by large evergreen broadleaved forests, and small scale fields from cropping and shifting cultivation practices.
(IV) Mekong Lowlands. The Mekong Lowlands are similar to the Mekong Highlands, though higher proportions of cropland and deciduous vegetation is found.
(V) Intensive Cultivation Region. This region is characterized by large portions of cultivated areas. Often rice is grown.
(VI) Tonle Sap/Mekong Delta. Intensive cultivation is also found here, along with swamp and mangrove forests, and aquaculture.

These six regions, as illustrated also in Fig. 6.1, were further used in this research. Only pixels which lie inside the MB (grey outline in Fig. 6.1) were used for the analyses, even if satellite imagery contained more data.

6.3 Data and Methodology

The gridded MODIS LST dataset from the Terra satellite (MOD11A2, version 005 – https://lpdaac.usgs.gov/products/modis_products_table/mod11a2) is the main data source for this analysis. Data from 2000 to 2012 is thereby considered. The dataset contains daytime and nighttime data in 1 km resolution and is in sinusoidal projection. LST values are given as the average of all clear-sky LSTs during 8-days periods. The LST of the daily input product are calculated using emissivities which are based on a land cover classification (http://www.icess.ucsb.edu/modis/LstUsrGuide/MODIS_LST_products_Users_guide_C5.pdf). For data before July 2001, the IGBP (International Geosphere-Biosphere Program) land cover product was used. For later data, the land cover product generated from MODIS data itself was used. Daily LSTs which are input into the 8-days average are produced using the generalized split-window algorithm (Wan and Dozier 1996). The accuracy of the daily products was found to be 1 K, however higher errors may occur for example in semi-arid and arid regions and at large viewing angles (http://landval.gsfc.nasa.gov/; Frey et al. 2012).

To facilitate understanding, the 8-day LST averages will be called LST_8day throughout this article. For convenience the LST_8day product has also been converted from K to °C in all figures and tables.

Although the MODIS product uses a cloud mask, there are still pixels in the daily products which are cloud contaminated, resulting in LST values that are too low (Frey and Künzer 2014). Leinenkugel et al. (2013a) analysed clouds in the MB and found generally high cloud cover during the SW Monsoon from May to mid-October, with 85–95 % of all pixels cloudy. During the NE Monsoon between November and March, average cloud cover was about 50 %. Frey and Künzer (2014) found that cloud gaps can influence the accuracy of the daily regional means, especially in months with high cloud cover. The contamination is reduced through the averaging process in the 8-day product, but there are still pixels whose values were too low. To reduce this effect only LST_8day pixels that were flagged as 'good quality' in the accompanying quality layer were used and an additional cloud screening – simplified from Neteler (2010) – was applied by the authors. This screening uses "lower boundary values" as a threshold for detecting invalid pixels in the LST_8day data. If any pixel is below the "lower boundary value" belonging to the appropriate 8-day period, year and region then the pixel is marked as missing value. The lower boundaries are calculated separately for each 8-day period and 300×300 windows based on quartiles, which are derived from all valid values in each respective 8-day period and window. To avoid extreme outliers in the lower boundary values, the time series of lower boundaries is smoothed with a 3×3 kernel and missing values are interpolated using spline interpolation.

The analysis of the dataset is organized as follows: In Sect. 6.4 the dataset is described quantitatively in space and time using various statistical measures, e.g. means, standard deviations, deviations from the long-term mean or variability. Afterwards in Sect. 6.5 the article discusses some factors that influence the spatial and temporal characteristics.

The quantitative description is arranged in Sects. 6.4.1, 6.4.2 and 6.4.3. In Sect. 6.4.1, the LST_8day time series are described using various statistical parameters and time series plots. The analysis is conducted region-wise using the physiographically homogeneous regions derived by Leinenkugel et al. (2013b). This was done because the spatial LST_8day distribution is also a function of surface characteristics and it is assumed that the regions form a suitable framework to distinguish different temporal and spatial patterns. The average LST_8day for a certain region is addressed as 'regional mean LST_8day'. It is the mean of all pixels within a region (e.g. all pixels of the region 'Tibetan Plateau'). The temporal resolution remains 8 days. Further, the magnitude of the diurnal cycle will be looked at. Generally, the incoming solar radiation is the major controlling factor for LST, leading to diurnal and annual cycles with peak LST around noon and in summer. While annual curves can be shown using the Terra MODIS LST time series, it is not possible to fully resolve the diurnal cycle relying only on MODIS LST data. Differences between daytime and nighttime values may not reflect the maximal diurnal magnitude, as acquisition times are not congruent with daily peak times. Acquisition times of the daytime scenes range around 10:30–11:30 local time, nighttime scenes are acquired around 22:00–23:00 local time. Nevertheless, the differences between daytime and nighttime values can be used to assess spatial patterns and temporal behaviour.

In the following Sect. (6.4.2), the study looks at deviations from the long-term mean. Deviations are the differences of actual LST_8day values and the 13-year mean of respective pixels. The 13-year mean is calculated as the mean of all

LST_8day values belonging to a certain 8-day period. The newly created 13-year mean dataset contains one image for each 8-day period, and will be addressed as LST_13year throughout this paper.

Section 6.4.3 analyses the variability of the deviations using the standard deviation of the yearly mean deviations. The variability is also calculated on pixel basis, separately for the daytime and the nighttime data.

Section 6.5 discusses how factors like land cover or surface height influence the distribution of surface temperatures. A layer with land use classes from Leinenkugel et al. (2013b) was utilized. Following classes were selected as combinations of the more differentiated classes from the authors above:

- Agricultural areas: 'Three season cultivated lands', 'Two season cultivated lands', 'Single season cultivated lands'
- Orchards: 'Single season orchards'
- Aqua culture: 'Aqua culture'
- Urban and built up lands: 'Urban and built up lands'
- Deciduous forests and shrublands: 'Deciduous broadleaf forests', 'Deciduous wood- and shrublands'
- Evergreen forests and shrublands: 'Evergreen broadleaf forests', 'Evergreen broadleaf wood- and shrublands', 'Evergreen needleleaf forests', 'Evergreen needleleaf wood- and shrublands'
- Swamps and mangroves: 'Swamp forests', 'Mangrove forests'
- Alpine grassland: 'Alpine grassland'

For a detailed analysis of LST_8day change in deforested areas, a forest change layer from Leinenkugel et al. (2014) was used. The layer was regridded from lat/lon to sinusoidal projection. This led to slight shifts in the location of a few polygons with a maximal shift of one pixel. To minimize possible misinterpretations, only polygons with a minimum area of 10 km^2 were chosen for the deforestation analysis.

Further, a digital elevation model (DEM) from the Shuttle Radar Topography Mission (SRTM) was applied in analyses requiring the topographic height. For the analysis the DEM was projected into the sinusoidal projection of the gridded MODIS data and resampled to 1 km resolution.

Topographic height data was also used to calculate the 'residual LST_8day'. Residual LST_8day was calculated by subtracting a modelled height-dependent LST, which was derived from regression calculations between the actual LST_8day and the SRTM over the whole MB. The residual LST_8day retains the temporal resolution. For some analyses also the 13 year mean (on 8 day basis) was calculated. This residual LST_13year is used to analyse the LST difference between the land use classes without the interfering elevation effect. Omitting this correction would distort the results mainly in the 'Transition zone' and 'Mekong Highlands' regions, as their land cover classes are not uniformly distributed. For example, agricultural areas in the Mekong Highland are found at lower elevations than forested areas, leading to higher LST in agricultural pixels due to the height effect.

Finally, the MODIS 16-day shortwave albedo product (MCD43B3) from 2000 to 2012 was used. This data was required for spatial analysis only. Therefore the 13-year time series was averaged, resulting in 13-year means for each 16-day period.

6.4 Quantitative Analysis of the LST_8day Dataset

6.4.1 Analysis of LST_8day Time Series

To analyse the temporal evolution of LST_8day, region-wise time series are plotted in Fig. 6.2. The magnitude of the annual cycle is strongest in the Tibetan Plateau, where the difference between the maxima and minima of the regional mean LST_8day may reach 30 °C in the daytime scenes and 26 °C in the nighttime scenes. The Transition Zone also shows strong seasonality. More southern regions exhibit smoother annual courses. The Tonle Sap and Mekong Delta region show minimum differences. The annual temperature range in regional means reaches 16 °C in the daytime scenes and only 10 °C in the nighttime scenes.

Generally, the Tibetan plateau shows the coldest regional mean LST_8day, exhibiting extremely low temperatures in the nighttime scenes. The nighttime minimum regional mean LST_8day reached −20 °C in the study period. Warmest nighttime regional mean LST_8day are found in the Tonle Sap and Mekong Delta region reaching 28 °C. In the daytime scenes the highest regional mean LST_8day are found in the Intensive Cultivation Region with regional mean values going up to 42 °C (outside plotted range in Fig. 6.2).

Table 6.1 shows the median values of all yearly minima and maxima, as well as average regional means.

The time series plotted in Fig. 6.2 show some negative outliers. An extreme example of such an outlier is found in the nighttime data of the 2003 summer in the Transition Zone. These outliers are due to cloud-influenced pixels that were not detected by the MODIS cloud mask and the post-seasonal cloud-screening. Such outliers may occur in periods with massive cloud occurrence. In such cases the lower boundary value is set to a fixed threshold which might be too low in respective cases. Outliers can also occur in regions with strong spatial LST gradients like in the transition zone, where the lower boundary may not be appropriate for all pixels. Apart from these outliers, the annual courses of LST_8day remain more or less constant over the time.

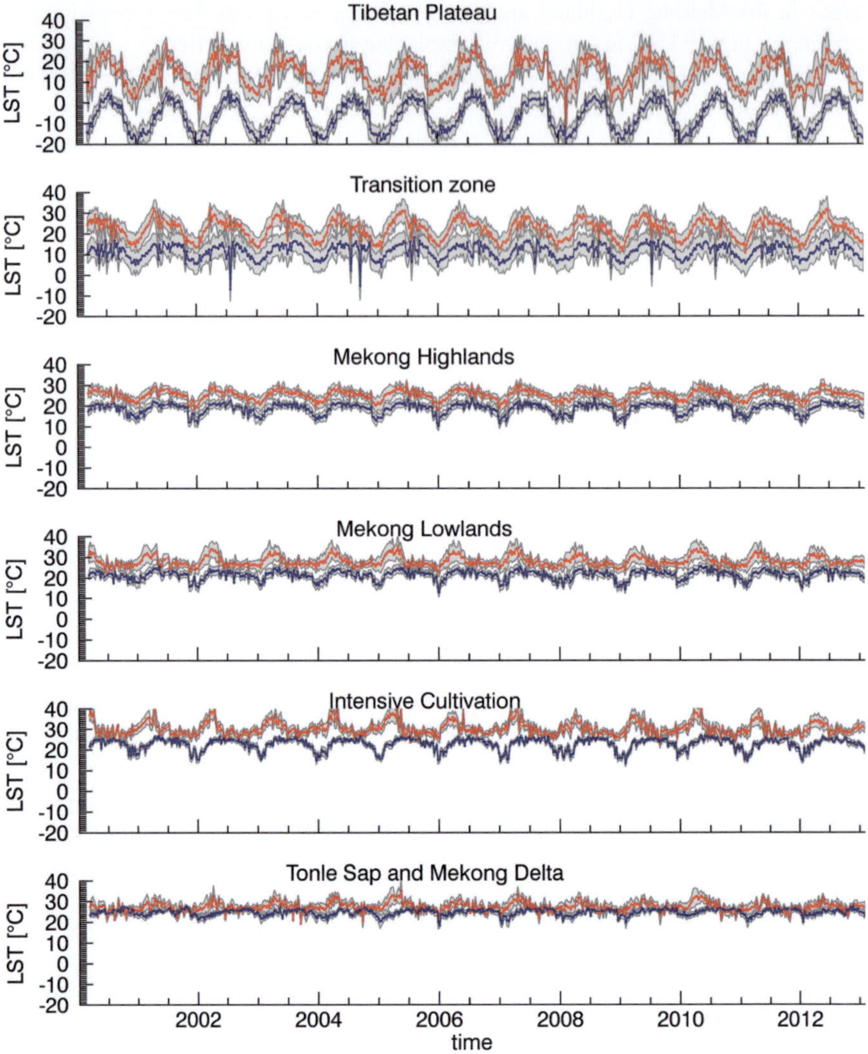

Fig. 6.2 Time series of daytime (*red*) and nighttime (*blue*) regional mean LST_8day for the six physiographic homogenous regions of the Mekong Basin. The grey background areas depict the standard deviation

6.4.2 Analysis of Deviations from the 13-Year Mean

Pixel-wise deviations from the calculated 13-years average (LST_13year) range from −11 °C to 6 °C for the daytime scenes and from −12 °C to 18 °C considering the nighttime scenes. Positive deviations stand for higher temperatures than average, negative deviations stand for lower than average. Mean regional absolute

Table 6.1 Daytime and nighttime statistical values from the LST_8day time series

	Daytime			Nighttime		
	Median of yearly maxima [°C]	Median of yearly minima [°C]	Mean value [°C]	Median of yearly maxima [°C]	Median of yearly minima [°C]	Mean value [°C]
Tibetan Plateau	21.3	5.2	14.0	3.0	−17.2	−7.0
Transition zone	28.9	14.1	22.0	15.4	6.1	11.4
Mekong Highlands	28.2	21.3	25.1	21.2	14.3	18.6
Mekong Lowlands	32.7	25.8	27.9	24.0	16.9	21.2
Intensive Cultivation Region	36.0	27.5	30.1	25.3	17.3	22.6
Tonle Sap and Mekong Delta	30.6	24.9	27.3	26.6	22.6	24.6

Table 6.2 Statistical parameters of the LST deviations (actual values minus 13-year average) for each of the six physiographically homogenous regions

	Maximum positive difference		Maximum negative difference		Mean absolute difference		Standard deviation of absolute difference	
	Day-time	Night-time	Daytime	Night-time	Daytime	Night-time	Daytime	Night-time
Tibetan Plateau	6.17	17.97	−10.85	−11.80	2.27	1.44	1.82	1.45
Transition Zone	3.99	4.70	−5.67	−5.51	0.98	0.96	0.88	0.92
Mekong Highlands	2.93	6.03	−4.09	−5.76	0.86	0.86	0.84	0.87
Mekong Lowlands	2.96	3.66	−3.70	−3.01	0.78	0.63	0.66	0.62
Intensive Cultivation Region	2.87	2.31	−5.33	−3.01	0.71	0.64	0.77	0.58
Tonle Sap and Mekong Delta	3.04	2.44	−2.87	−3.15	0.59	0.59	0.62	0.51

deviations range from 1 °C to 2 °C in the daytime scenes and are around 1 °C in the nighttime scenes (Table 6.2). The statistical parameters given in Table 6.2 are derived from LST_8day; the maximum positive difference e.g. thus represents the maximum difference found in all pixels and in all 8-day periods. The mean absolute difference gives evidence about the magnitude of the differences. All regions, except the Tibetan Plateau feature very similar values. Daytime differences are between ±3 °C and ±6 °C. Mean absolute differences and their standard deviations are all below 1 °C. However, the Tibetan Plateau stands out from the other regions having the largest regional deviations (up to 18 °C) as well as larger regional mean absolute differences (up to 2 °C). The high standard deviation (daytime 2 °C) points to a large temporal variability in the area.

For this analysis, yearly means were also calculated from the LST_8day deviations for the years 2000–2012. It was found that there are years with strong deviations like for example the daytime deviations in the year 2008. In other years LST_8day is similar to LST_13year, for example in nighttime in the year 2000. The deviations are also not distributed evenly over the study region. In some years there was a decoupling of the northern region, mainly the Tibetan Plateau, from the Mekong Highlands and Lowlands and the Intensive Cultivation and Tonle Sap/Mekong Delta regions, e.g. in the year 2005. In other years, the deviation pattern seem to be randomly distributed, e.g. in the year 2001 (data not shown). In summary, no regular pattern was found in the annual deviations. It is assumed that the natural variability of the weather is responsible for these deviations; however an influence of cloud detection errors cannot be excluded. Generally, it is found that inter-annual deviations tend to be larger in the winter months, when the NE Monsoon occurs between November and March. Lower deviations are found during the SW Monsoon in summer and autumn (May to mid-October). This finding is true for all regions except the Tibetan Plateau, where deviations are high throughout the year. Figures 6.4 and 6.5 show the regional mean deviations in the study period, separately for each month. For these figures, the maximum deviation was taken from all available 8-days periods per month. The figures show that the season-dependent deviation pattern is found in the daytime and nighttime images. Looking at the temporal evolution of the deviations, no significant trends could be detected in the data. 13 years is insufficient for this kind of analysis.

6.4.3 Analysis of the Variability

Generally, there is a much higher variability in the daytime scenes than in the nighttime scenes. Only some small parts in the Mekong Highlands show some higher values in the nighttime variability. In the daytime strong variability is found in the Tibetan Plateau region with its alpine grasslands, as well in the regions with large cropland areas (Intensive Cultivation Region, Mekong Lowlands). High variability also occurs at the southern coast of Vietnam where wetlands mix with aquaculture. Regions with low variability in the daytime scenes are the Transition Zone, most parts of the Mekong Highlands, where evergreen forests cover the Earth's surface, and the Tonle Sap Lake in Cambodia (Figs. 6.4 and 6.5).

Some of the regions with high daytime LST variability in the Lower Mekong Basin strongly agree with areas which underwent permanent tree cover loss between 2000 and 2011 (delineated by Leinenkugel et al. 2014). These areas are

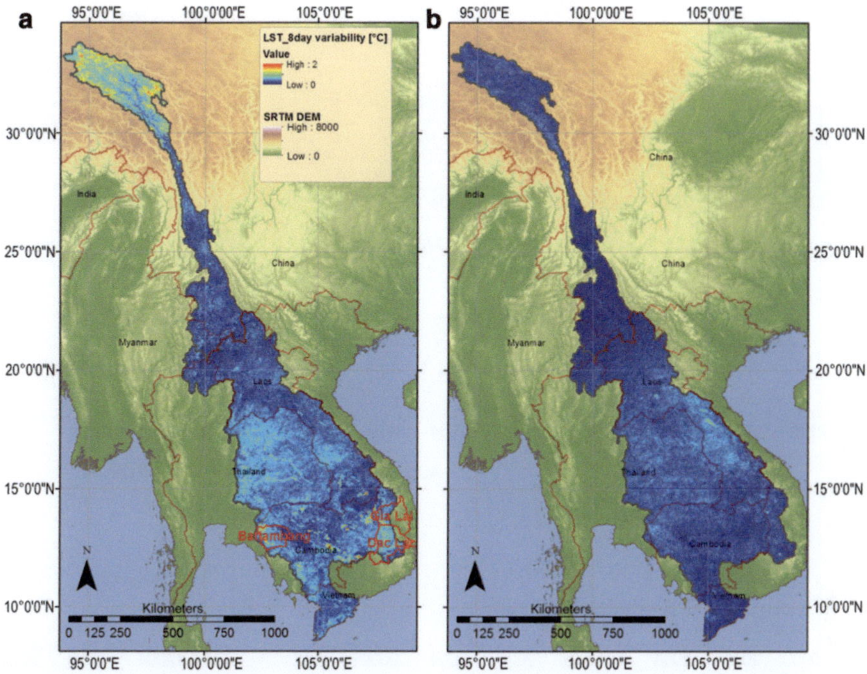

Fig. 6.3 LST variability in the MB for (**a**) the daytime and (**b**) the nighttime images. The variability is expressed as the standard deviation of all mean annual deviations. Maximal variability is 4.8 °C, for visualization reasons however, the scale bar just goes until 2 °C

mainly in Cambodia. For example, great clusters of deforestation which also show high LST variability are found in Batdambang, in Gia Lai, and in Dac Lac province (see also Fig. 6.3). There are also other small spots of high variability in Cambodia which might point to additional deforestation activities. The Khorat Plateau in Thailand also shows high daytime LST variability, yet, according to Leinenkugel et al. (2014) only few areas are affected there by deforestation. The effect of deforestation on LST_8day is discussed further in Sect. 6.5 (Figs. 6.4 and 6.5).

6.5 Influencing Factors

Solar irradiation is the main driver of LST. It mainly determines the diurnal and annual shapes of LST curves. Deviations from such ideal curves arise mainly from spatially varying surface characteristics and manifest themselves in spatial patterns. As such, the LST_8day distribution in the MB shows a distinct spatial pattern, which is well correlated with the topography of the area. On a large spatial scale, the elevation of the MB decreases towards the south, whereas the temperatures increase (Fig. 6.6). At small scale, a similar effect can be observed in the deep Mekong River

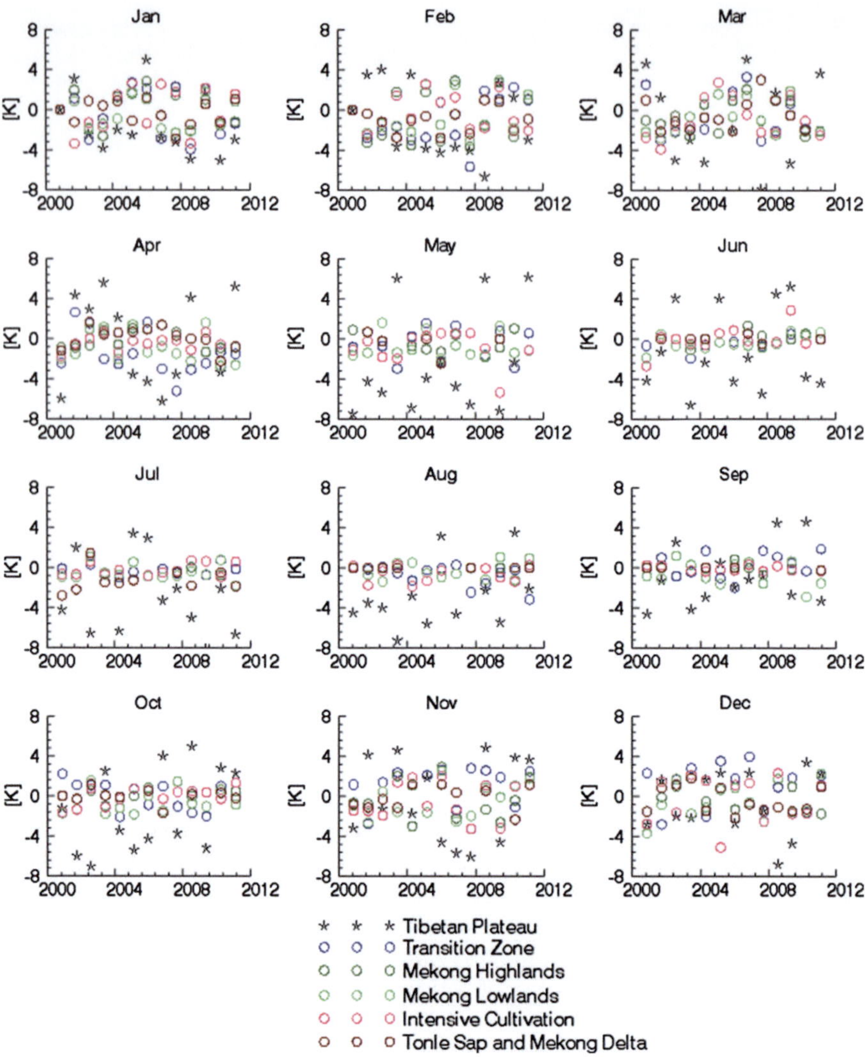

Fig. 6.4 Maximum of mean daytime deviations from the 13-year average LST (LST_13year), separately for the six physiographically homogenous regions

valley in the transition zone, where steep temperature gradients are observed along the slopes. This effect can be seen for example in the daytime scene of July (Fig. 6.6b). This effect is also present in the other months and in nighttime image of Fig. 6.6 but is not very visible due to the chosen colour stretch for the images of Fig. 6.6 (for further information see also Frey and Kuenzer 2014). Correlation coefficients r^2 between the topographic height and the individual 8-day periods in the LST_13years dataset range between 0.42 and 0.89 for the daytime scenes and between 0.95 and 0.97 for the nighttime scenes. Low correlation values are only

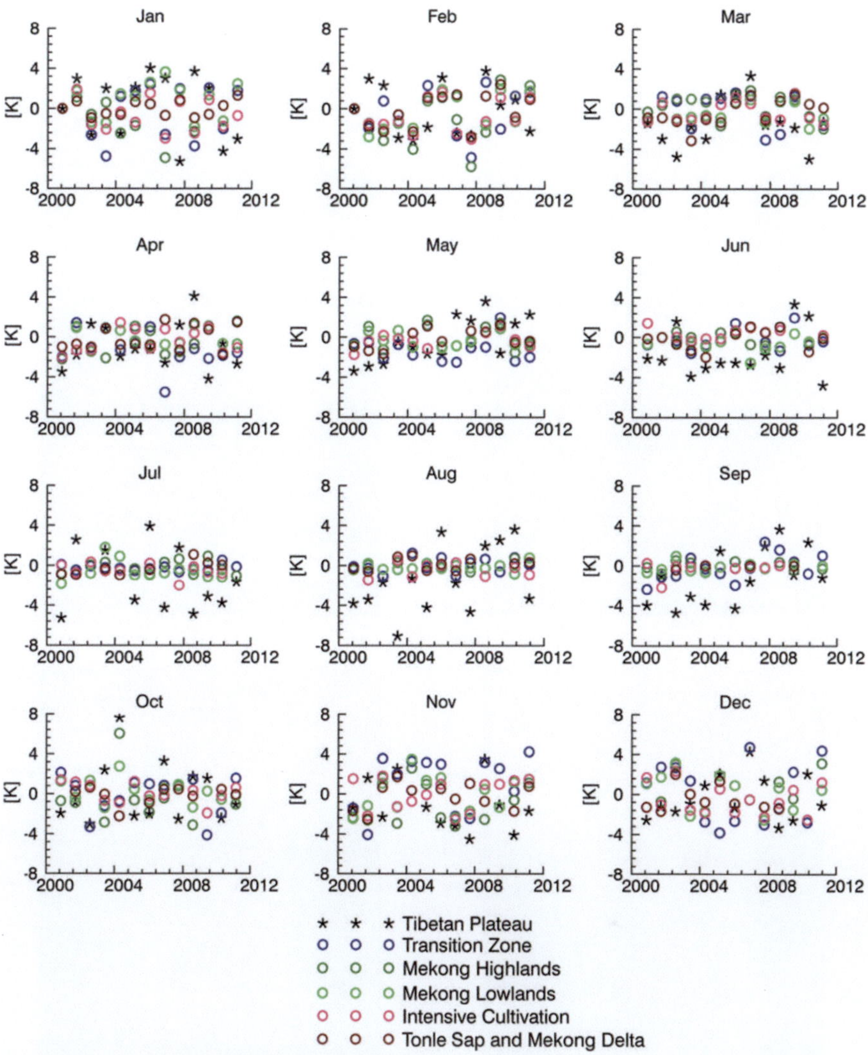

Fig. 6.5 Maximum of mean nighttime deviations from the 13-years average LST, separately for the six physiographic homogenous regions

found in the daytime summer scenes, where other aspects of the topography (slope and aspect), and surface characteristics like shortwave albedo, longwave emissivity, and thermal inertia exhibit a stronger influence on the spatial distribution of LST. These dependencies are nonlinear. For example the shortwave albedo influences the LST by controlling the solar irradiation energy available for heating. Areas with low albedo generally heat up more strongly than areas with high albedo. In the northern regions (Tibetan Plateau) this correlation can be found in the data, especially when there are snowy or glaciated areas. Soil moisture does also

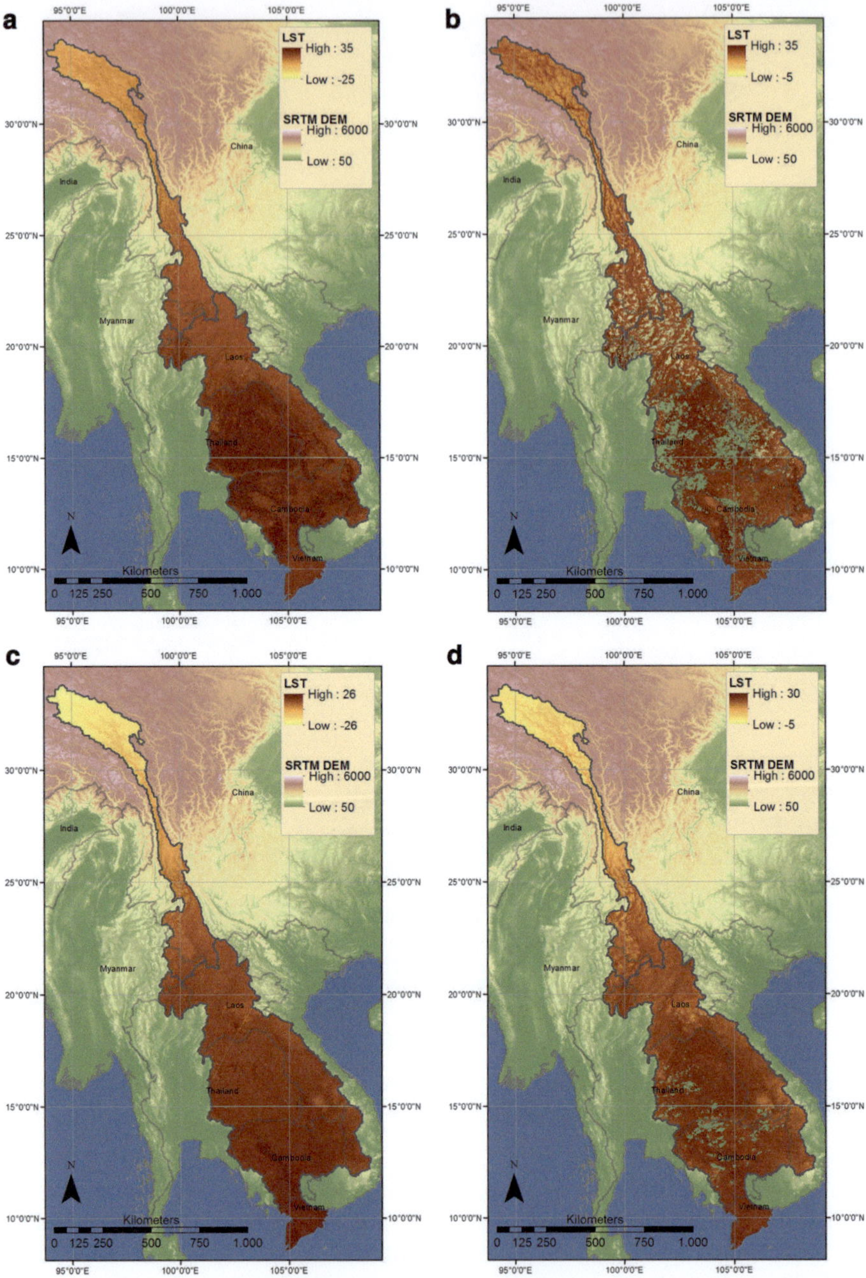

Fig. 6.6 13-year mean LST (**a**) January daytime, (**b**) July daytime, (**c**) January nighttime, (**d**) July nighttime. In the July scenes (rainy season), there are many cloud gaps. In some areas, no good cloud-free LST value was available in the whole 13-year period

influence LST: dry areas may show higher temperatures than wet areas, as no evaporative cooling takes place. Costa-Cabral et al. (2008) found that soil moisture in the MB is often coupled with land cover. Agricultural areas show highest soil moisture values- lowest values are found in grassland and woodland areas, when no antecedent high precipitation has taken place. However, the Intensive Cultivation Region generally has higher temperatures than its wooded neighbouring regions, even after subtracting the influence of the topography. Only in the nighttime scenes some winter months showed a pattern consistent with a cooling influence from soil moisture. Another factor, namely the latitude, adds to the correlation between LST and topography. Temperatures usually get warmer when moving towards the equator due to enhanced solar irradiation and this effect is observed in this dataset. Finally, Fig. 6.6 shows the high thermal capacity of water bodies. The Tonle Sap Lake in western Cambodia, which is the largest freshwater lake in Southeast Asia, can be distinguished by its cooler LST_8day in the daytime scenes and its warmer LST_8day in the nighttime scenes (black arrow in Fig. 6.6a).

Land use classes encompass regions with similar surface properties. It is therefore straightforward to assume LST variability within a land use class would be smaller than the variations amongst different classes. A relationship between land use and LST pattern is confirmed by analysing the temporal evolution of the residual LST_13year. Figure 6.7 shows the long-term mean residual LST_8day (=residual LST_13year) in its annual course separated for the six regions and for 10 land use classes (Leinenkugel et al. 2013b). The residual LST_13year does not show the usual annual curve, as this cancels out in the calculation. Variations within one class and one region are due to changing surface properties during the annual course within one class. But also the limited accuracy of the height model and other influencing factors like influence of undetected clouds, slope and exposition may have an impact.

Figure 6.7 shows that there is a substantial difference in daytime regional residual LST_13year between forested and agricultural areas, especially in the regions Transition zone, Mekong Highlands, Mekong Lowlands, and Intensive Cultivation Region. The largest differences occurred in the Mekong Lowlands and the Intensive Cultivation Region where the residual LST_13year in forested areas is up to 4 °C cooler than the agricultural areas. In the region Tonle Sap and Mekong Delta however, no such difference is found. The region Tibetan Plateau does not include agricultural areas, so no such comparison was made. The nighttime data did not reveal such differences between the land use classes. Only the two classes Evergreen forests and shrublands and Alpine grasslands on the Tibetan Plateau showed considerable differences.

LST is an important variable in the surface energy budget and thus in the local microclimate (Frey et al. 2012). For the regional climate therefore, differences between agricultural and forested areas may have substantial impacts. As already stated, Cambodia has large areas where deforestation has taken place during the last 10 years. As a consequence, LST_8day has increased in these areas. Figure 6.8 shows the residual LST_8day after elevation correction over time for only those pixels that underwent deforestation during the last years (deforestation mask:

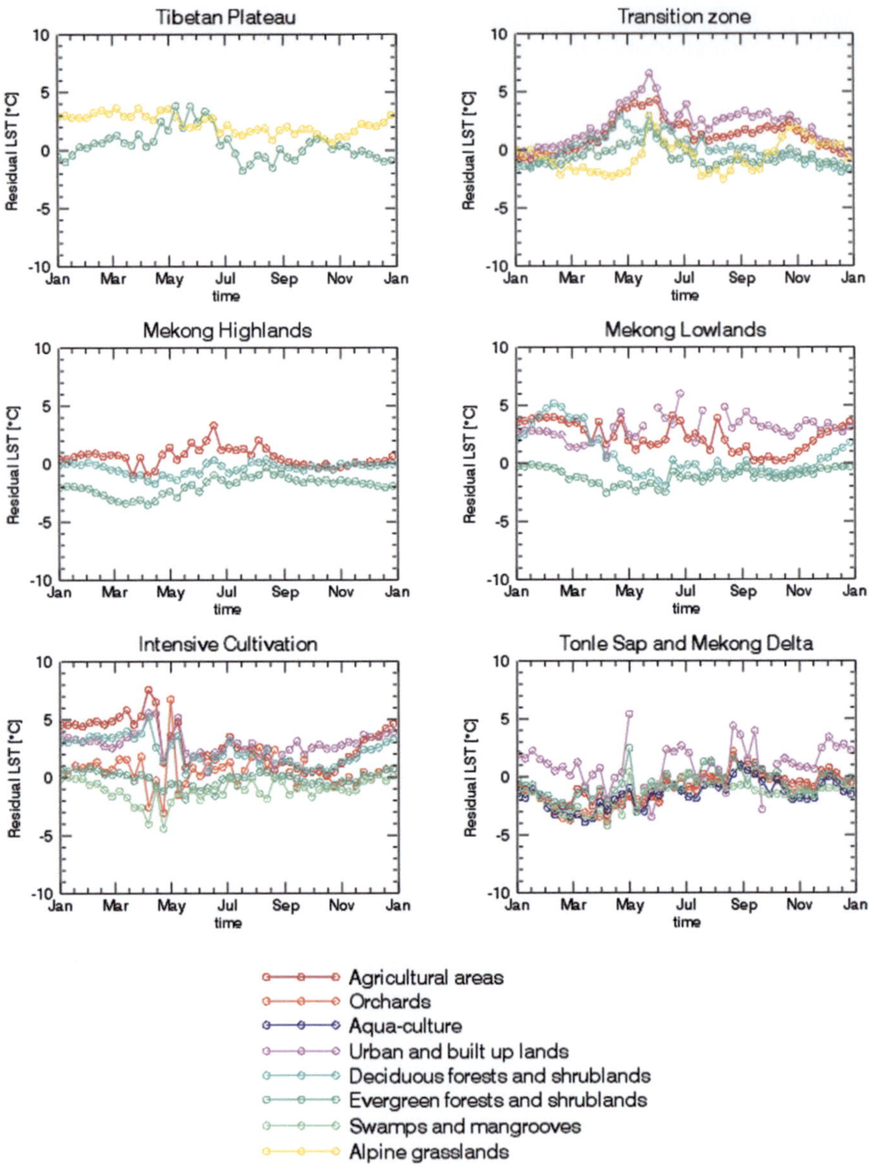

Fig. 6.7 Residual LST_13year separated for the six regions of the MB and land use classes

Leinenkugel et al. 2014). The number of deforested pixels per region was more than 500 in each of the regions. It must be noted that the deforestation mask did not only flag pixels whose area was completely cleared. The majority of the pixels are only partially affected by deforestation (personal communication, P. Leinenkugel). The residual LST_8day increase is therefore expected to be even higher on completely

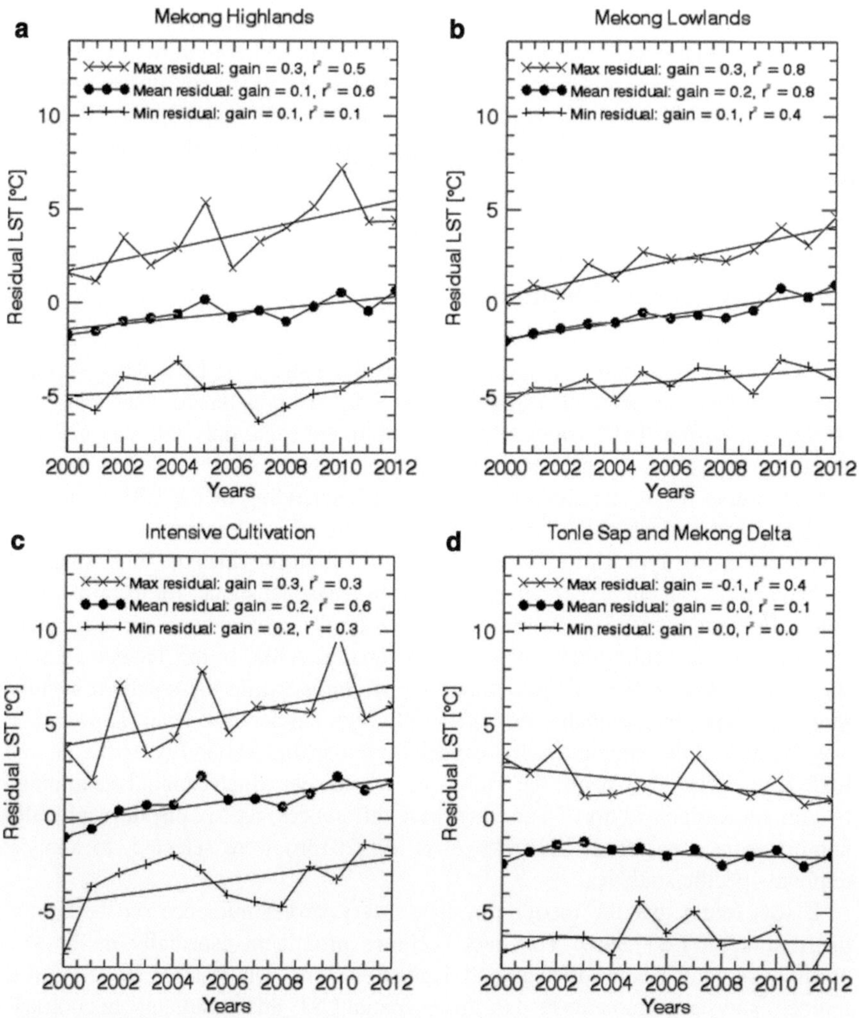

Fig. 6.8 Daytime residual LST_8day (Max, Mean, Min) for areas that underwent deforestation since 2001. (**a**) Mekong Highlands, (**b**) Mekong Lowlands, (**c**) Intensive Cultivation Region, (**d**) Tonle Sap and Mekong Delta

deforested pixels. Annual mean, maximum and minimum residual LST_8day values, as well as respective gain and correction coefficients are given in Fig. 6.8. A positive growth is shown for the Mekong Highland and Mekong Lowland and the Intensive Cultivation Region with increases of 0.1–0.2 °C per year. The Tonle Sap/Mekong Delta region however shows no change in the mean LST residuals, and even a continuous cooling of maximum values. Considering that agriculture in this region involves heavy irrigation and seasonal flooding, this finding is plausible. The observed warming is not equal for all months, but higher during the dry season and lower during the wet season. The growth of the mean values for the dry months of January and February for example, is 0.3 °C per year for the deforested pixels in the Mekong Highland, Mekong Lowland and Intensive Cultivation Region. For the wet months of August and September the growth is 0.0 or 0.1 for aforementioned regions. It should be mentioned that this growth is directly related to the deforestation and does not point to a long-term climate trend. The other pixels do not show such warming.

6.6 Discussion and Summary

In this study the temporal evolution and spatial pattern of LST_8day (8 days average of land surface temperature) of the MB (Mekong Basin) were analysed. MODIS data from the Terra satellite were used. As such, only one day and one nighttime measurement were available per pixel. Data that resolve the whole diurnal course would certainly enhance the understanding of the LST regime in the MB. Nevertheless, the MODIS Terra LST data allowed extracting spatial and temporal patterns typical for the MB. Solar irradiation provides the main shape of diurnal and annual curves. The magnitude of the seasonal signal, for example, gets much lower towards the south as mean temperatures increase due to more constant and higher solar zenith angles towards the equator. As such, the Tibetan Plateau shows large seasonality and low mean temperatures, while the southern regions show very smooth seasonality on a generally high temperature level. However, it was found that the magnitude and especially the spatial variability of the annual LST_8day curves vary inside the study area to an extent which cannot be explained by climatic variables only. To analyse these differences, the six physiographically homogeneous areas from Leinenkugel et al. (2013b) were selected, to analyse regional-specific statistics.

If was found that the topography has a very strong influence on the spatial distribution of LST_8day. This fact becomes prominent especially in the two northern regions Tibetan Plateau and Transition Zone, where large elevation differences and steep gradients lead to strong spatial LST_8day gradients. In contrary, the southern regions show lower intra-regional LST_8day gradients, where the topography is more even. The relation between topography and LST_8day was analysed quantitatively. At night 95–97 % of the LST_8day distribution could be explained by the topographic height. In daytime the influence was less and varied from 89 % to 42 % with low r^2 in the summer months. Wang et al. (2013) confirm the influence of the topography on spatial temperature pattern by analysing station air temperature data of a region mostly congruent with our Transition Zone. However, the sparseness of *in situ* stations in this region does not allow reproduction the spatial patterns that were found in this study. Xu et al. (2013) also agrees on the strong influence of topography on LST for the Tibetan Plateau, as well as the large intra-annual gradients. Latter finding is also confirmed by Salama et al. (2012).

The Tibetan Plateau also stands out in its variability (= standard deviation of the monthly deviation from the long-term mean), featuring a much higher variability than the other regions. This is especially true in the time of the SW Monsoon (May to mid-October), when deviations are lower than in the rest of the year in all regions except the Tibetan Plateau. Generally, the daytime scenes show higher variability than the nighttime scenes. In the daytime scenes also agricultural areas feature a high variability, especially areas which underwent deforestation and transformation from forest to agriculture.

Apart from elevation land use also plays an important role in explaining temporal and spatial LST pattern. Urban and built-up lands generally feature higher temperatures than their natural counterparts. This is a well-known effect proved in many Surface Urban Heat Island studies (e.g. Heldens et al. 2013; Parlow et al. 2014; Peng et al. 2012). Not only man-made materials influence LST, but different natural surfaces exhibit large differences, with the soil water content playing a major role. Swamps and mangroves for example exhibit clearly lower LST_8day than deciduous forests and shrublands. Interestingly, agricultural areas also feature higher LST_8day than forests and shrublands in the Transition Zone, the Mekong Lowlands and the Intensive Cultivation Region. The relation between land use and LST is confirmed by Xu et al. (2013) for the Tibetan Plateau. While they also found strong spatial LST gradients dependent on the topography, they showed that forested areas and grassland have higher mean LST than water but lower LST than bare land. Also the intra-annual gradients vary amongst the different land uses. Julien et al. (2006) investigated changes in LST and NDVI (Normalized Difference Vegetation Index) over Europe and confirmed different temporal evolutions of LST for different land cover types.

As LST_8day distribution varies amongst the different land use classes, land use change over time will also lead to changes in LST_8day in many cases. This was shown on the example of deforestation, which is carried out on a massive scale in Cambodia, where forest is cleared for extending the agricultural fields. As a result, LST_8day increased strongly over these areas. This effect results from processes which are independent from climate change related temperature increases as predicted e.g. by Eastham et al. 2008. This deforestation effect will likely add to the projected warming due to climate change, leading to even higher temperatures over the affected areas. Apart from this, a series of impacts are possible. This local temperature enhancement may increase the effects of climate change like a shifting of the vegetation growth periods through increased air temperatures (Eastham et al. 2008), and influence future rice yields (Chinvanno et al. 2008). Higher temperatures – coupled with heat waves – may also lead to increased mortality rates (Tan et al. 2010) or droughts, which frequently occur in the Lower Mekong Basin (Son et al. 2012). The local enhancement of LST_8day however, may also alter climate projections through potentially increased cloud cover and local triggering of rain events. Increased cloud cover in turn would reduce the regional warming. As such, the long-term effects of deforestation on local climate are subject to further research. The authors recommend including future predictions of land cover changes in climate simulations especially in areas, where deforestation is taking place. For extracting significant long-term trends it would be worth using longer time series of LST, e.g. from the AVHRR (Advanced Very High Resolution Radiometer) sensor.

6.7 Conclusions

The Mekong Basin (MB) recently came into the focus of various stakeholders, decision makers, and researchers, as the region undergoes rapid development (deforestation, hydropower, and urbanisation) which leads to significant land cover changes. The climatic and physiographic description of the area is therefore not only beneficial for future research like hydrologic and climatic modelling, but also for bodies working on land use planning and regulatory measures in affected countries.

The Mekong Basin is a region with strongly varying physiographic conditions. This reflects in the spatial and temporal distribution of LST (land surface temperature) of the region. This was shown quantitatively in six physiographic homogeneous regions of the Mekong Basin, which show varying mean LST as well as intra-annual and diurnal LST magnitudes. However, some of the regions show also very strong intra-regional gradients which could largely be explained by topography and land use. The topography explains a large portion of the spatial variability, especially in the area where strong height gradients exist. Besides the spatial variability, temporal variability is also found in the data, which itself varies spatially: some regions show larger temporal variability than others. This might be induced by changing weather patterns, but also through land use change. The latter is an important factor for understanding LST distribution and seasonality of LST. Land use changes influence the LST and therefore the microclimate. As such, anthropogenic land use change activities like deforestation and transformation into agriculture, lead to higher LST. This effect adds to regional climate change effects and may alter some of the local predictions. To account for this effect in regional climate models, it is recommended to include predictions of land cover change into simulation models in this region. This study has shown that it is possible to characterize the detailed spatial and temporal LST regime of the MB using a 13-year MODIS time series. The spatial distribution of LST can be assessed in high detail. With recently developed new approaches to convert LST to air temperatures (Lazzarini et al. 2014; Peón et al. 2014), this dataset may even help to overcome the problem with the lack of available *in situ* stations in this region. The authors encourage using longer time series together with a resolving of the diurnal course to deepen the understanding of the LST regime in the MB and to be able to detect climatic trends.

Acknowledgements The authors thank the German Ministry of Education and Research, BMBF, for funding the German–Vietnamese WISDOM project (www.wisdom.eoc.dlr.de). Further, the authors thank P. Leinenkugel for providing the physiographic regions, the land cover dataset, and the deforestation layer of the Mekong Basin used in this research.

References

Chinvanno S, Boulidam S, Inthavong T, Souvannalath S, Lersupavithnapa B, Kerdsuk V, Thuan NTH (2008) Climate risks and rice farming in the Lower Mekong River Basin. In: Leary N, Conde C, Kulkarni J, Nyong A, Adejuwon J, Barros V, Burton I, Lasco R, Pulhin J (eds) Climate change and vulnerability. Earthscan climate. Routledge, London

Costa-Cabral MC, Richey JE, Goteti G, Lettenmaier DP, Feldkötter C, Snidvongs A (2008) Landscape structure and use, climate, and water movement in the Mekong River basin. Hydrol Process 22:1731–1746

Eastham J, Mpelasoka F, Mainuddin F, Ticehurst C, Dyce P, Hodgson G, Ali R, Kirby M (2008) Mekong river basin water resources assessment: impacts of climate change. CSIRO: Water for a Healthy Country National Research Flagship

Frey CM, Kuenzer C (2014) Land surface temperature dynamics in the Upper Mekong Basin derived from MODIS time series. Int J Remote Sens 35(8):2780–2798

Frey CM, Kuenzer C, Dech S (2012) Quantitative comparison of the operational NOAA-AVHRR LST product of DLR and the MODIS LST product V005. Int J Remote Sens 33:7165–7183

Guillevic PC, Bork-Unkelbach A, Gottsche FM, Hulley G, Gastellu-Etchegorry J-P, Olesen FS, Privette JL (2013) Directional viewing effects on satellite land surface temperature products over sparse vegetation canopies – a multisensor analysis. IEEE Geosci Remote Sens Lett 10 (6):1464–1468

Gupta A (2009) Geology and landforms of the Mekong Basin. In: Campbell IC (ed) The Mekong. Biophysical environment of an international river basin. Academic, San Diego, pp 29–51

Heldens W, Taubenböck H, Esch T, Heiden U, Wurm M (2013) Analysis of surface thermal patterns in relation to urban structure types: a case study for the city of Munich. In: Kuenzer C, Dech S (eds) Thermal infrared remote sensing – sensors, methods, applications. Remote sensing and digital image processing series 17. Springer, Netherlands, pp 475–495

Jin ML, Dickinson RE (2010) Land surface skin temperature climatology: benefitting from the strengths of satellite observations. Environ Res Lett 5:1–13

Julien Y, Sobrino JA, Verhoef W (2006) Changes in land surface temperatures and NDVI values over Europe between 1982 and 1999. Remote Sens Environ 103(1):43–55

Kuenzer C, Dech S (2013) Theoretical background of thermal infrared remote sensing. In: Kuenzer C, Dech S (eds) Thermal infrared remote sensing – sensors, methods, applications. Remote sensing and digital image processing series 17. Springer, Netherlands, pp 1–26

Kuenzer C, Campbell I, Roch M, Leinenkugel P, Vo QT, Dech S (2012) Understanding the impacts of hydropower developments in the context of upstream-downstream relations in the Mekong River Basin. Sustain Sci 11:1–20

Kuenzer C, Guo H, Huth J, Leinenkugel P, Li X, Dech S (2013) Flood mapping and flood dynamics of the Mekong delta: ENVISAT-ASAR-WSM based time series analyses. Remote Sens 5:687–715

Lazzarini M, Marpu PR, Eissa Y, Ghedira H (2014) Toward a near real-time product of air temperature maps from satellite data and in situ measurements in arid environments. IEEE J Sel Top Appl Earth Obs Remote Sens 7(7):3093–3104

Leinenkugel P, Kuenzer C, Dech S (2013a) Comparison and enhancement of MODIS cloud mask products for Southeast Asia. Int J Remote Sens 34:2730–2748

Leinenkugel P, Kuenzer C, Oppelt N, Dech S (2013b) Characterisation of land surface phenology and land cover based on moderate resolution satellite data in cloud prone areas – a novel product for the complete Mekong Basin. Remote Sens Environ 136:180–198

Leinenkugel P, Wolters M, Oppelt N, Kuenzer C (2014) Tree cover and forest cover dynamics in the Mekong Basin from 2001 to 2011. Remote Sens Environ 158(3):376–392

Merchant CJ (2013) Thermal remote sensing of sea surface temperature. In: Kuenzer C, Dech S (eds) Thermal infrared remote sensing, remote sensing and digital image processing, 17. Springer, Netherlands, pp 287–313

Neteler M (2010) Estimating daily land surface temperatures in mountainous environments by reconstructed MODIS LST data. Remote Sens 2:333–351

Parlow E, Vogt R, Feigenwinter C (2014) The urban heat island of Basel – seen from different perspectives. Die Erde – J Geogr Soc Berl 145(1–2):96–110

Peng S, Piao S, Ciais P, Friedlingstein P, Ottle C, Bréon F-M, Nan H, Zhou L, Myneni RB (2012) Surface urban heat island across 419 global big cities. Environ Sci Technol 46(2):696–703

Peón J, Recondo C, Calleja JF (2014) Improvements in the estimation of daily minimum air temperature in peninsular Spain using MODIS land surface temperature. Int J Remote Sens 35 (13):5148–5166

Salama MS, Van der Velde R, Zhong L, Ma Y, Ofwono M, Su Z (2012) Decadal variations of land surface temperature anomalies observed over the Tibetan Plateau by the Special Sensor Microwave Imager (SSM/I) from 1987 to 2008. Clim Chang 114:769–781

Shrestha B, Babel MS, Maskey S, van Griensven A, Uhlenbrook S, Green A, Akkharath I (2013) Impact of climate change on sediment yield in the Mekong River Basin: a case study of the Nam Ou basin, Lao PDR. Hydrol Earth Syst Sci 17:1–20

Sobrino JA, Julien Y (2013) Time series corrections and analyses in thermal remote sensing. In: Kuenzer C, Dech S (eds) Thermal infrared remote sensing, remote sensing and digital image processing 17. Springer, Netherlands, pp 267–285

Son NT, Chen CF, Chen CR, Chang LY, Minh VQ (2012) Monitoring agricultural drought in the Lower Mekong Basin using MODIS NDVI and land surface temperature data. Int J Appl Earth Obs Geoinf 18:417–427

Tan J, Zheng Y, Tang X, Guo C, Li L, Song G, Zhen X, Yuan D, Kalkstein AJ, Li F, Chen HHH (2010) The urban heat island and its impact on heat waves and human health in Shanghai. Int J Biometeorol 54:75–84

Tian F, Qiu GY, Yang YH, Xiong YJ, Wang P (2012) Studies on the relationships between land surface temperature and environmental factors in an inland river catchment based on geographically weighted regression and MODIS data. IEEE J Sel Top Appl Earth Obs Remote Sens 5(3):687–698

Vinnikov KY, Yu Y, Goldberg MD, Tarpley D, Romanov P, Laszlo I, Chen M (2012) Angular anisotropy of satellite observations of land surface temperature. Geophys Res Lett 39(23)

Wan Z, Dozier J (1996) A generalized split-window algorithm for retrieving land-surface temperature from space. IEEE Trans Geosci Remote Sens 34:892–905

Wang X, Liang P, Wu F (2013) Analysis of regional temperature variation characteristics in the Lancang River Basin in southwestern China. Quat Int 33:198–206

Xu Y, Shen Y, Wu Z (2013) Spatial and temporal variations of land surface temperature over the Tibetan Plateau based on harmonic analysis. Mt Res Dev 33(1):85–94

Chapter 7
TIMESAT: A Software Package for Time-Series Processing and Assessment of Vegetation Dynamics

Lars Eklundh and Per Jönsson

Abstract Large volumes of data from satellite sensors with high time-resolution exist today, e.g. Advanced Very High Resolution Radiometer (AVHRR) and Moderate Resolution Imaging Spectroradiometer (MODIS), calling for efficient data processing methods. TIMESAT is a free software package for processing satellite time-series data in order to investigate problems related to global change and monitoring of vegetation resources. The assumptions behind TIMESAT are that the sensor data represent the seasonal vegetation signal in a meaningful way, and that the underlying vegetation variation is smooth. A number of processing steps are taken to transform the noisy signals into smooth seasonal curves, including fitting asymmetric Gaussian or double logistic functions, or smoothing the data using a modified Savitzky-Golay filter. TIMESAT can adapt to the upper envelope of the data, accounting for negatively biased noise, and can take missing data and quality flags into account. The software enables the extraction of seasonality parameters, like the beginning and end of the growing season, its length, integrated values, etc. TIMESAT has been used in a large number of applied studies for phenology parameter extraction, data smoothing, and general data quality improvement. To enable efficient analysis of future Earth Observation data sets, developments of TIMESAT are directed towards processing of high-spatial resolution data from e.g. Landsat and Sentinel-2, and use of spatio-temporal data processing methods.

L. Eklundh (✉)
Department of Physical Geography and Ecosystem Science,
Lund University, Lund, Sweden
e-mail: Lars.Eklundh@nateko.lu.se

P. Jönsson
Group for Materials Science and Applied Mathematics,
Malmö University, Malmö, Sweden

C. Kuenzer et al. (eds.), *Remote Sensing Time Series*, Remote Sensing and Digital Image Processing, DOI 10.1007/978-3-319-15967-6_7

7.1 Introduction

Satellite-derived time-series data help us understand interactions of terrestrial vegetation dynamics with climate and the carbon cycle, and their trends over time (Keenan et al. 2014). Using efficient processing methods for analyzing existing remotely sensed time-series data is important for monitoring and mapping vegetation dynamics, thereby contributing to improved understanding of the global climate system. We will in this chapter present and describe one available tool, named TIMESAT, for processing time-series of satellite sensor data to enable meaningful data extraction for modeling vegetation dynamics.

The first time-series of satellite imagery for studies of dynamic Earth processes were made available from weather satellites. It was a series of satellites launched by the American National Oceanic Administration (NOAA) that generated daily data covering the entire Earth, and enabled the generation of global near-real time vegetation data. The first of these weather satellites to have bands suitable for vegetation mapping was NOAA-6, carrying an improved Advanced Very High Resolution Radiometer (AVHRR) sensor (Zhu et al. 2012). Though the sensor generated data at coarse spatial resolution (approx. 1×1 km resampled into a 4×4 km global product) the value of the data for global vegetation monitoring soon became evident. A series of data products based on the Normalized Difference Vegetation Index (NDVI), computed from the NOAA channels 1 and 2, were developed and were used for studying the temporal dynamics of global land vegetation (Justice et al. 1985; Townshend and Justice 1986). These were the NOAA Pathfinder data set (James and Kalluri 1994), the University of Maryland GIMMS data set (Tucker et al. 2005), and the recent, improved GIMMS (Global Inventory Modeling and Mapping Studies) NDVI(3 g) data set (Jiang et al. 2013). These data sets contain global images of NDVI from 1981 onwards at a time step of 10–15 days and a spatial resolution of ca 8×8 km. This temporal and spatial resolution is adequate for studying seasonal and interannual dynamics of vegetation biomes. Hence, several studies from the mid-1980's and onwards have demonstrated how the information can be used for better understanding of vegetation dynamics as well as aiding land cover classifications (Defries and Townshend 1994; Running et al. 1994). In parallel, the increasing supply of high-spatial resolution data from sensors with 10–30 m resolution (e.g. Landsat and SPOT), and later on the development of satellites generating data at meter resolution (e.g. IKONOS, Quickbird, and Worldview), led to much of the technical development focusing on methods for classifying and quantifying high-resolution data. Hence, the development of time-series methodology in remote sensing was initially slow. However, it expanded quickly towards the beginning of the 2000's with the need to process large volumes of time-series data from the Terra and Aqua MODIS (Advanced Very High Resolution Radiometer) sensors at 250 m spatial resolution.

The interest in developing the TIMESAT software package arose from a need to manage time-series data in remote sensing in order to help tackle problems related to global change and monitoring of vegetation resources. The research community was interested in solving a range of questions related to time-series: is vegetation in the world's drylands changing; can satellite data be used for issuing early warnings of drought and famine; how does NDVI respond to changes in environmental driving forces such as rainfall and temperature; can satellites be used for monitoring carbon uptake from the vegetation; are growing seasons changing; and how does vegetation respond to climate change? In fact, these and many other related questions have been the focus of a large body of research during the last 30–40 years. TIMESAT is just one of many approaches for data processing and extraction of phenological information from Earth observation time-series data.

The background, theory and some future issues related to TIMESAT are described in the remaining sections.

7.2 Handling Remotely Sensed Time-Series Data – Assumptions and Some General Problems

The study of vegetation seasonality from space is based on two fundamental assumptions. The first assumption is that the optical data correctly model biophysical vegetation properties (such as leaf area index (LAI), green biomass, or fractional absorbed photosynthetically active radiation (FAPAR)). Unfortunately this assumption is not perfectly satisfied. It is true that many of the commonly used vegetation indices are empirically related to biophysical vegetation properties, but they are also affected by several other processes and disturbances. For example, even when ignoring the effects of clouds, angular effects and the atmosphere, the popular vegetation indices NDVI and Enhanced Vegetation Index (EVI) do not only respond to vegetation variations, but are both very sensitive to e.g. snow or moisture-induced background variations (Huete et al. 2002). Sensor degradation and drift in satellite overpass times (particularly evident with the NOAA satellites) are other examples of factors affecting the data reliability. All these influences cause ambiguity in the interpretation of the signal, affecting the information value and our ability to interpret the extracted seasonality data.

The second assumption is that the temporal signal from the vegetation is smooth. The canopy leaf mass, and the bulk of pigmentation and leaf water strongly dominate the optical signal from vegetation; these all tend to vary relatively consistently with time in a seasonal pattern. The variation can be slow (e.g. coniferous evergreen forest), or rapid (e.g. semi-arid grasslands), however it is not random. On the other hand, some short-term variations do occur in vegetation, e.g. due to light saturation and plant stress, which may lead to short-term variations in chlorophyll fluorescence that adds to the apparent reflectance. Though this addition is generally small for broad wavelength bands it may add substantially to certain wavelengths; up to 10–25 % at 685 nm and 2–6 % at 740 nm (Campbell et al. 2008). Also reflectance at 631 nm can change rapidly with variations in photosynthetic radiation-use efficiency (Gamon et al. 1997). Naturally, also some vegetation disturbances, caused by insect infestations, storms, and fires, can lead to rapid decline in canopy foliage which affects the reflectance. Overall, however, the seasonal canopy signal tends to change smoothly in a seasonal perspective, particularly in comparison with the many disturbing factors that may change rapidly from image to image: the atmosphere, clouds, angular variations due to different viewing and illumination angles, and geometric inaccuracies. Figure 7.1 shows daily MODIS NDVI data for 3 years from a coniferous forest site in southern Sweden, illustrating the noise in these data. The data in Fig. 7.1a seem to be more or less fully made up of noise, and the seasonal variation is quite difficult to discern.

To transform the noisy data into an understandable signal, a number of processing steps are necessary. These steps may include removing cloud interference by applying cloud masks (often based on thresholds in visible and thermal wavelengths), removing atmospheric absorption and scattering effects, and applying methods for correcting bi-directional illumination and viewing effects in the data. Employing a perfect set of physically based methods would be the ideal way of generating correct time-series data. However, with thousands of images having to be corrected it is usually necessary from a practical point of view to clean up the data using simple and rapid methods. One of these methods is maximum-value compositing, in which data over a short time-period (8–15 days) are scanned, and the maximum NDVI value retained to represent the time period (Holben 1986). The method has proven to be surprisingly effective in reducing noise in NDVI data, since cloud, atmospheric absorption, background color variations, etc., tend to lower the NDVI values. The result of 8-day maximum-value compositing applied to the coniferous MODIS data is seen in Fig. 7.1b.

A further way of managing noise is to use the quality flags, e.g. MODIS QA (MODIS Quality assessment), which are delivered with many remotely sensed products today, and which indicate the reliability of each observation. Though these flags are useful for removing doubtful data they are not easily applied in a more quantitative sense for improving the quality of the time-series.

Returning to Fig. 7.1, it can be seen from Fig. 7.1b that the maximum-value compositing has not been able to remove all the noise. Several observations of doubtful quality remain, and they have a clear negative bias. Thus, in most cases it is necessary to smooth the time-series data further using filters or other smoothing functions before extracting seasonality data. In doing so, the methods should take the negative noise bias into account, and should be able to handle missing data. Figure 7.1c shows the result of applying a smoothing function in the TIMESAT software package. This has resulted in a smooth curve that fits to the upper envelope of the data points. More information about data smoothing is given in the sections below.

Once a smooth data set has been generated it is possible to extract growing seasons and phenological parameters. Since vegetation indices are affected by a range of different processes (compare discussion above), their biophysical meaning is sometimes vague, and extraction of phenological parameters becomes somewhat subjective. Given this uncertainty it is not possible to define universal thresholds for defining the beginning and end of growing seasons. A further complication is that

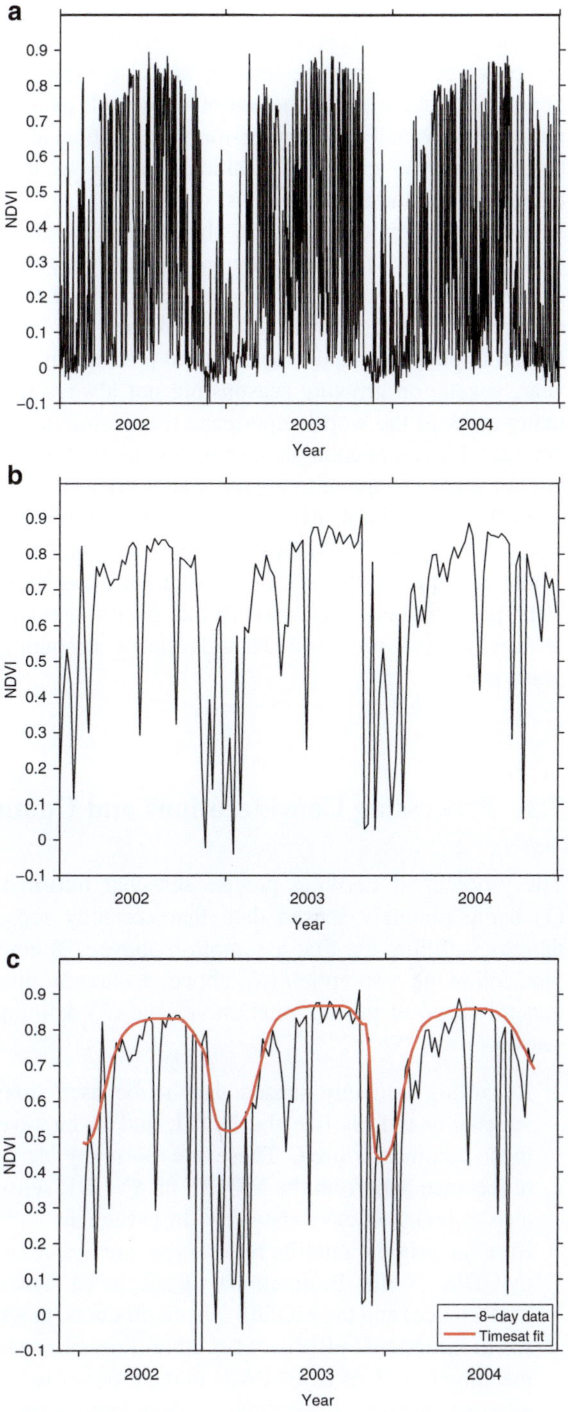

Fig. 7.1 (**a**) Daily NDVI data from MODIS for a mixed coniferous forest pixel in central Sweden; (**b**) 8-day maximum-value composite data for the same pixel; (**c**) smooth representation of the 8-day data from the same pixel using an upper-envelope weighted asymmetric Gaussian fit in TIMESAT

the time periods of most rapid shift in the sensed signal often coincide with meteorological changes. In tropical drylands, for example, the rainy season marks the onset of the growing season, but also brings cloudiness that affects signal quality; in cold areas the period of snow melt overlaps with the leaf development phase. Additionally, the understorey vegetation in many climate zones develops before the tree canopies, making it hard to use remotely sensed data for distinguishing the two processes.

Another complicating factor when mapping growing seasons is that, though the seasons normally follow an annual rhythm, they do not necessarily occur within single calendar years. In the Southern Hemisphere the growing season may begin in one year and end the year after. Although we are used to describing annually repeating phenomena, like agricultural production, with statistics for each calendar year, vegetation growing seasons are not always well suited to this. In addition, many areas of the world experience two (sometimes even three) growing seasons per year. Hence, phenological statistics should preferably not be reported per year, but per season (relative to a fixed starting date).

Last, but not least, it is necessary to consider the huge, and rapidly growing, storages of digital Earth Observation data available. For example, processing the whole of Africa at 250 $\times$ 250 m resolution using MODIS 8-day data for the 2000–2013 period means that roughly 523.6 billion points have to be analyzed; it is obviously necessary to use fast and reliable computing algorithms when estimating seasonality.

7.3 Processing Considerations and Common Methods

The problem of deriving precise seasonal information consists of three parts: (1) using remotely sensed data that correctly represent vegetation phenology, thereby fulfilling the first assumption above; (2) employing a smoothing method that, following assumption two above, accurately filters noise without altering the general shape of the seasonal curve; and (3) defining parameters of the growing season.

1. Regarding remotely sensed data to be used, maximum-value composites of vegetation indices like the NDVI, and in later years the EVI, have been the most commonly used. These are normally derived from top-of-atmosphere reflectance data from the MODIS or AVHRR sensors. However, there is reason also to focus on other data sets. In particular, higher-order products developed from the original satellite reflectances are important, such as the MODIS NBAR (MODIS Nadir Bidirectional Reflectance Distribution Function Adjusted Reflectance) and the MODIS albedo products, in which data have been corrected for bi-directional effects. Also other derived products with a clear biophysical meaning (e.g. LAI or FAPAR) make it easier to interpret the resulting seasonal parameters from a vegetation phenology point of view (provided that the

products accurately model these parameters). The development of new and improved biophysically relevant data sets is a highly active and relevant research field. For example, a recently developed plant phenology index (PPI), which is linearly related to green LAI, has strong potential for more accurately mapping of vegetation phenology than the traditionally used indices (Jin and Eklundh 2014).

2. A variety of smoothing methods have been developed and tested. Fourier series were among the first methods to be tested for extracting seasonality information from remotely sensed imagery (van Dijk et al. 1987; Menenti et al. 1993; Olsson and Eklundh 1994). The parameters of the harmonic functions contain useful information about the timing of seasons and the number of growing seasons per year. However, the method is inflexible when modeling individual years; Fourier series are better suited to data with less interannual variability than is often seen for remotely sensed time-series data. Another line of development is the use of various temporal filters for smoothing the time-series data. One early method was the best index slope extraction (BISE) (Viovy et al. 1992). In this method the upper envelope of the time-series is extracted by connecting the upper-most data points in a sliding window. The method is based on the principle of minimizing noise by consistently selecting the highest NDVI values; however in doing so it neglects the fact that also positive noise, e.g. due to angular effects, is present in the data. Also other smoothing filters have been used, e.g. the 4352H filter (van Dijk et al. 1987) and median filters (Reed et al. 1994). More recently various functions have been fitted to data: asymmetric Gaussian functions (Jönsson and Eklundh 2002), logistic functions (Zhang et al. 2003; Jönsson and Eklundh 2004; Fisher et al. 2006), and spline functions (Bradley et al. 2007; Hermance et al. 2007). Also wavelet transforms have been shown to be useful (Sakamoto et al. 2005; Lu et al. 2007; Campos and Di Bella 2012). In general, the choice of smoothing method is related to the type of input data and the desired result. If data are relatively smooth and the aim is to preserve variations on the seasonal curve, local filtering methods can be employed. If data are very noisy it might be necessary to enforce a general seasonal shape on the data by employing a more global type of function (e.g. asymmetric Gaussian or logistic function).
3. Regarding the extraction of phenological parameters, some different methods have been used. Most are based on absolute or relative thresholds of the seasonal amplitude. Others are purely mathematical parameters (inflexion points or derivatives of different order). Common for all these methods is that they seldom are based on any biological or physical understanding of the phenological process, but rather on empirical relationships. A more elaborate method, based on fitting shape models to smoothed data, yielded high fidelity for crop phenological parameters (Sakamoto et al. 2010). The choice of method cannot be separated from the type of input data or the fitting method used. For example, methods based on derivatives should not be used with data that are not very smooth.

It can be questioned whether it is possible to define a single set of smoothing and parameter extraction methods that will work across all different ecosystems and with all different types of remotely sensed data. White et al. (2009) made an extensive study including ten different methods for estimating spring phenology across the United States, concluding that the different methods did not behave consistently, and that, in their study, there was no rational basis for selecting one method over the other. Considerable inter-method differences were also documented by Cong et al. (2013). It is likely that bias and random errors due to cloud interference lead to temporally and spatially varying performance of different smoothing and filtering methods (Chen et al. 2013). Furthermore, phenological parameters are generally difficult to assess, as ground data have large variability and are often observed over small areas. Hence, it is in general very difficult to assess the reliability of processing methods; achieving smoothness is one thing, accurately depicting true vegetation variations is not necessarily the same. Intermediate-scale canopy data from phenocams and near-ground spectral sensors serve an important means of understanding and validating satellite-derived phenological parameters (Richardson et al. 2007; Eklundh et al. 2011; Hufkens et al. 2012).

7.4 The TIMESAT Approach

7.4.1 Processing Principles

TIMESAT has been developed with flexibility in mind, and is thus not oriented towards any specific data source or format. Hence, users are required to pre-process data before the actual TIMESAT processing can begin. Depending on the data source different preparation steps may be necessary, e.g. converting image data into the binary formats used in TIMESAT, organizing images in time stacks with equal time step, preparing lists of file names, and converting quality information into rank units that can be processed by TIMESAT. The actual TIMESAT processing consists of a series of steps: (1) computing the trend in the data using the Seasonal Trend decomposition by Loess (STL) method (Cleveland et al. 1990); (2) pre-filtering of data, in which extreme outliers and pixels with too few data points are removed; (3) computing a coarse seasonal fit to de-trended data based on sinusoidal harmonics to determine the number and approximate location of growing seasons; (4) smoothing the data using either of three different methods: adaptive Savitzky-Golay filter, asymmetric Gaussian or double logistic functions; (5) computing seasonal parameters for each extracted season; and (6) generating output data in the form of single-pixel data or images. The output includes smoothed data for each time step, and seasonality parameters for each identified growing season. The processing is controlled from a graphical user interface in which the necessary settings are determined based on visual control of sample time-series from the image data stack. When suitable settings have been determined, the full image data

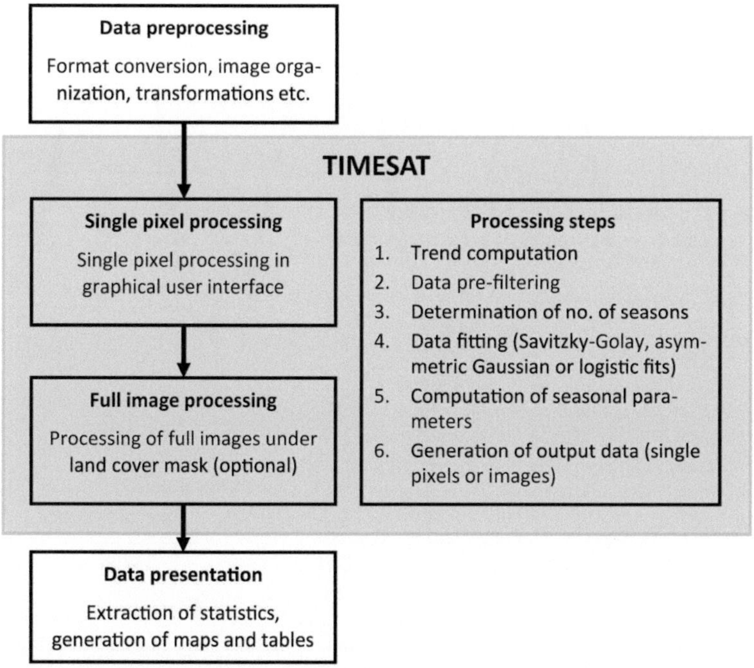

Fig. 7.2 Principle of TIMESAT work flow. The *white boxes* show tasks done outside of TIMESAT, whereas the *grey boxes* show functionality in TIMESAT. The processing steps 1–6 are done for single pixels as well as for full images, and are further described in the text

can be processed. A summary of the processing steps is presented in Fig. 7.2; we also refer to the TIMESAT manual for more detailed information (Eklundh and Jönsson 2012).

All data values to be processed have an associated weight which can be derived from product quality flag data or from STL. In the subsequent processing the weights can be modified if the user wishes to fit data to the upper envelope. This is done by reducing the weights of data points below the fitted functions, in up to a maximum of three iterations. All data fitting is done using weighted least squares, which means that short data gaps are handled without interpolation.

The first smoothing method implemented in TIMESAT was based on asymmetric Gaussian functions (Jönsson and Eklundh 2002). The method consists of seasonal functions fitted piecewise to the data and merged to a global continuous data series. Subsequently, double logistic functions and Savitzky-Golay filtering were added to TIMESAT (Jönsson and Eklundh 2004). An example of the results of running the three smoothing methods in TIMESAT are shown in Fig. 7.3. It can be seen that the Gaussian and logistic functions are very smooth and global in nature. They are most useful when data are very noisy and the user wishes to enforce a bell-shaped pattern on the data. The Gaussian functions adapt somewhat better than the logistic functions to flat peaks, otherwise the two methods are very similar. The Savitzky-Golay method, on the other hand, filters the data and follows local

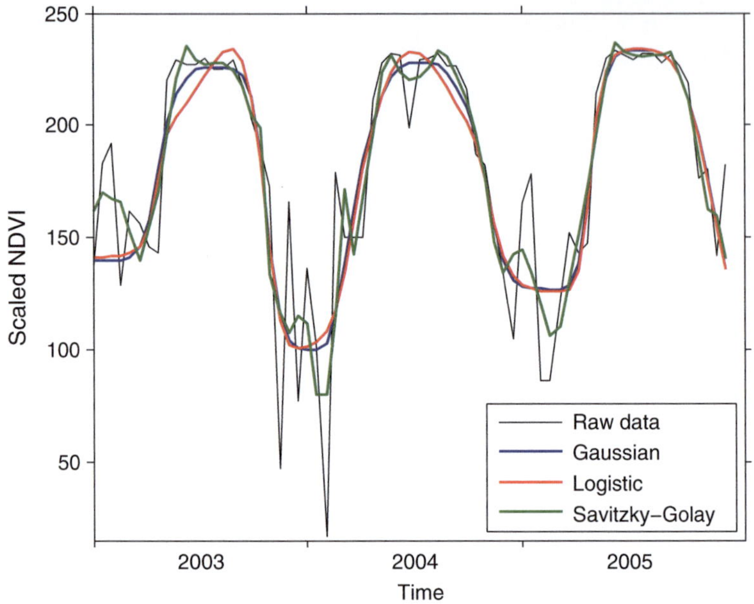

Fig. 7.3 TIMESAT fits to MODIS 16-day NDVI data from a deciduous forest in S. Sweden

variations in the seasonal curve more closely. The Savitzky-Golay implementation in TIMESAT is adaptive in that it iteratively tightens the search window in order to capture very rapid increase or decrease in the data. This is useful when monitoring e.g. semi-arid grasslands, where the ground can green-up in the course of a few days, leading to a very rapid increase in vegetation index data. Smoothing very noisy data requires an increased search window, which in turn can produce some artefacts. Therefore, Savitzky-Golay filtering is best used with data that is not extremely noisy.

Hird and McDermid (2009) showed that the methods in TIMESAT have good performance, balancing the ability to reduce noise and maintain the signal integrity.

Several of the methods in TIMESAT require the user to make individual settings, e.g. controlling the degree of smoothing or the envelope fitting. In small areas it might be enough to do this once, but for large areas with diverse land cover it might be necessary to define different settings for different areas. In order to maintain flexibility it is possible to store several groups of settings, and then apply these to different areas in the image, controlled by e.g. a land cover map.

After smoothing the data, TIMESAT proceeds to compute phenological parameters. The user determines thresholds for defining the start and end of seasons (absolute values or fractions of the amplitude), and the following parameters are then computed for each season: times of start and end of season; length of season; base level; time of midpoint; maximum value; amplitude; rates of increase and decrease; and large and small integrals. Examples of two phenological parameters, mapped for West Africa from AVHRR data, are shown in Fig. 7.4. The definition

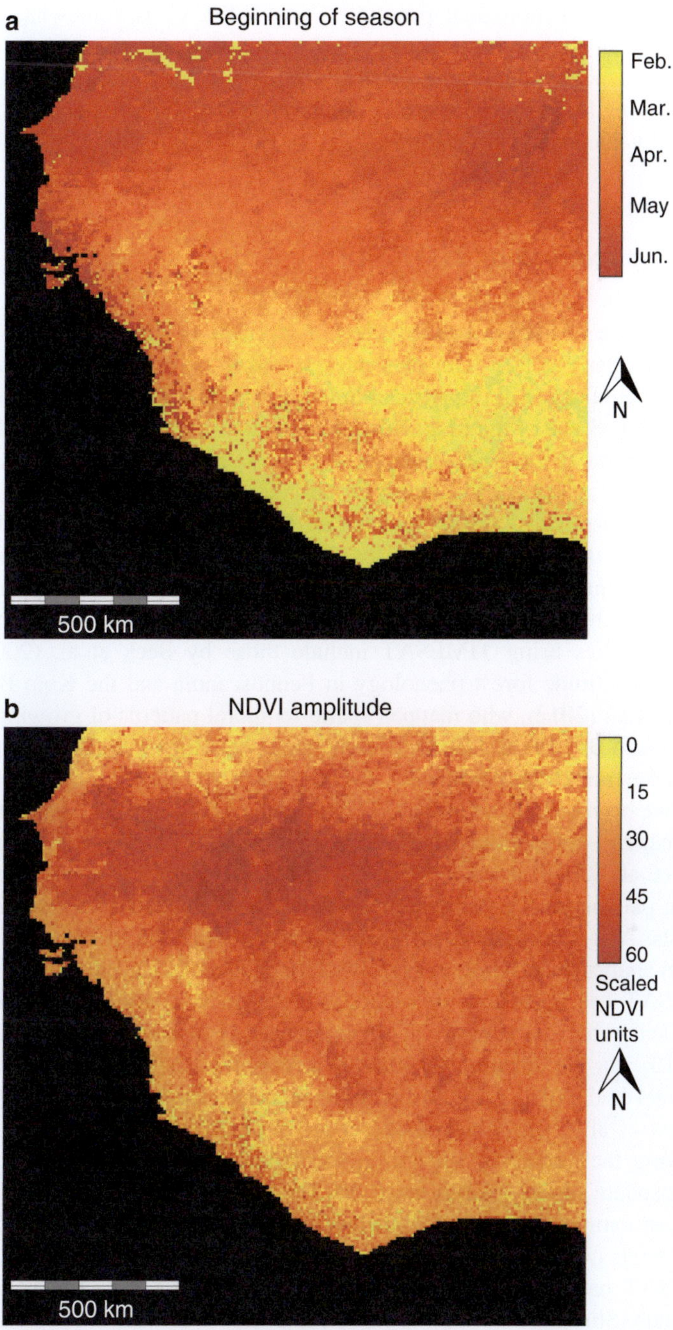

Fig. 7.4 TIMESAT seasonality parameters computed from a Gaussian-smoothed NOAA AVHRR data set over West Africa 1999; (**a**) start of season; (**b**) amplitude of season

and selection of phenological parameters in TIMESAT is somewhat arbitrary. Though several studies have shown that many of them make sense from an ecosystem perspective and are empirically related to inter-seasonal variations in climatic driving forces, more research is clearly needed to more precisely establish their actual value and ecological meaning.

7.4.2 Applications of TIMESAT

TIMESAT has been used in a wide variety of applications since the first version was written in the early 2000's. Our own interest was initially focused on mapping of environmental changes in the African Sahel using the AVHRR data records from 1982 till today. Some of the first evidence of the increasing greenness in the Sahel, from the droughts in the 1980's, was presented by Eklundh and Olsson (2003); this increase was subsequently linked to variations in climate drivers (Hickler et al. 2005; Olsson et al. 2005; Seaquist et al. 2009). In these studies TIMESAT was primarily used for computing seasonal amplitudes and integrals of NDVI. However, Heumann et al. (2007) also studied the changes in other phenological parameters in the Sahel, like the start and end of the growing seasons. Other phenology studies using TIMESAT include those by Beck et al. (2007), who mapped high-latitude forest phenology in Fennoscandia and the Kola Peninsula, O'Connor et al. (2012), who mapped spatio-temporal patterns of growing seasons on Ireland, and Boyd et al. (2011), who mapped phenology in S. England using the MERIS terrestrial chlorophyll index (MTCI; Dash and Curran 2007). Other case studies have been conducted in the US (Zhao et al. 2013), Europe (Han et al. 2013), South America (van Leeuwen et al. 2013), and in Arctic areas (Zeng et al. 2013). Jönsson et al. (2010) used TIMESAT while demonstrating the difficulties in extracting phenological parameters from MODIS NDVI data over boreal coniferous forests. Also disturbances in phenological patterns due to insect infestations have been analyzed (Eklundh et al. 2009; Olsson et al. 2012; Buma et al. 2013).

TIMESAT has been used in several studies on vegetation classification and phenological characterization of ecosystems (Tottrup et al. 2007; Clark et al. 2010; van Leeuwen et al. 2010; Wessels et al. 2011; Zhang et al. 2013a; Leinenkugel et al. 2013). It has furthermore been used for fire and fire risk modeling (Verbesselt et al. 2006; Veraverbeke et al. 2010; Le Page et al. 2010), and for investigating the impact of vegetation variability on predictability of a coupled land-atmosphere model (Weiss et al. 2012). In agriculture, TIMESAT has been used for estimation of sow dates (Lobell et al. 2013) and for mapping of abandoned cropping fields (Alcantara et al. 2012).

TIMESAT has been used for estimating diurnal air temperature from MSG SEVIRI data (Stisen et al. 2007), to study expansion of the thermal growing season and associated change in the biospheric carbon uptake (Barichivich et al. 2012), and to study the impact of extreme precipitation on reduction of terrestrial ecosystem production (Zhang et al. 2013b).

An important application field of TIMESAT is data smoothing to improve signal quality: Improved MODIS data quality has, when calibrating models with eddy-covariance flux tower data and other environmental data, led to generally better possibilities to estimate carbon fluxes (Olofsson and Eklundh 2007; Olofsson et al. 2007, 2008; Sjöström et al. 2009, 2011; Schubert et al. 2010, 2012; Tang et al. 2013). TIMESAT has also been used for data quality improvement with MODIS and AVHRR satellite products (Fensholt and Proud 2012), and for smoothing of GIMMS NDVI(3G) data for high northern latitudes (Barichivich et al. 2013). Data quality improvement is also the reason for using TIMESAT in an improved reprocessed version of the global MODIS LAI data set for land surface and climate modeling (Yuan et al. 2011).

7.5 Future Perspectives

We currently have over three decades of global AVHRR data, and over one decade of MODIS data from the Terra and Aqua satellites available. New satellites will continue to extend these time series into the future. As the data records grow, using them for studying impacts of climate and human action on the environment will be possible with increased confidence. This will increase the demand for the data, and call for further improving the methods for time-series data management and for exploiting the data, e.g. to extract linear and non-linear trends (Verbesselt et al. 2010; Jamali et al. 2014, 2015).

Earth observation is now taking an important step into a new era, with growing archives of time-series data at high spatial resolution. The release of the Landsat archive into the open domain has opened up for a range of new applications (Wulder et al. 2012); several new methods for exploiting these data, particularly for forest monitoring, are being developed (e.g. Huang et al. 2010; Kennedy et al. 2010; Zhu et al. 2012).

The next leap will be taken with the ESA Sentinel-2 satellites, to be launched in 2015 and 2017, generating Earth observation data at 10 m resolution with a 5-day interval. This will present both enormous opportunities and challenges. First, the high spatial resolution will mean that data validation against field measurements will be much improved compared with the 250–1,000 m data presently used. At this high resolution it will be possible to monitor vegetation at the scale of individual forest stands rather than at the ecosystem scale. Second, the high time resolution will mean that it will be possible to model seasonality more accurately than is possible with Landsat, SPOT or the other existing high-resolution sensors. The nature of data will present many new challenges, such as irregular time steps; hence new methods for gap-filling, smoothing and data fusion will have to be explored. Modification of TIMESAT to enable analysis of high-resolution data from Sentinel-2 is ongoing (Eklundh et al. 2012). Third, the new satellites will generate enormous volumes of data, calling for high-performance computing methods for processing all the data. A version of TIMESAT for parallel computing has been developed, showing almost linear scaling with the number of processors.

A further line of development is the integration of spatial and temporal dimensions. We have previously seen that incorporating the spatial domain will increase the significance in estimation of trend parameters across time (Bolin et al. 2009). It is likely that noise in time-series data can be reduced when estimating seasonal trajectories by extending the analysis into the spatial domain. Hence, we are currently exploring spline based methods in TIMESAT that can smooth the data across both time and space (Eklundh and Jönsson 2013).

Remote sensing science has come a long way towards extraction of environmentally meaningful time-series data during the last 10–20 years. With the new data types being released, and new and efficient processing methods being developed, Earth observation is now being accepted as an established and accurate tool for analyzing the Earth and its changes.

TIMESAT can freely be downloaded from http://www.nateko.lu.se/TIMESAT.

References

Alcantara C, Kuemmerle T, Prishchepov AV, Radeloff VC (2012) Mapping abandoned agriculture with multi-temporal MODIS satellite data. Remote Sens Environ 124:334–347

Barichivich J, Briffa KR, Osborn TJ, Melvin TM, Caesar J (2012) Thermal growing season and timing of biospheric carbon uptake across the Northern Hemisphere. Glob Biogeochem Cycles 26:GB4015

Barichivich J, Briffa KR, Myneni RB, Osborn TJ, Melvin TM, Ciais P, Piao S, Tucker C (2013) Large-scale variations in the vegetation growing season and annual cycle of atmospheric CO_2 at high northern latitudes from 1950 to 2011. Glob Chang Biol 19:3167–3183

Beck PSA, Jönsson P, Hogda KA, Karlsen SR, Eklundh L, Skidmore AK (2007) A ground-validated NDVI dataset for monitoring vegetation dynamics and mapping phenology in Fennoscandia and the Kola peninsula. Int J Remote Sens 28:4311–4330

Bolin D, Lindström J, Eklundh L, Lindgren F (2009) Fast estimation of spatially dependent temporal vegetation trends using Gaussian Markov random fields. Comput Stat Data Anal 53:2885–2896

Boyd DS, Almond S, Dash J, Curran PJ, Hill RA (2011) Phenology of vegetation in Southern England from Envisat MERIS terrestrial chlorophyll index (MTCI) data. Int J Remote Sens 32:8421–8447

Bradley BA, Jacob RW, Hermance JF, Mustard JF (2007) A curve fitting procedure to derive inter-annual phenologies from time series of noisy satellite NDVI data. Remote Sens Environ 106:137–145

Buma B, Pugh ET, Wessman CA (2013) Effect of the current major insect outbreaks on decadal phenological and LAI trends in southern Rocky Mountain forests. Int J Remote Sens 34:7249–7274

Campbell PKE, Middleton EM, Corp LA, Kim MS (2008) Contribution of chlorophyll fluorescence to the apparent vegetation reflectance. Sci Total Environ 404:433–439

Campos AN, Di Bella CM (2012) Multi-temporal analysis of remotely sensed information using wavelets. J Geogr Inf Syst 4:383–391

Chen W, Foy N, Olthof I, Latifovic R, Zhang Y, Li J, Fraser R, Chen Z, McLennan D, Poitevin J, Zorn P, Quirouette J, Stewart HM (2013) Evaluating and reducing errors in seasonal profiles of

AVHRR vegetation indices over a Canadian northern national park using a cloudiness index. Int J Remote Sens 34:4320–44343
Clark ML, Aide TM, Grau HR, Riner G (2010) A scalable approach to mapping annual land cover at 250 m using MODIS time series data: a case study in the Dry chaco ecoregion of South America. Remote Sens Environ 114:2816–2832
Cleveland RB, Cleveland WS, McRae JE, Terpenning I (1990) STL: a seasonal-trend decomposition procedure based on loess. J Off Stat 6:3–73
Cong N, Wang T, Nan H, Ma Y, Wang X, Myneni RB, Piao S (2013) Changes in satellite-derived spring vegetation green-up date and its linkage to climate in China from 1982 to 2010: a multimethod analysis. Glob Chang Biol 19:881–891
Dash J, Curran PJ (2007) Evaluation of the MERIS terrestrial chlorophyll index (MTCI). Adv Space Res 39:100–1104
Defries RS, Townshend JRG (1994) NDVI-derived land cover classifications at a global scale. Int J Remote Sens 15:3567–3586
Eklundh L, Jönsson P (2012) TIMESAT 3.1 software manual. Lund University, Lund
Eklundh L, Jönsson P (2013) A new spatio-temporal smoother for extracting vegetation seasonality with TIMESAT. The 35th international symposium on remote sensing of environment, 22–26 Apr 2013, Beijing, China
Eklundh L, Olsson L (2003) Vegetation index trends for the African Sahel 1982–1999. Geophys Res Lett 30:1430–1433
Eklundh L, Johansson T, Solberg S (2009) Mapping insect defoliation in Scots pine with MODIS time-series data. Remote Sens Environ 113:1566–1573
Eklundh L, Jin H, Schubert P, Guzinski R, Heliasz M (2011) An optical sensor network for vegetation phenology monitoring and satellite data calibration. Sensors 11:7678–7709
Eklundh L, Sjöström M, Ardö J, Jönsson P (2012) High resolution mapping of vegetation dynamics from Sentinel-2. In: Proceedings of the first Sentinel-2 preparatory symposium, 23–27 Apr 2012, Frascati, Italy
Fensholt R, Proud SR (2012) Evaluation of earth observation based global long term vegetation trends – comparing GIMMS and MODIS global NDVI time series. Remote Sens Environ 119:131–147
Fisher JI, Mustard JF, Vadeboncoeur MA (2006) Green leaf phenology at Landsat resolution: scaling from the field to the satellite. Remote Sens Environ 100:265–279
Gamon JA, Serrano L, Surfus JS (1997) The photochemical reflectance index: an optical indicator of photosynthetic radiation-use efficiency across species, functional types, and nutrient levels. Oecologia 112:492–501
Han Q, Luo G, Li C (2013) Remote sensing-based quantification of spatial variation in canopy phenology of four dominant tree species in Europe. J Appl Remote Sens 7:073485
Hermance JF, Jacob RW, Bradley BA, Mustard JF (2007) Extracting phenological signals from multiyear AVHRR NDVI time series: framework for applying high-order annual splines with roughness damping. IEEE Trans Geosci Remote Sens 45:3264–3276
Heumann BW, Seaquist JW, Eklundh L, Jönsson P (2007) AVHRR derived phenological change in the Sahel and Soudan, Africa, 1982–2005. Remote Sens Environ 108:385–392
Hickler T, Eklundh L, Seaquist J, Smith B, Ardö J, Olsson L, Sykes M, Sjöström M (2005) Precipitation controls Sahel greening trend. Geophys Res Lett 32:L21415
Hird JN, McDermid GJ (2009) Noise reduction of NDVI time series: an empirical comparison of selected techniques. Remote Sens Environ 113:248–258
Holben BN (1986) Characteristics of maximum-value composite images from temporal AVHRR data. Int J Remote Sens 7:1417–1443
Huang C, Goward SN, Masek JG, Thomas N, Zhu Z, Vogelmann JE (2010) An automated approach for reconstructing recent forest disturbance history using dense Landsat time series stacks. Remote Sens Environ 114:183–198

Huete A, Didan K, Miura T, Rodriguez EP, Gao X, Ferreira LG (2002) Overview of the radiometric and biophysical performance of the MODIS vegetation indices. Remote Sens Environ 83:195–213
Hufkens K, Friedl M, Sonnentag O, Braswell BH, Milliman T, Richardson AD (2012) Linking near-surface and satellite remote sensing measurements of deciduous broadleaf forest phenology. Remote Sens Environ 117:307–3321
Jamali S, Seaquist J, Eklundh L, Ardö J (2014) Automated mapping of vegetation trends with polynomials using NDVI imagery over the Sahel. Remote Sens Environ 141:79–89
Jamali S, Eklundh L, Jönsson P, Seaquist J, Ardö J (2015) Detecting changes in vegetation trends using time series segmentation. Remote Sens Environ 156:182–195
James ME, Kalluri SNV (1994) The pathfinder AVHRR land data set: an improved coarse resolution data set for terrestrial monitoring. Int J Remote Sens 15:3347–3363
Jiang N, Zhu W, Zheng Z, Chen G, Fan D (2013) A comparative analysis between GIMSS NDVIg and NDVI3g for monitoring vegetation activity change in the Northern Hemisphere during 1982–2008. Remote Sens 5:4031–4044
Jin H, Eklundh L (2014) A physically based vegetation index for improved monitoring of plant phenology. Remote Sens Environ 152:512–525
Jönsson P, Eklundh L (2002) Seasonality extraction by function fitting to time-series of satellite sensor data. IEEE Trans Geosci Remote Sens 40:1824–1832
Jönsson P, Eklundh L (2004) TIMESAT – a program for analysing time-series of satellite sensor data. Comput Geosci 30:833–845
Jönsson AM, Eklundh L, Hellström M, Bärring L, Jönsson P (2010) Annual changes in MODIS vegetation indices of Swedish coniferous forests in relation to snow dynamics and tree phenology. Remote Sens Environ 114:2719–2730
Justice CO, Townshend JRG, Holben BN, Tucker CJ (1985) Analysis of the phenology of global vegetation using meteorological satellite data. Int J Remote Sens 6:1271–1318
Keenan TF, Gray J, Friedl MA, Toomey M, Bohrer G, Hollinger DY, Munger JW, O'Keefe J, Schmid HP, Wing IS, Yang B, Richardson AD (2014) Net carbon uptake has increased through warming-induced changes in temperate forest phenology. Nat Clim Chang 4:598–604
Kennedy RE, Yang Z, Cohen WB (2010) Detecting trends in forest disturbance and recovery using yearly Landsat time series: 1. LandTrendr – temporal segmentation algorithms. Remote Sens Environ 114:2897–2910
Le Page Y, Oom D, Silva JMN, Jönsson P, Pereira JMC (2010) Seasonality of vegetation fires as modified by human action: observing the deviation from eco-climatic fire regimes. Glob Ecol Biogeogr 19:575–588
Leinenkugel P, Kuenzer C, Oppelt N, Dech S (2013) Characterisation of land surface phenology and land cover based on moderate resolution satellite data in cloud prone areas – a novel product for the Mekong basin. Remote Sens Environ 136:180–198
Lobell DB, Ortiz-Monasterio JI, Sibley AM, Sohu VS (2013) Satellite detection of earlier wheat sowing in India and implications for yield trends. Agric Syst 115:137–143
Lu X, Liu R, Liu J, Liang S (2007) Removal of noise by wavelet method to generate high quality temporal data of terrestrial MODIS products. Photogramm Eng Remote Sens 73:1129–1139
Menenti M, Azzali S, Verhoef W, van Swol R (1993) Mapping agroecological zones and time lag in vegetation growth by means of Fourier analysis of time series of NDVI images. Adv Space Res 13:233–237
O'Connor B, Dwyer E, Cawkwell F, Eklundh L (2012) Spatio-temporal patterns in vegetation start of season across the island of Ireland using the MERIS Global Vegetation Index. ISPRS J Photogramm Remote Sens 68:79–94
Olofsson P, Eklundh L (2007) Estimation of absorbed PAR across Scandinavia from satellite measurements. Part II: modeling and evaluating the fractional absorption. Remote Sens Environ 110:240–251

Olofsson P, Eklundh L, Lagergren F, Jönsson P, Lindroth A (2007) Estimating net primary production for Scandinavian forests using data from Terra/MODIS. Adv Space Res 39:125–130

Olofsson P, Lagergren F, Lindroth A, Lindström J, Klemedtsson L, Kutsch W, Eklundh L (2008) Towards operational remote sensing of forest carbon balance across northern Europe. Biogeosciences 5:817–832

Olsson L, Eklundh L (1994) Fourier transformation for analysis of temporal sequences of satellite imagery. Int J Remote Sens 15:3735–3741

Olsson L, Eklundh L, Ardö J (2005) A recent greening of the Sahel – trends, patterns and potential causes. J Arid Environ 63:556–566

Olsson PO, Jönsson AM, Eklundh L (2012) A new invasive insect in Sweden – physokermes inopinatus – tracing forest damage with satellite based remote sensing. For Ecol Manag 285:29–37

Reed BC, Brown JF, VanderZee D, Loveland TR, Merchant JW, Ohlen DO (1994) Measuring phenological variability from satellite imagery. J Veg Sci 5:703–714

Richardson AD, Jenkins JP, Braswell BH, Hollinger DY, Ollinger SV, Smith ML (2007) Use of digital webcam images to track spring green-up in a deciduous broadleaf forest. Oecologia 152:323–334

Running SW, Loveland TR, Pierce LL (1994) A vegetation classification logic based on remote sensing for use in global biogeochemical models. Ambio 23:77–81

Sakamoto T, Yokozawa M, Toritani H, Shibayama M, Ishitsuka N, Ohno H (2005) A crop phenology detection method using time-series MODIS data. Remote Sens Environ 96:366–374

Sakamoto T, Wardlow BD, Gitelson AA, Verma SB, Suyker AE, Arkebauer TJ (2010) A two-step filtering approach for detecting maize and soybean phenology with time-series MODIS data. Remote Sens Environ 114:2146–2159

Schubert P, Eklundh L, Lund M, Nilsson M (2010) Estimating northern peatland CO_2 exchange from MODIS time series data. Remote Sens Environ 114:1178–1189

Schubert P, Lagergren F, Aurela M, Christensen T, Grelle A, Heliasz M, Klemedtsson L, Lindroth A, Pilegaard K, Vesala T, Eklundh L (2012) Modeling GPP in the Nordic forest landscape with MODIS time series data – comparison with the MODIS GPP product. Remote Sens Environ 126:136–147

Seaquist JW, Hickler T, Eklundh L, Ardö J, Heumann B (2009) Disentangling the effects of climate and people on Sahel vegetation dynamics. Biogeosciences 6:469–477

Sjöström M, Ardö J, Eklundh L, El-Tahir BA, El-Khidir HAM, Hellström M, Pilesjö P, Seaquist J (2009) Evaluation of satellite based indices for gross primary production estimates in a sparse savanna in the Sudan. Biogeosciences 6:129–138

Sjöström M, Ardö J, Arneth A, Cappelaere B, Eklundh L, de Grandcourt A, Kutsch WL, Merbold L, Nouvellon Y, Scholes B, Seaquist J, Veenendaal EM (2011) Exploring the potential of MODIS EVI for modeling gross primary production across African ecosystems. Remote Sens Environ 115:1081–1089

Stisen S, Sandholt I, Norgaard A, Fensholt R, Eklundh L (2007) Estimation of diurnal air temperature using MSG SEVIRI data in West Africa. Remote Sens Environ 110:262–274

Tang XG, Wang X, Wang ZM, Liu DW, Jia MM, Dong ZY, Xie J, Ding Z, Wang HR, Liu XP (2013) Influence of vegetation phenology on modelling carbon fluxes in temperate deciduous forest by exclusive use of MODIS time-series data. Int J Remote Sens 34:8373–8392

Tottrup C, Schultz Rasmussen M, Eklundh L, Jönsson P (2007) Mapping fractional forest cover across the highlands of mainland Southeast Asia using MODIS data and regression tree modelling. Int J Remote Sens 28:23–46

Townshend JRG, Justice CO (1986) Analysis of the dynamics of African vegetation using the normalized difference vegetation index. Int J Remote Sens 7:1435–1445

Tucker CJ, Pinzon JE, Brown ME, Slayback D, Pak EW, Mahoney R, Vermote E, El Saleous N (2005) An extended AVHRR 8-km NDVI data set compatible with MODIS and SPOT vegetation NDVI data. Int J Remote Sens 26:4485–5598 van Dijk A, Callis SL, Sakamoto CM, Decker WL (1987) Smoothing vegetation index profiles: an alternative method for reducing radiometric disturbance in NOAA/AVHRR data. Photogramm Eng Remote Sens 53:1059–1067

van Leeuwen WJD, Davison JE, Casady GM, Marsh SE (2010) Phenological characterization of desert sky island vegetation communities with remotely sensed and climate time series data. Remote Sens 2:388–415

van Leeuwen WJD, Hartfield K, Miranda M, Meza FJ (2013) Trends and ENSO/AAO driven variability in NDVI derived productivity and phenology alongside the Andes mountains. Remote Sens 5:1177–1203

Veraverbeke S, Lhermitte S, Verstraeten WW, Goossens R (2010) The temporal dimension of differenced Normalized Burn Ratio (dNBR) fire/burn severity studies: the case of the large 2007 Peloponnese wildfires in Greece. Remote Sens Environ 114:2548–2563

Verbesselt J, Jönsson P, Lhermitte S, van Aardt J, Coppin P (2006) Evaluating satellite and climate data-derived indices as fire risk indicators in savannah ecosystems. IEEE Trans Geosci Remote Sens 44:1622–1632

Verbesselt J, Hyndman R, Newnham G, Culvenor D (2010) Detecting trend and seasonal changes in satellite image time series. Remote Sens Environ 114:106–115

Viovy N, Arino O, Belward AS (1992) The Best Index Slope Extraction (BISE): a method for reducing noise in NDVI time-series. Int J Remote Sens 13:1585–1590

Weiss M, Hurk B, Haarsma R, Hazeleger W (2012) Impact of vegetation variability on potential predictability and skill of EC-Earth simulations. Clim Dyn 39:2733–2746

Wessels K, Steenkamp K, von Maltitz G, Archibald S (2011) Remotely sensed vegetation phenology for describing and predicting the biomes of South Africa. Appl Veg Sci 14:49–66

White MA, De Beurs KM, Didan K, Inouye DW, Richardson AD, Jensen OP, O'keefe J, Zhang G, Nemani RR, Van Leeuwen WJD, Brown JF, De WITA, Schaepman M, Lin X, Dettinger M, Bailey AS, Kimball J, Schwartz MD, Baldocchi DD, Lee JT, Lauenroth WK (2009) Intercomparison, interpretation, and assessment of spring phenology in North America estimated from remote sensing for 1982–2006. Glob Chang Biol 15:2335–2359

Wulder MA, Masek JG, Cohen WB, Loveland TR, Woodcock CE (2012) Opening the archive: how free data has enabled the science and monitoring promise of Landsat. Remote Sens Environ 122:2–10

Yuan H, Dai Y, Xiao Z, Ji D, Shangguan W (2011) Reprocessing the MODIS Leaf Area Index products for land surface and climate modelling. Remote Sens Environ 115:1171–1187

Zeng H, Jia G, Forbes BC (2013) Shifts in Arctic phenology in response to climate and anthropogenic factors as detected from multiple satellite time series. Environ Res Lett 8:035036

Zhang X, Friedl MA, Schaaf CB, Strahler AH, Hodges JCF, Gao F, Reed BC, Huete A (2003) Monitoring vegetation phenology using MODIS. Remote Sens Environ 84:471–475

Zhang MQ, Guo HQ, Xie X, Zhang TT, Ouyang ZT, Zhao B (2013a) Identification of land-cover characteristics using MODIS time series data: an application in the Yangtze river estuary. PLoS One 8:e70079

Zhang Y, Moran MS, Nearing MA, Ponce Campos GE, Huete AR, Buda AR, Bosch DD, Gunter SA, Kitchen SG, McNab WH, Morgan JA, McClaran MP, Montoya DS, Peters DPC, Starks PJ (2013b) Extreme precipitation patterns and reductions of terrestrial ecosystem production across biomes. J Geophys Res Biogeosci 118:2169–8961

Zhao J, Wang Y, Hashimoto H, Melton FS, Hiatt SH, Zhang H, Nemani RR (2013) The variation of land surface phenology from 1982 to 2006 along the Appalachian trail. IEEE Trans Geosci Remote Sens 51:2087–2095

Zhu Z, Woodcock CE, Olofsson P (2012) Continuous monitoring of forest disturbance using all available Landsat imagery. Remote Sens Environ 122:75–91

Chapter 8
Assessment of Vegetation Trends in Drylands from Time Series of Earth Observation Data

Rasmus Fensholt, Stephanie Horion, Torbern Tagesson, Andrea Ehammer, Kenneth Grogan, Feng Tian, Silvia Huber, Jan Verbesselt, Stephen D. Prince, Compton J. Tucker, and Kjeld Rasmussen

Abstract This chapter summarizes approaches to the detection of dryland vegetation change and methods for observing spatio-temporal trends from space. An overview of suitable long-term Earth Observation (EO) based datasets for assessment of global dryland vegetation trends is provided and a status map of contemporary greening and browning trends for global drylands is presented. The vegetation metrics suitable for per-pixel temporal trend analysis is discussed, including seasonal parameterisation and the appropriate choice of trend indicators. Recent methods designed to overcome assumptions of long-term linearity in time series analysis (Breaks For Additive Season and Trend(BFAST)) are discussed. Finally, the importance of the spatial scale when performing temporal trend analysis is introduced and a method for image downscaling (Spatial and Temporal Adaptive Reflectance Fusion Model (STARFM)) is presented.

R. Fensholt (✉) • S. Horion • T. Tagesson • A. Ehammer
K. Grogan • F. Tian • K. Rasmussen
Section of Geography, Department of Geosciences
and Natural Resource Management, Faculty of Science,
University of Copenhagen, Copenhagen K, Denmark
e-mail: rf@ign.ku.dk

S. Huber
DHI GRAS, Copenhagen K, Denmark

J. Verbesselt
Laboratory of Geo-Information Science and Remote Sensing,
Wageningen University, Wageningen, The Netherlands

S.D. Prince
Department of Geographical Sciences, University of Maryland,
College Park, MD, USA

C.J. Tucker
NASA Goddard Space Flight Center, Greenbelt, MD, USA

C. Kuenzer et al. (eds.), *Remote Sensing Time Series*, Remote Sensing and Digital Image Processing, DOI 10.1007/978-3-319-15967-6_8

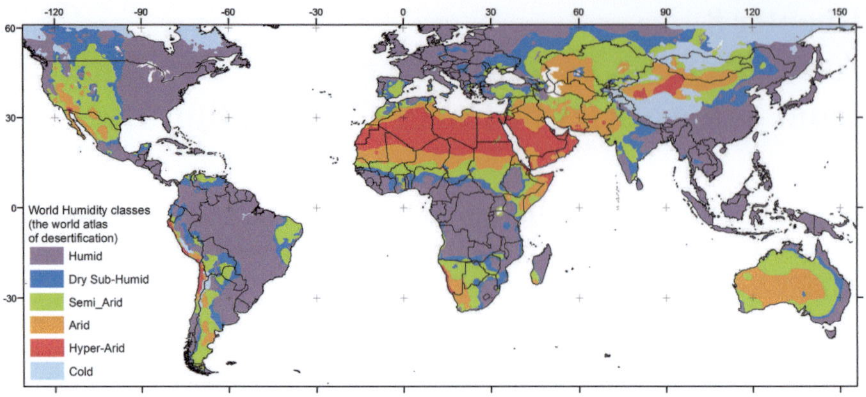

Fig. 8.1 World humidity classes (World Atlas of Desertification, United Nations Environment Programme 1997). Drylands are classified into four ranges of the Aridity Index (AI, the ratio between the mean annual precipitation and the mean annual potential evaporation): Hyperarid, AI < 0.05 (7.5 % of global land area); Arid, 0.05 < AI < 0.20 (12.1 % land); Semi-Arid, 0.20 < AI < 0.50 (17.7 % land); Dry subhumid, 0.50 < AI < 0.65 (9.9 % land)

8.1 Introduction

Dryland areas, defined as those areas of the world where water is an important limitation for plant growth (Fig. 8.1), have become the subject of increased interest due to the impacts of current global changes and concern for the sustainability of human lifestyles. Dryland ecosystems provide vital services, such as food, grazing, as well as energy and forestry products. Following the severe drought in the early 1970s in the Sahel region, the UN spurred an intensive interest in the issue of dryland degradation/desertification, most prominently marked by the UN Conference on Desertification in 1977, followed by the UN Convention to Combat Desertification (emerging from the Rio-conference in 1992). This fuelled a significant increase in the scientific interest to provide an improved understanding of both climatic and anthropogenic factors involved in the dynamics of drylands in general and in the Sahel in particular – a region from where several of the examples in the following two chapters are drawn. This chapter considers the concepts of long-term EO-based vegetation monitoring and discusses some of the most widely used EO (Earth Observation) datasets and methods for studying dryland vegetation trends.

8.2 Detection of Dryland Vegetation Change from Space

8.2.1 Towards EO-Based Monitoring of Vegetation Productivity

Vegetation indices (VI) measure the differential absorption and scattering by green leaf material of solar radiation in the visible waveband (0.4–0.7 μm, although only red wavelengths and near infrared wavebands (0.8–1 μm) are generally used in VIs). This difference is attributable to the absorption of visible radiation by chlorophyll and its accessory pigments while near infrared, although slightly absorbed, is strongly scattered by the complex air-water interfaces between cells within a leave (Gates et al. 1965). Thus the more green leaves in the field of view, the less visible and the more near infrared radiation are measured by a space borne sensor. The most widely used VI is the Normalized Difference Vegetation Index (NDVI):

$$NDVI = \frac{NIR - RED}{NIR + RED} \tag{8.1}$$

where RED and NIR are the surface reflectances in the red and near infrared, respectively (Tucker 1979).

Several factors other than leaf absorption and scattering also affect red and near infrared reflectance. These include factors related to the canopy/soil surface itself such as canopy structure, leaf angle distribution, non-green stems, branches and dead plant material (both attached and on the soil surface), the soil surface and also to factors related to sun-target-sensor geometric configurations governed by changes in solar zenith angle and sensor view angle. Different soils and litter reflect differently in the visible and near infrared, so the reflectance of sparse vegetation can be significantly affected by the non-green components (Huete et al. 1985). Because of the central role of vegetation cover as an indicator of the land surface condition in drylands, time series of vegetation indices are the main tool for trend detection using remote sensing (e.g. Prince and Goward 1995; Tucker et al. 1991).

Net primary production (NPP) is defined as the net carbon gain by plants and is the process whereby vegetation biomass is produced. Traditionally, Earth Observation based production efficiency modelling (PEM) has been based on the light use efficiency (LUE) concept originally proposed by (Monteith 1972), who considered biomass accumulation (g/m^2) as an ongoing process correlated with the amount of photosynthetically active radiation (PAR) absorbed or intercepted by green foliage (APAR in MJ/m^2). NPP depends on, among other factors, the fraction of photosynthetically active radiation (FPAR) absorbed by the canopy, for which VIs can be used as proxies (Myneni et al. 1995; Sellers 1985). Measurements and modelling have shown that NDVI is linearly related to FPAR (e.g. Asrar et al. 1984); however the relation is found to be biome specific (Fensholt et al. 2004; Goward and Huemmrich 1992). While several other vegetation indices have been proposed, to date, the only VI available globally for a long period (1981 to present) is the NDVI (Barreto-Munoz 2013; Pinzon and Tucker 2014). Properties of vegetation, in addition to FPAR that have been estimated from NDVI or related indices include Leaf Area Index (LAI), chlorophyll concentration, above-ground biomass, vegetation productivity, fractional vegetation cover, vegetation water content, and leaf nitrogen (Thenkabail et al. 2012). The properties with a close biophysical relationship to the absorption and scattering of solar radiation in the visible and near infrared (such as FAPAR, LAI and chlorophyll content) are generally better estimated than more indirectly related properties such as total biomass which requires additional data input not presently available from long-term EO time series data (Fensholt et al. 2006).

A single observation of NDVI cannot be used to retrieve quantitative information of leaf biomass, productivity or phenology. However, for annual crops and herbaceous vegetation in drylands, the integral of repeated NDVI measurements throughout the growing season is correlated with annual NPP (Paruelo et al. 1997; Prince 1991; Tucker et al. 1985). Coarse spatial resolution satellites can observe the same place on the ground almost daily so even during periods of cloud cover measurements at 10–15 day intervals can be achieved in most drylands. It should be noted that NDVI and all other vegetation indices based on RED and NIR cannot directly measure change in the species composition of vegetation (for instance a shift between palatable and unpalatable species) but provides a measure of photosynthetic vigour.

8.2.2 Building Consistent Long-Term Time Series

Long-term and consistent satellite data records are needed to monitor and quantify inter-annual trends in vegetation. The processing must be such that it does not inadvertently remove trends in data. Several global and regional NDVI time series have been created from the sequence of Advanced Very High Resolution Radiometer (AVHRR) instruments carried on the NOAA (National Oceanic and Atmospheric Administration) satellites from 1981 (NOAA-7) to present. 12 AVHRR sensors have been launched since 1981 and at least 7 of these have been used for constructing the long-term NDVI time series. Apart from the AVHRR record, there are no other global NDVI data with such a long record. Only since 1995, other instruments, often with improved sensor attributes, started to make NDVI time series measurements. Thus, for the purpose of trend detection, there is a trade-off between the length of the record and the quality of the observations. The AVHRR instrument is also part of the MetOp series of satellites from the EUMETSAT Polar System (EPS); but even though the AVHRR data stream is expected to continue for several years these data might be replaced by newer sensor systems (see below) that can be added to the AVHRR time series with appropriate harmonization (Pedelty et al. 2007).

The AVHRR was originally intended for visual analysis of meteorological conditions, and not for applications to vegetation studies (Cracknell 2001). The potential of AVHRR for vegetation studies came later (1981), when modifications to optimize the sensor for meteorological purposes (from NOAA-7) inadvertently provided RED and NIR data suitable for the derivation of VIs. Nevertheless, there are aspects of AVHRR data that, while not seriously affecting their meteorological uses, are not ideal for vegetation trend studies. These include post-launch degradation in sensor calibrations and drift in the satellite overpass times which have significant effects on VI time series, irrespective of ground conditions. Other limitations include the spectral properties of the sensors themselves (primarily the NIR sensor response function). Numerous investigations have evaluated NDVI continuity and have proposed inter-sensor translation methods among the different AVHRR instruments as well as between AVHRR instruments and newer sensors, including the Moderate Resolution Imaging Spectroradiometer (MODIS), the Système Pour l'Observation de la Terre – vegetation (SPOT-VGT), the Sea-viewing Wide Field-of-view Sensor (SeaWiFS), the Medium Resolution Imaging Spectrometer (MERIS) and the Visible/Infrared Imager Radiometer Suite (VIIRS) on the Suomi NPOESS series (e.g. Brown et al. 2006; Swinnen and Veroustraete 2008). It remains a challenge, however, to produce long-term and consistent vegetation index time series across the sequence of multiple sensor systems with their different spectral responses, spatial resolutions, swath widths and orbiting geometries.

Several archives of AVHRR data exist for which various types of processing have been applied to the satellite data in order to minimize unwanted effects, which are either specific to the satellite sensor used or inseparable from measurements made from a satellite, such as sun-sensor geometry and the effects of the atmosphere on reflectances. The aim of the processing is to estimate the NDVI that would be measured just above the vegetation surface, with all other significant variables normalized. Corrections and methods used in the processing have evolved through time and continue to be improved. Some archives have been reprocessed back to 1981 several times, each time using improved techniques and including newer sensor systems, while others continue to use the methods adopted at the inception of the data series. Some processing details of available global coverage archives are summarized in Table 8.1.

Sensor calibration is no longer a source of significant error (El Saleous et al. 2000). The greatest remaining sources of error in AVHRR estimates of surface NDVI are caused by absorption and scattering by atmospheric components (Nagol et al. 2009). Detection of partial cloud cover and hence the decision of whether to use or omit data is also a major source of uncertainty. Seasonal solar zenith angle variations and the effects of viewing vegetation at different angles (bi-directional reflectance distribution function (BRDF)) are further sources of error. As the atmospheric correction schemes improve, the BRDF effect becomes more pronounced (Nagol et al. 2009), thereby increasing the bias towards the selection of NDVI values from the forward scatter viewing direction (Fensholt et al. 2010; Huete et al. 2002). The selection of the maximum NDVI over a compositing period, so called maximum value compositing, can be effective by selecting the least hazy and nearest to nadir observations, but this is at the expense of any multiple, good observations in the compositing period. A series of physically-based corrections based upon those developed for the MODIS, SPOT VGT and VIIRS sensors have been suggested as a means to improve AVHRR NDVI data (Pedelty et al. 2007). The LTDR (Land Long Term Data Record), the VIP (Vegetation Index and Phenology Earth Science Data Record) and the GIMMS (Global Inventory Monitoring and Modeling System) data processing (Table 8.1) apply several or all of these new correction schemes as means of providing a seamless and consistent

Table 8.1 Characteristics of available AVHRR data sets

Data source	Level 1b (Raw-top of atmosphere)	NASA GIMMS3g	Long-term Data Record (LTDR) (v3 continuation from Modis)	Vegetation Index and Phenology Earth Science Data Record (VIP)
Spatial extent	Global	Global	Global	Global
Data provided	Channels 1–5	NDVI, solar zenith angle, channels 4 and 5	NDVI, Surface Reflectance (channel 1,2,3). TOA brightness temperature (channel 3,4,5) Sza, Vza, Raa, QA	NDVI, EVI2, Red, NIR, QA, Vza, Sza, Raa, Rank
Corrections				
Rayleigh	No	No	Yes	Yes
Ozone	No	No	Yes	Yes
Water vapour	No	No	Yes	Yes
Aerosols	No	Volcanic dust periods	Yes	AVHRR: No (Only volcanic)
				MODIS: Yes (Vermote et al. 2002)
Solar Zenith angle	No	Yes (Pinzon et al. 2005)	Yes	Yes
BRDF	No	No	Yes	No
Cloud screening	No	Temperatures from AVHRR channel 5	CLAVR (Stowe 1991)	AVHRR: CLAVR (Stowe 1991)
			LTDR v3 cloud mask (Vermote et al. 2009)	Additional Cloud screening by the LTDR Group (Pedelty et al. 2007) and (Barreto-Munoz 2013)
				MODIS: MOD35 Algorithm (Frey et al. 2008)

Sensor calibration	No	Rao and Chen (1995), Los (1998), and NOAA	Vermote and Kaufman (1995) and Vermote and Saleous (2006)	AVHRR: Rao and Chen 1995, Additional Calibration of AVHRR (Vermote et al. 2002)
				MODIS (Wu et al. 2008)
Spatial resolution	From 1.1 km at nadir	0.083 deg (~9.25 km at equator)	0.05 deg (~5 km at equator)	0.05 deg (~5.6 km at equator)
Length of archive	1981-present	1981–2012	LTDR AVHRR: 1981–1999	1981–2013
			LTDR Modis: 2000–2012	
Estimated geolocation error	–	0.083 deg (~9.25 km at equator)	0.025 deg (~2.5 km at equator)	<1.1 km
Compositing period	None	Bi-monthly	Daily	Daily, 7-day, 15-day, and monthly intervals
Where to get data	http://www.class.ncdc.noaa.gov	Contact: Pinzon and Tucker	http://ltdr.nascom.nasa.gov/ltdr/ltdr.html	http://vip.arizona.edu/viplab_data_explorer.php
				https://lpdaac.usgs.gov/user_community/nasa_measures
Key citations	Kidwell (1991)	Tucker et al. (2005)	Pedelty et al. (2007)	Didan (2010)
		Pinzon and Tucker (2014)		Barreto-Munoz (2013)

Note: *Sza* solar zenith angle, *Vza* viewing zenith angle, *Raa* relative azimuth angle, *QA* quality assessment sensor independent record of land surface vegetation. Upcoming European products (geoland2/Copernicus) time series products also from the AVHRR data stream are expected in the near future.

In the GIMMS project (Tucker et al. 2005), a NDVI time series was produced from different AVHRR sensors and accounting for various deleterious effects, such as calibration loss, orbital drift, atmospheric dust emitted by volcanic eruptions, etc. The latest version of the GIMMS NDVI data set spans the period July 1981 to December 2011 and is termed NDVI3g (third generation GIMMS NDVI from AVHRR) (Pinzon and Tucker 2014). A special issue "Monitoring Global Vegetation with AVHRR NDVI3g Data (1981–2011)" in the open access journal Remote Sensing (Myneni and Pinzón 2013) aims at an improved understanding of variability, long-term trends and changes in vegetation on our planet over the past 30 years using this new, improved dataset. Furthermore, long-term datasets of both FPAR and LAI have recently been produced from the AVHRR data record such as LAI3g and FPAR3g, both based on the NDVI3g dataset (Zhu et al. 2013), and the AVHRR based Global LAnd Surface Satellite (GLASS) LAI (Zhao et al. 2013).

Nevertheless, the AVHRR instruments lack the additional channels needed to make corrections as accurate as those possible for SeaWIFS, ENVISAT MERIS, Aqua/Terra MODIS and SPOT-VGT. Recently, successors of moderate resolution sensors applicable for vegetation monitoring have been launched like Proba-V and VIIRS instrument on the board the Suomi National Polar-orbiting Partnership (NPP). Future satellite missions capable of continuation of vegetation time series include Sentinel-3 developed by ESA (European Space Agency) as part of the Global Monitoring for Environment and Security (GMES) program. The first Sentinel-3 satellite is planned for launch in 2017.

Notwithstanding the caution needed in the interpretation of satellite measurements of VIs discussed above, these data are invaluable for detecting trends in important traits of vegetation. In fact, there are no other data sets as spatially and temporally comprehensive for the past 30+ years.

8.3 Greening or Browning of Global Drylands

Many reports of declining vegetation productivity in arid and semi-arid lands using field measurements have been published over the last decades, a recent and authoritative example being the Millennium Ecosystem Assessment (Adeel and World Resources Institute 2005). However, recent publications based on the use of EO-data have shown a more nuanced picture with both declines and increases (Bai et al. 2008; Beck et al. 2011; Fensholt et al. 2012), and a large body of literature focusing on increases in vegetation in the Sahel since the 1990s, the so-called 'greening of the Sahel' (e.g. Anyamba and Tucker 2005; Fensholt et al. 2013; Hellden and Tottrup 2008; Hickler et al. 2005; Olsson et al. 2005; Prince et al. 2007). These publications have been based on a variety of different EO datasets and they use slightly different methods but the common denominator is the use of data originating from the suite of AVHRR sensors on-board the NOAA satellites.

8.3.1 Trend Estimation

Long-term changes in vegetation greenness have been reported in many regional to global scale analyses (Donohue et al. 2009; Hellden and Tottrup 2008; Herrmann et al. 2005; Nemani et al. 2003; Zhao and Running 2010). Long-term trend analyses are often performed using a linear model providing the slope coefficient of an ordinary least squares regression (OLS) between the values of each pixel over time and a perfectly linear series. The outcome of a trend analysis however depends on several factors including data (data choice, data period and quality), choice of vegetation metric and the nonlinearity of process in relation to the selected trend indicator.

Since time series of vegetation metrics often do not meet parametric assumptions of normality and homoscedasticity (Hirsch and Slack 1984) median trend (Theil-Sen) procedure has also been applied to vegetation trend calculation (Fensholt and Rasmussen 2011; Fensholt et al. 2013). The Theil-Sen (TS) procedure is a rank-based test based on nonparametric statistics that calculates the slope and intercept of the time series and is therefore particularly effective for trend in noisy series of multiple estimates of the slopes derived from all pairs of observations (Hoaglin et al. 2000). Because it is based on the median, approximately 29 % of the samples can be unrelated noise but, owing to the TS analysis, have no impact on the statistic (Hoaglin et al. 2000). The TS approach is known to be robust against seasonality, non-normality, heteroscedasticity and temporal autocorrelation (at both intra- and inter annual scale) (Alcaraz-Segura et al. 2010; Hirsch and Slack 1984; Vanbelle and Hughes 1984) and has been recommended to be used for analysing time series of vegetation (de Beurs and Henebry 2005).

However, vegetation does not necessarily develop linearly over longer time scales and trends might often be better characterised by breaking the series into several different phases of alternating greening and/or browning periods. To accommodate for this, a trend breaks analysis procedure (Breaks For Additive Seasonal and Trend (BFAST)) has been developed (Verbesselt et al. 2010a, b, 2012) that enables the detection of change in trend within the time series. Examples of time series analyses based upon these different statistical approaches will be given in the Sect. 8.4 and in the following chapter.

8.3.2 Global Trends in Vegetation Greenness

A study of changes in dryland NDVI based on GIMMSg data from 1982 to 2007 (Fensholt et al. 2012) (Fig. 8.2) shows that, despite the regional nature of observed

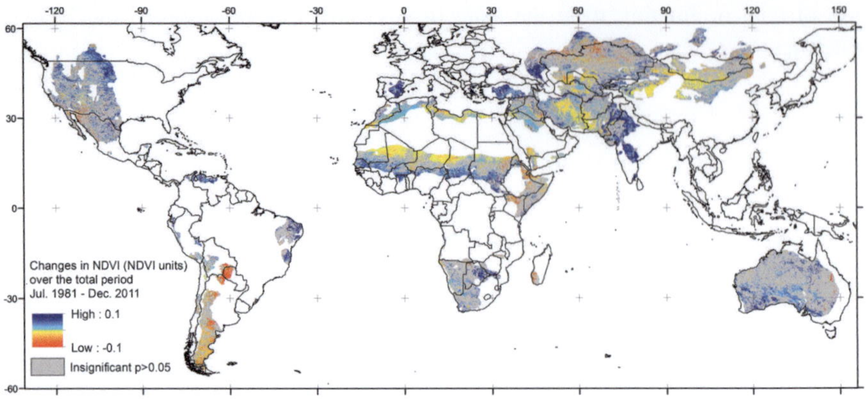

Fig. 8.2 Changes in NDVI for dryland areas (hyperarid not included) from 1981 to 2011. Only pixels with a statistically significant trend ($p < 0.05$) are shown

increases/decreases in NDVI, there was a noticeable trend of greening of the semi-arid zone across the globe (66 % of the pixels had a positive trend). 36 % of the semi-arid pixels were characterized by significant change values (both positive and negative) at the 0.05 significance level and out of the pixels with a significant positive/negative trend 77 % were characterized by positive trends. The changes in NDVI across the globe varied between regions and many regions with high rates of change (>0.05 NDVI units over the period of analysis) can be observed (Fig. 8.2; deep blue colours). The average slope value was +0.015 (NDVI units), indicating an overall greening for semi-arid areas across the globe over the 27 years. A recent study by Cook and Pau (2013) of global pasture land (spatially overlapping with global drylands) based on GIMMS3g (1982–2008) show statistically significant ($p < 0.05$) linear temporal trends in 23 % of all pasture pixels, with the vast majority of these areas showing positive trends. This suggests that degradation of pasture lands is not a globally widespread phenomenon, and in line with much of the terrestrial biosphere, there have been widespread increases in pasture productivity over the last 30 years.

8.3.3 Refining Vegetation Metrics for Long-Term Trends

When performing long-term trend analysis for dryland areas the specific vegetation metric to be used varies amongst scholars. Several studies focussing on the Sahelian region have discussed the most appropriate EO-based characterisation of the annual vegetation growth (see below). An annual summed NDVI value can be applied. However, for areas of pronounced seasonality it is generally accepted that the satellite signal derived from the dry season often introduces non-vegetation signals to long-term trend analyses. Thus, trends based on VIs derived only from the growing season should be used. Studies using a predefined set of months covering the period of growth (for a given region) can be used (Anyamba and Tucker 2005; Fensholt et al. 2009; Hellden and Tottrup 2008; Prince and Justice 1991; Wessels et al. 2012). However, if a temporal shift in the growing season occurs over time, this will influence such NDVI metric and thereby also impact the trend that is likely to decrease because part of the growing season is not anymore included in the analysis. By using software for parameterisation of the growing season (e.g. TIMESAT, (Jonsson and Eklundh 2002, 2004), MODIS MCD12Q2 (Ganguly et al. 2010; Zhang et al. 2006) or PHENOLO (Ivits et al. 2012)) vegetation metrics related to NDVI amplitude (the difference between NDVI dry season value and maximum value), length of growing season (includes the estimation of the onset and end of growing season) or the growing season integral (the NDVI integral for the period between onset and end of growing season) can be determined and used for long-term trend analysis (de Jong et al. 2011; Eklundh and Olsson 2003; Fensholt and Proud 2012; Fensholt et al. 2013; Heumann et al. 2007; Olsson et al. 2005).

The per-pixel temporal trend in GIMMS3g NDVI for the Sahel 1982–2010, using TIMESAT on the annual NDVI sum and the growing season integral are shown in Fig. 8.3. The slope expresses NDVI change over the total period of analysis and only NDVI slope values significant at the 95 % level ($p < 0.05$) are shown.

Annual NDVI sums (Fig. 8.3a) generally show a positive trend with an average NDVI slope for Sahel of +0.01 (Table 8.1). 33.7 % of the Sahelian pixels analyzed had a significant trend (30.7 % positive and 3.0 % negative). NDVI based on the seasonal integral (Fig. 8.3b) show a more pronounced positive trend in NDVI: 75 %

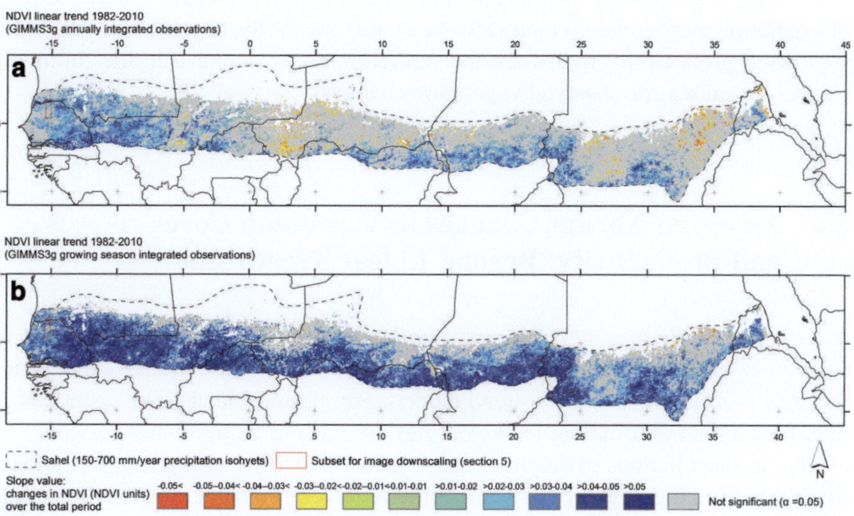

Fig. 8.3 GIMMS3g NDVI linear trend 1982–2010 based on (**a**) annually integrated NDVI and (**b**) growing season NDVI integral (so-called small integral) estimated from TIMESAT parameterization

of the Sahelian pixels analyzed show a significant trend (74.6 % positive and 0.4 % negative – only few pixels in Niger and Sudan were characterised by negative slope values).

Clearly, the percentage of Sahelian pixels characterised by a significant change in vegetation greenness depends on the NDVI metric used. Also in a study by Olsson et al. (2005) a much more widespread greening of the Sahel was found when calculated from seasonal NDVI integrals as compared to trend estimates calculated from the seasonal NDVI amplitude. This was explained by the possible saturation of the NDVI signal, thereby rendering the NDVI seasonal amplitude less sensitive to detection of changes for the greener parts of dryland areas. Seasonal maximum NDVI (closely related to seasonal amplitude) has however also been used as a metric for seasonal productivity to avoid the per-pixel length of season determination required for a per-pixel growing season integral (Fuller 1998; Jeyaseelan et al. 2007). The NDVI metric used for trend analysis should therefore be examined carefully taking into consideration vegetation density of the area studied. In cases where the scope of the analysis is global (including both areas with and without seasonality) a widely used approach is to estimate trends based on annual NDVI sums (or averages) rather than integrating only the signal from the growing season (Bai et al. 2008).

Which vegetation metric to be used for trend estimates of vegetation productivity in drylands is therefore not straight forward in the absence of direct *in situ* validation of the satellite derived time series. However, an alternative approach that could provide an indication of the appropriate vegetation metric to use for a given area would be to conduct per-pixel correlations with rainfall (which for most dryland areas is considered the primary constraint to vegetation growth) (Fensholt et al. 2012; Nemani et al. 2003). The highest correlation between rainfall and a set of vegetation metrics would then serve as a guideline for the best choice of metric if vegetation productivity trends are the desired analysis output (see the following chapter on causes for observed vegetation changes).

8.4 Assessing Abrupt Changes in Vegetation Cover and Productivity: Beyond Linear Trends Analysis

In recent years, there is a growing recognition that degradation and disturbance of vegetation traits cannot be sufficiently captured using traditional time series methods such as singular linear trend models. For instance, land cover disturbances (e.g. fires, insect defoliation, land clearing) often occur abruptly and may only be evident in short periods in the time series. Likewise, slower changes (e.g., biomass extraction from logging) will only affect the vegetation signal during the time of the ongoing disturbance and may change again later in the series. For such reasons it is argued that trends may be better captured by separating the series into individual segments, capturing specific vegetation conditions or stages of degradation through time. These approaches have only become possible as time series cover adequate numbers of years.

A method that can reveal these changes in trend within a time series analysis is Breaks For Additive Seasonal and Trend (BFAST) (Verbesselt et al. 2010a). BFAST integrates the decomposition of time series into seasonal, trend, and noise components with detection of abrupt changes, or breakpoints, within a time series using an additive model. The general model can be denoted as:

$$\mathrm{Yt} = \mathrm{Tt} + \mathrm{St} + \mathrm{et} \quad (\mathrm{t} = 1, \ldots, \mathrm{n}) \tag{8.2}$$

where Y_t is the observed data at time t, T_t is the trend component, S_t is the seasonal component, and e_t is the noise component. BFAST provides information on the number, timing, magnitude and direction of breakpoints in both the seasonal and trend components.

Figure 8.4 illustrates the decomposition of a MODIS 250 m (MOD13Q1) time series from a pixel covering a dry deciduous forest ecosystem. In this example there was a fire disturbance in 2006, followed by recovery from 2008. These distinct events are captured in the trend component by two breakpoints, and a statistically significant positive trend segment from 2008 to 2013. There was no break in the seasonal component as the land cover and phenology remained the same after the disturbance. However, if a change in land cover occurs (e.g. grassland to agriculture) a break in the seasonal component is also expected.

BFAST phenological and trend components have been used successfully on regional scales to detect abrupt land cover change in forest/grassland ecosystems (Verbesselt et al. 2010a, b) and on global scales to identify trend changes in vegetation greening and browning (de Jong et al. 2011, 2013). Furthermore, the BFAST approach has been adapted for near real-time monitoring of ecosystem disturbances such as drought (Verbesselt et al. 2012).

8.5 The Importance of Spatial Resolution for Vegetation Trend Analysis

Environmental variability in the temporal domain is affected by the spatial scale of the data and so analyses conducted at different spatial scales may give different results. Studies based on field or high spatial resolution remotely sensed data will typically be site-based, in contrast to studies using moderate to coarse spatial resolution satellite data. Analyses at different scales are often needed to reveal important information about drivers of observed changes, as shown in a recent EO-based case study from northern Burkina Faso (Rasmussen et al. 2014). The time series of moderate resolution (MODIS 250 m resolution) data, as compared to coarse resolution (AVHRR GIMMS3g resolution) data, identified local patterns of temporal trends pointing towards local, anthropogenic effects on vegetation

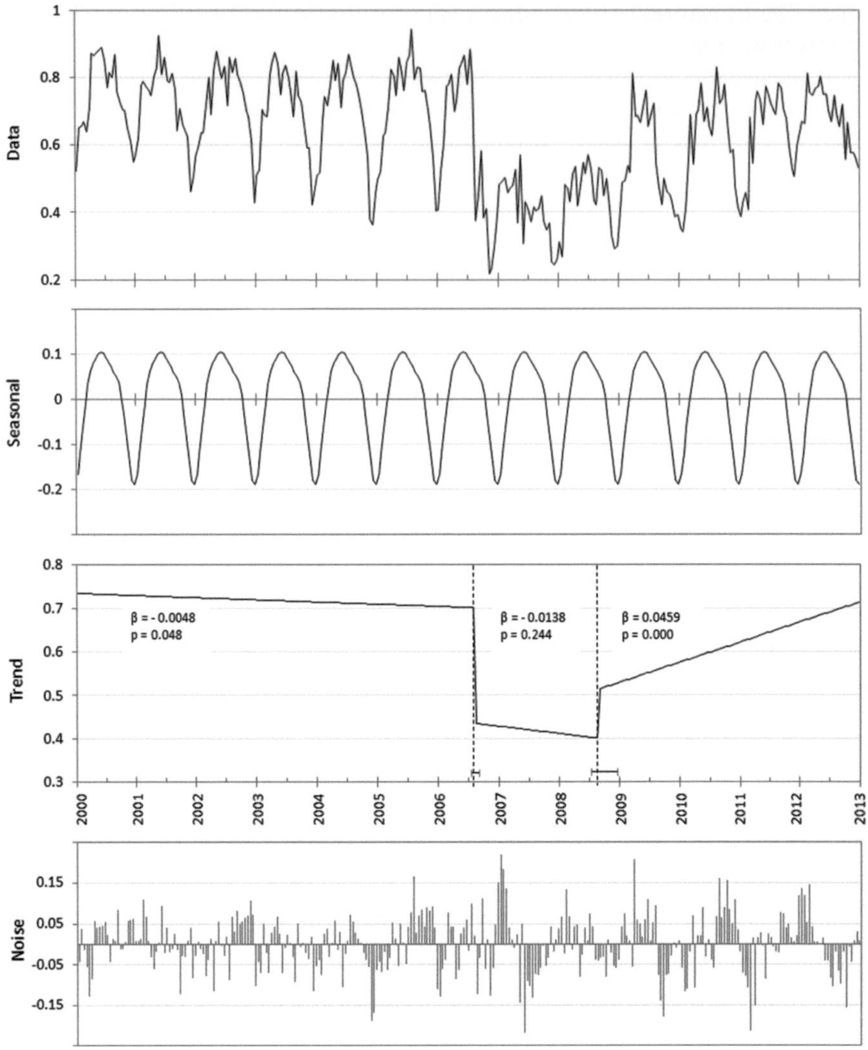

Fig. 8.4 An example of BFAST using MODIS 250 m 16- daytime series (MOD13Q1) for a pixel covering a seasonal dry deciduous forest. The data are decomposed into seasonal, trend, and noise components. Two breakpoints were detected in the trend component

productivity, while the regional pattern, detected from GIMMS3g data were attributed to changes in precipitation. In a field study over the last three decades, Herrmann and Tappan (2013) found an overall reduction in woody species richness, a loss of large trees, an increasing dominance of shrubs, and a shift towards more arid-tolerant, Sahelian species from ground observations in Senegal, despite the greening observed from the GIMMS NDVI data. Since this colonizing vegetation is of little economic value, it has been referred to as "green desert" by the local population in Sahel (Lykke et al. 2004).

8.5.1 Downscaling Vegetation Dynamics by Fusing Multi-temporal MODIS and Landsat Data

Current satellite capabilities create a trade-off between spatial and temporal resolutions. This is not a fundamental property; rather it is the result of too few satellites to cover the globe at adequate temporal frequencies. Landsat data captures land surface changes at high spatial resolution (30 m), which helps gaining insight to vegetation changes at that scale. However, the 16-day revisit cycle and frequent cloud contamination pose limitations to the use of Landsat data for trend and phenology analysis in general. In contrast, MODIS data provide near-daily observations which, by compositing (Holben 1986) substantially reduces cloud contamination but at coarser spatial resolutions (250–1,000 m). Downscaling to obtain images with both high spatial and high temporal resolution has been attempted by fusing data from Landsat and MODIS-like sensor systems (Atkinson 2013).

The Spatial and Temporal Adaptive Reflectance Fusion Model (STARFM) (Gao et al. 2006) has been developed for this purpose and used for studies related to vegetation phenology (Hilker et al. 2009; Hwang et al. 2011; Tian et al. 2013). STARFM assumes that a coarse-resolution MODIS pixel can be aggregated from the corresponding fine-resolution Landsat pixels when neglecting geo-location errors and differences in atmospheric correction. Following this assumption, for a homogenous pixel at the MODIS resolution, Landsat surface reflectance can be obtained by adding an error value (caused by differing bandwidth and solar geometry) to the MODIS surface reflectance. Suppose the land cover type and system errors do not change over time; it will be possible to predict the Landsat surface reflectance on a specific day (t_n) by blending one or more pairs of Landsat and MODIS surface reflectance images based on day (t_0) and the MODIS surface reflectance image acquired on day t_n. However, the relationships between MODIS and Landsat reflectance are complicated by the fact that MODIS observations may be mixed pixels, and land cover type and Bidirectional Reflectance Distribution Function (BRDF) may change over time. To alleviate this problem, STARFM employed a moving window to incorporate additional information from neighboring Landsat and MODIS image pixels using a weighting function defined by a combination of spectral, temporal and spatial differences among the given Landsat and MODIS scenes. With STARFM, daily Landsat-like data can be generated, thus allowing robust trend and phenology analysis at a finer spatial resolution.

The performance of STARFM is illustrated from an example of a semi-arid area in Senegal, West Africa characterized by a short and irregular growing season (July to October). The study area includes parts of the Senegal River in the north extending over one degree of latitude to a region in the south with more abundant vegetation. Three cloud free Landsat TM images and eleven MODIS 16-day composite nadir BRDF-adjusted reflectance products (MCD43A4, combining information from MODIS Terra and Aqua sensors) were used as inputs for STARFM. Eight Landsat-like images were generated and formed an image series during the growing season in 2010, along with the three Landsat TM images (Fig. 8.5). It is noticeable that cloud contamination cannot be eliminated in the

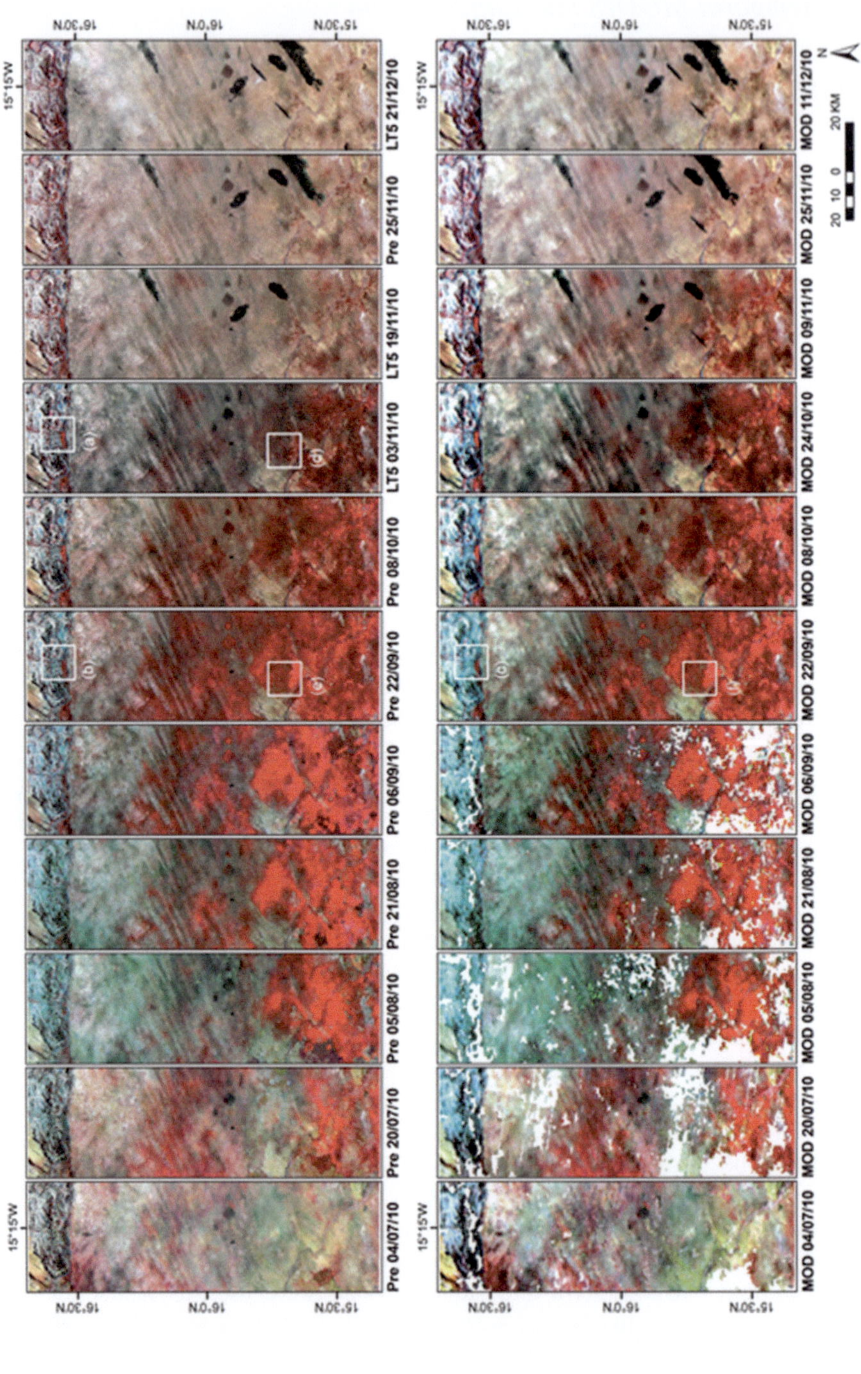

Fig. 8.5 Comparison between STARFM generated (*upper row*) and MODIS (MCD43A4, *lower row*) image series during the growing season in 2010. Observed Landsat images are labeled in *blue*, and cloudy pixels are shown in *white*. The dates of MODIS images are the first day of the 16-day composition period. All images are shown in the R-G-B composition of near infrared-red-blue bands. Areas in *black boxes* are enlarged in Fig. 8.6

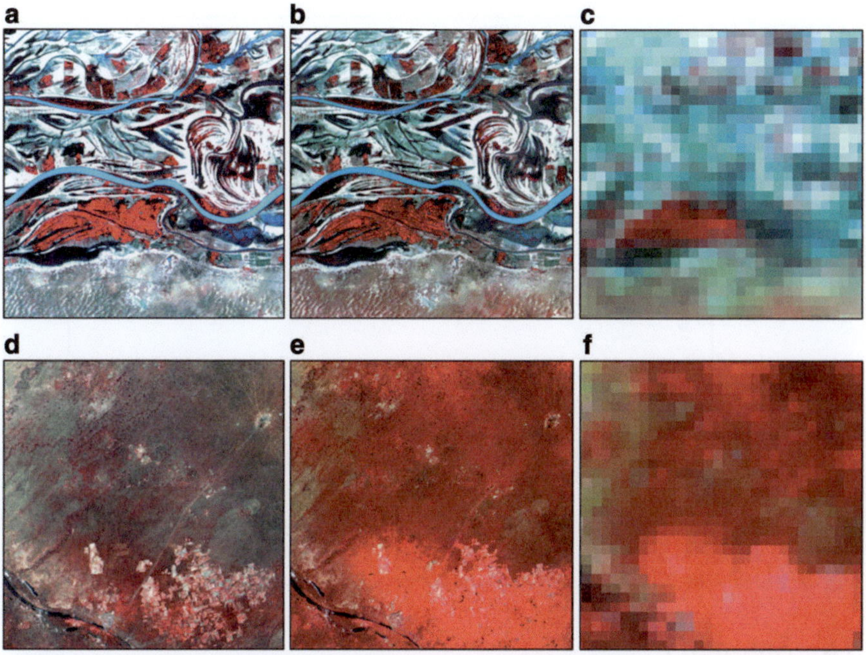

Fig. 8.6 Enlarged areas (from image subsets shown in Fig. 8.5) for better comparison of observed Landsat images (**a**, **d**), STARFM generated images (**b**, **e**) and MODIS images (**c**, **f**). The *upper row* (**a–c**) shows a subset of the Senegal River, while the *lower row* (**d–f**) shows an area of mixed savanna grassland and small scale agriculture

rainy season even with a 16-day composition of MODIS (lower row in Fig. 8.5). However, the STARFM generated images (upper row in Fig. 8.5) show high consistency with the corresponding MODIS images in terms of the overall pattern and the changing phenology of the vegetation were well captured by predicted image series. Figure 8.6 shows more detailed information of the image subset areas (shown by black boxes in Fig. 8.5) and the images generated (Fig. 8.6b, e) adequately combined the fine spatial pattern in Landsat (Fig. 8.6a, d) and the reflectance information in MODIS (Fig. 8.6c, f). NDVI images were produced using the NIR and Red bands for each of the images. Figure 8.7 shows that mean NDVI values of observed Landsat images fit well with the NDVI series of STARFM generated images and that both are marginally lower than mean MODIS NDVI. However the phenological change over time is identical in the MODIS and STARFM generated time series.

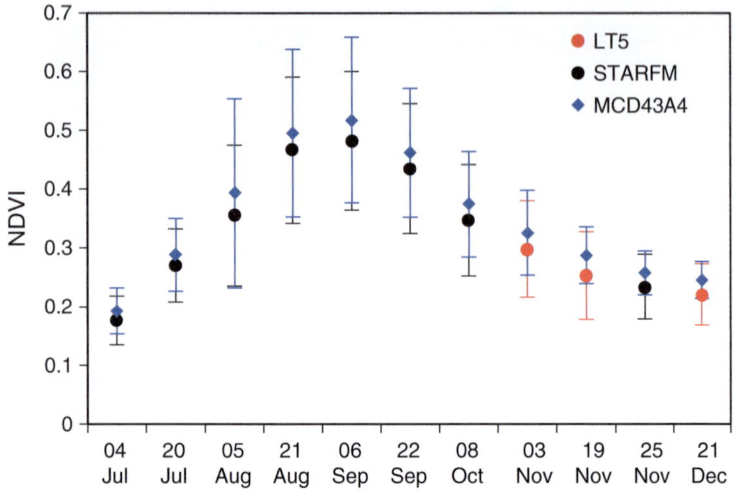

Fig. 8.7 Mean values of the NDVI images calculated from near infrared and red bands of the observed Landsat, MODIS and STARFM generated data in 2010. Cloudy pixels were excluded from the calculation. The *error bars* represent the corresponding standard deviations

8.6 Summary

Dryland ecosystems provide vital services, such as food, grazing, as well as energy and forestry products and have become the subject of increased interest due to the impacts of current global changes and concern for the sustainability of human lifestyles. This chapter gave a short overview of satellite sensor systems and approaches for detection of long-term dryland vegetation change at the global scale. Long-term Earth Observation (EO) datasets based on the Advanced Very High Resolution Radiometer (AVHRR) instruments carried on the NOAA satellites (1981 to present) are widely used for monitoring changes in dryland vegetation. The Normalized Difference Vegetation Index (NDVI) calculated from the red and near-infrared spectral bands of the AVHRR sensors has been found to be related to vegetation greenness or photosynthetic vigour and is widely used for regional to global scale vegetation trend studies due to the availability datasets covering more than three decades. The AVHRR instruments suffer from the lack of onboard calibration for its visible to short wave infrared channels and lack the additional channels needed to make corrections as accurate as those possible for later sensor systems like Sea-WIFS, ENVISAT MERIS, Aqua/Terra MODIS and SPOT-VGT. However, large research effort has been put into the production of seamless and intercalibrated AVHRR time-series data using a combination of mature and tested algorithm and the best available polar-orbiting satellite data from the past to the present. Recently also biophysical variables like FAPAR (fraction of photosynthetically active radiation) and LAI (leaf area index) have been produced from AVHRR data stream.

A status map of greening and browning trends for global drylands is presented based on the GIMMSg (Global Inventory Monitoring and Modeling System) data from 1982 to 2007. The map indicates an overall greening for semi-arid areas across the globe over the 27 years. For areas of pronounced seasonality it is generally accepted that the satellite signal derived from the dry season often introduces non-vegetation signals to long-term trend analyses and the vegetation metrics suitable for per-pixel temporal vegetation trend analysis (including seasonal parameterisation and the appropriate choice of trend indicators) was discussed.

In recent years there has been a growing recognition that degradation and disturbance of vegetation traits cannot be sufficiently captured using traditional time series methods such as singular linear trend models and it is argued that trends may be better captured by separating the series into individual segments, capturing specific vegetation conditions or stages of degradation through time. A method designed to overcome assumptions of long-term linearity in time series analysis (Breaks For Additive Season and Trend (BFAST)) is presented. The strengths of the BFAST approach is illustrated by the decomposition of a MODIS 250 m time series (2000–2013) from a pixel covering a dry deciduous forest ecosystem. In this example there was a fire disturbance in 2006, followed by recovery from 2008 and these distinct events were captured by the BFAST as two time-series breakpoints, and a statistically significant positive trend segment from 2008 to 2013.

Finally, the importance of the spatial scale when performing temporal trend analysis is introduced. Environmental variability in the temporal domain is affected by the spatial scale of the data and so analyses conducted at different spatial scales may give different results and point towards different drives of the observed changes. A method for image downscaling (Spatial and Temporal Adaptive Reflectance Fusion Model (STARFM)) is presented as a way to obtain images with both high spatial and high temporal resolution by fusing data from Landsat and MODIS-like sensor systems. The performance of STARFM is illustrated from an example of a semi-arid area in Senegal, West Africa characterized by a short and irregular growing season and with STARFM, daily Landsat-like data can be generated, thus allowing robust trend and phenology analysis at a finer spatial resolution.

Acknowledgements This book chapter is written within the frame of the project entitled Earth Observation based Vegetation productivity and Land Degradation Trends in Global Drylands. The project is funded by the Danish Council for Independent Research (DFF) Sapere Aude programme.

References

Adeel Z, World Resources Institute (2005) Ecosystems and human well-being: desertification synthesis: a report of the Millennium Ecosystem Assessment. World Resources Institute, Washington, DC

Alcaraz-Segura D, Chuvieco E, Epstein HE, Kasischke ES, Trishchenko A (2010) Debating the greening vs. browning of the North American boreal forest: differences between satellite datasets. Glob Chang Biol 16:760–770
Anyamba A, Tucker CJ (2005) Analysis of Sahelian vegetation dynamics using NOAA-AVHRR NDVI data from 1981–2003. J Arid Environ 63:596–614
Asrar G, Fuchs M, Kanemasu ET, Hatfield JL (1984) Estimating absorbed photosynthetic radiation and leaf-area index from spectral reflectance in wheat. Agron J 76:300–306
Atkinson PM (2013) Downscaling in remote sensing. Int J Appl Earth Obs Geoinf 22:106–114
Bai ZG, Dent DL, Olsson L, Schaepman ME (2008) Proxy global assessment of land degradation. Soil Use Manag 24:223–234
Barreto-Munoz A (2013) Multi-sensor vegetation index and land surface phenology earth science data records in support of global change studies: data quality challenges and data explorer system. Ph.D. thesis, The University of Arizona, Tucson, AZ, USA
Beck HE, McVicar TR, van Dijk AIJM, Schellekens J, de Jeu RAM, Bruijnzeel LA (2011) Global evaluation of four AVHRR-NDVI data sets: intercomparison and assessment against Landsat imagery. Remote Sens Environ 115:2547–2563
Brown ME, Pinzon JE, Didan K, Morisette JT, Tucker CJ (2006) Evaluation of the consistency of long-term NDVI time series derived from AVHRR, SPOT-Vegetation, Sea-WiFS, MODIS, and Landsat ETM+ sensors. IEEE Trans Geosci Remote Sens 44:1787–1793
Cook BI, Pau S (2013) A global assessment of long-term greening and browning trends in pasture lands using the GIMMS LAI3g dataset. Remote Sens 5:2492–2512
Cracknell AP (2001) The exciting and totally unanticipated success of the AVHRR in applications for which it was never intended. Calibration Charact Satell Sens Accuracy Deriv Phys Parameters 28:233–240
de Beurs KM, Henebry GM (2005) Land surface phenology and temperature variation in the International Geosphere-Biosphere Program high-latitude transects. Glob Chang Biol 11:779–790
de Jong R, de Bruin S, de Wit A, Schaepman ME, Dent DL (2011) Analysis of monotonic greening and browning trends from global NDVI time-series. Remote Sens Environ 115:692–702
de Jong R, Verbesselt J, Zeileis A, Schaepman ME (2013) Shifts in global vegetation activity trends. Remote Sens 5:1117–1133
Didan K (2010) Multi-satellite earth science data record for studying global vegetation trends and changes. In: Proceedings of the 2010 international geoscience and remote sensing symposium, Honolulu, HI, USA, 25–30 July 2010, IEEE International, pp 25–30
Donohue RJ, McVicar TR, Roderick ML (2009) Climate-related trends in Australian vegetation cover as inferred from satellite observations, 1981–2006. Glob Chang Biol 15:1025–1039
Eklundh L, Olsson L (2003) Vegetation index trends for the African Sahel 1982–1999. Geophys Res Lett 30:1430. doi:10.1029/2002GL016772, 8
El Saleous NZ, Vermote EF, Justice CO, Townshend JRG, Tucker CJ, Goward SN (2000) Improvements in the global biospheric record from the Advanced Very High Resolution Radiometer (AVHRR). Int J Remote Sens 21:1251–1277
Fensholt R, Proud SR (2012) Evaluation of earth observation based global long term vegetation trends – comparing GIMMS and MODIS global NDVI time series. Remote Sens Environ 119:131–147
Fensholt R, Rasmussen K (2011) Analysis of trends in the Sahelian 'rain-use efficiency' using GIMMS NDVI, RFE and GPCP rainfall data. Remote Sens Environ 115:438–451
Fensholt R, Sandholt I, Rasmussen MS (2004) Evaluation of MODIS LAI, fAPAR and the relation between fAPAR and NDVI in a semi-arid environment using in situ measurements. Remote Sens Environ 91:490–507
Fensholt R, Sandholt I, Rasmussen MS, Stisen S, Diouf A (2006) Evaluation of satellite based primary production modelling in the semi-arid Sahel. Remote Sens Environ 105:173–188
Fensholt R, Rasmussen K, Nielsen TT, Mbow C (2009) Evaluation of earth observation based long term vegetation trends – intercomparing NDVI time series trend analysis consistency of Sahel

from AVHRR GIMMS, Terra MODIS and SPOT VGT data. Remote Sens Environ 113:1886–1898

Fensholt R, Sandholt I, Proud SR, Stisen S, Rasmussen MO (2010) Assessment of MODIS sun-sensor geometry variations effect on observed NDVI using MSG SEVIRI geo-stationary data. Int J Remote Sens 31:6163–6187

Fensholt R, Langanke T, Rasmussen K, Reenberg A, Prince SD, Tucker C, Scholes RJ, Le QB, Bondeau A, Eastman R, Epstein H, Gaughan AE, Hellden U, Mbow C, Olsson L, Paruelo J, Schweitzer C, Seaquist J, Wessels K (2012) Greenness in semi-arid areas across the globe 1981–2007 – an Earth Observing Satellite based analysis of trends and drivers. Remote Sens Environ 121:144–158

Fensholt R, Rasmussen K, Kaspersen P, Huber S, Horion S, Swinnen E (2013) Assessing land degradation/recovery in the African Sahel from long-term earth observation based primary productivity and precipitation relationships. Remote Sens 5:664–686

Frey RA, Ackerman SA, Liu YH, Strabala KI, Zhang H, Key JR, Wang XG (2008) Cloud detection with MODIS. Part I: Improvements in the MODIS cloud mask for collection 5. J Atmos Ocean Technol 25:1057–1072

Fuller DO (1998) Trends in NDVI time series and their relation to rangeland and crop production in Senegal, 1987–1993. Int J Remote Sens 19:2013–2018

Ganguly S, Friedl MA, Tan B, Zhang XY, Verma M (2010) Land surface phenology from MODIS: characterization of the Collection 5 global land cover dynamics product. Remote Sens Environ 114:1805–1816

Gao F, Masek J, Schwaller M, Hall F (2006) On the blending of the Landsat and MODIS surface reflectance: predicting daily Landsat surface reflectance. IEEE Trans Geosci Remote Sens 44:2207–2218

Gates DM, Keegan HJ, Schleter JC, Weidner VR (1965) Spectral properties of plants. Appl Opt 4 (1):11–20

Goward SN, Huemmrich KF (1992) Vegetation canopy Par absorptance and the normalized difference vegetation index – an assessment using the Sail model. Remote Sens Environ 39:119–140

Hellden U, Tottrup C (2008) Regional desertification: a global synthesis. Glob Planet Chang 64:169–176

Herrmann SM, Tappan GG (2013) Vegetation impoverishment despite greening: a case study from central Senegal. J Arid Environ 90:55–66

Herrmann SM, Anyamba A, Tucker CJ (2005) Recent trends in vegetation dynamics in the African Sahel and their relationship to climate. Glob Environ Chang Hum Policy Dimens 15:394–404

Heumann BW, Seaquist JW, Eklundh L, Jonsson P (2007) AVHRR derived phenological change in the Sahel and Soudan, Africa, 1982–2005. Remote Sens Environ 108:385–392

Hickler T, Eklundh L, Seaquist JW, Smith B, Ardo J, Olsson L, Sykes MT, Sjostrom M (2005) Precipitation controls Sahel greening trend. Geophys Res Lett 32:L21415. doi:10.1029/2005GL024370

Hilker T, Wulder MA, Coops NC, Seitz N, White JC, Gao F, Masek JG, Stenhouse G (2009) Generation of dense time series synthetic Landsat data through data blending with MODIS using a spatial and temporal adaptive reflectance fusion model. Remote Sens Environ 113:1988–1999

Hirsch RM, Slack JR (1984) A nonparametric trend test for seasonal data with serial dependence. Water Resour Res 20:727–732

Hoaglin DC, Mosteller F, Tukey JW (2000) In: Wiley classics library (ed) Understanding robust and exploratory data analysis. Wiley, New York

Holben BN (1986) Characteristics of maximum-value composite images from temporal Avhrr data. Int J Remote Sens 7:1417–1434

Huete AR, Jackson RD, Post DF (1985) Spectral response of a plant canopy with different soil backgrounds. Remote Sens Environ 17:37–53

Huete A, Didan K, Miura T, Rodriguez EP, Gao X, Ferreira LG (2002) Overview of the radiometric and biophysical performance of the MODIS vegetation indices. Remote Sens Environ 83:195–213

Hwang T, Song CH, Bolstad PV, Band LE (2011) Downscaling real-time vegetation dynamics by fusing multi-temporal MODIS and Landsat NDVI in topographically complex terrain. Remote Sens Environ 115:2499–2512

Ivits E, Cherlet M, Toth G, Sommer S, Mehl W, Vogt J, Micale F (2012) Combining satellite derived phenology with climate data for climate change impact assessment. Glob Planet Chang 88–89:85–97

Jeyaseelan AT, Roy PS, Young SS (2007) Persistent changes in NDVI between 1982 and 2003 over India using AVHRR GIMMS (Global Inventory Modeling and Mapping Studies) data. Int J Remote Sens 28:4927–4946

Jonsson P, Eklundh L (2002) Seasonality extraction by function fitting to time-series of satellite sensor data. IEEE Trans Geosci Remote Sens 40:1824–1832

Jonsson P, Eklundh L (2004) TIMESAT – a program for analyzing time-series of satellite sensor data. Comput Geosci 30:833–845

Kidwell K (1991) NOAA polar orbital data user's guide. NOAA NESDIS technical report. USDoC, Washington, DC

Los SO (1998) Estimation of the ratio of sensor degradation between NOAA AVHRR channels 1 and 2 from monthly NDVI composites. IEEE Trans Geosci Remote Sens 36:206–213

Lykke AM, Kristensen MK, Ganaba S (2004) Valuation of local use and dynamics of 56 woody species in the Sahel. Biodivers Conserv 13:1961–1990

Monteith JL (1972) Solar-radiation and productivity in tropical ecosystems. J Appl Ecol 9:747–766

Myneni RB, Pinzón JE (2013) Special issue "Monitoring Global Vegetation with AVHRR NDVI3g Data (1981–2011)". Remote sensing special issue. http://www.mdpi.com/journal/remotesensing/special_issues/monitoring_global. Accessed 30 Nov 2014

Myneni RB, Hall FG, Sellers PJ, Marshak AL (1995) The interpretation of spectral vegetation indexes. IEEE Trans Geosci Remote Sens 33:481–486

Nagol JR, Vermote EF, Prince SD (2009) Effects of atmospheric variation on AVHRR NDVI data. Remote Sens Environ 113:392–397

Nemani RR, Keeling CD, Hashimoto H, Jolly WM, Piper SC, Tucker CJ, Myneni RB, Running SW (2003) Climate-driven increases in global terrestrial net primary production from 1982 to 1999. Science 300:1560–1563

Olsson L, Eklundh L, Ardo J (2005) A recent greening of the Sahel – trends, patterns and potential causes. J Arid Environ 63:556–566

Paruelo JM, Epstein HE, Lauenroth WK, Burke IC (1997) ANPP estimates from NDVI for the Central Grassland Region of the United States. Ecology 78:953–958

Pedelty J, Devadiga S, Masuoka EB, Pinzon JM, Tucker CJ, Vermote EF et al (2007) Generating a long-term land data record from the AVHRR and MODIS instruments. In: Proceedings of the 2007 international geoscience and remote sensing symposium, 2007. IGARSS 2007, IEEE International, pp 1021–1025

Pinzon JE, Tucker CJ (2014) A non-stationary 1981–2012 AVHRR NDVI3g time series. Remote Sens 6:6929–6960

Pinzon JE, Brown ME, Tucker CJ (2005) Hilbert-Huang transform and its applications. In: Huang NE, Shen SSP (eds) Interdisciplinary mathematical sciences, vol 5. World Scientific, Singapore/Hackensack/London

Prince SD (1991) A model of regional primary production for use with coarse resolution satellite data. Int J Remote Sens 12:1313–1330

Prince SD, Goward SN (1995) Global primary production: a remote sensing approach. J Biogeogr 22:815–835

Prince SD, Justice CO (1991) Coarse resolution remote-sensing of the Sahelian environment – editorial. Int J Remote Sens 12:1137–1146

Prince SD, Wessels KJ, Tucker CJ, Nicholson SE (2007) Desertification in the Sahel: a reinterpretation of a reinterpretation. Glob Chang Biol 13:1308–1313

Rao CRN, Chen J (1995) Intersatellite calibration linkages for the visible and near-infrared channels of the advanced very high-resolution radiometer on the Noaa-7, Noaa-9, and Noaa-11 spacecraft. Int J Remote Sens 16:1931–1942

Rasmussen K, Fensholt R, Fog B, Rasmussen LV, Yanogo I (2014) Explaining NDVI trends in northern Burkina Faso. Geografisk Tidsskr Dan J Geogr 114:17–24

Sellers PJ (1985) Canopy reflectance, photosynthesis and transpiration. Int J Remote Sens 6:1335–1372

Stowe LL (1991) Cloud and aerosol products at Noaa Nesdis. Glob Planet Chang 90:25–32

Swinnen E, Veroustraete F (2008) Extending the SPOT-VEGETATION NDVI time series (1998–2006) back in time with NOAA-AVHRR data (1985–1998) for southern Africa. IEEE Trans Geosci Remote Sens 46:558–572

Thenkabail PS, Lyon JG, Huete A (2012) Hyperspectral remote sensing of vegetation. CRC Press, Boca Raton

Tian F, Wang Y, Fensholt R, Wang K, Zhang L, Huang Y (2013) Mapping and evaluation of NDVI trends from synthetic time series obtained by blending Landsat and MODIS data around a coalfield on the loess plateau. Remote Sens 5:4255–4279

Tucker CJ (1979) Red and photographic infrared linear combinations for monitoring vegetation. Remote Sens Environ 8:127–150

Tucker CJ, Vanpraet CL, Sharman MJ, Vanittersum G (1985) Satellite remote-sensing of total herbaceous biomass production in the senegalese sahel – 1980–1984. Remote Sens Environ 17:233–249

Tucker CJ, Dregne HE, Newcomb WW (1991) Expansion and contraction of the Sahara desert from 1980 to 1990. Science 253:299–301

Tucker CJ, Pinzon JE, Brown ME, Slayback DA, Pak EW, Mahoney R, Vermote EF, El Saleous N (2005) An extended AVHRR 8-km NDVI dataset compatible with MODIS and SPOT vegetation NDVI data. Int J Remote Sens 26:4485–4498

Vanbelle G, Hughes JP (1984) Nonparametric-tests for trend in water-quality. Water Resour Res 20:127–136

Verbesselt J, Hyndman R, Newnham G, Culvenor D (2010a) Detecting trend and seasonal changes in satellite image time series. Remote Sens Environ 114:106–115

Verbesselt J, Hyndman R, Zeileis A, Culvenor D (2010b) Phenological change detection while accounting for abrupt and gradual trends in satellite image time series. Remote Sens Environ 114:2970–2980

Verbesselt J, Zeileis A, Herold M (2012) Near real-time disturbance detection using satellite image time series. Remote Sens Environ 123:98–108

Vermote E, Kaufman YJ (1995) Absolute calibration of Avhrr visible and near-infrared channels using ocean and cloud views. Int J Remote Sens 16:2317–2340

Vermote EF, Saleous NZ (2006) Calibration of NOAA16 AVHRR over a desert site using MODIS data. Remote Sens Environ 105:214–220

Vermote EF, El Saleous NZ, Justice CO (2002) Atmospheric correction of MODIS data in the visible to middle infrared: first results. Remote Sens Environ 83:97–111

Vermote E, Justice C, Csiszar I, Eidenshink J, Myneni R, Baret F, Masuoka E, Wolfe R, Devadiga S (2009) A terrestrial surface climate data record for global change studies. Eos Trans. AGU 90 (52), Fall Meet. Suppl., Abstract IN42A-08, 2009

Wessels KJ, van den Bergh F, Scholes RJ (2012) Limits to detectability of land degradation by trend analysis of vegetation index data. Remote Sens Environ 125:10–22

Wu A, Xiaoxiong X, Changyong C, Amit A (2008) Monitoring MODIS calibration stability of visible and near-IR bands from observed top-of-atmosphere BRDF-normalized reflectances over Libyan Desert and Antarctic surfaces. In: Proceedings of SPIE 7081, Earth Observing Systems XIII, 708113 August 28. doi:10.1117/12.795296

Zhang XY, Friedl MA, Schaaf CB (2006) Global vegetation phenology from Moderate Resolution Imaging Spectroradiometer (MODIS): evaluation of global patterns and comparison with in situ measurements. J Geophys Res. doi:10.1029/2006JG000217

Zhao MS, Running SW (2010) Drought-induced reduction in global terrestrial net primary production from 2000 through 2009. Science 329:940–943

Zhao X, Liang SL, Liu SH, Yuan WP, Xiao ZQ, Liu Q, Cheng J, Zhang XT, Tang HR, Zhang X, Liu Q, Zhou GQ, Xu S, Yu K (2013) The Global Land Surface Satellite (GLASS) remote sensing data processing system and products. Remote Sens 5:2436–2450

Zhu ZC, Bi J, Pan YZ, Ganguly S, Anav A, Xu L, Samanta A, Piao SL, Nemani RR, Myneni RB (2013) Global data sets of vegetation Leaf Area Index (LAI)3g and Fraction of Photosynthetically Active Radiation (FPAR)3g derived from Global Inventory Modeling and Mapping Studies (GIMMS) Normalized Difference Vegetation Index (NDVI3g) for the period 1981 to 2011. Remote Sens 5:927–948

Chapter 9
Assessing Drivers of Vegetation Changes in Drylands from Time Series of Earth Observation Data

Rasmus Fensholt, Stephanie Horion, Torbern Tagesson, Andrea Ehammer, Kenneth Grogan, Feng Tian, Silvia Huber, Jan Verbesselt, Stephen D. Prince, Compton J. Tucker, and Kjeld Rasmussen

Abstract This chapter summarizes methods of inferring information about drivers of global dryland vegetation changes observed from remote sensing time series data covering from the 1980s until present time. Earth observation (EO) based time series of vegetation metrics, sea surface temperature (SST) (both from the AVHRR (Advanced Very High Resolution Radiometer) series of instruments) and precipitation data (blended satellite/rain gauge) are used for determining the mechanisms of observed changes. EO-based methods to better distinguish between climate and human induced (land use) vegetation changes are reviewed. The techniques presented include trend analysis based on the Rain-Use Efficiency (RUE) and the Residual Trend Analysis (RESTREND) and the methodological challenges related to the use of these. Finally, teleconnections between global sea surface temperature (SST) anomalies and dryland vegetation productivity are illustrated and the associated predictive capabilities are discussed.

R. Fensholt (✉) • S. Horion • T. Tagesson • A. Ehammer
K. Grogan • F. Tian • K. Rasmussen
Section of Geography, Department of Geosciences
and Natural Resource Management, Faculty of Science,
University of Copenhagen, Copenhagen K, Denmark
e-mail: rf@ign.ku.dk

S. Huber
DHI GRAS, Copenhagen K, Denmark

J. Verbesselt
Laboratory of Geo-Information Science and Remote Sensing,
Wageningen University, Wageningen, The Netherlands

S.D. Prince
Department of Geographical Sciences, University of Maryland,
College Park, MD, USA

C.J. Tucker
NASA Goddard Space Flight Center, Greenbelt, MD, USA

C. Kuenzer et al. (eds.), *Remote Sensing Time Series*, Remote Sensing and Digital Image Processing, DOI 10.1007/978-3-319-15967-6_9

9.1 Introduction

The United Nations Convention to Combat Desertification (UNCCD) definition of desertification, or dryland degradation (used synonymously) is:

> land degradation in arid, semi-arid and dry sub-humid areas resulting from various factors, including climatic variations and human activities" followed by "land degradation" means reduction or loss, in arid, semi-arid and dry sub-humid areas, of the biological or economic productivity and complexity of rainfed cropland, irrigated cropland, or range, pasture, forest and woodlands resulting from land uses or from a process or combination of processes, including processes arising from human activities and habitation patterns (UNCCD homepage, www.unccd.int).

This definition implies that change in vegetation productivity is a key indicator (but not the only one) of land degradation. Furthermore, vegetation productivity is of great economic importance because crop and livestock production is the most essential economic activity in many arid and semi-arid regions. Moreover, primary production is an important element in dryland key supporting ecosystem services, as defined by the Millennium Ecosystem Assessment (MEA) Desertification Synthesis (Adeel and World Resources Institute 2005). Therefore, spatially and temporally consistent, long-term data on changes and trends in vegetation productivity are of great interest for the assessment of environmental conditions and their trends in dryland regions. Earth Observation (EO) satellite data provide the only suitable means of temporally and spatially consistent global scale data, covering the last three decades (Prince 2002).

According to Adeel and World Resources Institute (2005), at least 10–20 % of drylands are already degraded and a recent publication from the UNCCD (UNCCD-secretariat 2013) states that global assessments indicate an increase in the percentage of highly degraded land area from 15 % in 1991 to 25 % by 2011. Many reputable sources rank desertification among the greatest environmental challenges today and a major impediment to meeting basic human needs in drylands (MEA & UNCCD). It is, however, also underlined that more elaborate studies are needed to identify where the problems occur and what is their true extent. This chapter introduces different EO-based methods for monitoring indicators of land degradation and to gain insight into the driving mechanisms of observed changes in vegetation productivity.

9.2 Inferring Causes for Observed Changes

Trends in vegetation productivity may be related to climatic as well as non-climatic causes of change (e.g. management), and it is obviously of great policy relevance to better understand the drivers and causal mechanisms of observed productivity trends. There is good correspondence between EO-based vegetation dynamics and precipitation in most dryland areas (Fig. 9.1) which is not surprising since

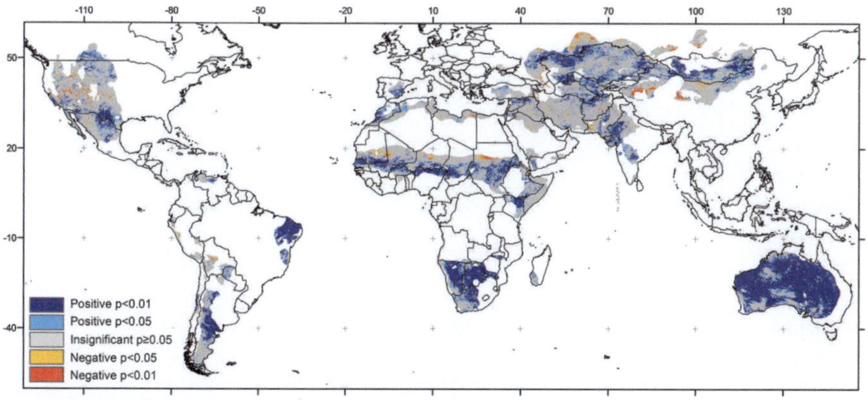

Fig. 9.1 Significance of linear correlation between annual integrated GIMMS3g NDVI and annual summed CMAP precipitation 1982–2010 for dryland areas (hyper-arid not included). CMAP precipitation has been resampled to match the spatial resolution of the GIMMS3g NDVI

vegetation growth is primarily water constrained in these areas (Nemani et al. 2003). However, large dryland regions of non-significant correlation between rainfall and vegetation growth can also be observed.

Different datasets of precipitation exist for continental to global scale analysis based on a combination of rain gauge measurements and a variety of different satellite observations (Huffman et al. 2009; Huffman et al. 2007; Xie and Arkin 1997). Three different products have been used in this chapter (GPCP (Global Precipitation Climatology Project), CMAP (CPC Merged Analysis of Precipitation) and TRMM (Tropical Rainfall Measuring Mission)) and are summarized in Table 9.1.

It has been shown that dryland areas across the globe, on average have experienced an increase in greenness during the satellite record, from 1981 till present (Fig. 9.2, previous chapter). However, similar increases in greenness over the last three decades in the same or different regions may have widely different explanations (Fensholt et al. 2012; Mao et al. 2013) including driving mechanisms of both climate and human induced changes in land use and land cover. Mao et al. (2013) estimated satellite-derived relative change in annual LAI (leaf area index) from the years 1982 to 2009 at the global scale and found a South-to-North asymmetry in the trends coinciding with trends in temperature over the same period. Precipitation patterns were found to decrease this asymmetric-latitudinal LAI trend, with strong local effects. By combining EO data analysis with model simulations it was found that positive and negative vegetation trends in dryland areas were primarily driven by changes in climate, with positive trends dominating. de Jong et al. (2013a) used an additive spatial model with 0.5° resolution, including climate-associated effects and influence of other factors such as land use change to separate possible drivers of observed changes. They attributed just above 50 % of the spatial variance in global productivity to changes in climate variables.

Table 9.1 Precipitation datasets used in this chapter and their main characteristics

Satellite product	CPC Merged Analysis of Precipitation (CMAP)	Global Precipitation Climatology Project (GPCP)	Tropical Rainfall Measuring Mission (TRMM)
Spatial resolution	2.5 × 2.5°	2.5 × 2.5°	0.25 × 0.25°
Spatial coverage	Global	Global	Latitude: 50 N – 50 S
			Longitude: 180 W – 180 E
Temporal resolution	Monthly	Monthly	Aggregated to monthly (from 3 hourly)
Temporal coverage	1979–present	1979–present	1998–present
Sensors included	GPCC rain gauge,	SSM/I emission,	SSM/I,
	IR-based GOES precipitation index,	SSM/I scattering,	Advanced Microwave Scanning Radiometer for Earth Observing System (AMSR-E),
	OLR precipitation index, Microwave Sounding	IR-based Goddard Earth Observing System (GEOS) precipitation index,	Advanced Microwave Sounding Unit-B (AMSU-B),
	Unit (MSU),	Television and Infrared	Infrared (IR) data from the international constellation of geosynchronous earth orbit (GEO) satellites, Gauge,
	SSM/I scattering,	Observation Satellite Operational Vertical Sounder (TOVS)-based estimates,	GPCC,
	SSM/I emission,	Outgoing longwave radiation (OLR) precipitation index, GPCC,	Climate Assessment and Monitoring System (CAMS).
	National Centers for Environmental Prediction–National Center for Atmospheric Research (NCEP–NCAR) reanalysis.	Global Historical Climate Network (GHCN, produced by NOAA) and CAMS.	
Download address	ftp://ftp.cpc.ncep.noaa.gov/precip/cmap/monthly	http://www1.ncdc.noaa.gov/pub/data/gpcp/v2/sat_gauge_precip	http://daac.gsfc.nasa.gov/data/datapool/TRMM/
References	Xie and Arkin (1997)	Adler et al. (2003)	Huffman et al. (2007)

The relative importance of precipitation, air temperature and incoming solar radiation for vegetation growth across the globe has been mapped by (Nemani et al. 2003) showing regions of different climatic dominant limiting factors. Areas primarily constrained by precipitation occupied approximately 50 % of the global

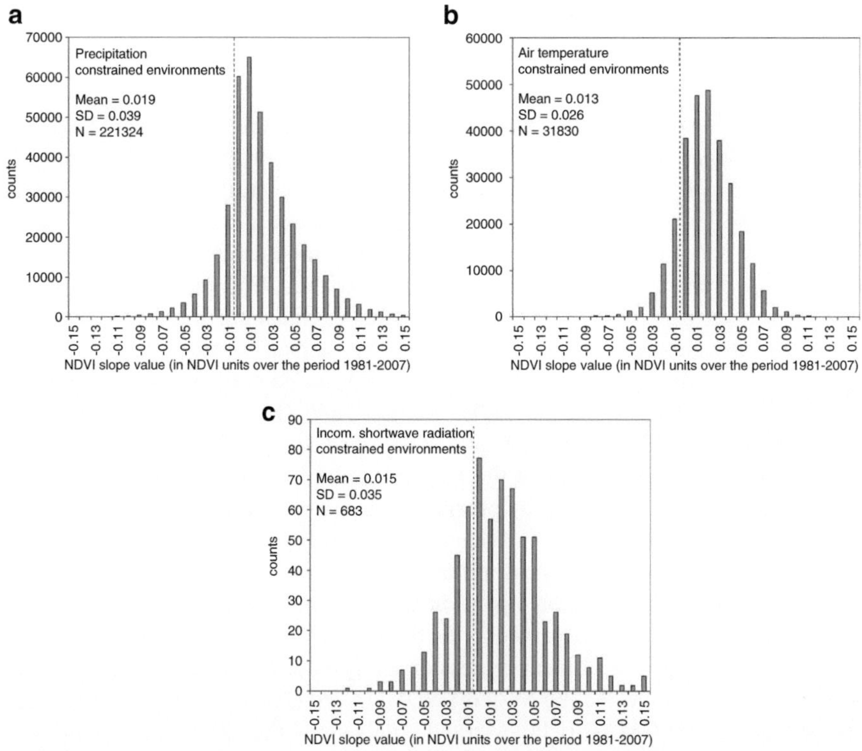

Fig. 9.2 Histograms of the NDVI slope in semi-arid areas from July 1981 to December 2007 in environments constrained by, (**a**) precipitation, (**b**) air temperature and (**c**) incoming shortwave radiation. Dashed vertical line represents NDVI trend values of 0 (NDVI units over the total period 1981–2007). Note the different scale on the y-axis value for each sub-plot due to the different number of pixels in each category

semi-arid areas, 7 % by air temperature and <1 % by incoming shortwave radiation (the remaining 42 % of semi-arid pixels were not characterized by a single predominant driver) (Fensholt et al. 2012). The NDVI trend coefficients of these three categories of potential climatic constraints to plant growth for semi-arid areas across the globe (Fig. 9.2a–c) were found to be positive on average for all three constraints (mean NDVI trend coefficients of 0.019, 0.013, and 0.015 for precipitation, air temperature, and incoming shortwave radiation, respectively). This implies that current generalizations, claiming that land degradation is ongoing in dryland areas worldwide (Adeel and World Resources Institute 2005; UNCCD Secretariat 2013) are not supported by the most recent satellite based analysis of vegetation greenness (being closely related to the key indicator of biological productivity).

9.2.1 Precipitation Controlling Observed Vegetation Changes in the Sahel

The Sahel is one of the world's largest dryland areas bordering the Sahara Desert to the north. Sahel has been referred to as the region of largest global rainfall anomalies during the last century (Nicholson 2000), suffering from recurrent droughts and large inter-annual variations in vegetation productivity. The grasslands of the Sahel constitute the basis for livestock production and the livelihoods of millions of people. Since the 'Sahel drought' of the 1970s and early 1980s, this zone has been described as a hotspot of land degradation, threatened both by recurrent droughts (Nicholson 2000) and by human overuse, e.g., through overgrazing (Hulme 2001; Lamb 1982) which is in contrast to more recent EO findings (Anyamba and Tucker 2005; Eklundh and Olsson 2003; Fensholt and Rasmussen 2011; Herrmann et al. 2005; Prince et al. 1998; Rasmussen et al. 2001). The productivity of the semi-natural grasslands of the Sahel is to a considerable extent controlled by precipitation. Recent analyses of trends in precipitation based on rain gauge measurements (Lebel and Ali 2009), as well as on global precipitation datasets (Fensholt and Rasmussen 2011; Fensholt et al. 2013; Huber et al. 2011) show that precipitation has increased in the Sahel since the mid-1980s. Thus the greening, observed in the field and by use of time series of satellite images, is not surprising. Linear regression analysis of GIMMS3g (Global Inventory Monitoring and Modeling System) NDVI (Normalized Difference Vegetation Index) against the CMAP (CPC Merged Analysis of Precipitation) precipitation (Xie and Arkin 1997) was conducted by Fensholt et al. (2013) for the period 1982–2010 (Fig. 9.3). An overall strong linear correlation between growing season integrated NDVI and precipitation is observed for the Sahel with 65.1 and 47.7 % of the pixels analysed being significantly positively correlated ($p < 0.05$ and 0.01, respectively).

9.2.2 Assessing Drivers of Observed Changes Based on Rain-Use Efficiency

If the greening in drylands is predominantly an effect of increased precipitation, it could be argued that this may disguise continued degradation caused by other factors, such as excessive cultivation and overgrazing. Over the last decades several studies have attempted to eliminate the effect of rainfall change (by a normalization procedure) on biological productivity, to better isolate the impact of non-rainfall related changes, e.g. human impacts (Evans and Geerken 2004; Prince et al. 1998; Wessels et al. 2007). This is sought to be captured by the concept of Rain-Use Efficiency (RUE), defined as the ratio of ANPP (aboveground net primary productivity) to annual precipitation (Le Houérou 1984, 1989; Prince et al. 1998). Consequently, changes in RUE have been suggested as an integral measure for evaluating

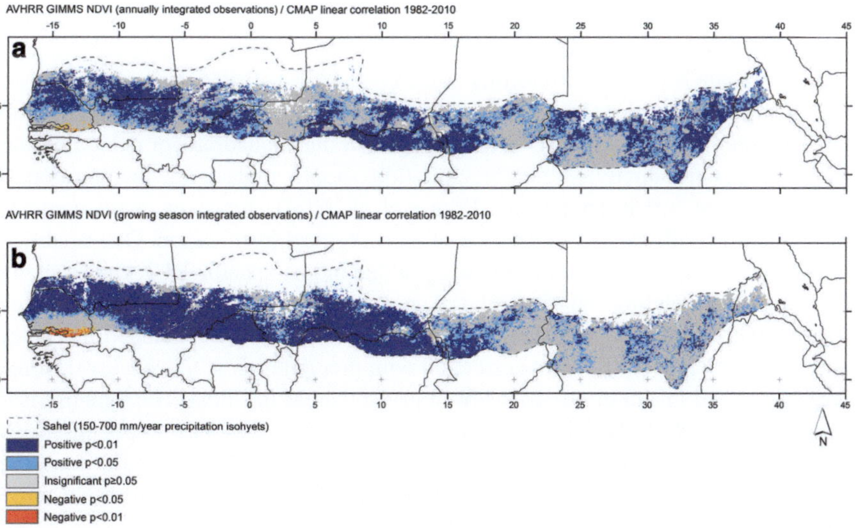

Fig. 9.3 Significance of correlation between GIMMS3g NDVI and precipitation from CMAP (Table 9.1) 1982–2010 for (**a**) Annually integrated NDVI (**b**) growing season integrated NDVI. CMAP precipitation has been resampled to match the spatial resolution of the GIMMS3g NDVI

land degradation and desertification and a number of authors have attempted to assess non-precipitation related land degradation – or the reverse – in global drylands. Time series of RUE have been estimated wholly or partly from satellite remote sensing or using only ground measurements (Bai et al. 2008; Hein and de Ridder 2006; Hein et al. 2011; Prince et al. 1998, 2007).

The basic assumption involved in the use of RUE is that NPP (net primary productivity) is proportional to (or at least linearly related to, see below) precipitation in the absence of human-induced land degradation. If this assumption of proportionality does not hold, the normalization for precipitation, which is the basis for the use of RUE is not successful (Prince 2002) and the use of RUE to detect non-precipitation related land degradation will become biased by changes in precipitation. Several papers have questioned this proportionality. It is well known that for increasing amounts of rainfall, the importance of water availability will at some point decrease (Prince et al. 2007) violating the assumption of proportionality between productivity and precipitation. An important question is whether the transition from water being the primary constraint for vegetation growth into other factors such as nutrients and incoming solar radiation, is observed for dryland areas. Using ground data from a variety of semi-arid rangelands in the Sahel and elsewhere, (Hein and de Ridder (2006), Hein et al. (2011)), as well as Hein (2006) argued that at high precipitation levels RUE will tend to decrease in dryland areas, because other production factors than water availability become limiting. The interval of annual precipitation in which proportionality may be assumed is debated and varies with vegetation, soil and climate. Hein et al. (2011) cited Breman and

Dewit (1983) for the statement that the proportionality breaks down at/above around 300 mm of rainfall per year.

Hein and de Ridder (2006) further argued that RUE will also decrease in areas of very low precipitation because most of the precipitation will evaporate and thus not be available for vegetation. Thus, they suggested that a quadratic or cubic relationship between productivity and precipitation should replace the assumption of proportionality for dryland areas. Prince et al. (2007) however challenged this interpretation with respect to the lack of an ecological justification. Other publications based on *in situ* measurements suggest that biome-specific RUE values should be applied depending on the rainfall regime (Huxman et al. 2004; Paruelo et al. 1999; Ruppert et al. 2012). However, Hu et al. (2010) concluded that inter-annual variation in RUE is not correlated with precipitation at the site level from a large dataset of *in situ* observations from dryland areas in China. The hypothesis of a constant of RUE for different species/rainfall regimes has implications for interpreting values of EO-based (Earth Observation) RUE in both the temporal and spatial domain since RUE values might not be directly inter-comparable across space for drylands receiving different amounts of rainfall (Prince et al. 1998). Also, if the amount of rainfall for a given pixel changes towards wetter or dryer conditions over time, this will have implications for the interpretation of RUE if a non-proportional relation between productivity and precipitation exists.

Based on annually integrated NDVI and annual precipitation (Fensholt and Rasmussen (2011), Fensholt et al. (2013)) demonstrated that for most pixels in the Sahel there is no proportionality, but sometimes a linear relation between ΣNDVI (seasonal or annual) and annual precipitation exists (as in Fig. 9.3). Proportionality is mathematically defined as the relationship of two variables whose ratio is constant, and unless the linear relationship between the vegetation metric and precipitation crosses the origin (0,0) of the Cartesian coordinate plane, proportionality is not obtained. It is argued that this lack of proportionality undermines the general use of satellite-based RUE time series as a means of identifying non-precipitation related land degradation (Fensholt et al. 2013), Veron et al. (2005). The specific data pre-processing of EO-based metrics for vegetation productivity have implications for the proportionality between productivity and precipitation and will therefore impact on the degradation/recovery assessment results obtained when using RUE. Fensholt et al. (2013) studied the sensitivity of the RUE approach to the EO-based proxies used (Fig. 9.4). Annually summed AVHRR GIMMS3g NDVI was shown to be linearly related to annual precipitation but no proportionality was found, thereby making a normalisation impossible (the inability of RUE to normalise for variability in precipitation is obvious from a remaining high per-pixel temporal correlation between RUE and precipitation). The results show significant negative trends in RUE (Fig. 9.4a) (primarily western and central Sahel). If substituting annually summed AVHRR GIMMS3g NDVI with a different vegetation productivity metric (the growing season integrated NDVI) in the RUE calculation (Fig. 9.4b), proportionality between productivity and precipitation was attained for the majority of pixels in the Sahel allowing for a successful use of RUE (no correlation between RUE and precipitation). The use of a growing season

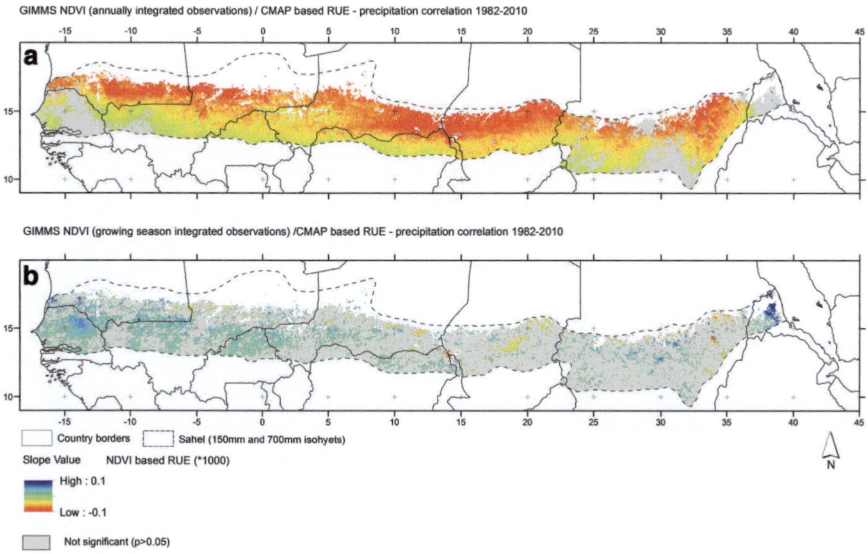

Fig. 9.4 RUE linear trends 1982–2010 based on (**a**) Annual sums of AVHRR GIMMS3g NDVI. (**b**) Growing season integrals of AVHRR GIMMS3g NDVI. Both productivity estimates are divided by annual sums of precipitation from GPCP (Table 9.1) to obtain Rain-Use Efficiency (RUE). GPCP precipitation has been resampled to match the spatial resolution of the GIMMS3g NDVI

integral of NDVI, which does not violate proportionality, produces very different results in trends of RUE, with the majority of pixels being characterised by significant positive RUE trends across the entire Sahelian belt. Clearly, this example illustrates that widely different conclusions concerning drivers of observed changes in vegetation trends and land degradation in the Sahel may be obtained depending on the vegetation parameterization approach used for the RUE analysis. Care must be taken that the assumed precipitation normalisation is in fact successful; otherwise trends in RUE will be nothing but a simple reflection of the trend in the precipitation dataset or perhaps other factors controlling NPP.

A different use of RUE as a measure of land degradation has been suggested by (del Barrio et al. 2010). The RUE values for each site and date were rescaled according to the upper and lower bounds of the VI (vegetation index)/precipitation point scatter to calculate the performance of RUE for a given landscape location to a reference potential conditions (i.e. maximum RUE observed) for this landscape type. However, the reference values depend on the actual observations, and assume that some areas are in their potential condition and others are fully degraded. It could also be that the RUE of a given pixel as compared to a reference landscape will be dependent on local soil variability and topographic conditions.

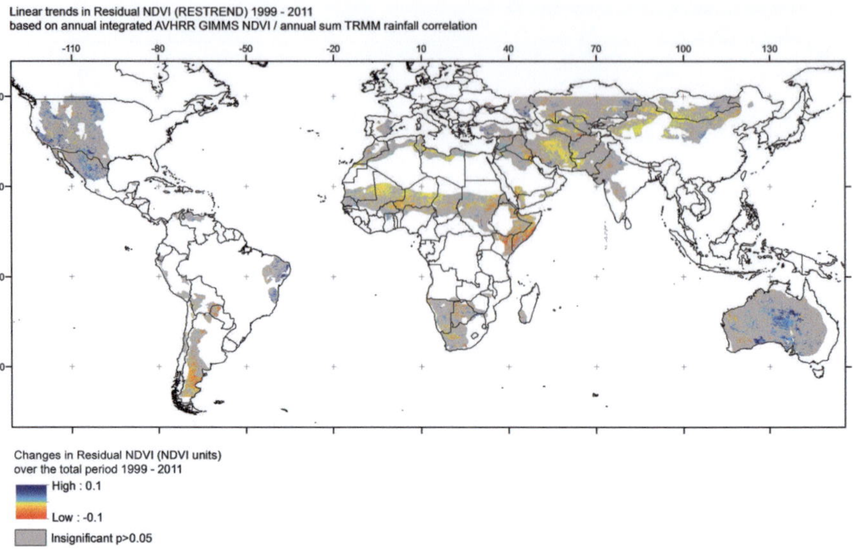

Fig. 9.5 Linear trends in Residual NDVI (RESTREND) 1999–2011. The residuals were estimated from linear regressions between annual integrated AVHRR GIMMS3g NDVI and annual summed TRMM (Table 9.1) rainfall. TRMM precipitation has been resampled to match the spatial resolution of the GIMMS3g NDVI

9.2.3 Assessing Drivers of Observed Changes Using the Residual Trends Productivity Approach

A different approach, called Residual Trend Analysis (RESTREND), has been developed in an attempt to distinguish rainfall-related variations and trends from human-induced land degradation (Archer 2004; Evans and Geerken 2004; Wessels et al. 2007). Following this method, per-pixel ΣNDVI (seasonal or annual) is regressed against annual precipitation, as with RUE, and then residuals are calculated for each site/time point from the best-fit linear regression for all sites. These residuals are then plotted against time to detect any temporal trends in deviations from the potential (as estimated by the best-fit regression). Just as with RUE, RESTREND seeks to expose factors other than precipitation, including a human-induced change (Herrmann et al. 2005; Huber et al. 2011; Wessels et al. 2007) that affect NPP.

In Fig. 9.5 AVHRR GIMMS3g NDVI was regressed against satellite-measured precipitation data from TRMM (Tropical Rainfall Measuring Mission; latitudinal coverage: 50 N-50S) from 1999 to 2011 and then used in a RESTREND analysis. Mixed patterns of increasing and decreasing trends of residual NDVI can be observed, with large negative values in eastern Africa and southern America and mainly positive trends in Australia and northern America.

9.2.4 Limitations/Challenges for RUE and RESTREND Approaches

Instead of assuming proportionality or linearity between precipitation and productivity for the use of RUE as being criticised by Hein and de Ridder (2006) and Hein et al. (2011) it was suggested by Fensholt et al. (2013) to restrict the analysis applicable to the RUE approach to regions or pixels for which proportionality can be shown to exist from remotely sensed data. This allows for maintaining the basic simple notion of RUE (as formulated by Le Houérou (1984)) as a means of normalizing for the effect rainfall on vegetation productivity and also helps in defining the limits within which RUE should be applied, i.e. to regions where rainfall is the primary constraint to vegetation growth. For a given pixel, however, in the case of severe ongoing land degradation in the middle part of the time series being studied, the linearity between rainfall and productivity may decline. This may be captured in the RUE time series as gradual changes that may reverse over time involving a trend break. Hence, if one applies strict statistical criteria at the per-pixel level there is a risk of excluding pixels from the analysis that are in fact the ones showing signs of human-induced land degradation. It is therefore suggested to apply the statistical requirement of a significant correlation between precipitation and vegetation productivity to be fulfilled at the regional level by a zonation/stratification of the per-pixel relation.

Also the use of the RESTREND approach for assessing human induced influence of vegetation changes is based on linearity between rainfall and productivity. For pixels for which a high linear correlation between ΣNDVI (annual/seasonal) and annual precipitation exists, meaningful estimations based on the RESTREND technique is feasible. If, however, for a given pixel a weak relation between ΣNDVI and annual rainfall exists, this approach is of little use, because the uncertainty caused by estimating the NDVI residuals increases proportionally. As pointed out by Wessels et al. (2012) this is likely to happen for a scenario, as above, where human-induced land degradation starts in the middle of a time series. A simulated degradation intensity $\geq$20 % was shown to cause an otherwise strong relationship between NDVI and rainfall to break down, thereby making the RESTREND an unreliable indicator of land human induced degradation.

A way to minimize the effects of fitting only one linear regression for the whole time-series is the identification of gradual or abrupt changes in the RUE time series using change detection method such as Breaks For Additive Seasonal and Trend (BFAST) (Verbesselt et al. 2010a, b). As described in the previous chapter, the basic principle of the BFAST algorithm is the decomposition of a time series into seasonal, trend, and remainder components, coupled with the detection of abrupt changes in both the trend and seasonal components. BFAST enables the detection of trend changes within EO time series assuming that nonlinearity can be approximated by piecewise linear models. This type of analysis can provide valuable information on the occurrence of trend changes, as well as on the timing and

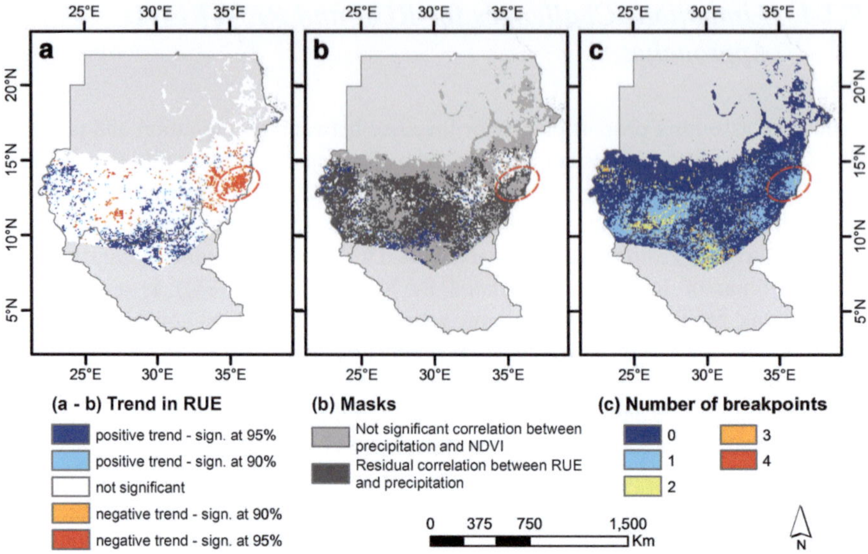

Fig. 9.6 (**a**) Direction and significance of 1982–2011 trends in RainUse Efficiency derived from the GIMMS3g NDVI and the GPCP yearly totals for dryland areas of Sudan. Non vegetated areas were masked out (*light grey*). (**b**) As in (a) but superimposed by pixels being masked due to lack of correlation between rainfall and NDVI (*medium grey*) and residual correlation between RUE and rainfall (*dark grey*). (**c**) Number of break points in Rain-Use Efficiency identified by BFAST between 1982 and 2011

magnitude of related break points in the time series (de Jong et al. 2012, 2013b; Verbesselt et al. 2012). Land degradation assessment based on a joint analysis of both long-term trends and abrupt changes in precipitation and vegetation time series should therefore be more accurate as they will not be solely based on diagnosis of long-term linear changes in ecosystem efficiencies but will also use more accurate information on potential abrupt changes observed either in climate or in the vegetation traits.

An example of the application of BFAST to address the issues of the RUE and RESTREND approaches for land degradation assessment is an analysis for Sudan (Fig. 9.6). Sudan is characterised by widespread and rapidly accelerating environmental degradation, which is sufficiently severe to be amongst the factors triggering tensions and conflicts (United Nations Environment Programme. 2007). This example is based on the growing-season NDVI integral derived from the GIMMS3g archive (1981 to the present) used as proxy for vegetation productivity and annual precipitation from the Global Precipitation Climatology Project (GPCP) (Table 9.1). Figure 9.6a shows the trends in RUE without taking into consideration that there are large areas of the semi-arid Sudan where the preconditions for using RUE are not fulfilled (lack of linearity between vegetation productivity and precipitation and/or residual correlations between RUE and precipitation are observed) (Fig. 9.6b). However, breaks in the RUE time series detected by the BFAST

(Fig. 9.6c) indicate that for many pixels (e.g. the region highlighted with a red circle) the linearity assumption on which the RUE approach is based upon is not fulfilled because of a distinct breakpoint within the period of analysis. Therefore the rejection of these pixels based on a too strict statistical criteria of linearity may actually lead to disregarding some of the regions that are most vulnerable and most seriously hit by land degradation.

9.3 Assessment of the Roles of Climate on Anomalies of Dryland Vegetation Productivity

9.3.1 Combining Dynamic Global Vegetation Models and EO Data

Recent studies based on process-based modelling approaches (Dynamic Global Vegetation Models; DGVM's) have attempted to disentangle the climate and human effects on the Sahelian greening (Hickler et al. 2005; Seaquist et al. 2009) and greening at the global scale (Mao et al. 2013). The use of DGVM's like the LPJ (Lund-Potsdam-Jena) (Sitch et al. 2003) allows studying the causes for current and historical variability and trends in vegetation productivity of global drylands when comparing against time series of EO data for the same period (Hickler et al. 2005). DGVM's provide the potential vegetation properties and modelling includes atmospheric CO_2 fertilization, nitrogen/phosphorous deposition and land use and land cover change (not accommodated in EO-based Light Use Efficiency (LUE) approaches) as well as dryland resilience in the context of disturbance processes from human influence like bush fires. Discrepancies between modelled and EO-based observed productivity have therefore been used as means of inferring information of drivers of changes (Seaquist et al. 2009). Combining process-based ecosystem models with high-temporal resolution remote sensing using data assimilation offers an interesting way forward adding insights about the patterns and mechanisms driving observed vegetation dynamics at these spatial scales; yet it remains an underutilized avenue of research (Seaquist et al. 2012).

9.3.2 Sea Surface Temperature and Vegetation Productivity Teleconnections

Vegetation productivity across different dryland regions is known to be affected locally by changes in precipitation as discussed in the previous sections. The causes of inter-annual precipitation variability has also been related to variability in regional climate driven by SST patterns. In the African Sahel, the reasons for the large inter-annual and decadal fluctuations in rainfall are still not entirely understood, but early works by (Folland et al. (1986), Lamb (1978), Palmer 1986) found a relationship (teleconnection) with regional and global SST conditions. Sahelian precipitation and SST patterns have been related to the ENSO (El Nino Southern Oscillation) and NAO (North Atlantic Oscillation) (Biasutti et al. 2008; Palmer 1986; Shanahan et al. 2009; Ward 1998). Relationships between precipitation and SST have been found also in the Pacific (Caminade and Terray 2010; Janicot et al. 1998; Mohino et al. 2011), the Indian Ocean (Bader and Latif 2003; Giannini et al. 2003; Lu 2009) and the Mediterranean (Philippon et al. 2007; Raicich et al. 2003; Rowell 2003).

In the Sahel, the importance of SST on precipitation is still unclear. While several studies have reported limited correlations (Anyamba and Eastman 1996; Anyamba and Tucker 2005; Anyamba et al. 2001; Philippon et al. 2007; Propastin et al. 2010), others have shown stronger relationships (Camberlin et al. 2001; Oba et al. 2001; Ward 1998). Oba et al. (2001), attributed large parts of the inter-annual variation of vegetation productivity during the 1980s to the NAO. Wang (2003), on the other hand, did not find a consistent relationship. Other studies (e.g. Brown et al. (2010)) have found significant relationships individually between the Pacific Decadal Oscillation (PDO) and two phenological metrics of NDVI (start of season and seasonal integrated NDVI) in West Africa but a limited influence of the Indian Ocean Dipole (IOD). However, although Williams and Hanan (2011), found the IOD and the Multivariate ENSO Index (MEI) to be related to rainfall individually, (when taken together) interacting effects of the two indices removed the correlations.

Direct relationships between SST and vegetation measurements from AVHRR time series have also been demonstrated. For example, Huber and Fensholt (2011) studied the direct correlations between the Sahelian dryland vegetation variability and large-scale ocean–atmosphere phenomena causing changes in SST patterns. It was concluded that over the last 3 decades, significant correlations existed between global climate indices/SST anomalies and Sahelian productivity, however with different characteristics in western, central and eastern Sahel. Whereas the vegetation productivity in the western Sahel could be associated with SST for large oceanic areas of the Pacific, the Atlantic as well as the Indian Ocean (Fig. 9.7), for the eastern Sahel only small areas in the Atlantic were found to be significantly related to dynamics in NDVI.

Overall, these large scale climate indices and especially SST anomalies for specific ocean areas were found to have predictive power expressed by a statistically significant relation between northern latitude winter/spring SSTs and summer vegetation productivity in the Sahel (Fig. 9.8). This time lag of several months could be of immense importance for forecasting annual vegetation productivity in this region and possibly in other dryland areas across the globe, home to the world's poorest populations.

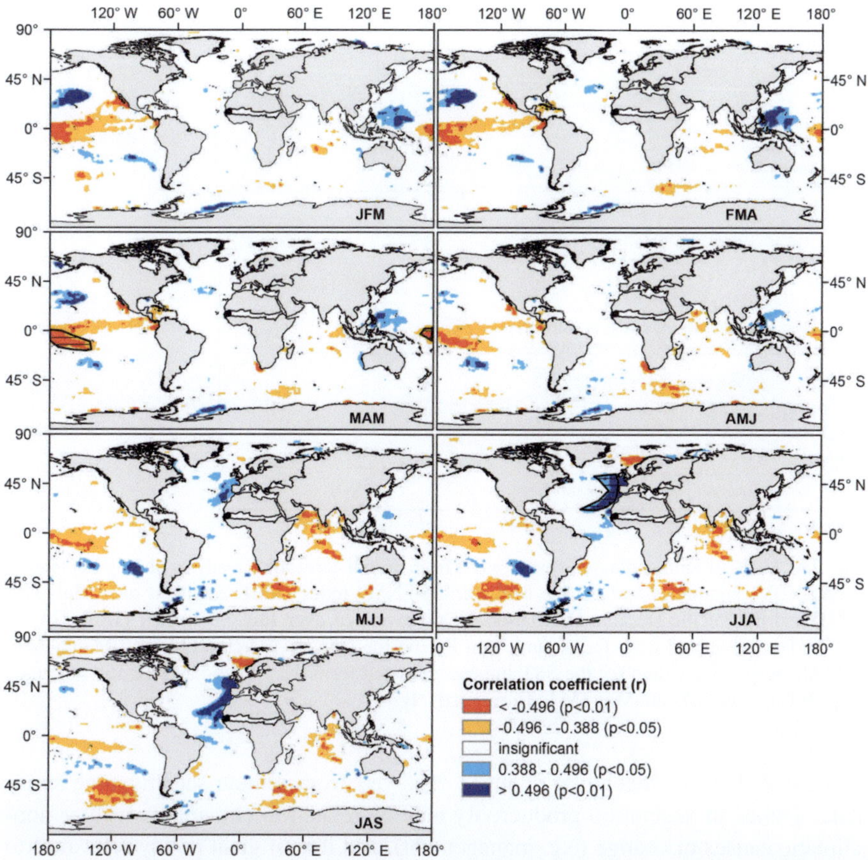

Fig. 9.7 Maps of significant correlation coefficients ($p < 0.05$) between the Sahel NDVI anomaly index (based on July–September NDVI (JAS)) for the West African Sahel sub-region and mean SST anomalies from 1982 to 2007 for different intra-annual time lags (e.g., correlation between JAS NDVI anomalies and JFM (January–March) SST anomalies)

9.4 Summary

The United Nations Convention to Combat Desertification (UNCCD) definition of desertification (degradation in dryland areas) implies that change in vegetation productivity is a key indicator (but not the only one) of land degradation. Spatially and temporally consistent, long-term data on vegetation productivity is therefore of great interest for the assessment of changes in environmental conditions in dryland regions and Earth Observation (EO) satellite data provide the only suitable means of consistent monitoring of changes at the global scale.

Current generalizations, claiming that land degradation is ongoing in dryland areas worldwide are not supported by recent satellite based analysis of vegetation and this chapter introduced some of the most widely used methods of inferring

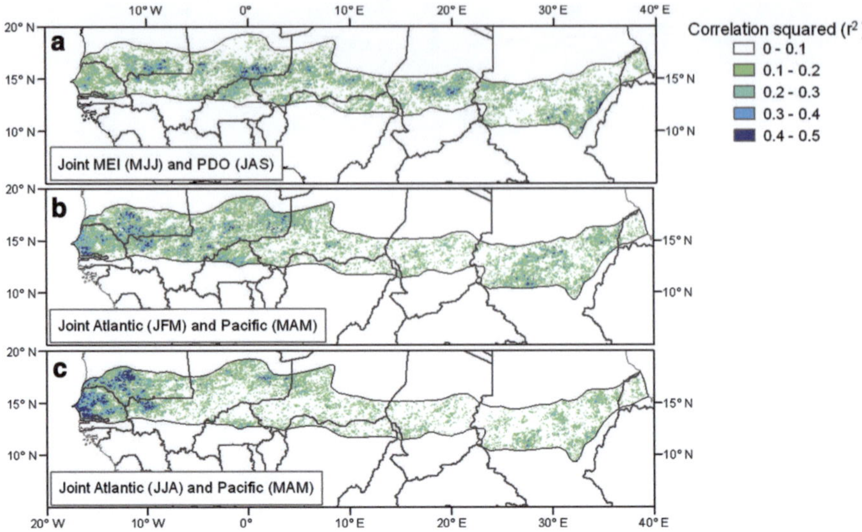

Fig. 9.8 Maps of joint explained variance (r^2) from partial correlation analysis of July-September (JAS) NDVI anomalies and (**a**) the Multivariate ENSO Index (MEI) averaged over May–July (MJJ) and the Pacific Decadal Oscillation (PDO) averaged over July–September (JAS), (**b**) the SST indices extracted from the Atlantic and Pacific for January–March (JFM) and March–May (MAM), respectively, and (**c**) the SST indices extracted from the Atlantic and Pacific for June–August (JJA) and March–May (MAM), respectively

drivers of dryland vegetation changes observed from remote sensing time series data. Trends in vegetation productivity may be related to climatic as well as non-climatic causes of change (e.g. management), and it is of great policy relevance to better understand the drivers and causal mechanisms of observed productivity trends. However, one of the main challenges in dryland vegetation research remains resolving and disentangling the impact from climate and human induced land use change respectively. A strong coupling between EO-based vegetation dynamics and precipitation and/or temperature was found in most dryland areas but also large regions of non-significant correlation between rainfall/temperature and vegetation growth was observed pointing towards human influence on vegetation from changes in land use practices.

The Sahel (being one of the world's largest dryland areas) has suffered from recurrent droughts and large inter-annual variations in vegetation productivity over recent decades. Sahel was used here as a showcase for two interrelated methods of detecting the impact of non-rainfall related changes on vegetation; the concept of RUE (Rain-Use Efficiency) and the RESTREND approach. Both approaches however are based on the assumption of a strong per-pixel linear relationship between rainfall and productivity (over time) that might be compromised in the case of escalating land degradation during the period under study. Rather than fitting only one per-pixel linear regression for the whole time-series, it is suggested here to combine a change detection method such as BFAST (Breaks For Additive Seasonal and Trend) (see previous chapter) for identification of time series breakpoints in combination with RUE/RESTREND approaches to overcome the problem of the assumption of long-term rainfall/vegetation linearity that might be incompatible with the manifestation of degradation.

Finally, global sea surface temperature (SST) anomalies (caused by large-scale ocean–atmosphere phenomena) were shown to be teleconnected to regional scale vegetation productivity in the Sahel, thereby being important for an improved understanding of inter-annual changes in the Sahelian dryland productivity. Large scale climate indices and especially SST anomalies for specific ocean areas were found to have predictive power for the vegetation productivity in the Sahel and the existence of a time lag of several months between SST anomalies and vegetation productivity provides an important basis for forecasting annual vegetation productivity in this region and possibly in other dryland areas across the globe.

Acknowledgements This book chapter is written within the frame of the project entitled Earth Observation based Vegetation productivity and Land Degradation Trends in Global Drylands. The project is funded by the Danish Council for Independent Research (DFF) Sapere Aude programme.

References

Adeel Z, World Resources Institute (2005) Ecosystems and human well-being: desertification synthesis: a report of the millennium ecosystem assessment. World Resources Institute, Washington, DC

Adler RF, Huffman GJ, Chang A, Ferraro R, Xie PP, Janowiak J, Rudolf B, Schneider U, Curtis S, Bolvin D, Gruber A, Susskind J, Arkin P, Nelkin E (2003) The version-2 global precipitation climatology project (GPCP) monthly precipitation analysis (1979-present). J Hydrometeorol 4:1147–1167

Anyamba A, Eastman JR (1996) Interannual variability of NDVI over Africa and its relation to El Nino Southern Oscillation. Int J Remote Sens 17:2533–2548

Anyamba A, Tucker CJ (2005) Analysis of Sahelian vegetation dynamics using NOAA-AVHRR NDVI data from 1981–2003. J Arid Environ 63:596–614

Anyamba A, Tucker CJ, Eastman JR (2001) NDVI anomaly patterns over Africa during the 1997/98 ENSO warm event. Int J Remote Sens 22:1847–1859

Archer ERM (2004) Beyond the "climate versus grazing" impasse: using remote sensing to investigate the effects of grazing system choice on vegetation cover in the eastern Karoo. J Arid Environ 57:381–408

Bader J, Latif M (2003) The impact of decadal-scale Indian Ocean sea surface temperature anomalies on Sahelian rainfall and the North Atlantic Oscillation. Geophys Res Lett 30:2169. doi:10.1029/2003GL018426

Bai ZG, Dent DL, Olsson L, Schaepman ME (2008) Proxy global assessment of land degradation. Soil Use Manag 24:223–234

Biasutti M, Held IM, Sobel AH, Giannini A (2008) SST forcings and Sahel rainfall variability in simulations of the twentieth and twenty-first centuries. J Clim 21:3471–3486

Breman H, Dewit CT (1983) Rangeland productivity and exploitation in the Sahel. Science 221:1341–1347

Brown ME, de Beurs K, Vrieling A (2010) The response of African land surface phenology to large scale climate oscillations. Remote Sens Environ 114:2286–2296

Camberlin P, Janicot S, Poccard I (2001) Seasonality and atmospheric dynamics of the teleconnection between African rainfall and tropical sea-surface temperature: Atlantic vs. ENSO. Int J Climatol 21:973–1005

Caminade C, Terray L (2010) Twentieth century Sahel rainfall variability as simulated by the ARPEGE AGCM, and future changes. Clim Dyn 35:75–94

de Jong R, Verbesselt J, Schaepman ME, de Bruin S (2012) Trend changes in global greening and browning: contribution of short-term trends to longer-term change. Glob Chang Biol 18:642–655

de Jong R, Schaepman ME, Furrer R, De Bruin S, Verburg PH (2013a) Spatial relationship between climatologies and changes in global vegetation activity. Glob Chang Biol 19:1953–1964

de Jong R, Verbesselt J, Zeileis A, Schaepman ME (2013b) Shifts in global vegetation activity trends. Remote Sens 5:1117–1133

del Barrio G, Puigdefabregas J, Sanjuan ME, Stellmes M, Ruiz A (2010) Assessment and monitoring of land condition in the Iberian Peninsula, 1989–2000. Remote Sens Environ 114:1817–1832

Eklundh L, Olsson L (2003) Vegetation index trends for the African Sahel 1982–1999. Geophys Res Lett 30(8)

Evans J, Geerken R (2004) Discrimination between climate and human-induced dryland degradation. J Arid Environ 57:535–554

Fensholt R, Rasmussen K (2011) Analysis of trends in the Sahelian ‘rain-use efficiency’ using GIMMS NDVI, RFE and GPCP rainfall data. Remote Sens Environ 115:438–451

Fensholt R, Langanke T, Rasmussen K, Reenberg A, Prince SD, Tucker C, Scholes RJ, Le QB, Bondeau A, Eastman R, Epstein H, Gaughan AE, Hellden U, Mbow C, Olsson L, Paruelo J, Schweitzer C, Seaquist J, Wessels K (2012) Greenness in semi-arid areas across the globe 1981–2007 – an Earth observing satellite based analysis of trends and drivers. Remote Sens Environ 121:144–158

Fensholt R, Rasmussen K, Kaspersen P, Huber S, Horion S, Swinnen E (2013) Assessing land degradation/recovery in the African Sahel from long-term earth observation based primary productivity and precipitation relationships. Remote Sens 5:664–686

Folland CK, Palmer TN, Parker DE (1986) Sahel rainfall and worldwide sea temperatures, 1901–85. Nature 320:602–607

Giannini A, Saravanan R, Chang P (2003) Oceanic forcing of Sahel rainfall on interannual to interdecadal time scales. Science 302:1027–1030

Hein L (2006) The impacts of grazing and rainfall variability on the dynamics of a Sahelian rangeland. J Arid Environ 64:488–504

Hein L, de Ridder N (2006) Desertification in the Sahel: a reinterpretation. Glob Chang Biol 12:751–758

Hein L, de Ridder N, Hiernaux P, Leemans R, de Wit A, Schaepman M (2011) Desertification in the Sahel: towards better accounting for ecosystem dynamics in the interpretation of remote sensing images. J Arid Environ 75:1164–1172

Herrmann SM, Anyamba A, Tucker CJ (2005) Recent trends in vegetation dynamics in the African Sahel and their relationship to climate. Glob Environ Change-Hum Policy Dimens 15:394–404

Hickler T, Eklundh L, Seaquist JW, Smith B, Ardo J, Olsson L, Sykes MT, Sjostrom M (2005) Precipitation controls Sahel greening trend. Geophys Res Lett 32(21)

Hu ZM, Yu GR, Fan JW, Zhong HP, Wang SQ, Li SG (2010) Precipitation-use efficiency along a 4500-km grassland transect. Glob Ecol Biogeogr 19:842–851

Huber S, Fensholt R (2011) Analysis of teleconnections between AVHRR-based sea surface temperature and vegetation productivity in the semi-arid Sahel. Remote Sens Environ 115:3276–3285

Huber S, Fensholt R, Rasmussen K (2011) Water availability as the driver of vegetation dynamics in the African Sahel from 1982 to 2007. Glob Planet Chang 76:186–195

Huffman GJ, Adler RF, Bolvin DT, Gu GJ, Nelkin EJ, Bowman KP, Hong Y, Stocker EF, Wolff DB (2007) The TRMM multisatellite precipitation analysis (TMPA): Quasi-global, multiyear, combined-sensor precipitation estimates at fine scales. J Hydrometeorol 8:38–55

Huffman GJ, Adler RF, Bolvin DT, Gu G (2009) Improving the global precipitation record: GPCP version 2.1. Geophys Res Lett. 36:L17808. doi:10.1029/2009GL040000

Hulme M (2001) Climatic perspectives on Sahelian desiccation: 1973–1998. Glob Environ Change-Hum Policy Dimens 11:19–29

Huxman TE, Smith MD, Fay PA, Knapp AK, Shaw MR, Loik ME, Smith SD, Tissue DT, Zak JC, Weltzin JF, Pockman WT, Sala OE, Haddad BM, Harte J, Koch GW, Schwinning S, Small EE, Williams DG (2004) Convergence across biomes to a common rain-use efficiency. Nature 429:651–654

Janicot S, Harzallah A, Fontaine B, Moron V (1998) West African monsoon dynamics and eastern equatorial Atlantic and Pacific SST anomalies (1970–88). J Clim 11:1874–1882

Lamb PJ (1978) Large-scale tropical Atlantic surface circulation patterns associated with sub-Saharan weather anomalies. Tellus 30:240–251

Lamb PJ (1982) Persistence of sub-Saharan drought. Nature 299:46–48

Le Houérou HN (1984) Rain use efficiency – a unifying concept in arid-land ecology. J Arid Environ 7:213–247

Le Houérou HN (1989) The grazing land ecosystems of the African Sahel. Springer, Berlin/New York

Lebel T, Ali A (2009) Recent trends in the Central and Western Sahel rainfall regime (1990–2007). J Hydrol 375:52–64

Lu J (2009) The dynamics of the Indian Ocean sea surface temperature forcing of Sahel drought. Clim Dyn 33:445–460

Mao JF, Shi XY, Thornton PE, Hoffman FM, Zhu ZC, Myneni RB (2013) Global latitudinal-asymmetric vegetation growth trends and their driving mechanisms: 1982–2009. Remote Sens 5:1484–1497

Mohino E, Janicot S, Bader J (2011) Sahel rainfall and decadal to multi-decadal sea surface temperature variability. Clim Dyn 37:419–440

Nemani RR, Keeling CD, Hashimoto H, Jolly WM, Piper SC, Tucker CJ, Myneni RB, Running SW (2003) Climate-driven increases in global terrestrial net primary production from 1982 to 1999. Science 300:1560–1563

Nicholson SE (2000) The nature of rainfall variability over Africa on time scales of decades to millenia. Glob Planet Chang 26:137–158

Oba G, Post E, Stenseth NC (2001) Sub-Saharan desertification and productivity are linked to hemispheric climate variability. Glob Chang Biol 7:241–246

Palmer TN (1986) Influence of the Atlantic, Pacific and Indian Oceans on Sahel Rainfall. Nature 322:251–253

Paruelo JM, Lauenroth WK, Burke IC, Sala OE (1999) Grassland precipitation-use efficiency varies across a resource gradient. Ecosystems 2:64–68

Philippon N, Jarlan L, Martiny N, Camberlin P, Mougin E (2007) Characterization of the interannual and intraseasonal variability of West African vegetation between 1982 and 2002 by means of NOAA AVHRR NDVI data. J Clim 20:1202–1218

Prince SD (2002) Spatial and temporal scales of measurement of desertification. In: Stafford-Smith M, Reynolds JF (eds) Global desertification: do humans create deserts? Dahlem University Press, Berlin, pp 23–40

Prince SD, De Colstoun EB, Kravitz LL (1998) Evidence from rain-use efficiencies does not indicate extensive Sahelian desertification. Glob Chang Biol 4:359–374

Prince SD, Wessels KJ, Tucker CJ, Nicholson SE (2007) Desertification in the Sahel: a reinterpretation of a reinterpretation. Glob Chang Biol 13:1308–1313

Propastin P, Fotso L, Kappas M (2010) Assessment of vegetation vulnerability to ENSO warm events over Africa. Int J Appl Earth Obs Geoinform 12:S83–S89

Raicich F, Pinardi N, Navarra A (2003) Teleconnections between Indian monsoon and Sahel rainfall and the Mediterranean. Int J Climatol 23:173–186

Rasmussen K, Fog B, Madsen JE (2001) Desertification in reverse? Observations from northern Burkina Faso. Global Environ Change-Hum Policy Dimens 11:271–282

Rowell DP (2003) The impact of Mediterranean SSTs on the Sahelian rainfall season. J Clim 16:849–862

Ruppert JC, Holm A, Miehe S, Muldavin E, Snyman HA, Wesche K, Linstadter A (2012) Meta-analysis of ANPP and rain-use efficiency confirms indicative value for degradation and supports non-linear response along precipitation gradients in drylands. J Veg Sci 23:1035–1050

Seaquist JW, Hickler T, Eklundh L, Ardo J, Heumann BW (2009) Disentangling the effects of climate and people on Sahel vegetation dynamics. Biogeosciences 6:469–477

Seaquist J, Lehsten V, Lindeskog M, Eklundh L (2012) Analyzing vegetation dynamics by combining remote sensing with process-based ecosystem models. In: Workshop on temporal analysis of satellite images, Mykonos, Greece

Shanahan TM, Overpeck JT, Anchukaitis KJ, Beck JW, Cole JE, Dettman DL, Peck JA, Scholz CA, King JW (2009) Atlantic forcing of persistent drought in West Africa. Science 324:377–380

Sitch S, Smith B, Prentice IC, Arneth A, Bondeau A, Cramer W, Kaplan JO, Levis S, Lucht W, Sykes MT, Thonicke K, Venevsky S (2003) Evaluation of ecosystem dynamics, plant geography and terrestrial carbon cycling in the LPJ dynamic global vegetation model. Glob Chang Biol 9:161–185

UNCCD Secretariat (2013) A stronger UNCCD for a land-degradation neutral world, Issue Brief, Bonn, Germany

United Nations Environment Programme (2007) Sudan: post-conflict environmental assessment. United Nations Environment Programme, Nairobi

Verbesselt J, Hyndman R, Newnham G, Culvenor D (2010a) Detecting trend and sea-sonal changes in satellite image time series. Remote Sens Environ 114:106–115

Verbesselt J, Hyndman R, Zeileis A, Culvenor D (2010b) Phenological change detection while accounting for abrupt and gradual trends in satellite image time series. Remote Sens Environ 114:2970–2980

Verbesselt J, Zeileis A, Herold M (2012) Near real-time disturbance detection using satellite image time series. Remote Sens Environ 123:98–108

Veron SR, Oesterheld M, Paruelo JM (2005) Production as a function of resource availability: slopes and efficiencies are different. J Veg Sci 16:351–354

Wang GL (2003) Reassessing the impact of North Atlantic Oscillation on the sub-Saharan vegetation productivity. Glob Chang Biol 9:493–499

Ward MN (1998) Diagnosis and short-lead time prediction of summer rainfall in tropical North Africa at interannual and multidecadal timescales. J Clim 11:3167–3191

Wessels KJ, Prince SD, Malherbe J, Small J, Frost PE, VanZyl D (2007) Can human-induced land degradation be distinguished from the effects of rainfall variability? A case study in South Africa. J Arid Environ 68:271–297

Wessels KJ, van den Bergh F, Scholes RJ (2012) Limits to detectability of land degradation by trend analysis of vegetation index data. Remote Sens Environ 125:10–22

Williams CA, Hanan NP (2011) ENSO and IOD teleconnections for African ecosystems: evidence of destructive interference between climate oscillations. Biogeosciences 8:27–40

Xie PP, Arkin PA (1997) Global precipitation: a 17-year monthly analysis based on gauge observations, satellite estimates, and numerical model outputs. Bull Am Meteorol Soc 78:2539–2558

Chapter 10
Land Surface Phenology in a West African Savanna: Impact of Land Use, Land Cover and Fire

Ursula Gessner, Kim Knauer, Claudia Kuenzer, and Stefan Dech

Abstract Phenological change and variation have become increasingly relevant topics in global change science due to recognition of their importance for ecosystem functioning and biogeophysical processes. Remote sensing time series offer great potential for assessing phenological dynamics at landscape, regional and global scales. Even though a number of studies have investigated phenology, mostly with a focus on climatic variability, we do not yet have a detailed understanding of phenological cycles and respective biogeographical patterns. This is particularly true for biomes like the tropical savannas, which cover approximately one eighth of the global land surface. Savannas are often characterized by high human population density and growth, one example being the West African Sudanian Savanna. The phenological characteristics in these regions can be assumed to be particularly influenced by agricultural land use and fires, in addition to climatic variability. This study analyses the spatio-temporal patterns of land surface phenology in a Sudanian Savanna landscape of southern Burkina Faso based on time series of the Moderate Resolution Spectroradiometer (MODIS), and on multi-temporal Landsat data. The analyses focus on influences of fire, land use, and vegetation structure on phenological patterns, and disclose the effects of long-term fire frequency, as well as the short-

U. Gessner (✉) • C. Kuenzer
German Remote Sensing Data Center, DFD, Earth Observation Center, EOC, German Aerospace Center, DLR, Oberpfaffenhofen, Germany
e-mail: ursula.gessner@dlr.de; claudia.kuenzer@dlr.de

K. Knauer
German Remote Sensing Data Center, DFD, Earth Observation Center, EOC, German Aerospace Center, DLR, Oberpfaffenhofen, Germany

Remote Sensing, Institute of Geology and Geography, University of Wuerzburg, Wuerzburg, Germany
e-mail: kim.knauer@dlr.de

S. Dech
German Remote Sensing Data Center, DFD, German Aerospace Center, DLR, Wessling, Germany

Institute for Geography and Geology, University of Wuerzburg, Wuerzburg, Germany

C. Kuenzer et al. (eds.), *Remote Sensing Time Series*, Remote Sensing and Digital Image Processing, DOI 10.1007/978-3-319-15967-6_10

term effects of burning on the vegetation dynamics observed in the following growing season. Possibilities of further improvements for remote sensing based analyses of land surface phenology are seen in using earth observation datasets of increased spatial and temporal resolution as well as in linking phenological metrics from remote sensing with actual biological events observed on the ground.

10.1 Introduction

Phenology has become an increasingly relevant topic in global change science (e.g. Rosenzweig et al. 2007). It addresses "the timing of recurrent biological events, the causes of their timing with regard to biotic and abiotic forces, and the interrelation among phases of the same or different species" (Lieth 1974). During the last decades, changes in phenology have been observed in different regions of the world and could frequently be related to climatic variability and change or to alterations in land cover and land use practices (e.g. de Beurs and Henebry 2004; Heumann et al. 2007; Richardson et al. 2013). Phenological changes are of particular relevance as they may impact a number of ecosystem functions such as ecosystem productivity and carbon sequestration (e.g. Churkina et al. 2005; Richardson et al. 2010), reproductive patterns (e.g. Ramírez and Briceño 2011) and consumer-resource interactions (e.g. Kerby et al. 2012). Given the close link between phenology and resource availability for herbivores in many regions of the world, phenological variations can even impact land use decisions, such as long-distance movements of livestock (Butt et al. 2011). In addition to ecological aspects, phenology is closely linked to land-atmosphere fluxes of water and energy as it determines seasonal variations of biogeophysical land surface properties such as albedo (Ryu et al. 2008), leaf area index, or surface roughness length (Blanken et al. 1997).

Plant phenology is analyzed at the scale of individual plants or vegetation communities and is usually based on field assessments. Land surface phenology in contrast is studied at landscape, regional and global scales, where earth observation delivers information on the seasonal variations of the vegetated land surface (de Beurs and Henebry 2005; Henebry and de Beurs 2013). Space-borne, medium resolution (250 m–1 km) sensors provide daily to bi-weekly data of vegetation indices for large areas, and currently cover up to three decades of time series. This makes remote sensing an indispensable basis for many studies of phenology and phenological changes.

Tropical savannas cover approximately one eighth of the global land surface and more than half of the African continent (Scholes and Archer 1997). The land surface phenology of savannas has been studied at various scales, mainly focusing on the influence of climatic variability (e.g. Archibald and Scholes 2007; Butt et al. 2011; Heumann et al. 2007; Ma et al. 2013; Wagenseil and Samimi 2006). However, a detailed understanding of the phenological cycles of savanna ecosystems and their biogeographical patterns is still missing, particularly at landscape and regional scales (Ma et al. 2013). Furthermore, in regions like West Africa, where population density and growth are high, the phenological characteristics of savannas can be particularly influenced, in addition to climatic variability, by agricultural land use and fires (e.g. Devineau 1999; Devineau et al. 2010).

The aim of this study is to analyze the spatio-temporal patterns of phenology in a typical region of the West Sudanian Savanna in southern Burkina Faso. A special focus is on investigating the interrelations of phenological patterns with fire, land use and vegetation structure. It is assumed that fire can have both long-term, and direct, short-term effects on land surface phenology. Long-term effects might be observable, when the average frequency of fires over long time periods influences average phenological characteristics. Such effects may be due to shifts in species or growth form composition as a result of the long-term fire repetition rates (e.g. Hoffmann 1998). On the other hand there might be direct, short-term effects of fire events on the phenological development of the subsequent growing season. These effects could be, for example, resulting from the removal of dead biomass that gives way for new sprouting, or from potential fire damage that hinders or delays vegetation development. This study is based on optical remote sensing time series of the Moderate Resolution Spectroradiometer (MODIS) with a spatial resolution of 250 m, on the MODIS burned area product MCD45A1 with spatial resolution of 500 m, and on multi-date Landsat spectral data with a spatial resolution of 30 m. Spatio-temporal information on phenology, land use, land cover and fire occurrence is derived from these datasets, and analyzed for a 14-year period from 2000 to 2013.

10.2 Study Region

The study region is located in southern Burkina Faso, just north of the national border with Ghana and covers an area of approximately 15,400 km^2 (Fig. 10.1, 2°23.5′W-1°10.25′W/11°58.75′N-10°58.4′N). It belongs to the tributary area of the White Volta and includes the northern part of the Sissili catchment. The topography is relatively flat with elevations between 250 m and 400 m above sea level. Pronounced seasonal alterations between a monsoonal rainy season and a dry season characterize the region. The mean annual precipitation ranges between 800 and 980 mm and falls mainly between May/June and September/October. Temperatures are relatively constant throughout the year with monthly averages between 26 and 30 °C. The study region is part of the Sudanian Savanna zone (White 1983) with different types of savanna vegetation ranging from open grassland savannas to closed woodland savannas and small patches of forests, mainly along rivers. The vegetation is characterized by alternating periods of vegetation growth during the rainy season and reduced vegetation activity during the dry season. Large parts of the study area are intensively used, mainly for the cultivation of crops, and partly for grazing. A further intensification and expansion of land use can be expected for the future years, given the high annual population growth of 2.9 % in Burkina Faso (The World Bank 2013), in a society where currently more than 90 % of the population work in the field of agriculture (FAO 2013). Some parts of the study region are protected areas, most importantly the Nazinga Game Ranch,

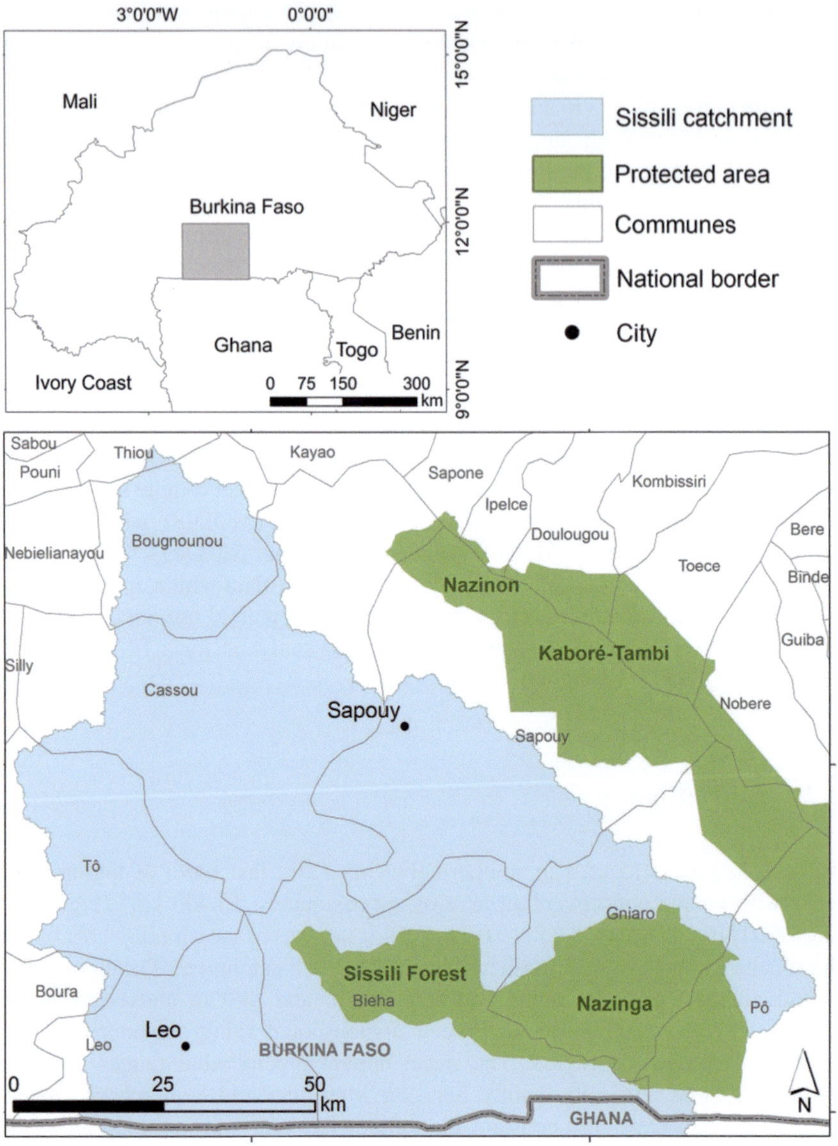

Fig. 10.1 Overview over the study region and its location in southern Burkina Faso

the Kaboré-Tambi (K.T.) National Park, the Sissili classified forest and the Nazinon reserved forest (Fig. 10.1). Regular fires are a typical feature of the study region and have a key influence on the vegetation structure of savannas (Rueth 2010; Wardell et al. 2004). While some fires in West Africa have natural causes, the majority of burnings are human-induced and often closely related to land use practices (Goldammer and Ronde 2004; Gornitz and NASA 1985). In the study region, four fire seasons can be distinguished, the very early fire season in October, the early fire season in November and December, the late fire season in January and February, and the very late fire season in March and April (Rueth 2010). Controlled fire management is found within the protected areas. In major parts of the Nazinga Game Ranch for example, controlled early burning is undertaken. The aim is to prevent the more destructive late fires, to provoke a fresh grass flush for grazing animals, and to improve the visibility of wildlife for tourists. On the other hand, fire has been completely excluded by the park management from some parts of the southern Nazinga Game Ranch for more than 30 years, in order to study the long-term effect of fire suppression on the savanna ecosystem.

10.3 Material and Methods

10.3.1 Time Series of Land Surface Phenology

Phenological information was derived from time series of the MODIS Enhanced Vegetation Index (EVI, Eq. 10.1). This index considers the difference in blue and red reflectances for estimating influences of the atmosphere, and minimizes soil-brightness related variations (Huete et al. 1999, 2002). Compared to the frequently used Normalized Difference Vegetation Index (NDVI), EVI is more sensitive in areas of dense vegetation and is able to reduce canopy background effects and aerosols (Huete et al. 2002; Justice et al. 1998).

$$EVI = \frac{\rho_{NIR} - \rho_{RED}}{\rho_{\mathrm{NIR}} + \mathrm{C}_1\rho_{\mathrm{RED}} - \mathrm{C}_2\rho_{\mathrm{BLUE}} + \mathrm{L}} * G \tag{10.1}$$

where ρ_x = fully or partially atmospheric-corrected surface reflectances
L = canopy background adjustment factor (L = 1)
C_1, C_2 = coefficients of the aerosol resistance term ($C_1 = 6$; $C_2 = 7.5$)
G = scaling factor (G = 2.5)

EVI time series were extracted from the MODIS product MOD13Q1 that delivers 16-day composites at a spatial resolution of 250 m. Despite the mentioned advantages of EVI, this global vegetation index product can still contain low quality values that are for example due to unfavorable observation geometries or due to remaining atmospheric effects. The latter is particularly relevant for situations of high and persistent cloud coverage, which are typical for the rainy season in the study area. With the aim to reduce such atmospheric and geometry related effects, all pixels were excluded from further processing that did not meet one or more of the three following conditions: (1) assignment of 'good' or 'marginal' in the MODIS quality information layer 'pixel reliability', (2) reflectance above 0.09 in the blue band, and (3) view zenith angle below 60°. In case these conditions excluded more than 18 composites of 1 year, the criteria were iteratively attenuated until at least 5 valid observations were available. This was done by keeping all data that met at least two, or finally at least one of the three conditions. In case even one

Table 10.1 Phenological metrics derived from EVI time series after Jönsson and Eklundh (2002)

Metric		Description
Start of season	(SOS)	Date for which EVI has increased to a level of 0.25 for the first time during the annual cycle
Rate of increase	(RIN)	Linear slope between the points where the time series reaches the 20 % and 80 % levels of its annual amplitude respectively
Maximum EVI	(MAX)	Largest EVI of the annual cycle
Large EVI integral	(LIN)	Integral of the fitted function from growing season start to growing season end (date at which EVI falls below a level of 0.25 for the first time during the annual cycle)

Table 10.2 Landsat data used for the classification of land cover and land use changes

Date	Sensor
18 November 1986	TM (Landsat-5)
3 November 2001	ETM+ (Landsat-7)
27 October 2013	OLI (Landsat-8)

All datasets were provided by USGS and cover WRS path 195, row 52

single criterion resulted in too few observations, the original data were kept. Time series were assembled for the period 2001–2013, and data gaps resulting from the removal of low quality composites were linearly interpolated.

The EVI time series were smoothed using the adaptive Savitzky-Golay filtering approach with a moving window width of 6 (96 days). Phenological metrics for the 13 years were derived from the fitted EVI time series as defined in Table 10.1. For Savitzky-Golay fitting and for the derivation of phenological metrics, the software tool TIMESAT was used (Eklundh and Jönsson 2012; Jönsson and Eklundh 2004; Jönsson and Eklundh 2002).

10.3.2 Land Cover and Land Use Changes

Baseline information on current land cover and land use as well as on the historic development of agricultural areas in the study region was derived from Landsat data. Nearly cloud free TM (Thematic Mapper), ETM+ (Enhanced Thematic Mapper Plus), and OLI (Operational Land Imager) data were available for the years 1986, 2001 and 2013 at a spatial resolution of 30 m (Table 10.2). These datasets, provided by the United States Geological Survey (USGS), did not need any further geometric correction as they showed satisfactory agreement with GPS records taken in the field. Calibration and atmospheric correction was done using Atcor-2 (Richter 1996) for flat terrain which is suitable for the flat topography in the study region.

Based on in-situ field plots taken in October and November 2013 and on very high resolution (0.6 m), pansharpened QuickBird data of October 2013, training areas of seven major land cover/use classes were defined for the year 2013. The considered classes are *forest* (tree cover > 65 %), *closed woodland* (woody cover 40–65 %), *open woodland and shrubland* (woody cover 15–40 %), *grassland* (woody cover 5–15 %, grass cover > 15 %), *bare to sparsely vegetated areas* (vegetative cover < 15 %), *agriculture* and *water*. In addition, training areas for agricultural and non-agricultural land in 1986 and 2001 could be identified directly from the respective Landsat data. As very high resolution imagery and in-situ information were missing for these years, a further subdivision of the non-agricultural vegetation types was not possible. Based on the training information, the Landsat data were classified with a combined random forest and maximum likelihood approach.

10.3.3 Time Series of Burned Areas

Spatial distribution, timing, and frequency of fires in the study area were derived from time series of the MODIS burned area product MCD45A1 (Justice et al. 2006). This dataset is based on MODIS Terra and Aqua observations and provides information on burned areas as well as the approximate date of burning at a spatial resolution of 500 m. The MCD45A1 approach identifies fire-affected areas as sudden changes in daily MODIS reflectance time series by considering bi-directional reflectance variations. Further details on the MODIS burned area mapping algorithm can be found in Justice et al. (2006) and Roy et al. (2002). The approximate date of burning was extracted from MCD45A1 for the years 2000–2012. From this information, fire frequency and fire seasonality were determined considering very early and early fires occurring between October and December as well as late and very late fires occurring between January and April.

10.4 Results

10.4.1 Land Cover Types and Land Use History

The land cover and land use map of 2013, based on OLI data, visualizes the spatial patterns of land cover types in the study region (Fig. 10.2). The protected areas (Fig. 10.1) are characterized by heterogeneous patterns of the semi-natural vegetation types closed woodland, open woodland and shrubland as well as grassland. Forests occur in smaller patches and frequently follow courses of rivers (gallery forests). Agriculturally used areas dominate outside the protected zones, intermingled by fragments of woodlands and forests, and in the northern part of the Nazinon reserved forest.

The history of clearing for agriculture since the mid-1980s is shown in Fig. 10.3a–c. While previous to our study period, the overall extent of agricultural

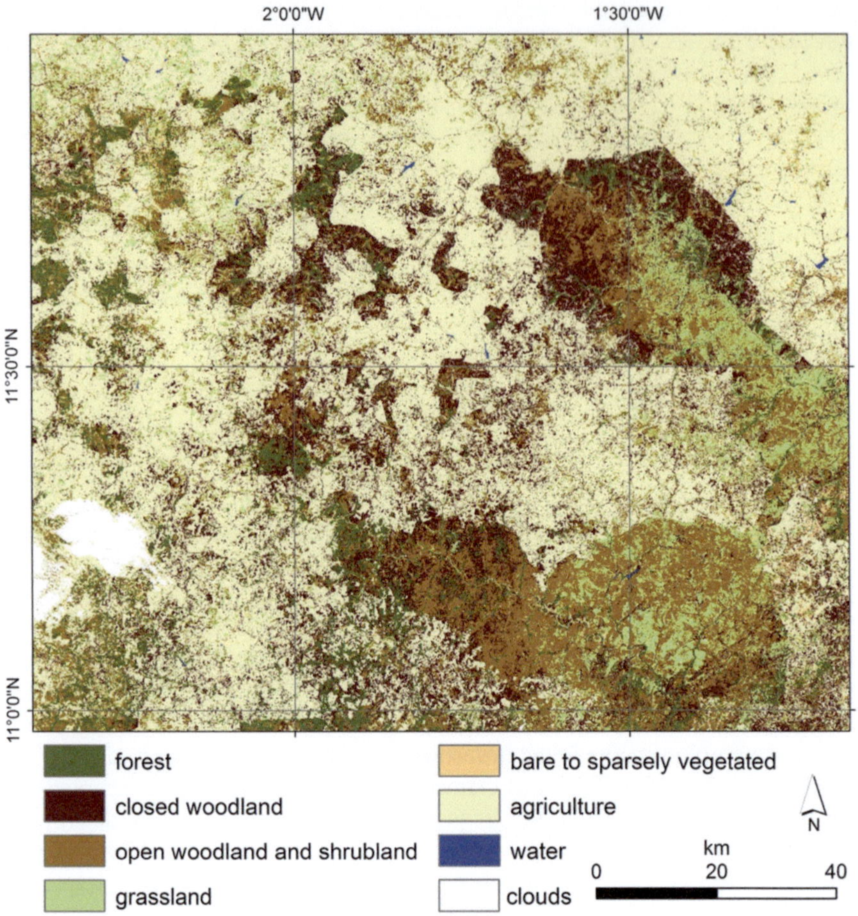

Fig. 10.2 Land cover and land use in the year 2013, based on OLI data

area has been relatively stable (1986: 3,869 km^2; 2001: 3,804 km^2), it has increased to approximately double the size of 2001 within the last 12 years (2013: 7,724 km^2). This increase in agricultural area since 2001 is marked in red color in Fig. 10.3d. Despite these enormous changes, some parts of the study region show only minor alterations with respect to agricultural land use over the last 28 years. As an example, the land to the northeast of K.T. National Park has been cultivated since the mid-1980s already (yellow color in Fig. 10.3d). The situation is also stable within most protected areas and in a buffer zone parallel to the northeastern border of K.T. National Park, where hardly any agricultural land use has been observed for the period of study. The considerable increase in cultivated area has thus concentrated on the land between and to the west of the protected areas (orange color in Fig. 10.3d). During the past 13 years, agriculture has more and more advanced

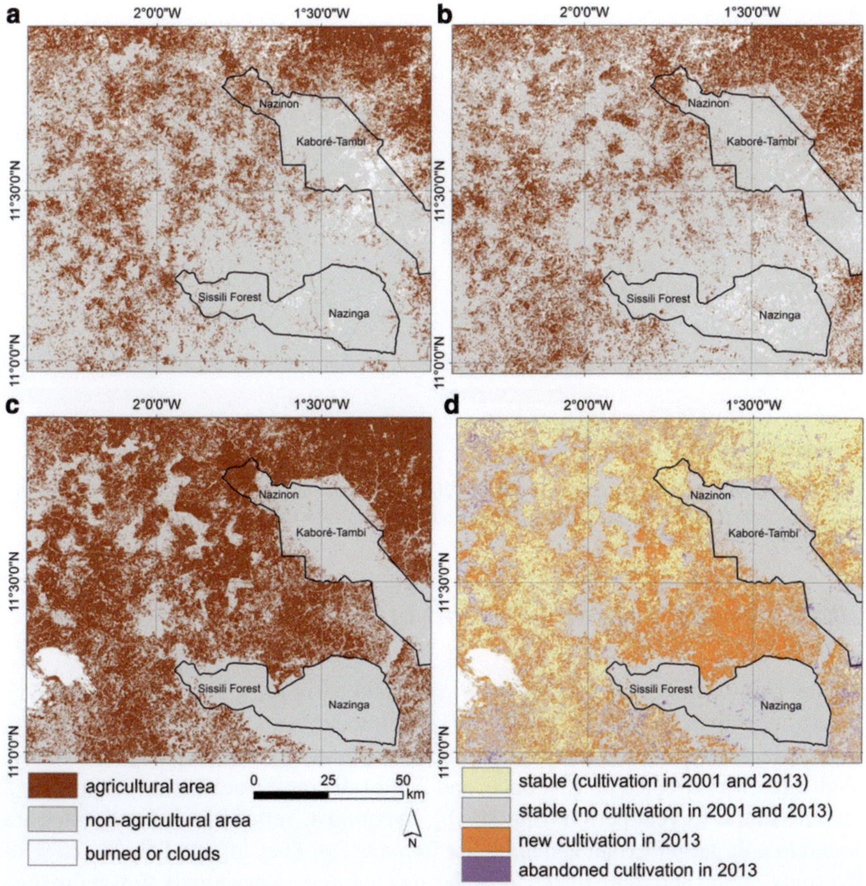

Fig. 10.3 Agricultural land use in (**a**) 1986, (**b**) 2001, and (**c**) 2013; for color coding see *bottom left*. (**d**) Changes in cultivated areas between 2001 and 2013 (**d**); for color coding see *bottom right*

towards the protected zones, resulting today in sharp fence line contrasts at the park borders.

Any areas affected by land use changes during the period of study (2001–2013) might disturb the intended phenological analyses and therefore need to be excluded from further investigations. The semi-natural savanna vegetation types existing in 2013 were not agriculturally used in 2001 (hardly any purple color in Fig. 10.3d), so that for these land cover types only minor changes during the period of study can be assumed. Considering the major transformations that have occurred since 2001 on the present cultivated land, only those agricultural areas that have existed in 2001 and in 2013 are considered in the following analyses.

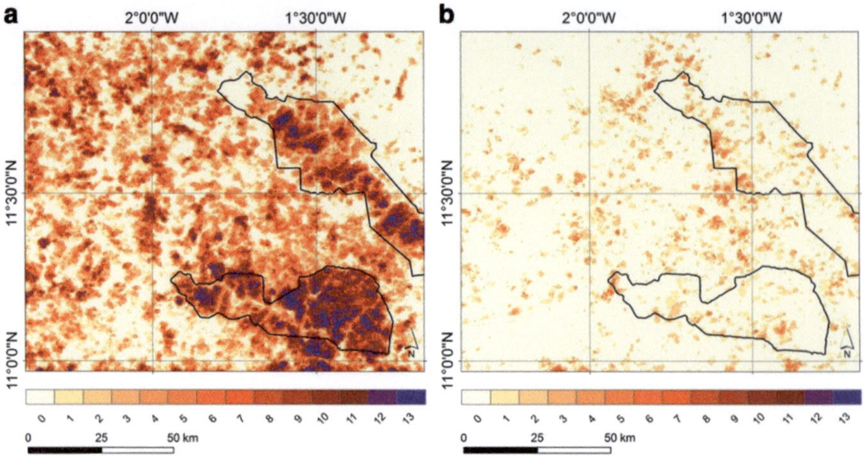

Fig. 10.4 Number of fires in a 13 year period (2000–2012) according to MOD45A1: (**a**) very early and early fires (October-December); (**b**) late and very late fires (January-April)

10.4.2 Spatio-Temporal Patterns of Fires

Figure 10.4 shows the number of very early and early (a) as well as late and very late (b) bushfires for the years 2000–2012, extracted from the MODIS burned area time series. Very late and late burnings are comparatively seldom; only 20 % of the study area was affected at least once, and 2 % of the region at least three times by a late fire in the 13 year period (Fig. 10.4b). In contrast, very early and early fires are found to a far larger extent and at higher frequencies. They affected almost 75 % of the study region and more than half of the area has experienced early fires more than three times between 2000 and 2012. Frequent burnings are concentrated in the protected areas where fires are usually found in intervals of 1–2 years. In the following analyses, only very early and early fires were considered as they are most typical for the study region.

When regarding the typical temporal fire patterns of different vegetated land use and land cover types in the study region, a relationship between fire frequency and vegetation structure can be observed (Fig. 10.5). With decreasing woody and increasing herbaceous components of land cover, fire frequency is increasing. While forests are on average affected by fire in only 3 out of 13 fire seasons, the fire frequency increases in more open woodland and shrubland types and reaches highest values for grasslands that burn in 11 out of 13 fire seasons, on average. Burning of agricultural areas in contrast was detected very rarely with a median value of 1 fire within 13 years.

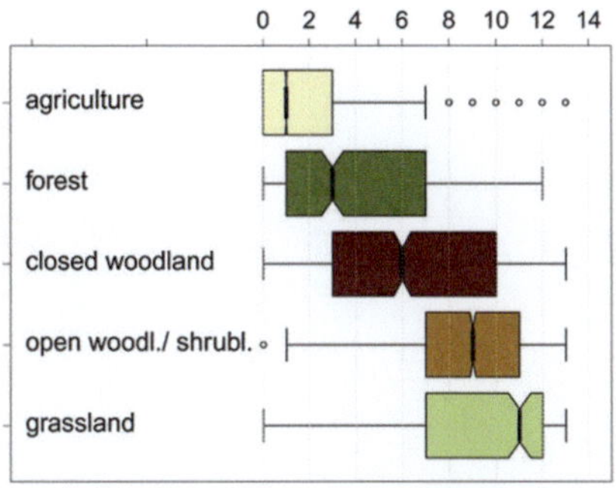

Fig. 10.5 Number of fires in the period 2000–2012, differentiated by land cover/use types. Only pure MODIS pixels that contain at least 80 % of one single land cover type are considered. Boxplot elements: *box* = values of 2nd and 3rd quartile; *horizontal line* = median; *whiskers* = lowest/ highest values; *circles* = outliers beyond 1.5 times the inter-quartile range; *notches* = indication of the 95 % confidence interval of the median after Chambers et al. (1983)

10.5 Spatio-Temporal Patterns of Phenology

10.5.1 Multi-annual Average Patterns of Phenology

The spatio-temporal patterns of phenological characteristics in the study region are illustrated in Figs. 10.6 and 10.7. Displayed are the 13 year average (left column) and variability (right column) of the start of season, the rate of increase in EVI, the maximum EVI and the EVI integral of the growing season (cf. Table 10.1). The statistics of the same phenological metrics, grouped by vegetated land cover/use classes, are displayed in Fig. 10.8. Here, the semi-natural land cover types are ordered from left (forests) to right (grasslands) according to their vegetation structure with decreasing woody and increasing herbaceous components.

Figures 10.6 and 10.7 show that the agricultural areas in the north and northeast of the study region which had been widely cleared for agriculture in 1986 already (cf. Fig. 10.3a), show distinct phenological characteristics when compared to the more recently cleared agricultural land in the center and west of the study region (cf. orange color in Fig. 10.3d). In the areas of longer agricultural land use in the north and northeast, greening starts later (Fig. 10.6, top left), the rate of seasonal increase in green vegetation is smaller (Fig. 10.6, bottom left), and both the maximum and integral of EVI indicate a lower productivity (Fig. 10.7, left column). In terms of inter-annual variability of phenological characteristics however, the behavior of all cultivated land in the study region is similar (Figs. 10.6 and 10.7, right columns).

With regard to the semi-natural land cover types, which are found mainly inside the protected areas, the spatial patterns (Figs. 10.6 and 10.7) and the statistical distributions (Fig. 10.8) reveal a general relationship between phenological characteristics and vegetation structure. The start of season tends to be later with

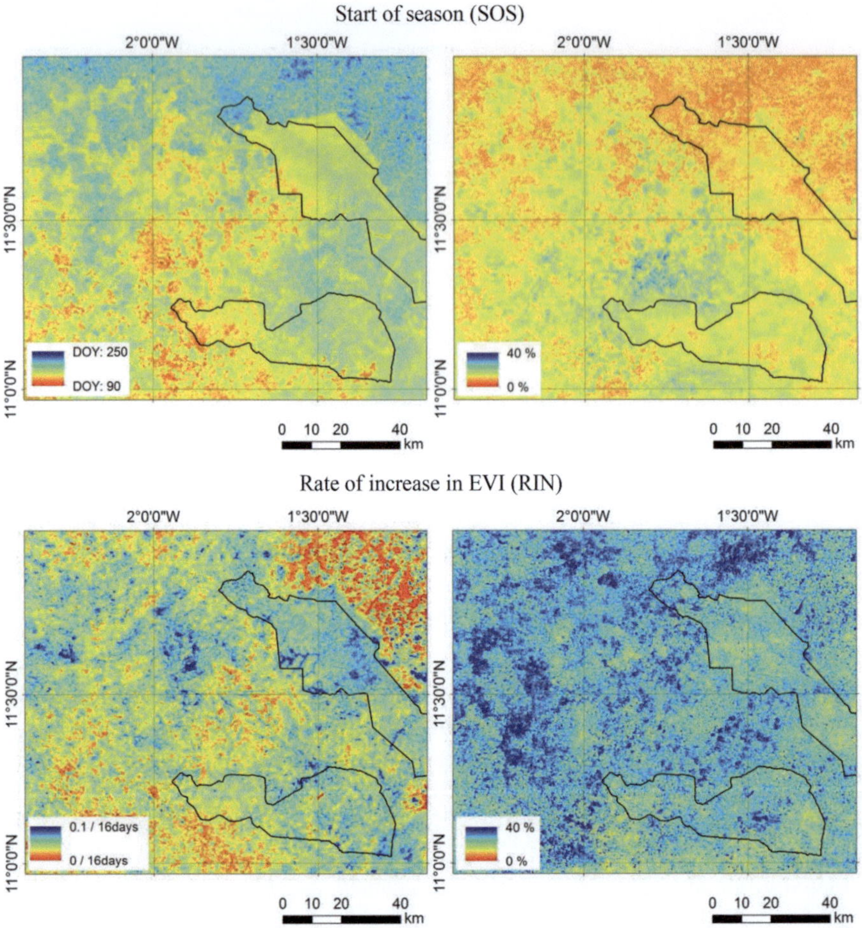

Fig. 10.6 Multi-annual (2001–2013) mean (*left column*) and coefficient of variation (*right column*) of start of season (*top row*) and rate of increase in EVI (*bottom row*). For the purpose of orientation, borders of protected areas are included in the maps (cf. Fig. 10.1)

decreasing woody and increasing herbaceous components with differences around 40 days between forests and grasslands. Likewise, the maximum and the integral of EVI are clearly related to vegetation structure, showing decreasing values with decreasing woody components. However, when considering the rate of increase in greenness, a dependency on vegetation structure is not as obvious. The statistical distributions (Fig. 10.8) show a decelerating green-up with decreasing woody cover from forests over closed woodlands to open woodlands/shrublands. For grasslands however the rate of increase in EVI is elevated again. Regarding the spatial patterns, the rate of increase is relatively high in the K.T. National Park as well as in the remaining forest and woodland fragments in the center of the study region

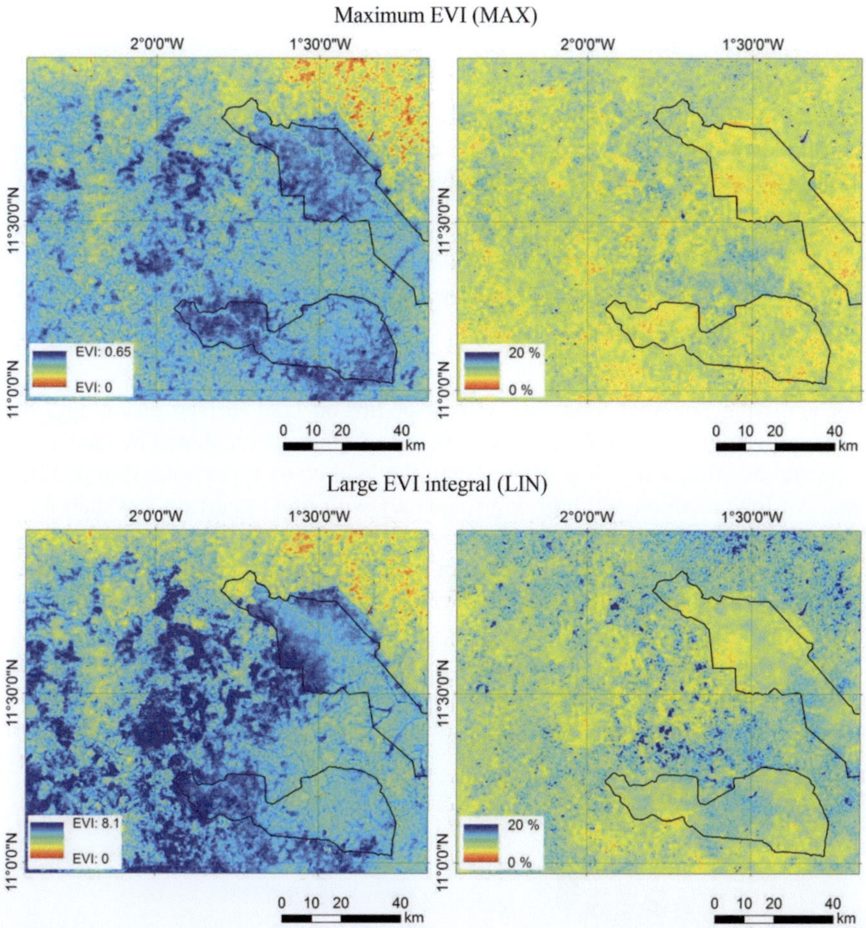

Fig. 10.7 Multi-annual (2001–2013) mean (*left column*) and coefficient of variation (*right column*) of maximum EVI value (*top row*) and large EVI integral (*bottom row*). For the purpose of orientation, borders of protected areas are included in the maps (cf. Fig. 10.1)

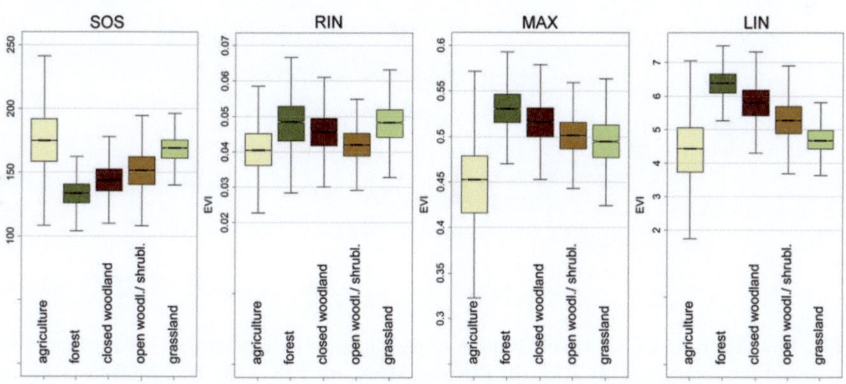

Fig. 10.8 Statistics of multi-annual (2001–2013) mean phenological metrics, grouped by vegetated land cover/use types. Only pure MODIS pixels that contain at least 80 % of one single land cover type are considered. Boxplot elements: *box* = values of 2nd and 3rd quartile; *horizontal line* = median; *whiskers* = lowest/highest values, except for outliers; *notches* = indication of the 95 % confidence interval of the median after Chambers et al. (1983)

(Fig. 10.6, bottom left). Noticeable is also the particularly strong seasonal EVI increase in the river valleys, e.g. in those of K.T. National Park, but for the closed woodlands in the Sissili forest and the riparian vegetation in the Nazinga Game Ranch, the rate of increase is comparatively low.

10.5.2 Effects of Fire Frequency on Phenological Characteristics

For assessing potential long-term effects of fire on land surface phenology, the 13-year averages of phenological metrics were further differentiated by land cover type and fire frequency (Fig. 10.9). Here, the land cover types were grouped into areas of low, medium and high fire frequency with fires occurring less than three times, between three and five times, and more than five times in 2000–2012. Agricultural areas were excluded from this analysis, as crop plants and their phenology are not directly affected by fire because sowing takes place after burning.

The differentiation into fire frequency classes reveals that frequent fires tend to delay the multiannual average start of season. This effect is stronger in grasslands and open woodlands/shrublands than in the tree dominated closed woodlands, and it

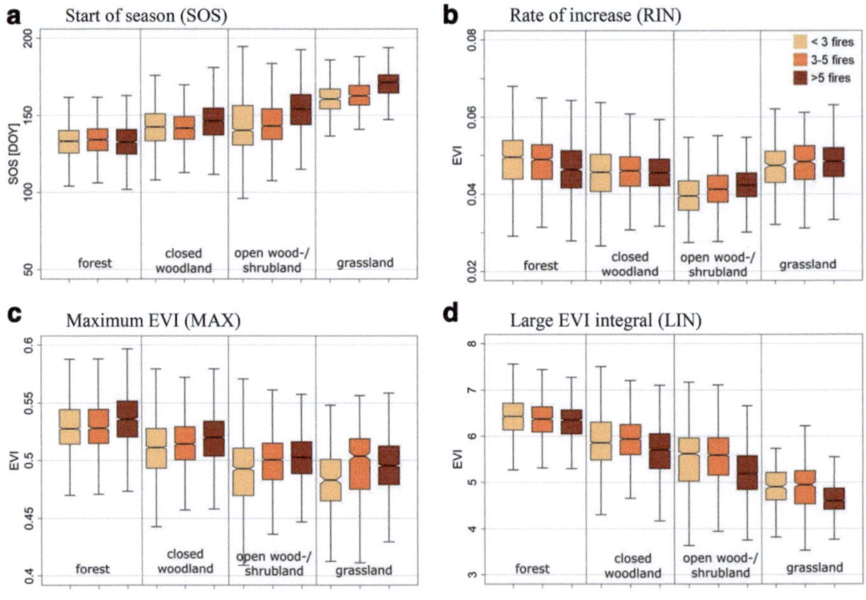

Fig. 10.9 Multi-annual (2001–2013) mean of phenological metrics differentiated by land cover type and fire frequency. (**a**) SOS, (**b**) RIN, (**c**) MAX, (**d**) LIN. Only pure MODIS pixels that contain at least 80 % of one single land cover type are considered. Boxplot elements are described in Fig. 10.8

is not observable in forests (Fig. 10.9a). The increase in greenness (Fig. 10.9b) decelerates in forests at high fire frequencies, while no significant effect could be observed in closed woodlands and grasslands. For open woodlands/shrublands in contrast, higher fire frequencies are associated with an accelerating increase in greenness. The seasonal maxima of EVI (Fig. 10.9c) generally rise with increasing fire frequencies, except for grasslands where intermediate fire frequencies lead to highest seasonal EVI maxima. There is no observable, significant effect of fire frequency on the EVI integral in forests while in the other semi-natural land cover classes, the integrated EVI is significantly smaller at high fire frequencies when compared to medium and low fire frequencies (Fig. 10.9d).

10.5.3 Effects of Fire Events on Phenological Characteristics

The direct, short-term effects of fire events on the phenology of the subsequent growing season are presented in Fig. 10.10. In this plot, land cover types and fire frequency classes are further differentiated into seasons with and without previous fire. As we are considering annual phenologies in this analysis, it needs to be assumed that fire effects on phenology are overlaid by rainfall effects in years of extraordinarily high and low precipitation. Therefore, only years with close to average rainfall amounts were considered here. Annual rainfall sums were assessed based on precipitation datasets of the Global Precipitation Climatology Centre (GPCC, Schneider et al. 2011a, b). Only those years that had experienced an annual rainfall within a range of one standard deviation around the multi-annual (2001–2013) mean were selected. The forest class could not be accounted for in this analysis due to its small areal coverage that resulted in too few data points within the once more subdivided and reduced groups.

In grasslands and open woodlands/shrublands the start of season (Fig. 10.10, first column) is observed slightly earlier (up to 8 days) when fire has occurred previously while for the more tree dominated closed woodlands this effect is not significant. For all considered classes, the average increase in greenness (Fig. 10.10, second column) is accelerated in growing seasons with previous fires. This effect is strongest in grasslands and closed woodlands with medium to low fire frequency. Similar patterns are found for the average maximum EVI (Fig. 10.10, third column) which is slightly higher in the growing seasons after burnings except for open woodlands/shrublands with high fire frequencies. The integral of growing season EVI (Fig. 10.10, fourth column) is weakly affected by previous fire occurrences with a tendency towards increased values after burnings in grasslands and open woodlands/shrublands. In closed woodlands, a significant effect could only be found at high fire frequencies where fires tend to lower the next year EVI integral.

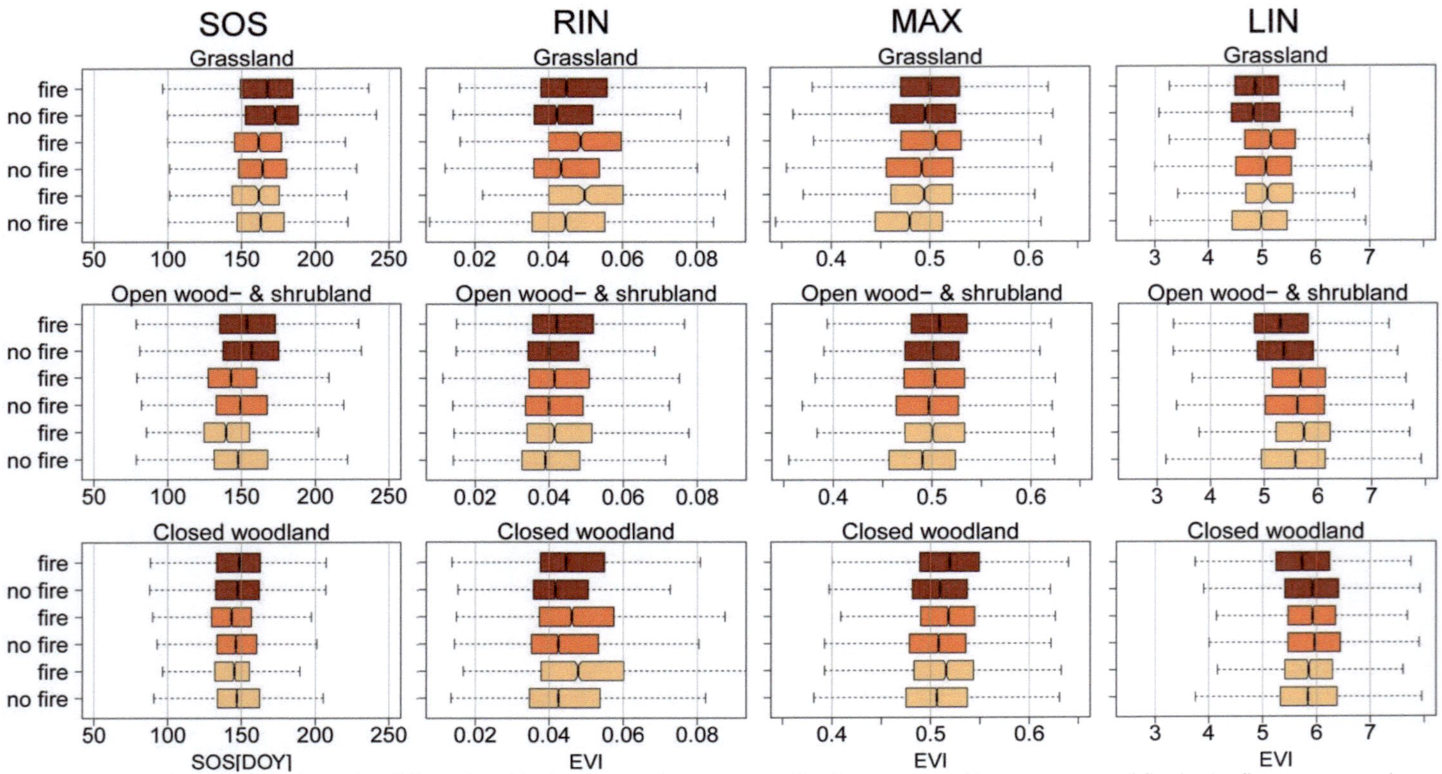

Fig. 10.10 Annual phenological metrics differentiated by land cover, by long-term fire frequency, and by occurrence of fire in the fire season previous to the respective growing season. Only pure MODIS pixels that contain at least 80 % of one single land cover type, and only close to average rainfall years (2004–2006, 2008–2009, 2011–2013) were considered. For color coding see Fig. 10.9. Boxplot elements are described in Fig. 10.8

10.6 Discussion and Conclusions

This study analyzed the influence of fire, land use, and vegetation structure on phenological patterns in a savanna region in southern Burkina Faso. For a 13-year period (2000–2012), the spatio-temporal patterns of land surface phenology were delineated from phenological metrics based on time series of MODIS EVI data. Multi-temporal Landsat data (1986–2013) was used for assessing current and historic land use patterns and vegetation structure. Time series of the MODIS burned area product were analyzed for identifying annual fire patterns and for assessing multi-annual spatio-temporal patterns of fire frequency.

The fire frequency in the study region was found to increase with the proportion of herbaceous components in savanna land cover types (Fig. 10.5). This finding is in agreement with other studies in West Africa and can be related to fine fuels from dry grasses that contribute to elevated fire frequencies, whereas leaf litter in tree dominated land cover types shows lower ignitability (e.g. Devineau et al. 2010; Hennenberg et al. 2006). Burned areas were hardly observed for agricultural fields even though these areas are known to be, at least to some extent, subject to fires as burning is used e.g. for clearing fields (Wardell et al. 2004).

A reason for this probable underestimation could be the fact that agricultural burnings usually occur on single fields that are too small for being detected as burnt areas at a spatial resolution of 500 m (Devineau et al. 2010). Moreover, a general dependency of land surface phenological metrics on vegetation structure was found in the presented study (Fig. 10.8). The shift towards an earlier start of season with increasing tree proportions can be explained by the fact that deciduous tree species in West Africa frequently start sprouting before the first rains (Bie et al. 1998) while grass development is restricted to the rainy season. The larger growing season integral of EVI in tree- and shrub-dominated land cover types can serve as an indicator for higher seasonal vegetation productivity in these areas, in dependence on what has been found in other studies for NDVI (cf. Budde et al. 2004; Lo Seen Chong et al. 1993). Similarly, the integrated EVI indicates that productivity is higher on agricultural areas that have been cleared since 2001 (center of the study area) than for areas that had been already cultivated in the mid-1980s (northeast of the study area).

Moreover, the presented analyses reveal certain effects of fire on phenological patterns. These can be subdivided into effects of long-term fire frequency, and into direct, short-term effects of fire events on the subsequent growing season. Here, several interesting aspects arise, e.g. in the context of the delayed start of season in grasslands and open woodlands/shrublands with increasing fire frequencies (Fig. 10.9a). At the first view, there seems to be a contradiction when comparing this long-term effect to the observed direct, short-term effect which implies that the start of season tends to be earlier in years of previous fire than in years without previous fire (Fig. 10.10). However, we assume that these findings can be explained by two separate effects. There might be a general shift in species composition with differing green-up phenologies due to long-term fire frequency on the one hand. For example Sawadogo et al. (2005) found in the Sudanian Savanna that frequent fires result in a long-term increase of annual and a long-term decrease of perennial grasses. On the other hand, as a direct, short-term effect, an enhanced sprouting of grasses can be assumed when the dry grass biomass of the previous growing season has been removed by fire. Reasons could be that a large detritus component on unburnt sites usually reduces light for emerging shoots, and alleviates tillering (Knapp and Seastedt 2014). However, the litter component on unburnt sites could also reduce bare soil evaporation which could, in a semi-arid environment, lead to an enhanced grass development. For a reliable understanding and interpretation of these findings, further research is needed that includes in-situ assessments in addition to remote sensing analyses and puts an additional focus on soil moisture.

Another interesting aspect is the observation that the average maximum EVI broadly increases with increasing fire frequency (Fig. 10.9c), while the integral of growing season EVI decreases (Fig. 10.9d). This shows that the peak of vegetation activity during the growing season tends to be more pronounced in areas of high fire frequency but, at the same time, the overall productivity, as indicated by the EVI integral, tends to be smaller. It can thus be concluded that investigating vegetation productivity based on single snapshots that are for example only covering the peak of the growing season, will lead to different results than when analyzing continuous time series of vegetation development.

This study allowed the detection of interrelationships between land surface phenology, land use, vegetation structure and fire in a typical West African Sudanian Savanna, based on remotely sensed time series of vegetation indices and burned area information as well as on multi-date land use/cover data. The spatial and temporal resolutions of land surface phenology and burned area time series were sufficient for delineating a number of relationships; however some of the observed tendencies were not statistically significant. Remotely sensed time series of higher spatial and temporal resolution might improve the validity of the analyses here. The use of near to daily time series of vegetation indices, for example from MODIS or from the upcoming Sentinel-3, might be a suitable option for increasing the temporal resolution. A simultaneously increased spatial resolution is a challenge that might be solved by advanced fusion techniques of daily medium resolution time series and high resolution acquisitions that cover only few dates per month (e.g. from Landsat or Sentinel-2). Furthermore, the additional consideration of moisture and its potential forcing on land surface phenology is likely to contribute to a better explanation of the interrelationships between fire occurrence, land use/cover and land surface phenology. Another variable that could be included in future studies is fire intensity as its influence on savanna vegetation is assumed to even exceed the influence of frequency (Ryan 2009). Furthermore, the combined analysis of land surface phenology based on remote sensing time series with plant phenology based on in-situ assessments is of great potential for further improving current phenological studies. Here, a major challenge will be to link remotely sensed phenological metrics with biological events observed in the field (Knauer et al. 2014).

Acknowledgements This study was funded by BMBF (German Federal Ministry of Education and Research) in the context of the project WASCAL (West African Science Service Center on Climate Change and Adapted Land Use) under FKZ 01LG1202D. We appreciate that NASA, the U.S. Geological Survey, and LP DAAC provide MODIS and Landsat data free of charge. We would also like to thank Lars Eklundh and Per Jönsson for the development and provision of TIMESAT and the anonymous reviewers for their valuable comments.

References

Archibald S, Scholes RJ (2007) Leaf green-up in a semi-arid African savanna – separating tree and grass responses to environmental cues. J Veg Sci 18:583–594

Blanken P, Black T, Yang P, Neumann H, Nesic Z, Staebler R, den Hartog G, Novak M, Lee X (1997) Energy balance and canopy conductance of a boreal aspen forest: partitioning overstory and understory components. J Geophys Res 102:28915–28927

Budde ME, Tappan G, Rowland J, Lewis J, Tieszen LL (2004) Assessing land cover performance in Senegal, West Africa using 1-km integrated NDVI and local variance analysis. J Arid Environ 59:481–498

Butt B, Turner MD, Singh A, Brottem L (2011) Use of MODIS NDVI to evaluate changing latitudinal gradients of rangeland phenology in Sudano-Sahelian West Africa. Remote Sens Environ 115:3367–3376

Chambers JM, Cleveland WS, Kleiner B, Tukey PA (1983) Graphical methods for data analysis. Wadsworth International, Duxbury Press, Boston

Churkina G, Schimel D, Braswell B, Xiao X (2005) Spatial analysis of growing season length control over net ecosystem exchange. Glob Chang Biol 11:1777–1787

De Beurs KM, Henebry GM (2004) Land surface phenology, climatic variation, and institutional change: analyzing agricultural land cover change in Kazakhstan. Remote Sens Environ 89:497–509

De Beurs KM, Henebry GM (2005) Land surface phenology and temperature variation in the International Geosphere-Biosphere Program high-latitude transects. Glob Chang Biol 11:779–790

de Bie S, Ketner P, Paasse M, Geerling C (1998) Woody plant phenology in the West Africa savanna. J Biogeogr 25:883–900

Devineau J (1999) Seasonal rhythms and phenological plasticity of savanna woody species in a fallow farming system (south-west Burkina Faso). J Trop Ecol 15:497–513

Devineau JL, Fournier A, Nignan S (2010) Savanna fire regimes assessment with MODIS fire data: their relationship to land cover and plant species distribution in western Burkina Faso (West Africa). J Arid Environ 74:1092–1101

Eklundh L, Jönsson P (2012) TIMESAT 3.1 Software manual

FAO (2013) FAOSTAT. http://faostat3.fao.org/home/index.html#HOME. Accessed 30 Nov 2013

Goldammer J, de Ronde C (2004) Wildland fire management handbook for Sub-Sahara Africa. Global Fire Monitoring Center

Gornitz V, NASA (1985) A survey of anthropogenic vegetation changes in West Africa during the last century – climatic implications. Clim Chang 7:285–325

Henebry GM, de Beurs KM (2013) Remote sensing of land surface phenology: a prospectus. In: Schwartz MD (ed) Phenology: an integrative environmental science. Springer Netherlands, Springer, Dordrecht, pp 385–411

Hennenberg KJ, Fischer F, Kouadio K, Goetze D, Orthmann B, Linsenmair KE, Jeltsch F, Porembski S (2006) Phytomass and fire occurrence along forest–savanna transects in the Comoé National Park, Ivory Coast. J Trop Ecol 22:303

Heumann BW, Seaquist JW, Eklundh L, Joensson P (2007) AVHRR derived phenological change in the Sahel and Soudan, Africa, 1982–2005. Remote Sens Environ 108:385–392

Hoffmann WA (1998) Post-burn reproduction of woody plants in a neotropical savanna: the relative importance of sexual and vegetative reproduction. J Appl Ecol 35:422–433
Huete A, Justice C, Leeuwen W van (1999) MODIS vegetation index (MOD 13) algorithm theoretical basis document. 129
Huete A, Didan K, Miura T, Rodriguez EP, Gao X, Ferreira LG (2002) Overview of the radiometric and biophysical performance of the MODIS vegetation indices. Remote Sens Environ 83:195–213
Jönsson P, Eklundh L (2002) Seasonality extraction by function fitting to time-series of satellite sensor data. IEEE Trans Geosci Remote Sens 40:1824–1832
Jönsson P, Eklundh L (2004) TIMESAT – a program for analyzing time-series of satellite sensor data. Comput Geosci 30:833–845
Justice CO, Vermote E, Townshend JRG, Defries R, Roy DP, Hall DK, Salomonson VV, Privette JL, Riggs G, Strahler A, Lucht W, Myneni RB, Knyazikhin Y, Running SW, Nemani RR, Wan Z, Huete AR, Leeuwen WV, Wolfe RE, Giglio L, Muller J-P, Lewis P, Barnsley MJ (1998) The moderate resolution imaging spectroradiometer (MODIS): land remote sensing for global change research. IEEE Trans Geosci Remote Sens 36:1228–1249
Justice C, Giglio L, Boschetti L, Roy D, Csiszar I, Morisette J, Kaufman Y (2006) Algorithm technical background document MODIS fire products
Kerby JT, Wilmers CC, Post E (2012) Climate change, phenology and the nature of consumer – resource interactions: advancing the match/mismatch hypothesis. In: Ohgushi T, Schmitz OJ, Holt RD (eds) Trait-mediated indirect interactions: ecological and evolutionary perspectives. Cambridge University Press, Cambridge, pp 508–525
Knapp AK, Seastedt TR (2014) Detritus accumulation limits productivity of Tallgrass Prairie. Bioscience 36:662–668
Knauer K, Gessner U, Dech S, Kuenzer C (2014) Remote sensing of vegetation dynamics in West Africa. Int J Remote Sens 35:6357–6396
Lieth H (1974) Purposes of a phenology book. In: Lieth H (ed) Phenology and seasonality modeling. Springer, Berlin/Heidelberg, pp 3–19
Lo Seen Chong D, Mougin E, Gastellu-Etchegorry JP (1993) Relating the Global Vegetation Index to net primary productivity and actual evapotranspiration over Africa. Int J Remote Sens 14:1517–1546
Ma X, Huete A, Yu Q, Coupe NR, Davies K, Broich M, Ratana P, Beringer J, Hutley LB, Cleverly J, Boulain N, Eamus D (2013) Spatial patterns and temporal dynamics in savanna vegetation phenology across the North Australian Tropical Transect. Remote Sens Environ 139:97–115
Ramírez N, Briceño H (2011) Reproductive phenology of 233 species from four herbaceous-shrubby communities in the Gran Sabana Plateau of Venezuela. AoB Plants 2011:1–17
Richardson AD, Black TA, Ciais P, Delbart N, Friedl MA, Gobron N, Hollinger DY, Kutsch WL, Longdoz B, Luyssaert S, Migliavacca M, Montagnani L, Munger JW, Moors E, Piao S, Rebmann C, Reichstein M, Saigusa N, Tomelleri E, Vargas R, Varlagin A (2010) Influence of spring and autumn phenological transitions on forest ecosystem productivity. Philos Trans R Soc Lond B Biol Sci 365:3227–3246
Richardson AD, Keenan TF, Migliavacca M, Ryu Y, Sonnentag O, Toomey M (2013) Climate change, phenology, and phenological control of vegetation feedbacks to the climate system. Agric For Meteorol 169:156–173
Richter R (1996) A spatially adaptive fast atmospheric correction algorithm. Int J Remote Sens 17:1201–1214
Rosenzweig C, Casassa G, Karoly DJ, Imeson A, Liu C, Menzel A, Rawlins S, Root TL, Seguin B, Tryjanowski P (2007) Assessment of observed changes and responses in natural and managed systems. In: Parry ML, Canziani OF, Palutikof JP, van der Linden P, Hanson CE (eds) Climate change 2007: impacts, adaptation and vulnerability. Contribution of working group II to the fourth assessment report of the intergovernmental panel on climate change. Cambridge University Press, Cambridge, pp 79–131

Roy DP, Lewis PE, Justice CO (2002) Burned area mapping using multi-temporal moderate spatial resolution data – a bi-directional reflectance model-based expectation approach. Remote Sens Environ 83:263–286

Rueth N (2010) Mapping bushfire distribution and burn severity in West Africa using remote sensing observations. 117

Ryan CM (2009) Carbon cycling, fire and phenology in a tropical savanna woodland in Nhambita. University of Edinburgh, Mozambique

Ryu Y, Baldocchi DD, Ma S, Hehn T (2008) Interannual variability of evapotranspiration and energy exchange over an annual grassland in California. J Geophys Res 113:D09104

Sawadogo L, Tiveau D, Nygård R (2005) Influence of selective tree cutting, livestock and prescribed fire on herbaceous biomass in the savannah woodlands of Burkina Faso, West Africa. Agric Ecosyst Environ 105:335–345. doi:10.1016/j.agee.2004.02.004

Schneider U, Becker A, Finger P, Meyer-Christoffer A, Rudolf B, Ziese M (2011a) GPCC monitoring product: near real-time monthly land-surface precipitation from rain-gauges based on SYNOP and CLIMAT data. http://dx.doi.org/10.5676/DWD_GPCC/MP_M_V4_100

Schneider U, Becker A, Finger P, Meyer-Christoffer A, Rudolf B, Ziese M (2011b) GPCC full data reanalysis version 6.0 at 0.5°: monthly land-surface precipitation from rain-gauges built on GTS-based and historic data. 10.5676/DWD_GPCC/FD_M_V6_050

Scholes RJ, Archer SR (1997) Tree-grass interactions in savannas. Annu Rev Ecol Syst 28:517–544

The World Bank (2013) Population growth. http://data.worldbank.org/indicator/SP.POP.GROW

Wagenseil H, Samimi C (2006) Assessing spatio-temporal variations in plant phenology using Fourier analysis on NDVI time series: results from a dry savannah environment in Namibia. Int J Remote Sens 27:3455–3471

Wardell DA, Nielsen TT, Rasmussen K, Mbow C (2004) Fire history, fire regimes and fire management in West Africa: an overview. In: Goldammer J, de Ronde C (eds) Wildland fire management handbook for Sub-Sahara Africa. Global Fire Monitoring Center, pp 350–374

White F (1983) The vegetation of Africa: a descriptive memoir to accompany the UNESCO/AETFAT/UNSO vegetation map of Africa. Nat Resour Res 20:1–356

Chapter 11
Assessing Rainfall-EVI Relationships in the Okavango Catchment Employing MODIS Time Series Data and Distributed Lag Models

Thomas Udelhoven, Marion Stellmes, and Achim Röder

Abstract Aboveground net primary productivity (ANPP) is limited by water availability especially in dry and desert regions, and many studies have linked ANPP to current and previous "effective" rainfall events. In this study a distributed lag model (DLM) was used to assess the impact of current and previous 16 day rainfall anomalies on the Enhanced Vegetation Index (EVI) as a proxy for ANPP in the Okavango catchment (South Africa). The two important aspects in using DLMs are the explained total ANPP variability by the rainfall regime and the duration of that dependency. The results indicate that more than 50 % of the Okavango Basin are sensitive towards current and previous rainfall anomalies. These regions are mainly restricted to the southern semi-arid parts of the catchment, whereas in the humid and sub-humid northern areas significant correlations were observed only locally. Here, the dominant land cover classes are shrub- and grassland, thornbush savannahs and mixed woodlands. The duration of significant rainfall-EVI dependencies ranges from concurrent anomalies to a time-shift of 3.5 months. A logistic regression model was applied to discriminate among the sensitive and non-sensitive areas in the basin in terms of possible physiogeographic covariates. The model was able to correctly classify ~80 % of the available pixels. Most relevant explanatory covariates were evaporation, elevation and land cover.

T. Udelhoven (✉) • M. Stellmes • A. Röder
Environmental Remote Sensing and Geoinformatics, Faculty of Regional and Environmental Sciences, University of Trier, Trier, Germany
e-mail: udelhoven@uni-trier.de

C. Kuenzer et al. (eds.), *Remote Sensing Time Series*, Remote Sensing and Digital Image Processing, DOI 10.1007/978-3-319-15967-6_11

11.1 Introduction

Vegetation strongly affects the characteristics of the Earth surface as it determines the exchange processes of energy and water between surface and the atmosphere. Water availability, and thus rainfall, is the main abiotic factor limiting plant production in many regions of the world. Although natural vegetation has developed a great capacity of physiological adaptation and resistance to long droughts and soil moisture below the theoretical wilting points (Kosmas 1999), precipitation is considered the primary limiting factor for plant growth in semi-humid and semi-arid areas (Wang et al. 2001, 2003; Karabulut 2003; Boer and Puigdefábregas 2003). Thus, especially in water-limited ecosystems significant positive correlations between the inter-annual variation of precipitation amount and aboveground net primary productivity (ANPP) have been found in many studies across a range of sites (Sala et al. 1988; Nicholson et al. 1990; Herrmann et al. 2005; Paruelo and Lauenroth 1995; Oesterheld et al. 2001; Udelhoven et al. 2008; Gessner et al. 2013; Peng et al. 2013).

There are several covariates that may affect the ANPP-rainfall relationship in semi-arid areas, with land cover being among the most important ones (Camberlin et al. 2007). For instance, south of the Sahara open grassland and cropland areas were found to be more sensitive towards the rainfall regime than woodlands and forest, while soil properties and soil types were only of minor importance. Temperature is an additional factor that affects ANPP (Balaghi et al. 2008). However, Udelhoven et al. (2008) found in the semi-arid regions of Spain that current and previous temperature anomalies have a smaller impact than rainfall anomalies. Many studies also suggest that the correlation between rainfall and ANPP is strongly determined by the degree of aggregation of the variables in the time dimension (Wang et al. 2003) and that biomass may lag behind rainfall by several weeks or months (Gessner et al. 2013). Consequently, in semi-arid environments vegetation greenness is more strongly correlated with soil moisture, which is determined by previous accumulated rainfall events, than with instantaneous rainfall (Herrmann et al. 2005).

The functional relation between monthly or annually aggregated biomass and rainfall data was found to be either log-linear (Davenport and Nicholson 1993), or linear (Wang et al. 2003; Evans and Geerken 2004). The establishment of such relationships has implications for modelling crop yield and analysing primary productivity in semi-arid regions (Gurgel and Ferreira 2003). Furthermore, knowledge about the vulnerability of a region toward climatic variability is useful in establishing early warning systems, to distinguish between climatic or human induced changes in surface conditions, and in decision support that addresses the exploitation of water resources by competing users.

Rainfall triggers a cascade of ecosystem responses that affects net primary productivity and soil respiration (Thomey et al. 2011; Parton et al. 2012), carbon fluxes (Huxman et al. 2004), plant aboveground net primary productivity (Li et al. 2013) and water use efficiency (Gao et al. 2011). According to Huxman et al. (2004) discrete precipitation pulses are the main factor for biological activities in North America's dry regions. Important covariates for primary production are soil physical properties, plant evapotranspiration, root zone soil moisture, vegetation type, surface temperature, irrigation activity and microclimate (Cihlar et al. 1991; Schultz and Halpert 1993; Kawabata et al. 2000; Li et al. 2013).

The efficiency of carbon assimilation of plants and how of producing biomass or grain yield after consuming water is described at plot scale by the concept of the water use efficiency (WUE), which reflects the balance between the production of biomass (kg of biomass produced or moles of CO_2 assimilated) and consumed water (m^3 of water used or moles of water transpired) (Tomás et al. 2013). The WUE can be measured at different spatial scales, from the leaf level ($WUE_{instantaneous}$) to the crop level (WUE_{yield} or 'water productivity') (Tambussi et al. 2007), or at different temporal scales from the instantaneous exchange of water vapour for carbon dioxide to the biomass accumulation at monthly scales (Tomás et al. 2013). At regional level the rain-use efficiency (RUE), which is defined as the ratio between vegetation productivity and annual precipitation, has been established as a measure to assess land condition (Le Houérou 1984; Fensholt and Rasmussen 2011). The rain-use efficiency concept allows identifying non-rainfall impacts on vegetation cover. It is based on the observation that RUE is stable over time and, thus, spatio-temporal changes in RUE reflect non-rainfall impacts on the biosphere, such as land degradation (Prince et al. 1998). Thus, one advantage of the RUE concept is that it has the potential to distinguish between human-induced and climatic (e.g. rainfall) triggered changes in vegetation cover. The drawback is that it relies on temporally aggregated (annual) rainfall data and thus, the direct or delayed impact of rainfall events on the ANPP cannot be directly analysed with this concept.

A variety of modelling concepts exists to describe the primary productivity response to rainfall in dry regions on different spatial and temporal scales. The Westoby-Bridges pulse-reserve hypothesis (Noy-Meir I 1973; Ogle and Reynolds 2004) addresses the response of various plant functional types (FTs, primarily annual plants) to pulses of precipitation and states that a rain event can trigger a production response (i.e. germination or growth) if a certain threshold is reached, such that the rainfall event becomes 'biologically important' or 'effective' (Ogle and Reynolds 2004). A significant rainfall event interacts with a plant's water-use patterns of utilizing soil moisture pulses at particular infiltration depths or durations (Zhao and Liu 2011). For instance, Li et al. (2013) found that in a desert ecosystem in China a threshold of a rainfall event of at least 5 mm is required to become ecologically significant. Robertson et al. (2009) and Peng et al. (2013) demonstrated for semi-arid grasslands that the effect of rainfall is different in the growing season, where precipitation affects ANPP directly, and the non-growing season, were precipitation only has an indirect lagged effect through an increase of soil moisture. In contrast, for temporally aggregated rainfall data an upper threshold seems to exist to which a significant relationships between ANPP and rainfall exists. For East Africa, Davenport and Nicholson (1993) found the rainfall threshold to be approximately 1,000 mm/year. Above this threshold other external factors, such as nutrient availability, dominate over rainfall effects, or the soil is water saturated for most time in the year so that water scarcity does not provide a dominant limitation for vegetation growth.

A drawback of the pulse-reserve hypothesis is that it does not assess the impact of potential delayed responses of vegetation to rainfall, and it does not account for explicit precipitation thresholds (Li et al. 2013). Furthermore, the pulse-reserve model was not developed for regional rainfall-biomass assessment. Ogle and Reynolds (2004) suggest the threshold-delay (T-D) model to overcome some of the limitations of the two-layer and pulse-reserve hypotheses. The T-D model uses six key parameters and integrates the ideas of resource partitioning, precipitation thresholds, and different plant FT strategies and responses, as well as potential delays in these responses. One limitation of this model is its difficulty to scale up from individual sites to the regional scale (Li et al. 2013).

With remote sensing techniques it is possible to obtain information about the impact of different rainfall regimes on the biosphere at different scales and to study interannual and seasonal changes in vegetation characteristics. The aim of this study is to analyse the effect of biweekly aggregated rainfall anomalies on ANPP at the regional scale for the Okavango Basin in Southern Africa. The area represents a highly complex socio-ecological system, where the variation in physiogeographic characteristics is reflected by different livelihood strategies. Besides locally important revenues from tourism and the beginning establishment of irrigation schemes, the majority of the population directly depend on ecosystem goods and services supplied from forest and savannah ecosystems, and on subsistence agriculture that is often practiced in a traditional slash and burn strategy (Röder et al. 2013). In this context, understanding responses to variation in rainfall event size and frequency will aid to distinguish between human and climatic induced changes in dry ecosystems and to assess how these ecosystems may change under future scenarios of more extreme precipitation regimes. To this end, a quantitative link between time-series of 16-day rainfall and Moderate Resolution Spectroradiometer (MODIS) data was established using a distributed lag model (DLM), with the goal of identifying regions that are sensitive towards current and previous rainfall anomalies. The identified regions were further analysed in relation to physio-geographical data such as dominant land cover terrain morphology and actual evaporation (aET).

11.2 Material and Methods

11.2.1 Study Area

The Okavango catchment (Fig. 11.1) is located in Southern Africa within the three countries Angola, Namibia and Botswana. The three prevalent landscape types comprise the Angolan highlands in the north, were altitudes of 1,600–1,800 m a.s.l. are reached, the lowlands of the Kalahari in the south-east and the inner Okavango

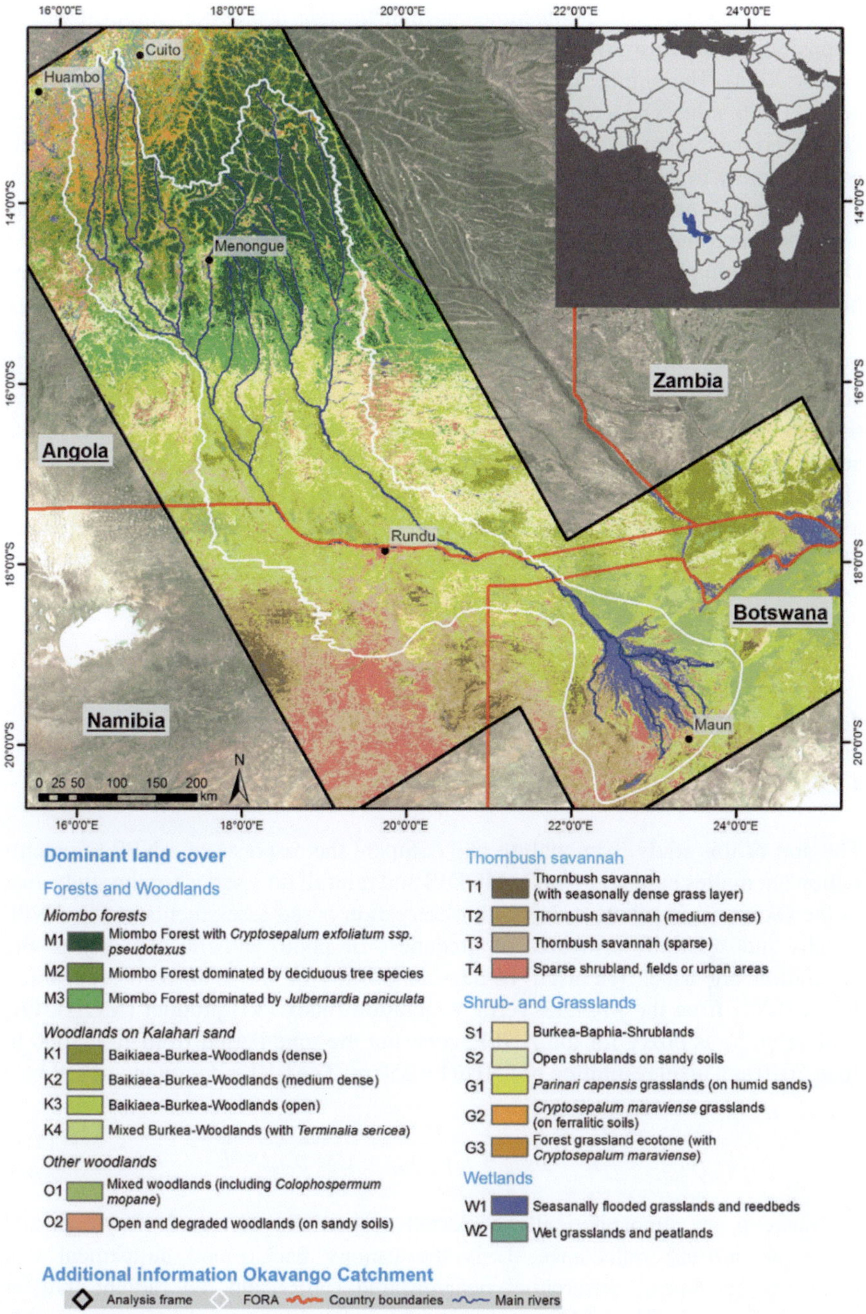

Fig. 11.1 Land cover map derived from 16-day MODIS EVI time series covering the observation period 2000–2011. Dominant land cover classes were produced by an unsupervised classification approach which was based on phenology metrics (Stellmes et al. 2013)

Delta in Botswana, which constitutes an indistinctive depression confined within extensions of the East African Rift Valley system (Wehberg and Weinzierl 2013; Weber 2013). The humid Angolan highlands receive an annual rainfall amount of up to 1,400 mm. The conditions change from semi-humid in the north of the Okavango catchment to semi-arid (<500 mm precipitation) in the south. Rainfall occurs mainly in the period from December to February.

The annual mean temperature is about 18 °C in the extreme northwest of the catchment and increases towards the south and achieves 24 °C in the Okavango Delta. Accordingly, potential evapotranspiration shows a similar trend und reaches highest rates during the summer (Weber 2013).

Figure 11.1 shows the dominant land cover classes in the Okavango catchment (Stellmes et al. 2013). The upper semi-humid catchment in the north is mostly covered by Miombo forests. These woodlands are traversed by river valleys feeding the tributaries of the Okavango River, which are dominated by open grasslands accompanied by dwarf shrubs. The middle reaches of the Okavango River are characterized by extensive woodlands on Kalahari sand. The Okavango Delta area is surrounded by mixed woodlands on the eastern side (including Terminalia and Mopane) while thornbush savannah represents the main vegetation type on the western side (Stellmes et al. 2013).

11.2.2 Data

11.2.2.1 Enhanced Vegetation Index (EVI)

The aim of this study is to analyse and compare the response of ANPP to precipitation anomalies based on 16-day MODIS and rainfall time series for the study area in the Okavango catchment. In Earth observation based assessments of the ANPP usually index based surrogates of greenness or photosynthetic capacity of the vegetation are used. We used 16-day-composites of the Enhanced Vegetation Index (EVI) from the MODIS-Terra Vegetation Index (VI) product (MOD13Q1, Collection 5) as proxy for the ANPP, covering the time period from July 2000 to June 2010 at a pixel resolution of 250 m by 250 m. The EVI is computed as follows:

$$EVI = G \frac{\rho_{NIR} - \rho_{red}}{\rho_{NIR} + C_1 \cdot \rho_{red} - C_2 \cdot \rho_{blue} + L} \tag{11.1}$$

Where ρ are atmospherically corrected or partially corrected (Rayleigh and ozone absorption) reflectance, L is the canopy background adjustment that addresses non-linear, differential near-infrared (NIR) and red radiant transfer through a canopy, C1 and C2 are coefficients related to aerosol correction and G is a gain factor. The blue band is used to remove residual atmosphere contamination caused by smoke and sub-pixel thin cloud. The EVI is reported to be sensitive to canopy structural variations, including canopy architecture and type, leaf area index (LAI), and plant physiognomy (Huete et al. 2002, 2006).

11.2.2.2 Precipitation and Actual Evaporation

Daily precipitation amounts were derived from the 3B42 (daily) product (Huffman et al. 2007). The data were provided by the NASA/Goddard Space Flight Center's Mesoscale Atmospheric Processes Laboratory and Precipitation Processing System (PPS), which develop and compute the multi-sensor precipitation data as a contribution to NASA's Tropical Rain Measurement Mission (TRMM). This mission provides comprehensive precipitation estimates and latent heating between 35°N and 35°S on a 0.25°× 0.25° grid. These data were resampled according to the temporal and spatial resolution of the MODIS EVI data. Gridded daily actual evaporation data was provided by the Climate Service Center, Helmholtz-Zentrum Geesthacht, Germany, and has been computed in a spatial resolution of 25 km × 25 km with the Regional Climate Model REMO (Jacob 2001) forced with the Global Circulation Model ECHAM for the domain of longitude 8°–32°, latitude 10°–27° and covering the period from 2000 to 2010, corresponding to the EVI time series.

Both, the precipitation as well as the actual evaporation data were aggregated corresponding to the 16-day EVI composites. The pixel size was resampled to 250 m using the nearest neighbor method to retain the MODIS resolution and hereby, reflect the landscape heterogeneity.

11.2.3 Methods

11.2.3.1 Data Pre-processing

Many studies have demonstrated that ecosystems in drylands have a "memory" of past precipitation events, which can last several weeks or months (Li et al. 2013; Schwinning et al. 2004; Udelhoven et al. 2008). Distributed lag models (DLM) are powerful statistical tools to assess such lagged effects between two or more time series (Wei 1990) and, thus, to analyse the impact of current and previous rainfall events on ANPP. The DLM used in this study was applied to regress seasonal rainfall anomalies on EVI anomalies, including time-lagged effects. A seasonal EVI anomaly is computed as

$$EVI_{diff(i)} = EVI_i - EVI_{mean(i)} \tag{11.2}$$

Where $EVI_{diff(i)}$ denotes an anomaly in season i, $EVI_{mean(i)}$ is the long-term mean in season i and EVI_i is the actual EVI in season i from a de-trended time series. Rainfall anomalies are computed in the same way. EVI anomalies can be caused either by a drought period (negative anomaly) or by above average rainfall events in

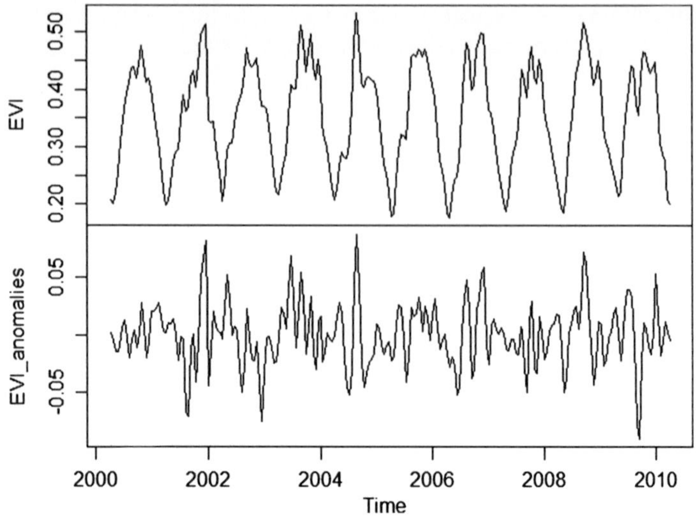

Fig. 11.2 Ten-year MODIS EVI time series and corresponding anomalies

a 16 day period (positive anomaly). Before applying Eq. 11.2 the series were de-trended for an unbiased estimation of the seasonal means. Figure 11.2 presents one example for an EVI time series and the related EVI anomalies.

11.2.3.2 Distributed Lag Models (DLM)

A distributed lag model (DLM) constitutes a multivariate statistical regression model for equidistant time series data in which current values of a dependent variable are predicted based on both the current values of an explanatory variable and lagged values of this explanatory variable. The DLM that was used in this study was used to model 16-day anomalies in the EVI in terms of current and previous rainfalls anomalies and takes on the general form

$$EVI_t = a + \sum_{i=0}^{i=\max} \beta_i \text{rainfall}_{t-i} + \varepsilon, \qquad (11.3)$$

where EVI and rainfall represent a seasonal (16-day) anomaly time series, β denotes the impulse response weight vectors and describes the weights assigned to past rainfall data values. The ε_t are the model errors and α is the constant term. The model order i addresses the number of lagged versions of the rainfall data. One lag corresponds to a time lag of 16 days. We considered a maximum lag of seven, corresponding to a maximal previous time shift of 3.5 months.

The 16-day anomalies still suffer from of temporal autocorrelation, which gives rise to some statistical problems if usual ordinary least-square (OLS) parameter estimation is used to solve Eq. 11.3. This leads to inappropriate assumptions of

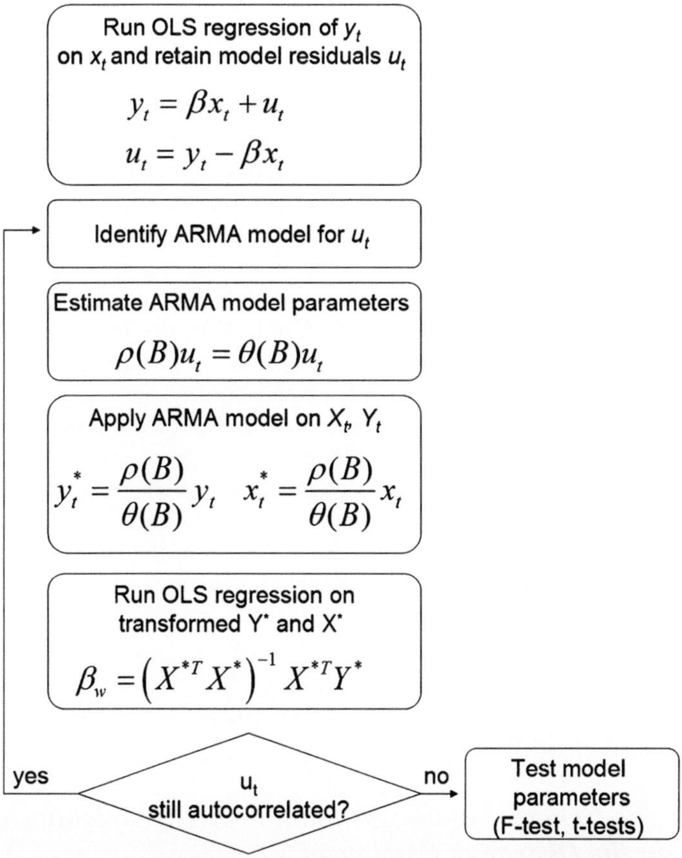

Fig. 11.3 Generalized least square parameter estimation for DLM in case of serial autocorrelation (Modified after Shumway and Stoffer 2000), where y and x are two time series, u denotes the autocorrelated error term, ρ denotes the vector of autoregressive parameters, θ the vector of moving average parameters, B is the backshift operator, β is the final vector of regression coefficients. Alternatively, any moving average component can be substituted by a higher order AR-model using the *stepwise autoregression* algorithm (Granger and Newbold 1986)

independent error terms, inflating standard errors of the parameters, and the tendency to overestimate the R^2 statistics due to the phenomenon of spurious correlation of two series, which might actually be unrelated to each other (Wei 1990; Chatfield 2004).

To remove serial autocorrelation and to achieve (weak) stationary conditions for OLS-regression analysis, the anomalies need to be pre-whitened. This can be achieved by temporal aggregation of the data unless autocorrelation fades away or by fitting a suitable autoregressive moving average (ARMA) model to the data and to keep the residuals. For DLM models, Liu and Hanssens (1982) and Shumway and Stoffer (2000) suggest to replace OLS-regression by an iterative generalized least square (GLS) approach for consistent and efficient parameter estimation in case of serial autocorrelation (Fig. 11.3). In a first step the two (autocorrelated) anomaly series are regressed to each other using standard OLS-parameter estimation, which typically results in autocorrelated model errors. Then a suitable ARMA model is fitted to these residuals. Alternatively, the stepwise autoregression algorithm proposed by Granger and Newbold (1986), that fits AR (p)-models of increasing order to these errors until the residual become white-noise, can be applied (Udelhoven et al. 2008). The two original rainfall and EVI anomaly series are then filtered using the identified AR- (or ARMA-) model and again regressed against each other using OLS- regression. The process is repeated until approximate confidence bounds for a white-noise process are achieved for the model residuals (Box et al. 1994), which is usually achieved already after the first iteration. According to Shumway and Stoffer (2000), the procedure converges toward the maximum likelihood solution under normality of the errors. The procedure is robust and easy to automate for big data archives.

Equation 11.3 was solved for each corresponding EVI and rainfall series and related F- and t-statistics were computed for the total model and for each (lagged) regression coefficient. A significant test statistic in a certain region indicates a significant impact of current or previous seasonal rainfall anomalies on EVI (or ANPP). To discuss identified vulnerable regions in the context of potential external covariates (such as climate conditions), logistic regression analysis was applied.

11.3 Results and Discussion

11.3.1 The General Significant Rainfall-ANPP Relationship in the Okavango Catchment

Figure 11.4 shows the mean annual rainfall and evapotranspiration maps for the Okavango catchment. A distinct rainfall gradient exists with high precipitation values in the north in the humid Angolan highlands with up to 1,400 mm, and low values (<500 mm) in the south in the lowlands of the Kalahari and the Okavango Delta. Corresponding evaporation values show an opposite spatial trend, being highest in the fringe of Kalahari Desert and the Okavango Delta in the south.

Figure 11.5 shows pairwise scatterplots of climate and additional spatial covariates. Mean annual rainfall is highly negatively correlated with actual evaporation. This demonstrates that in dry regions actual evaporation is mainly determined by available rainwater. The mean annual EVI shows the highest correlation with rainfall (positive relationship) and actual evaporation (negative relationship). Temperature is positively correlated with elevation and negatively with actual evaporation. The latter results from the fact that available rainwater is limited in semi-arid regions. Between rainfall and elevation a highly non-linear relationship can be observed which results from the difference of a mean annual rainfall gradient

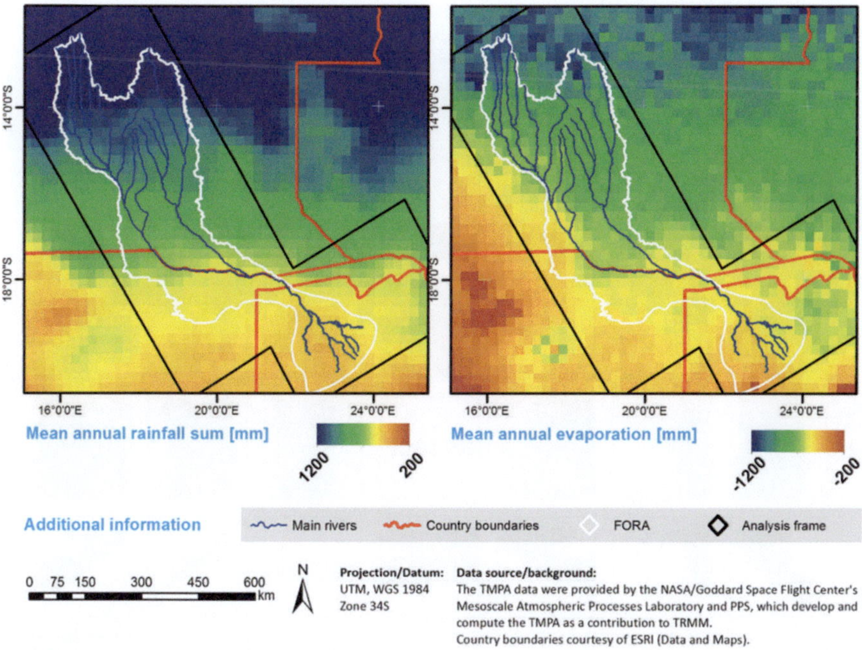

Fig. 11.4 Maps of mean annual rainfall and potential evapotranspiration in the Okavango catchment

being north to south orientated and a northwest to southeast gradient of elevation, with lowest altitudes occurring in the delta regions of the Okavango.

Figure 11.6 (right) illustrates the overall F-statistic of the DLM model which indicates whether an EVI pixel is vulnerable towards current and previous rainfall anomalies at the 5 % significance level. The respective multiple coefficient of the model (R^2) is shown in Fig. 11.6 (left). The F-statistic clearly suggests that large areas in the Okavango catchment are sensitive towards the rainfall regime, with responses of ANPP contingent on seasonal (16-day) levels. Sensitive areas are mainly located in the semi-arid southern region and they show a spatial structure that is similar to the rainfall gradient in Fig. 11.4. The R^2 values of the DLMs vary between 0.3 and 0.5 in these areas. Thus, although regionally up to 50 % of the variability in EVI anomalies can be explained by the rainfall regime, there remains an unexplained variability in ANPP that is attributed to noise and additional spatial covariates. According to Thiam (2003) additional possible determinants for variations in ANPP are factors such as deforestation, soil type, overgrazing, and agricultural land-use.

The Okvanango Delta is covered with seasonally flooded grasslands and reedbeds and does not show significant rainfall-EVI relationships. Here, soil water is not a limiting factor and according to Wehberg and Weinzierl (2013) two-thirds of the total area are seasonally inundated swamp (around 9,000 km^2) or permanently flooded.

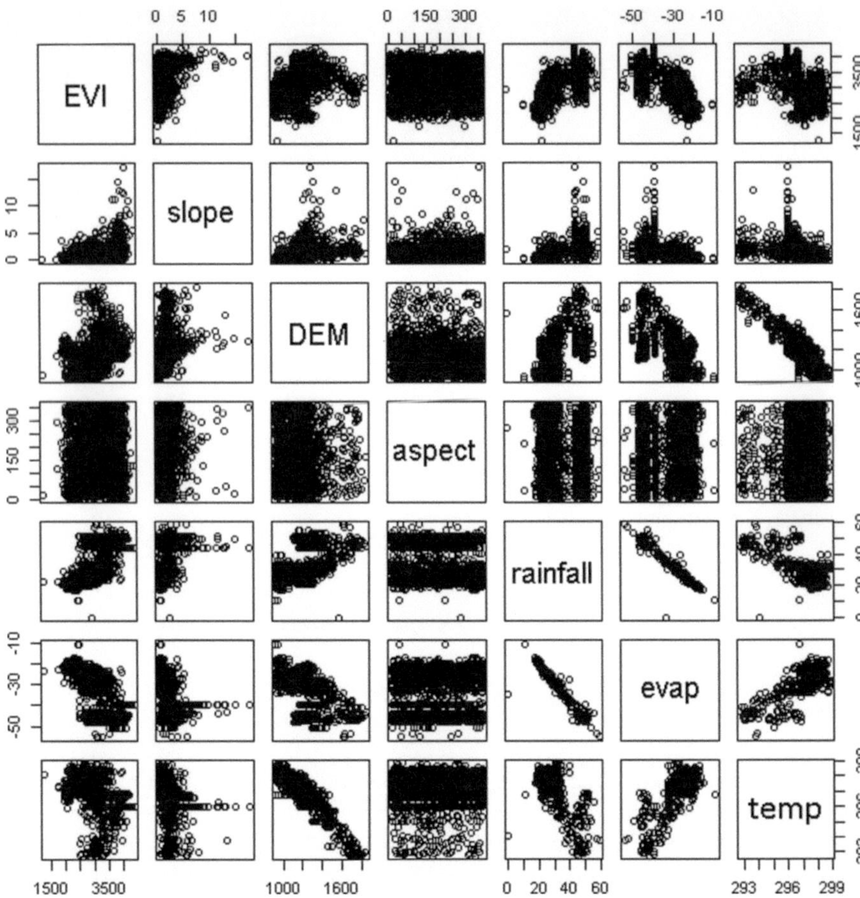

Fig. 11.5 Pairwise scatterplots of external variables (mean annual temperature, mean annual potential evapotranspiration, mean annual rainfall, aspect, slope, and elevation)

Significant areas in Fig. 11.6 represent potential climate-sensitive systems as the dependency of ANPP on rainfall conditions is high; in these regions, an increased high inter-annual or intra-annual rainfall variability is expected to go along with a high variability in ANPP. Climate change is likely to impose further variability on rainfall conditions in Africa. Even now, rainfed agriculture, upon which livelihoods of the majority of residents in the semi-arid regions in Africa are depending, is highly challenging, and climate change is expected to further reduce the length of the growing season, extend dry spells during the wet season and introduce higher variability in precipitation (Boko et al. 2007).

To discuss the identified significant vulnerable regions in the context of possible covariates a supervised classification was carried out using logistic regression to discriminate among the significant and non-significant pixels, according to the overall F-test ($\alpha = 0.01$) of the DLM. To this end 1,000 pixels each were randomly

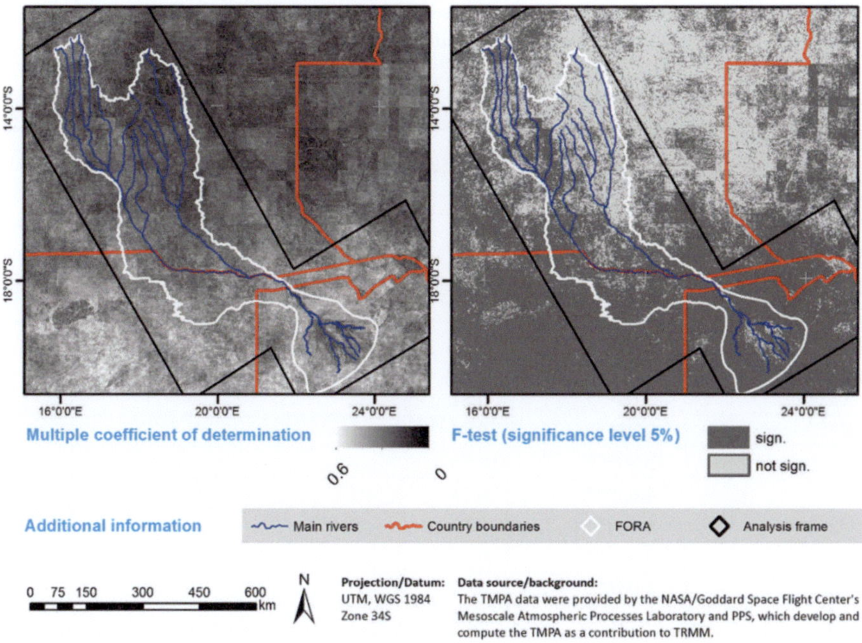

Fig. 11.6 Overall significance of the DLM (F-test, $\alpha = 1$ %) (*right*) and multiple R of the model (*left*)

selected from the sensitive and non-sensitive regions. Further 1,000 pixels from each class were used for model validation. As independent variables mean annual precipitation, temperature, evapotranspiration, slope, aspect, elevation (derived from Shuttle Radar Topography Mission (SRTM3) data courtesy of NASA) and land cover (see Fig. 11.1) were used as explanatory variables. The categorical land cover information was included as dummy-variable in the model. Table 11.1 shows the significant model coefficients and the result of the classification. From the calibration data set 81.05 % of all pixels were correctly classified as sensitive or non-sensitive. The total accuracy in the validation data set is 78.21 %, respectively.

The results suggest that among the external variables aET, elevation and temperature have a high significant prediction capacity to distinguish sensitive from insensitive regions. Annual rainfall was not found to be significant, as it is highly correlated with aET (compare Fig. 11.5). Also the effects of slope and aspect did not significantly contribute to a separation of sensitive and non-sensitive areas. The sign of the b-coefficients of evaporation is positive and negative for elevation, indicating a contrary influence of those variables. The odd ratios of aET and elevation show a similar pattern. A value greater than one indicates that as pET increases, the odds of a region being sensitive increases, too. Here, it should be remarked that actual ET is recorded with negative sign. Thus, low aET values appear "bigger" (small negative number) than high aET values (large negative number). Significant areas are thus located predominately in the low lands with

Table 11.1 Results from logistic regression to distinguish sensitive and non-sensitive regions (corresponding to significant EVI-rainfall relations ($\alpha = 1$ %) from DLM)

	B	SE(B)	95 % CI for odds ratio		
			Lower	Odds ratio	Upper
(Intercept)	368.83***	72.26			
Evap	0.16***	0.02	1.12	1.18	1.23
Temp	−1.20***	0.23	0.02	0.29	0.47
Elevation	−0.01***	0.01	0.99	0.99	1.00
Land cover					
M2	1.02	0.49			
M3	0.85	0.53			
K1	4.06***	0.62			
K2	3.64***	0.51			
K3	2.28***	0.46			
K4	3.89***	0.55			
O1	4.63***	0.74			
O2	1.86***	0.51			
T1	3.59***	0.76			
T2	4.05***	0.64			
T3	3.71***	0.60			
T4	4.33***	0.86			
S1	1.78***	0.46			
S2	2.82***	0.46			
G1	1.97*	0.80			
G2	1.71**	0.62			
G3	2.63***	0.51			
W1	0.45	0.55			
W2	2.35***	0.54			

Signif. codes: 0 '***' 0.001 '**' 0.01 '*' 0.05 '.' 0.1
Note. $R^2 = 0.428$ (Hosmer and Lemeshow), $R^2 = 0.447$ (Cox and Snell), $R^2 = 0.597$ (Nagelkerke)
Total accuracy in the calibration data set (n = 1,000 pixels): 81.05 %
Total accuracy in the validation data set (n = 1,000 pixels): 78.21 %

small aET levels and lower amounts of average precipitation. The positive sign of the variable mean annual temperature is reverse to the sign of aET, which was not expected due to the positive correlation of both variables (compare Fig. 11.5). It is interpreted as a significant statistical interaction effect between temperature and aET.

The indicator variables for land cover have a slightly different interpretation as for quantitative predictor variables. Since the land cover classes were coded as dummy variables, here the b-coefficients represent differences in the sensitivity of a particular land cover class compared to a reference class, here M1 (Miombo-Forest). For instance, woodland-class K1versus the reference class changes the log odds of being sensitive towards the rainfall regime by 4.06, which is highly significant. Although the high significance of these partial offsets might be also affected by the large sample size in the regression, which was necessary to capture the spatial heterogeneity in the study area, it nevertheless demonstrates that different plant communities show distinct individual responses to the rainfall regime (Li et al. 2013). The Miombo forest classes (M1, M2, and M3) are less sensitive. They are mainly located in the north and are associated with grasslands found in the valleys, forming the forest grassland ecotone. The plant root system of the Miombo tree species has easier access to the deeper soil water resources, such that single rainfall pulses have a lesser chance to become "effective".

According to the analysis, more sensitive classes are represented by the woodlands (classes K and O), thornbush savannah (classes T) and shrub and grassland (classes S and G) units in the semi-arid regions of the south. In the wetland categories, wet grasslands and peatlands (W2) have a significant partial intercept. This class is predominant in the Okavango Delta, which does otherwise not show any significant rainfall-EVI relationships. The Okavango Panhandle and Delta form a special ecosystem, where land cover is dominated by seasonally flooded grasslands and reedbeds (class W1) and is thus governed rather by inundation than by the amount of rainfall. The reason for the significance of W2 is probably a mixed pixel effect with neighbouring shrub and grassland areas which are associated with W2 (compare Fig. 11.1).

Figure 11.7 further differentiates characteristics of the sensitive areas within the individual land cover classes by showing the percentage of significant pixels in each class and considering the different lags in the model. This supports the results from the logistic regression models, whereupon shrub- and grasslands but also woodlands and thorn savannahs are highly sensitive towards rainfall anomalies. In these classes more than 80 % of the pixels are sensitive, whereas in the Miombo forests this contribution is only between eight (8 %, M1) and 30 % (M2). Between 40 % and 65 % of grass- and shrubland regions (classes S and G) are sensitive in the Okavango catchment. Although partial offsets of these classes are significant, the

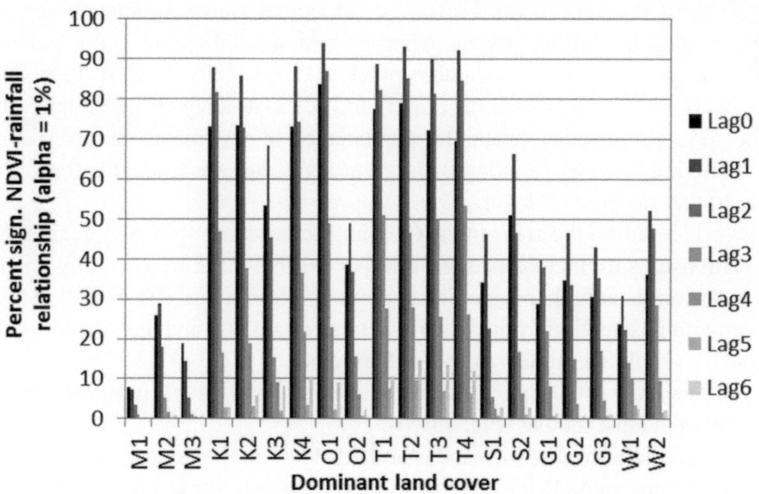

Fig. 11.7 Occurrence of significance DLM results in different land cover classes for different lags

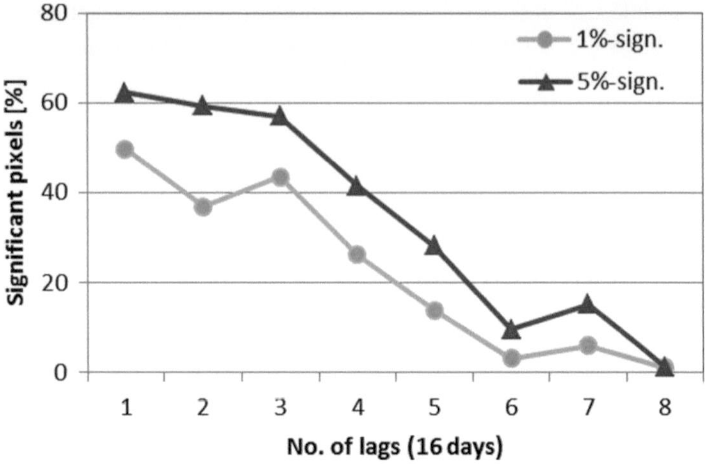

Fig. 11.8 Percentage of pixels with significant rainfall-EVI relationships at different lags in the Okavango catchment

coefficients in the logistic regression Table 11.1 are smaller compared to those of woodlands and thornbush savannahs.

11.3.2 Temporal Rainfall-ANPP Relationships at Specific Lags

Figure 11.8 depicts the percentage of significant pixels from the DLM for increasing lags considering two significance levels (1 % and 5 %). On the 1 % significance level 50 % of the area in the Okavango catchment shows significant sensitivity towards current rainfall anomalies. After a 1.5 month delay (lag 3) this percentage drops to almost half of the available pixels (25 %). Only 3 % of all pixels are sensitive after a lag of 2.5 months (lag 5) and 0.2 % at a lag of 3.5 months (lag 6). On the 5 % significance level, the percentages of significant pixels is slightly higher, ranging from 61 % at lag 0–0.8 % at lag 7, but the general pattern is quite similar to that on the 1 % level.

Figure 11.9 shows the areas in the Okavango catchment with significant rainfall-EVI relationships for increasing lags. It illustrates that there are generally no unique rainfall responses at higher lags. The common pattern rather shows EVI at a given pixel remaining sensitive from lag zero up to a certain lag, and then the influence of previously cumulated rainfall events is fading out.

To further analyse this effect, the mean annual rainfall levels in significant pixels were plotted against the lag (Fig. 11.10).

Both precipitation amounts and standard deviations decrease in regions that show significant rainfall-EVI relationships at higher lags. Lower precipitation amounts also bear the risk that periods between two “effective” rainfall events

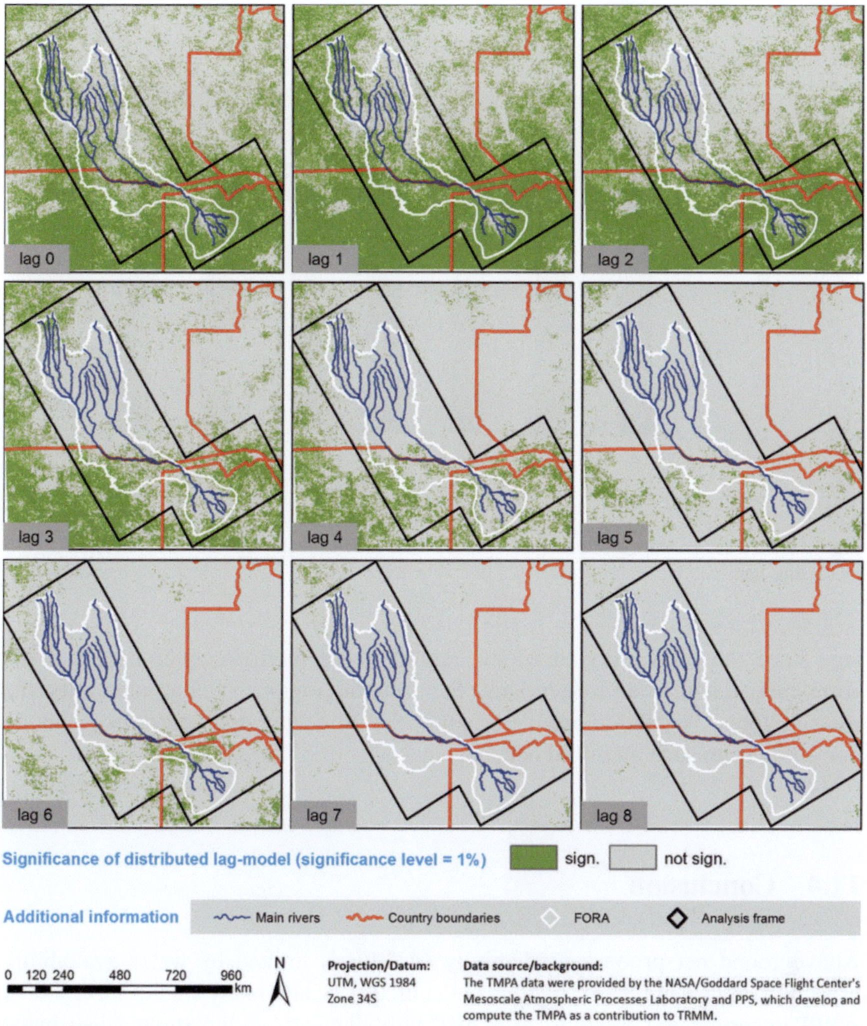

Fig. 11.9 Significance maps for the modelled rainfall-EVI relationships using a DLM considering different lags

are extended. The impact of an "average" rainfall anomaly remains until the next "effective" event that triggers another change in the EVI, according to the Westoby-Bridges pulse-reserve hypothesis. This results in extended significant rainfall-ANPP relationships. If no other "effective" rainfall occurs during the next months this anomaly will resist, unless water limited plant stress occurs or a new rainfall event triggers another EVI anomaly. In a similar way an extended drought period will negatively affect the ANPP, unless the vegetation recovers after a new biologically relevant rainfall event. On the other hand, if periods between two effective cumulated rainfall events are rather short or characterized by a large temporal variability, then the impact of the cumulated events on the EVI decreases at longer

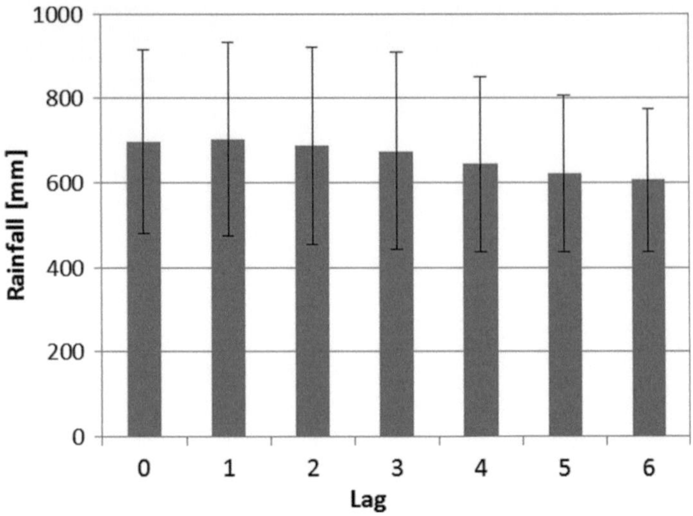

Fig. 11.10 Mean annual rainfall in areas with significant rainfall-EVI relations, for increasing lags

lags since the probability of a new rainfall pulse increases over time and the correlation declines at longer lags. Semi-humid regions, where especially the roots of forest vegetation are better connected to deeper soil water resources do not suffer from such a rainfall dependency.

11.4 Conclusion

Aboveground net primary productivity (ANPP) is limited by water availability especially in dry and desert regions (Li et al. 2013), and many studies have linked ANPP to current and previous "effective" rainfall events. In this study a distributed lag model (DLM) was used to assess the impact of previous 16-day accumulated rainfall anomalies on the Enhanced Vegetation Index (EVI) as proxy for the ANPP. The two important aspects in using DLMs are the total amount of ANPP variability that can be explained by the previous rainfall amounts and the duration of the dependency. Results demonstrate that in the Okavango catchment more than half of the area is sensitive towards anomalies in precipitation. These areas are mainly located in the semi-arid regions of the catchment, which correspond to the low lands in the south. Rainfall anomalies either directly affect ANPP, or the first significant effect is lagged up to 1 month after the anomaly. The impact of previous cumulated rainfall events remains significant until the next rainfall anomaly triggers a new change in ANPP. Consequently, the number of significant lags in the DLM is negatively correlated with total rainfall amount; a higher aridity entails the risk of longer dry periods between effective rainfall anomalies. Thus, on average rainfall anomalies in these areas have longer impacts on the vegetation than in regions with more frequent rainfall. Results further indicate that in dryland ecosystems different vegetation types have different EVI signals during dry spells and varied rates of response to the prevalent rainfall regime. This result is not surprising as different plant functional types can avoid competition, particularly by exploiting soil water resources during their growing season (Ogle and Reynolds 2004). Most sensitive land cover classes were shrub- and grassland, thornbush savannah and open woodlands.

In the semi-humid north of the catchment only locally significant rainfall-ANPP relationships were found. In these regions the effect of dry periods on vegetation is buffered through a higher available soil water content and the roots of the dominant Miombo forests have better access to deeper soil water layers. This reduces the dependency on former rainfall events.

For the respective populations, this means that livelihood strategies depending on the identified, most sensitive classes, will face further risks under predicted changing climate conditions. On the other hand, Miombo forests show a better buffering capability, which underlines their outstanding importance and makes their sustainable use a major requirement to land management schemes adapted to climate change.

Acknowledgements This study was partially conducted within the "Future Okavango" project funded by the German Federal Ministry of Education and Research (BMBF) in the frame of the funding measure "Sustainable Land Management". This support is gratefully acknowledged. The authors would also like to thank NASA/PPS for providing the MODIS imagery and the TRMM data, as well as the Climate Service Center, Helmholtz-Zentrum Geesthacht, especially Dr. Torsten Weber, for providing evaporation data.

References

Balaghi R, Tychon B, Eerens H (2008) Empirical regression models using NDVI, rainfall and temperature data for the early prediction of wheat grain yields in Morocco. Int J Appl Earth Obs Geoinf 10:438–452

Boer MM, Puigdefábregas J (2003) Predicting potential vegetation index values as a reference for the assessment and monitoring of dryland condition. Int J Remote Sens 24:1135–1141

Boko M, Niang I, Nyong A, Vogel C, Githeko A, Medany M, Osman-Elasha B, Tabo R, Yanda P (2007) Africa. In: Parry ML, Canziani OF, van der Linden PJ PJP, Hanson CE (eds) Climate change 2007. Impacts, adaptation and vulnerability. Contribution of working group II to the fourth assessment report of the intergovernmental panel on climate change: technical summary. Cambridge University Press, Cambridge, UK, pp 433–467

Box GEP, Jenkins GM, Reinsel GC (1994) Time series analysis, forecasting and control. Prentice-Hall, Englewood Cliffs

Camberlin P, Martiny N, Philippon N, Richard Y (2007) Determinants of the interannual relationships between remote sensed photosynthetic activity and rainfall in tropical Africa. Remote Sens Environ 106:199–216

Chatfield C (2004) The analysis of time series. An introduction. CRC Press, Boca Raton

Cihlar J, Laurend I, Dyer J (1991) Relation between normalized difference vegetation index and ecological variables. Remote Sens Environ 35:279–298

Davenport ML, Nicholson SE (1993) On the relation between rainfall and the normalized difference vegetation index for diverse vegetation types in East-Africa. Int J Remote Sens 14:2369–2389

Evans J, Geerken R (2004) Discrimination between climate and human-induced dryland degradation. J Arid Environ 57:535–554

Fensholt R, Rasmussen K (2011) Analysis of trends in the Sahelian 'rain-use efficiency' using GIMMS NDVI, RFE and GPCP rainfall data. Remote Sens Environ 115(2):438–451

Gao YZ, Chen Q, Lin S, Giese M, Brueck H (2011) Resource manipulation effects on net primary production, biomass allocation and rain-use efficiency of two semiarid grassland sites in Inner Mongolia, China. Oecologia 165:855–864

Gessner U, Naeimi V, Klein I, Kuenzer C, Klein D, Dech S (2013) The relationship between precipitation anomalies and satellite-derived vegetation activity in Central Asia. Glob Planet Chang 110:74–87

Granger CWJ, Newbold P (1986) Forecasting economic time series. Academic, Orlando

Gurgel HC, Ferreira NJ (2003) Annual and interannual variability of NDVI in Brazil and its connections with climate. Int J Remote Sens 24:3595–3609

Herrmann SM, Anyamba A, Tucker CJ (2005) Recent trends in vegetation dynamics in the African Sahel and their relationship to climate. Glob Environ Chang 15:394–404

Huete A, Didan K, Miura T, Rodriguez EP, Gao X, Ferreira LG (2002) Overview of the radiometric and biophysical performance of the MODIS vegetation indices. Remote Sens Environ 83(1–2):195–213

Huete AR, Didan K, Shimabukuro YE, Ratana P, Saleska SR, Hutyra LR, Yang W, Nemani RR, Myneni R (2006) Amazon rainforests green-up with sunlight in dry season. Geophys Res Lett 33:L06405

Huffman GJ, Adler RF, Bolvin DT, Gu G, Nelkin EJ, Bowman KP, Hong Y, Stocker EF, Wolff DB (2007) The TRMM multi-satellite precipitation analysis: quasi-global, multi-year, combined-sensor precipitation estimates at fine scale. J Hydrometeor 8:38–55

Huxman TE, Snyder KA, Tissue D, Leffler AJ, Ogle K, Pockman WT, Sandquist DR, Potts DL, Schwinning S (2004) Precipitation pulses and carbon fluxes in semiarid and arid ecosystems. Oecologia 141:254–268

Jacob D (2001) A note to the simulation of the annual and inter-annual variability of the water budget over the Baltic Sea drainage basin. Meteorog Atmos Phys 77:61–73

Karabulut M (2003) An examination of relationships between vegetation and rainfall using maximum value composite AVHRR-NDVI data. Turk J Bot 27:93–101

Kawabata A, Ichii K, Yamaguchi Y (2000) Global monitoring of interannual changes in vegetation activities using NDVI and its relationships to temperature and precipitation. Int J Remote Sens 22(7):1377–1382

Kosmas C (1999) The impacts of agriculture on desertification. In: Balabanis P et al (eds) Mediterranean desertification – research results and policy implications. Proceedings of the International Conference, 29 Oct–1 Nov 1996, Crete, Greece, pp 199–213

Le Houérou HN (1984) Rain use efficiency – a unifying concept in arid-land ecology. J Arid Environ 7:213–247

Li F, Zhao W, Liu H (2013) The response of aboveground net primary productivity of desert vegetation to rainfall pulse in the temperate desert region of Northwest China. PLoS ONE 8 (9):1–11

Liu LM, Hanssens DM (1982) Identification of multiple-input transfer function models. Commun Stat – Theory Method 11(3):297–314

Nicholson SE, Davenport ML, Malo AR (1990) A comparison of the vegetation response to rainfall in the Sahel and East Africa, using normalized difference vegetation index from NOAA AVHRR. Climate Change 17:209–241

Noy-Meir I (1973) Desert ecosystems: environment and producers. Annu Rev Ecol Evol Syst 4:25–51
Oesterheld M, Loreti J, Semmartin M, Sala OE (2001) Inter-annual variation in primary production of a semi-arid grassland related to previous-year production. J Veg Sci 12:137–142
Ogle K, Reynolds JF (2004) Plant responses to precipitation in desert ecosystems: integrating functional types, pulses, thresholds, and delays. Oecologia 141:282–294
Parton W, Morgan J, Smith D, Del Grosso S, Prihodko L, LeCain D, Kelly R, Lutz S (2012) Impact of precipitation dynamics on net ecosystem productivity. Glob Chang Biol 18:915–927
Paruelo JM, Lauenroth WK (1995) Regional patterns of normalized difference vegetation index in North American shrublands and grasslands. Ecology 76:1888–1898
Peng S, Piao S, Shen Z, Ciais P, Sun Z, Chen S, Bacour C, Peylin P, Chen A (2013) Precipitation amount, seasonality and frequency regulate carbon cycling of a semi-arid grassland ecosystem in Inner Mongolia, China: a modeling analysis. Agric For Meteorol 178–179:46–55
Prince SD, De Colstoun EB, Kravitz LL (1998) Evidence from rain-use efficiencies does not indicate extensive Sahelian desertification. Glob Chang Biol 4:359–374
Robertson TR, Bell CW, Zak JC, Tissue DT (2009) Precipitation timing and magnitude differentially affect aboveground annual net primary productivity in three perennial species in a Chihuahuan Desert grassland. New Phytol 181:230–242
Röder A, Stellmes M, Domptail S, Eschenbach A, Finckh M, Gröngröft A, Helmschrot J, Pröpper M, Schneibel A, Stoffels J (2013) Cumulative effects of policy and management actions on ecosystem services: challenges and methodological approaches in the future Okavango project. Biodivers Ecol 5:167–183
Sala OE, Parton WJ, Joyce LA, Lauenroth WK (1988) Primary production of the central grassland region of the USA. Ecology 69:40–45
Schultz PA, Halpert MS (1993) Global correlation of temperature, NDVI and precipitation. Adv Space Res 13:277–280
Schwinning S, Sala OE, Loik ME, Ehleringer JR (2004) Thresholds, memory, and seasonality: understanding pulse dynamics in arid/semi-arid ecosystems. Oecologia 141:191–193
Shumway RH, Stoffer DS (2000) Time series analysis and its applications. Springer, New York
Stellmes M, Frantz D, Finckh M, Revermann R (2013) Okavango basin – earth observation. In: Oldeland J, Erb C, Finckh M, Jürgens N (eds) Environmental assessments in the Okavango region. Biodiversity and ecology, vol 5., pp 23–27
Tambussi EA, Bort J, Araus JL (2007) Water use efficiency in C3 cereals under Mediterranean conditions: a review of physiological aspects. Ann Appl Biol 150(3):307–321
Thiam AK (2003) The causes and spatial patterns of land degradation risk in Southern Mauritania using multi-temporal AVHRR NDVI imagery and field data. Land Degrad Dev 14:133–142
Thomey ML, Collins SL, Vargas R, Johnson JE, Brown RF, Natvig DO, Friggens MT (2011) Effect of precipitation variability on net primary production and soil respiration in a Chihuahuan Desert grassland. Glob Chang Biol 17:1505–1515
Tomás M, Medrano H, Escalona JM, Martorell S, Pou A, Ribas-Carbó M, Flexas J (2013) Variability of water use efficiency in grapevines. Environ Exp Bot 103:148–157
Udelhoven T, Stellmes M, Del Barrio G, Hill J (2008) Modelling the NDVI – rainfall relationship in Spain (1989–1999) using distributed lag models. Int J Remote Sens 30(8):1961–1976
Wang J, Price KP, Rich PM (2001) Spatial patterns of NDVI in response to precipitation and temperature in the central Great Plains. Int J Remote Sens 22:1005–1027
Wang J, Rich PM, Price KP (2003) Temporal responses of NDVI to precipitation and temperature in the central Great Plains, USA. Int J Remote Sens 24:2345–2364
Weber T (2013) Okvango basin – climate. Biodivers Ecol 5:15–17
Wehberg J, Weinzierl T (2013) Okavango basin – physiogeographical setting. In: Oldeland J, Erb C, Finckh M, Jürgens N (eds) Environmental assessments in the Okavango region, vol 5, Biodiversity and Ecology., pp 11–13
Wei W (1990) Time series analysis. Univariate and multivariate methods. Addison-Wesley, Essex
Zhao W, Liu H (2011) Precipitation pulses and ecosystem responses in arid and semiarid regions: a review. J Appl Ecol 22:243–249

Chapter 12
Land Degradation in South Africa – A Degradation Index Derived from 10 Years of Net Primary Production Data

Markus Niklaus, Christina Eisfelder, Ursula Gessner, and Stefan Dech

Abstract Dry regions such as arid southern Africa are strained by unfavourable climatic conditions. Intensive land use as rangeland and for livestock farming leads to additional encroachment of these ecosystems. The consequence of this long-time stress is degradation in terms of loss of the vegetative cover and productivity. Albeit these are known facts there is still a lack of objectiveness in the long term assessment of degradation on a larger scale. We present a method of applying remote sensing time-series in a vegetation model that helps to fill this gap. The approach is based on time-series of the vegetative productivity computed by our vegetation model BETHY/DLR (Biosphere Energy Transfer Hydrology Model). The used data included SPOT-VGT LAI (Leaf area index) and ECMWF meteorology time-series for the period of 1999–2010. The trend-analysis of model output and climatic input results in a new land degradation index (LDI) that distinguishes between climatic and human-induced reduction of vegetative productivity.

12.1 Introduction

Arid or semi-arid areas are often affected by the process of land degradation, which is caused by different components from biophysical and socio-economic factors (Hoffman and Todd 2000). While the degradation of soils is mainly caused by wind and water erosion as well as soil acidification and salinization due to agricultural

M. Niklaus (✉) • C. Eisfelder • U. Gessner
German Remote Sensing Data Center, DFD, Earth Observation Center, EOC, German Aerospace Center, DLR, Oberpfaffenhofen, Germany
e-mail: markus.niklaus@web.de

S. Dech
German Remote Sensing Data Center, DFD, German Aerospace Center, DLR, Wessling, Germany

Institute for Geography and Geology, University of Wuerzburg, Wuerzburg, Germany

C. Kuenzer et al. (eds.), *Remote Sensing Time Series*, Remote Sensing and Digital Image Processing, DOI 10.1007/978-3-319-15967-6_12

land use. The degradation of vegetative cover is driven by either human activity as logging, agriculture, fire or mono-culturing or meteorological factors as the reduction in precipitation or the variation of temperature. Main factors for the regional decrease in vegetative productivity include the intense livestock farming (Perkins and Thomas 1993; Dougill et al. 1999). The assessment of these processes has already been subject of several scientific investigations especially in African regions (Abel and Blaikie 1989; Ringrose et al. 1999; Stringer and Reed 2007; Wessels et al. 2004, 2008; Knauer et al. 2014).

Most of these activities have been carried out either on a regional scale for particular biomes or for a single snapshot of the state of the vegetation. The assessment of land degradation on a national to sub-continental scale based on long term time-series would be an intense benefit, which could be achieved by using the advancements of modern vegetation models driven by remote sensing data.

In principle land degradation is correlated with a loss of biological productivity due to soil erosion, salinization, crusting and loss of soil fertility. Moreover, the vegetation cover and especially its biodiversity and its density (Le Houérou 1996) are influenced, what can be intensified by excessive land use. The carbon reservoirs in biomass and soil are sensitive indicators of degradation and the change in climate and environmental conditions. Thus, the land biomass was defined as one of the essential climate variables with high impact on the requirements of the UNFCCC (United Nations Framework Convention on Climate Change) (GCOS 2003). Changes on this reservoir cause a crucial feedback on climate and the balance of greenhouse gases (GTOS 2009). Net primary productivity (NPP), as the exchange of carbon between vegetation and atmosphere was identified as key variable to observe the ecological functionality and the lasting degradation processes (CGER 2000).

One way of rating the status of land degradation is to interpret the perception of land owners or agricultural extension officers. This was done by Hoffman and Todd (2000) from a survey in the magisterial districts of South Africa. From this they could elaborate a map showing the incidence of land degradation for the whole country, with KwaZulu-Natal, Limpopo and the Northern and Eastern Cape being mostly affected.

Wessels et al. (2008) presented a quantitative method for assessing human induced land degradation using remote sensing data. The authors tested the local NPP Scaling (LNS) method, for which the growth season NDVI (Normalized Difference Vegetation Index) sum (ΣNDVI) is calculated as a substitute for the vegetative productivity for each pixel. This value is then normalized to the highest values (90th percentile) of ΣNDVI observed in all pixels within the same Land Capability Unit (LCU). They concluded that the LNS method is a valuable tool for mapping land degradation at a regional scale.

Instead of using NDVI sums the productivity of the plants derived from SVAT (Soil Vegetation Atmosphere Transfer) models can be used to analyze the influences of environmental circumstances on the photosynthetic reactions of the plants. The carbon exchange between atmosphere and vegetation is determined at leaf level considering the energy and water balance. One model using this principle is BETHY/DLR (Biosphere Energy Transfer Hydrology Model) operated at the German Aerospace Center (DLR) (Knorr 1997; Knorr and Heimann 2001a, b; Wisskirchen 2005; Eisfelder et al. 2012, 2013). Meteorological and remote sensing based time-series are the drivers to compute the NPP in daily time steps. The analysis of the resulting time-series of NPP in arid and semi-arid areas is used to gather information about the process of land degradation. Additional analysis of climatic input parameters help to distinguish between climatic or human influences in this process.

12.2 The Study Area of Southern Africa

As region of interest for this study the area of southern Africa was chosen, including the countries Namibia, South Africa, Lesotho, Swaziland, Botswana, and Zimbabwe (see Fig. 12.1). The distribution of the Köppen-Geiger classification (Kottek et al. 2006) shows that the western part of the study area is categorized as steppe (BS) and desert (BW). The dry and hot climate of Namibia extends to Botswana and southern Zimbabwe and affects wide areas of the southern provinces in Mozambique. Warm-temperate climate zones (type C) can be found at the southern and southeastern coastal regions of South Africa, as well as at the higher altitudes of Lesotho and Swaziland in the east.

The classification of the different climates is based on the distribution of sums of precipitation and mean temperature (Tables 12.1 and 12.2). This climate classification from Kottek et al. (2006) is an update of the Köppen-Geiger classification (Köppen 1900; Geiger 1954) using monthly climate observations from meteorological stations provided by the Climatic Research Unit (CRU) of the University of East Anglia (Mitchell and Jones 2005) and a monthly precipitation data set provided by the Global Precipitation Climatology Centre (GPCC) (Beck et al. 2005). These datasets cover the period 1951–2000.

The climatic conditions of southern Africa can be seen in Fig. 12.2. Long time means are calculated from datasets of ECMWF (European Center for Medium Range Weather Forecasts) for the period 1989–2010. The combination of annual rainfall (Fig. 12.2a) and mean annual temperature (Fig. 12.2b) reflects the distribution of the Köppen-Geiger classification. Precipitation rates below 400 mm per year denote the western part of the region including the deserts Namib and Kalahari (<200 mm). Further east the precipitation rates increase more than 1,000 mm per year. The temperature shows a negative gradient from south to north with lowest temperatures below 10 °C in the higher altitudes of Lesotho and South Africa.

Variations of theses climatic parameters will be used to assess the influence of climatic changes in the regional processes of land degradation. Thus a differentiation can be inferred whether a decreased vegetative productivity is affected by climate change or human influence.

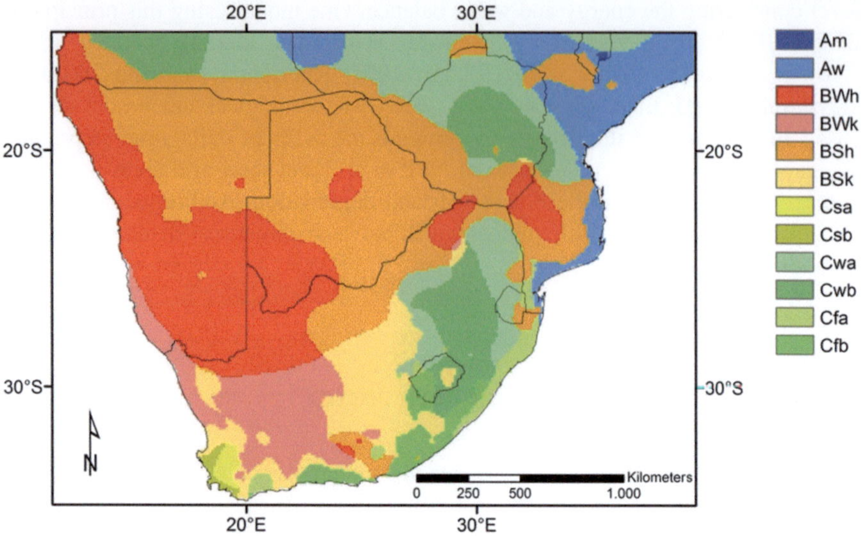

Fig. 12.1 Köppen-Geiger classification of climate types in the study area of southern Africa (For the description of the legend see Tables 12.1 and 12.2)

Table 12.1 Classification of main climates (A: equatorial, B: arid, C: warm temperate) considering characteristics of precipitation (*m* monsoonal, *S* steppe, *W* desert, *s* summer dry, *w* winter dry, *f* fully humid) for the study area following Köppen and Geiger (Köppen 1900; Geiger 1954; Kottek et al. 2006)

Climates	Description	Criteria
A	Equatorial	$T_{min} \geq +18$ °C
Am	Monsoonal	$P_{ann} \geq 25(100 - P_{min})$
Aw	Savanna	$P_{min} < 60$ mm in winter
B	Arid	$P_{ann} < 10P_{th}$
BS	Steppe	$P_{ann} > 5\ P_{th}$
BW	Desert	$P_{ann} \leq 5\ P_{th}$
C	Warm temperate	-3 °C $< T_{min} < +18$ °C
Cs	Warm summer dry	$P_{s,min} < P_{w,min}$, $P_{w,max} > 3\ P_{s,min}$ and $P_{s,min} < 40$ mm
Cw	Warm winter dry	$P_{w,min} < P_{s,min}$ and $P_{s,max} > 10\ P_{w,min}$
Cf	Fully humid	Neither Cs nor Cw

12.3 Method

12.3.1 NPP Time-Series from the Vegetation Model BETHY/DLR

We use the SVAT model BETHY/DLR to compute time-series of carbon uptake by vegetation (Knorr 1997; Roeckner et al. 2003). At the German Remote Sensing

Table 12.2 Division of temperature classes for the study area following Köppen and Geiger (Köppen 1900; Geiger 1954; Kottek et al. 2006)

Temperature classes	Description	Criteria
a	Hot summer	Mean temperature of warmest month higher than +22 °C
b	Warm summer	Mean temperature of warmest month lower than +22 °C, at least four months with mean temperature of +10 °C and higher
h	Hot arid	Mean annual temperature higher than +18 °C
k	Cold arid	Mean annual temperature lower than +18 °C

Data Center (DFD) of the DLR the model was modified by Wisskirchen (2005) for the use with time-series of meteorological and remotely sensed input data with higher resolution of up to 1 km. Now it is driven as BETHY/DLR on regional to national scales. The model mainly derives the carbon exchange between biosphere and atmosphere on basis of a photosynthetic parameterization. It also includes the water balance of the ecosystem considering precipitation, the soil water balance, uptake by the roots and the evapotranspiration of the plant.

The model needs meteorological and remote sensing based time series data. Air temperature at 2 m height, precipitation, wind speed at 10 m above ground and cloud cover are taken from the ECMWF. The daily mean of cloud cover over all three strata (high, medium and low) are used to calculate the fraction of photosynthetically active radiation (fPAR). The ECMWF provides the required data in a spatial resolution of $0.25° \times 0.25°$ (interpolated from station measurements) with a temporal resolution of every 6 h. The datasets are interpolated to time steps of 1 h. Leaf area index (LAI) and land cover (Global Land Cover, GLC2000) information are based on SPOT-VGT (Satellite Pour l'Observation de la Terre-Vegetation) and are used to describe the phenology. Since the model highly relies on LAI we need a long term, spatio-temporal continuous time-series of this parameter in model resolution. Here we use the product derived from SPOT-VGT satellite data, provided by Medias France for the years 1999–2003 and by Vito Belgium from 2004 onwards and reanalyze the data for gap filling using the method of harmonic analysis (Gessner et al. 2013).

Information on composition, stratification and thickness of the soils and their distribution are taken from the Harmonized World Soil Database (HWSD). This Dataset was created by the Food and Agriculture Organization (FAO) in collaboration with IIASA (International Institute for Applied Systems Analysis) (FAO et al. 2009). This data is aggregated from the digital soil map of the world from FAO-UNESCO, the European Soil Database (ESDB), the soil map of the Chinese Academy of Sciences and regional soil and terrain databases.

BETHY/DLR uses a two-flux scheme to approximate the radiation absorption in the canopy. A combined approach of Farquhar et al. (1980) and Collatz et al. (1992) describes the photosynthesis. Enzyme kinetics are parameterized on leaf level. For the special environment of southern Africa it is important to distinguish between C3 and C4 plants. C4 plants have different schemes of photosynthesis being better

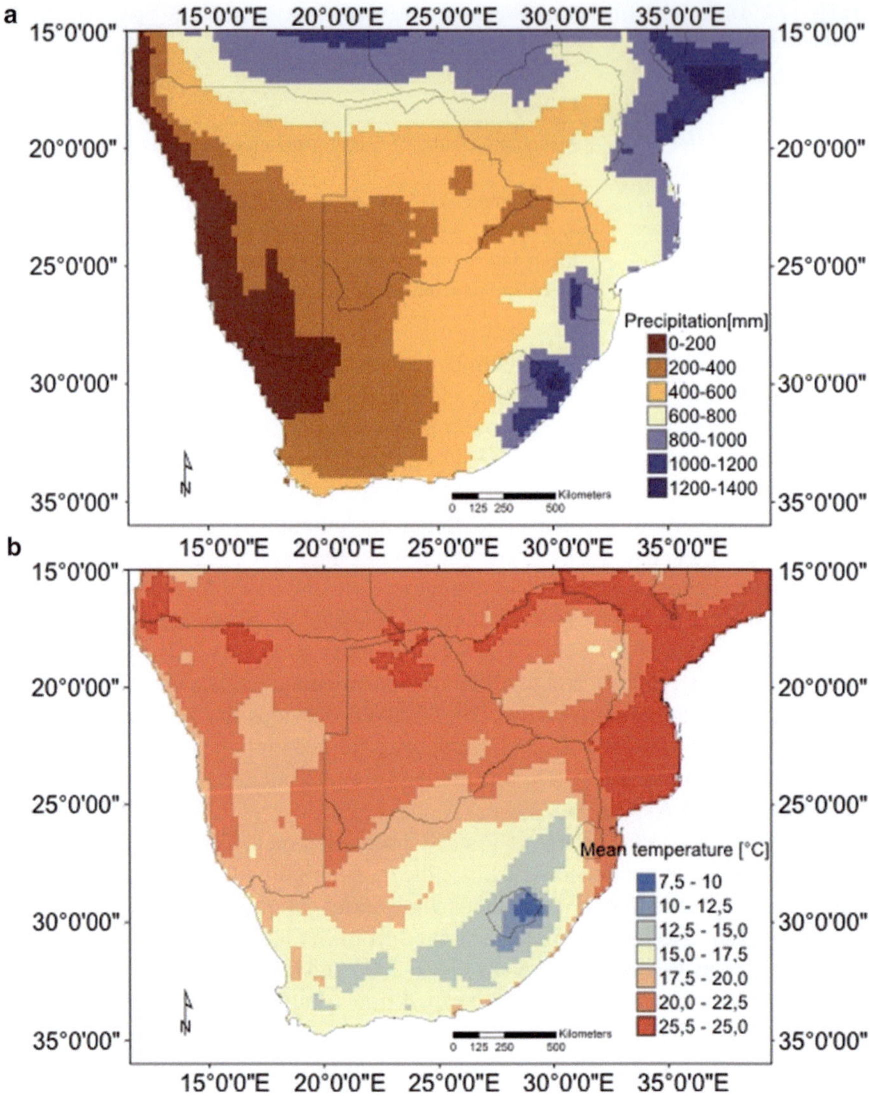

Fig. 12.2 Climatic conditions of the study area derived from time series of the ECMWF parameters precipitation (**a**) and temperature (**b**) calculated from the annual means of the period 1989–2010

adapted to dry and warm climate. The enzyme kinetics have significantly higher affinity for CO_2 than those of C3 plants. In a second step the photosynthetic rate is extrapolated from leaf to canopy level, taking into account both the canopy structure as well as the interaction between soil, atmosphere and vegetation. Stomatal and canopy conductance, evapotranspiration and soil water balance are included.

The output is given by time-series of the Gross Primary Productivity (GPP) and the autotrophic respiration where the NPP can be calculated as the difference of these two values. The results are computed in daily time steps with the spatial resolution and projection of the land cover classification (1 km × 1 km, latitude – longitude projection with WGS84 (World Geodetic System 1984) datum).

12.3.2 NPP Variations from Annual Sums

The results of our SVAT model BETHY/DLR are used to analyze the time-series of the productivity of the plants for wide areas in the region of southern Africa. From this, variations caused by natural or anthropogenic influences can be identified. This can be done by analyzing the development of annual accumulated NPP values over the period of years 1999/2000–2009/2010. This leads to a variation of productivity for each individual pixel. To identify influences of climate anomalies this variation is weighted by variations in time-series of climatic variables (temperature and rainfall). Hence it can be distinguished if positive or negative variations in vegetative productivity can be traced back to human activities or variability in climatic conditions.

In a first step the annual NPP sums for the 11 vegetative periods (99/00 to 09/10) are calculated. The vegetative periods are defined with the beginning in July (from Julien day 182) and the end in June of the following year (to 181st day of year). Figure 12.3 shows the time-series of daily NPP values for a single pixel (blue line) as model output. The pixel represents a grassland covered area in the South African province Northern Cape close to the border to Botswana (26.67S 21.93E). The daily values of the annual period are summed up to a cumulative NPP (red circles). From these annual values a temporal variation in vegetative productivity can be derived (blue line), for which this example shows a distinct negative development. The slope of this variation represents the mean annual loss of productivity and has a value of -1.4 gC $m^{-2}a^{-1}$ with a coefficient of determination of 0.67. So this pixel represents vegetation that has lost more than half (55 %) of its initial productivity (27.7 gC $m^{-2}a^{-1}$ in 99/00) in a period of 11 years.

This relative variation of NPP is shown in Fig. 12.4 for the study region. Blue colors represent a positive variation of productivity, red represents a negative variation. For studies of land degradation the negative variations regions are to be analyzed. In the study area single regions become apparent where there are high negative variations of about 100 % loss of initial productivity. NPP values at saturation can occur where an initially positive NPP becomes negative during the modeling period. These values can be attributed to be based on soil types, with poor water storage properties within the soil. Other soil types allow for a maximum rooting depth of few millimeters which again leads to an insufficient amount of accumulated water to supply the vegetation.

One of these peculiar observations in the northwest of Lesotho represents the soil type Lithosol with a horizon thickness of 10 cm. This soil type can commonly be

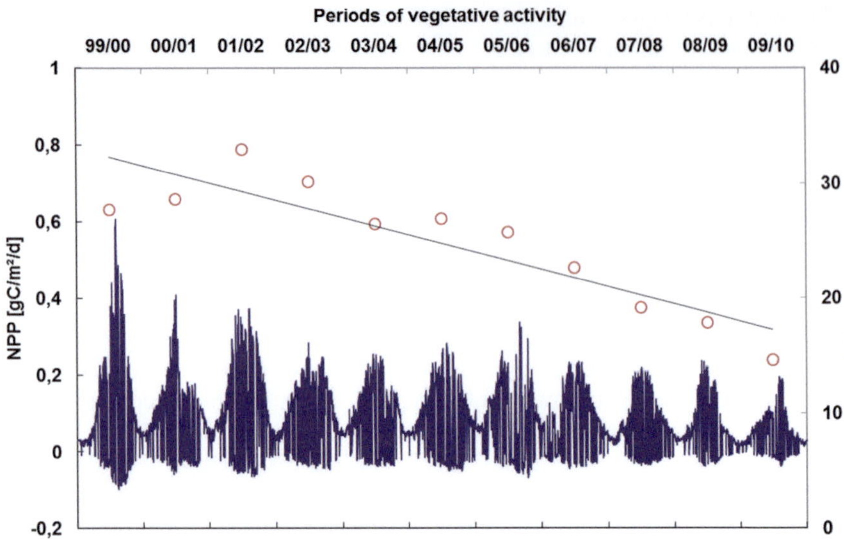

Fig. 12.3 Time-series of NPP from daily values (*blue line*) and annual sums (*red circles*) for the vegetative periods of 1999/2000–2009/2010

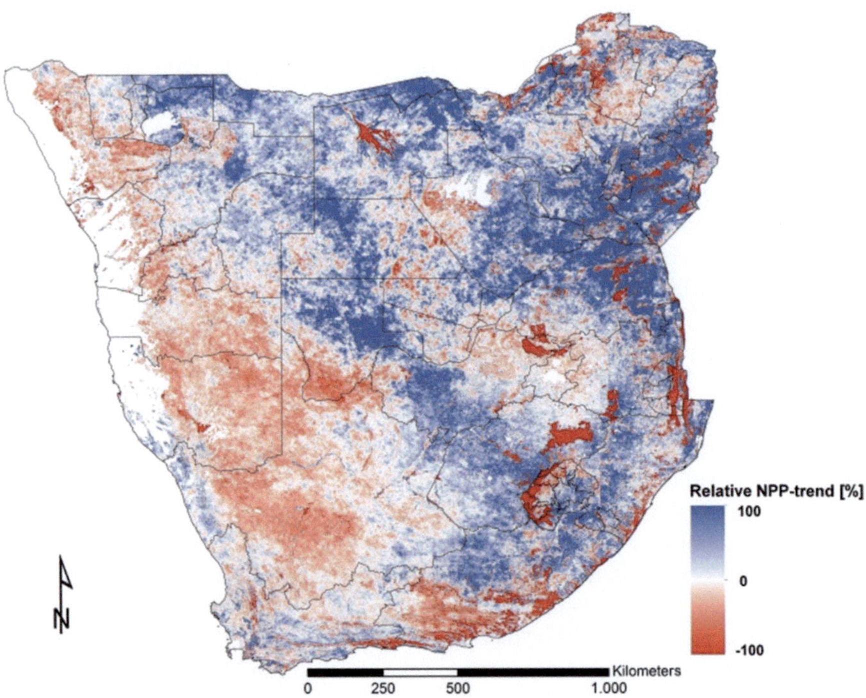

Fig. 12.4 Map of NPP-variations for the study area calculated from NPP sums of single vegetative periods

found in mountainous regions as thin layer on top of solid rocks. In contrast most of the soil types have a horizon thickness of 120 cm and more. From the soil water model the amount of water can be calculated, that can be used by the plant. Stress from insufficient water is simulated when only 25 % of this amount is left in the soil. For the Lithosol soil type the maximum amount is 2.4 cm water column, hence below 0.6 cm water column the plant experiences drought stress. Analyzing the precipitation rates of this pixel reveals a high variability over the modeling period with the amount of rain in the period 99/00–02/03 and 05/06 being up to 40 % higher than the mean value. The remaining periods show values of up to 30 % below mean where vegetation can experience stress almost the whole year. Hence the carbon assimilation is very low for these periods.

Other soils, e.g. Luvisols, have poor water storage properties with a high fraction of sand (>70 %), where water quickly percolates and gets unavailable for the vegetation. Additionally the density of vegetation cover defines the amount of water that is extracted from the soil. Translating the GLC2000 classes 2 and 3 (closed and open deciduous broadleaved forest) the fraction of cover is defined as 0.9 and 0.65 respectively. This leads to different variations in neighboring pixels when the fractional cover strongly varies on the same soil type with identical meteorological conditions. Exceeding the maximum water availability of the soil with having a denser vegetation cover can lead to negative variations, where the neighboring pixel is still getting enough water.

The negative development of the productivity is found in the dry grass- and shrubland areas in the west of the study area especially in Namibia and South Africa. Besides that the agricultural areas surrounding Pretoria and Johannesburg also show negative variations in a broad region. An even higher loss of productivity can be found in the surroundings of Port Elizabeth for forest areas, where the variation partially exceeds 100 %. This describes the change of the vegetation being a carbon sink towards a carbon source. Hence the calculated NPP reaches negative values, resulting from a greater amount of carbon being released to the atmosphere by maintenance respiration than the amount being assimilated during photosynthesis. The structures discussed before in Lesotho, in the South African province Neustaat, in Swaziland, or north of Johannesburg show the highest reduction in vegetative productivity at rates greater than 100 %. In contrast to the previous regions the equally high negative variation for the Okavango Delta can not only be attributed to the soil type. This region is particularly classified as regularly flooded marshland. This vegetation type has a very low electron transport rate ($J_m = 37$ μmol(CO_2) m^{-2} s^{-1}, (Knorr 1997)) hence the photosynthetic activity is highly susceptible to variations in the climatic conditions.

The primary vegetation type in the entire area is scrubland followed by grasslands. On the other hand, the distribution of the negative productivity variations indicates a higher susceptibility to degradation of the grassland vegetation. This is attributed to the intensive grazing land use in southern Africa. Due to grazing of large areas these ecosystems take considerable damage, which is reflected in the reduced productivity of the plants (Perkins and Thomas 1993). However, not the entire decline in the productivity can be claimed to the demands caused by human. Additional effects are especially caused by variations in rainfall.

12.3.3 Anomalies in Climatic Time-Series

It has to be determined whether a negative variation in productivity is attributable solely to the management of the natural area and no additional adverse climatic conditions have affected plant growth. Therefore variations for the time-series of precipitation and temperature are computed for the years 99/00–09/10 analogous to the NPP variations shown before.

For the pixel shown in Fig. 12.3 a negative variation of −16.8 mm a^{-1} ($R^2 = 0.35$) for annual precipitation and negative variation of −0.019 °C a^{-1} ($R^2 = 0.03$) for mean temperature was observed. These variations can help to identify if the vegetation is regressive despite or because having good climatic conditions.

The variation of the annual precipitation rate is negative for almost the whole study area (Fig. 12.5a). Exceptions are the western coast, with its desert areas of the Namib and the cape region of South Africa, and areas in the northeastern part of South Africa and the northeastern part of Zimbabwe. The variations are distinct with positive and negative direction, representing an increase or decrease of up to 50 % of the initial value in the period 99/00. The regions in southern Namibia and western South Africa having a dry climate but especially broad regions of central South Africa and Lesotho are affected by a significant decline in precipitation amount. This seems to have direct impact on the local ecosystem, as can be seen from the vegetative variation in Fig. 12.4.

The variation of the mean annual temperature shows a more heterogeneous distribution (Fig. 12.5b). Botswana as an example shows negative variations in temperature for almost the whole country and is surrounded, except for the north, by regions with positive variations up to +30 % (Lesotho and its northeastern surrounding) compared to the initial value in 99/00. These local peaks represent a raise of mean annual temperature of about 0.5 °C for the considered period of 11 years. The Namibian and South African coast in the west and the south again show negative variations. In the eastern part of the study area (Mozambique) the raise of temperature reaches up into the north also covering large parts of Zimbabwe. The coefficient of correlation of the linear fit especially for high temperature variations (>±15 %) has good values of 0.5 and higher.

An analysis of the anomalies of the single years of precipitation compared to the long time mean of 1989–2010 of ECMWF data revealed that the periods 99/00, 03/04 and 05/06 had a very high rate of precipitation. Extensive dry periods appeared for 04/05, 06/07, 08/09 and 09/10, pointing out the negative precipitation variations of Fig. 12.5a. The analysis of the mean annual temperature permanently shows positive anomalies for the periods 99/00–09/10 except for small regions in Botswana and Zimbabwe. Hence, the region of southern Africa is undergoing a

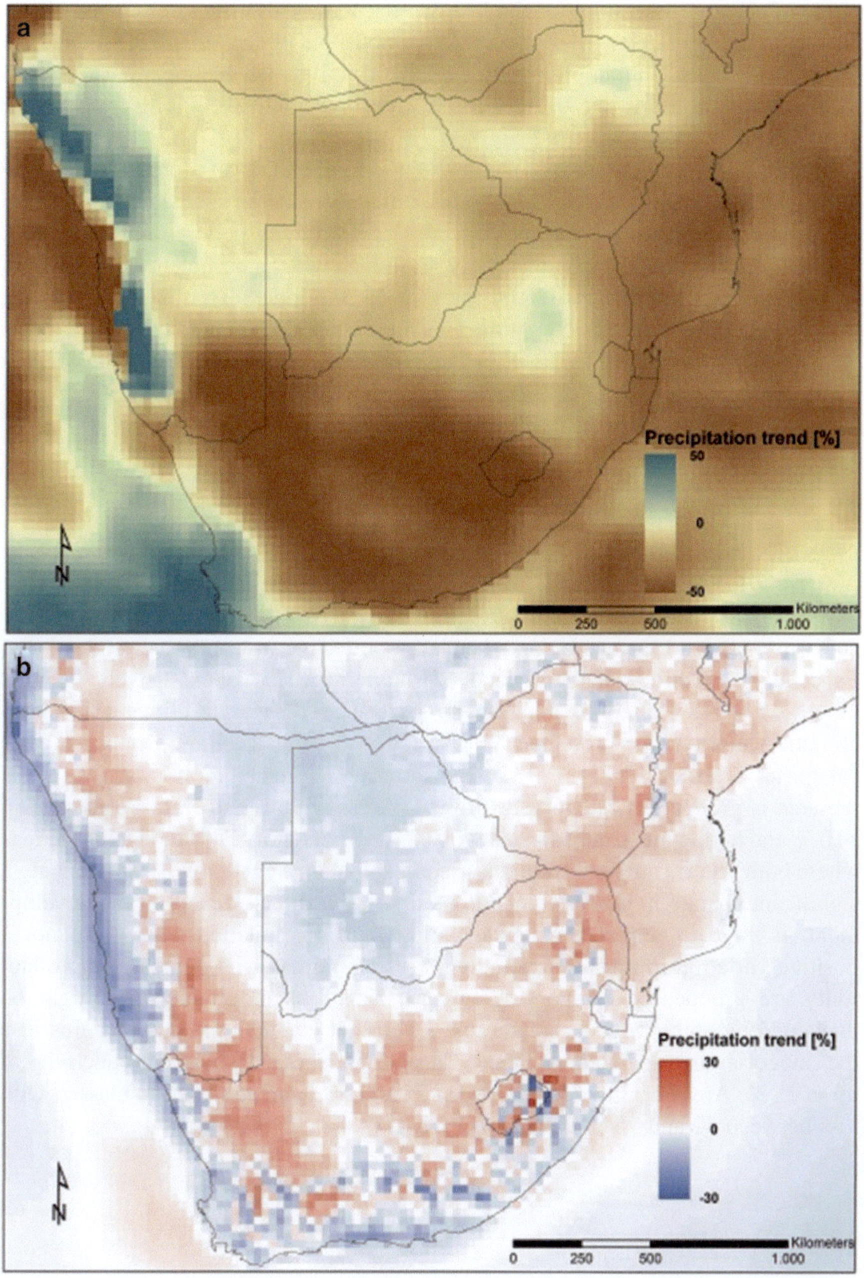

Fig. 12.5 Relative variation of the mean annual precipitation (**a**) and mean annual temperature (**b**) relating to the initial values of the period under observation 99/01–09/10

phase of rising temperature for the last two decades. This effect may on one hand lead to an increase in length of the growth periods, but on the other hand reduce the productivity due to limitations in photosynthetic activity when temperatures exceed a certain threshold, or due to decreasing water availability resulting from higher evaporation rates.

In the following definition of the degradation index only an increase of temperature and a decrease of precipitation rates will be considered. These are believed to be the most important drivers of vegetative productivity undergoing the highest changes due to climate variations. Only variations with a coefficient of correlation higher than 0.5 will be considered relevant for the index. Thus a climatological impact is assumed, if the temperature rises more than 10 % in the time period or the amount of rain is reduced by 30 %. Depending on the region this represents a rise in temperature of about 1–2 °C and a decrease in rainfall of 60–420 mm.

12.3.4 The Degradation Index

Assessing the land degradation using the model results of BETHY/DLR is supposed to consider climatic conditions. This can be helpful to determine if the cultivation of land surface by man has led to a decline in vegetative productivity. Especially the shrub- and grassland of southern Africa is extensively used as rangeland (Ross 1999; Fox and Rowntree 2001). To separate climatic and non-climatic variations the presented NPP-variations of Fig. 12.4 are grouped in four categories (Table 12.3, Fig. 12.6). If the variation in precipitation is below −30 % the area is grouped in category B (brown). Areas with a temperature variation of +10 % and higher are grouped in category C (red). Category D (blue) includes areas where both criteria are met. Here it is assumed, that the vegetation is experiencing a significant change in climatic conditions that has a direct impact on the development. If there is neither a negative variation in precipitation (<−30 %) nor a positive variation in mean temperature (>+10 %) areas with decrease in productivity are grouped in category A (green). This loss of productivity might be explained by anthropogenic influences. To further evaluate the grade of this loss the categories are separated in classes. Class 1 describes a decline of productivity of up to 25 %. Areas having a loss between 25 and 50 % are assigned to class 2, with loss between 50 and 75 % to class 3 and higher than 75 % to class 4.

12.4 Results

A direct influence of climate change can be seen for wide areas of the Namib Desert in southern Namibia and northwestern South Africa, which is classified as grassland. Here an extensive decline of about 50 % (LDI D2) is observed with local peaks of more than 75 % (LDI D4). Such areas can also be seen in Lesotho,

Table 12.3 Classification of the LDI using the color code of Fig. 12.6

ΔNPP	Non-climatic	ΔP<-30 %	ΔT>+10 %	ΔP<-30 % & ΔT>+10 %
<-75 %	A4	B4	C4	D4
-50 % to -75 %	A3	B3	C3	D3
-25 % to -50 %	A2	B2	C2	D2
0 % to -25 %	A1	B1	C1	D1

ΔNPP is the NPP-Variation, ΔP the precipitation variation and ΔT the temperature variation for the study period 99/00–09/10

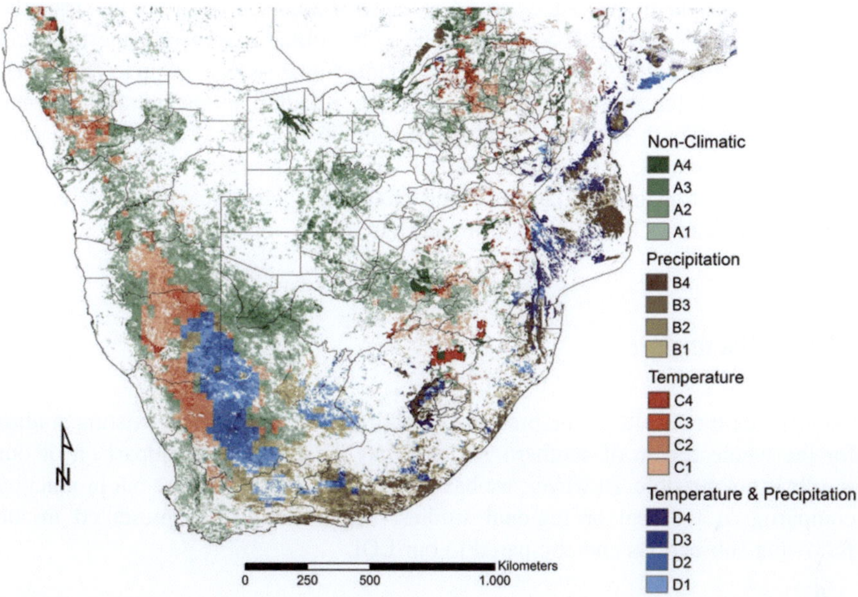

Fig. 12.6 Degradation index of the study area with the separation in categories A: non-climatic (*green*), B: significant decline of precipitation (*brown*), C: significant rise in temperature (*red*) and D: combined raise in temperature and decline of precipitation (*blue*). The *numbers* represent the grade of the NPP decline: 1: 0–25 %; 2: 25–50 %; 3: 50–75 %; 4: 75 % and higher

Swaziland and northern Zimbabwe, where generally an index of D4 is reached. They mostly coincide with the discussed soil types which typically have poor water storage capacities. This also can be seen for category B of which the highest values (LDI B4) also occur for these soil types. In these cases the problem of poor soil conditions facing decreasing water input from rainfall becomes apparent. Large regions grouped in category B include the forest areas at the southern coast of South Africa. These plants encounter loss of productivity of 75 % and more.

High increase of mean annual temperature may be the cause of the NPP decrease in wide areas of the Namib mainly in Namibia, but also in northwestern South Africa. An LDI of C2 and C3 is common for this region, but there also exist decreases of more than 75 % (LDI C4). The same classifications can be found in the west of the Etosha-pan in northern Namibia (shrub- and grassland) and in northern Zimbabwe (cultivated area). The LDI C4 in the South African province Freestate in turn can be explained by the properties of the soil type.

For the remaining regions no significant variations in climatic conditions can be observed (category A). These regions have a high potential of being intensively threatened by human cultivation. This can also be true for regions classified in categories B, C and D, but here it is assumed, that human impact solely is responsible for the changes in NPP. This is true for all negative NPP variations in Botswana with the highest index of A4 at the Okavango Delta. In South Africa the cultivated areas are affected adjacent to the metropolitan area of Johannesburg and Pretoria, but also in the Cape-region. Furthermore high loss rates of up to 75 % (LDI A3) are found for the grassland areas of the Kalahari at the border between Botswana and South Africa, partially exceeding 75 % (LDI A4). For the adjacent areas of category B, C and D in South Africa and Namibia a combination of overstraining from human land use and changes of climatic conditions can be assumed that causes the decline of productivity of the vegetation.

12.5 Discussion

To evaluate the results of the presented classification there are no existing studies for the whole region of southern Africa. Thus the consistent comparison of our results is not possible. However, we have the opportunity to analyze our product by comparing it to local or regional studies. These studies are presented in the following subsections and compared to our LDI.

12.5.1 Namibia

Strohbach (2001) describes degradation processes in the northern Oshikoto region of Namibia and the central region surrounding the capital Windhoek. This study uses the gradient of degradation developed by Bosch et al. (1987) and Bosch and van Rensburg (1987). This gradient observes the transition of a grass covered landscape to woody savannah vegetation due to intensive rangeland usage. In addition the loss of vegetated areas is relevant for many regions like the northern Oshikoto region west of the Etosha pan or the Oshana plain in the north (Strohbach 2000a, b). For both regions a LDI of A2 to A3 is classified as cultivated areas in the north and grassland in the west. For regions where the natural vegetation is replaced by other plants the change in productivity may not be represented with a

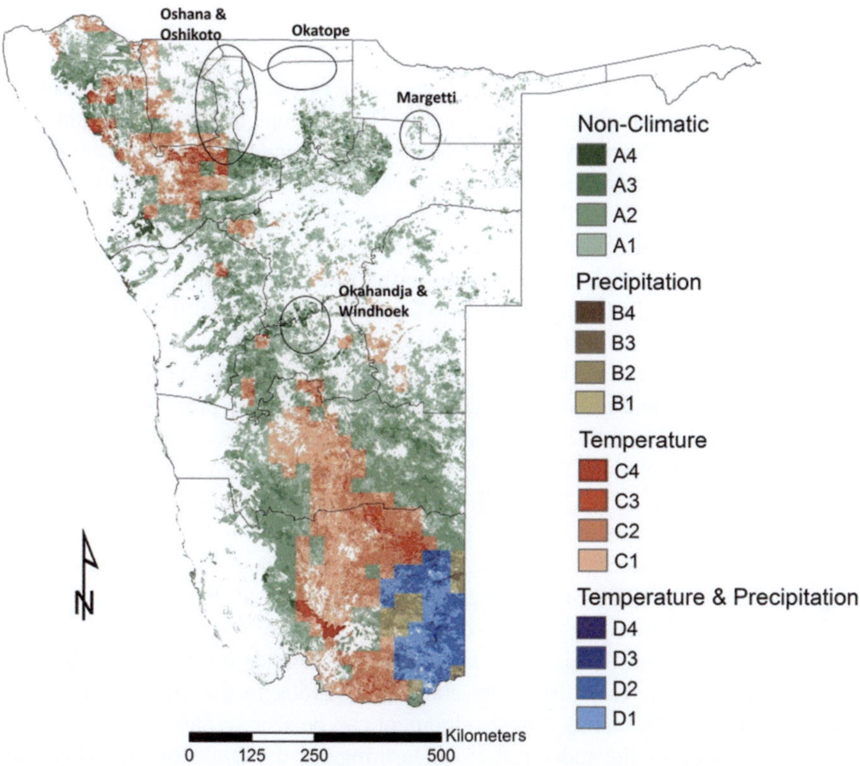

Fig. 12.7 Degradation index for Namibia with study regions of Strohbach (2000a, b, 2001) (*black markings*)

classification of the degradation index, as can be seen for the Okatope region in the northeast of the Etosha pan (Strohbach 2001).

In the Mangetti region a loss of shrubland vegetation is reported (Strohbach 2001) that can be affirmed by a LDI A2 in this region. Here also the land use as rangeland seems to be responsible for the loss of productivity. For central Namibia in the districts Okahandja and Windhoek a strong decline of grassland vegetation is observed for the period 1985–2001, where partially the whole vegetative soil cover is lost (Strohbach 2001). This also applies for good climatic conditions resulting in a LDI of up to A4 for the development in the period 99/00–09/10 for both regions. This is an obvious sign for the influence of the natural vegetation by intensive livestock farming (Fig. 12.7).

12.5.2 Botswana

Less than 5 % of the area of Botswana is suited for rain fed agriculture. Most of the land is used for livestock farming, whereas 71 % are operated by municipals and tribes, 23 % is state territory and the other 6 % are operated by commercial farms. 18 % of the area is rangeland and simultaneously identified as national parks and game reserves whereof conflicts arise progressively (Darkoh 1999).

The arid and semi-arid ecosystems, as they are in the Kalahari, obviously are extremely resilient due to the adaptation to the local climatic conditions. Though, the growing demand of rangeland, wood resources or agricultural land puts an immense stress on the vegetation the ecosystems can hardly cope with. These are the direct consequences of the faster growing population in these regions resulting in growing droves and intensifying cultivation of land. This problem seems to be persisting over the last decade, since all observed regions of the country are grouped in the category A (Fig. 12.8) and hence no climatic variations occurred in this period. Van Vegten (1981) already reported a reduction of grass cover in the southern Kgatleng district from 6–15 to 0–2 % caused by the local livestock farming. From our model results we get a LDI of A2 to A3 for cultivated areas and grassland (Fig. 12.8). Similar high values (partially A4), but spatially broader, are found for the Boteti region west of the Makgadikgadi salt pan. For this region Darkoh (1999) found obvious indications of land degradation and desertification over a period of 30 years from the end of the 1960s to the end of the 1990s. Following our model results this development can be seen to date.

The relevance of the industrial stock farming for the strain of the grass- and scrubland also was discussed by De Queiroz (1993). These studies in the Khutse wildlife reserve in the Kalahari showed that the decline of the local vegetation is in direct relation to the intensive grazing. The same was found by Dougill et al. (1999) for the vegetative development at the 'Uwe-Aboo-Farm' in the northeast. In this area the loss of mainly grassland vegetation was reported for the 1990s which is partly replaced by scrubland. At the persisting farms a high degradation index of A2 to A3 can be found for the period 1999–2010.

One specific feature in Botswana of course is the regularly flooded swamp region of the Okavango Delta. This area is classified with the highest LDI of A4. Hamandawana et al. (2007) used historical records, surveys regarding the environmental changes and satellite data (Landsat and Corona) to assess the land degradation in the delta covering the period 1860–2001. The analysis of the surveys reveals a decreasing amount of surface water, declining of the ground water level and the loss of pasture land. Negative precipitation variations, vegetation loss caused by floods and increasing demands by migrations to the region are accounted as reasons for the degradation. The assessment of the satellite images has also shown decline of the environmental status. A decrease of surface water by more than 12 % was observed before 1989. The tree cover reduced by 5 % between 1989 and 2001, whereas the scrubland increased by 6.6 % from 1967. A considerable decrease of about 44 % was found for grassland between 1967 and 2001. The main reason for

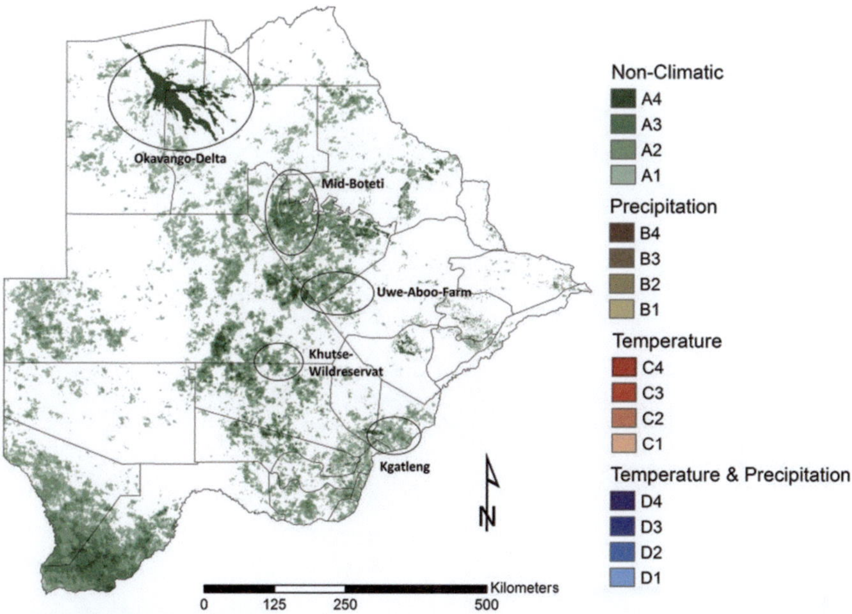

Fig. 12.8 Degradation index for Botswana with study regions of Van Vegten (1981), Darkoh (1999), and De Queiroz (1993) (*black markings*)

this development is the growth of population in the area of more than 50 % in the years between 1981 and 2001, causing an increase in livestock farming (Hamandawana et al. 2005). The results of the model presented in the degradation index show these variations to persist in the subsequent decade.

12.5.3 Zimbabwe

In Zimbabwe the Buhera District (marked in Fig. 12.9a) was analyzed by Mambo and Archer (2007) by comparing two sets of Landsat TM (Termal Mapper) and ETM (Enhanced Thematic Mapper) scenes of 1992 and 2002 to derive negative changes in the vegetative cover (Fig. 12.9c).

Both approaches locate degradation processes in similar regions. The locations of the degrading areas are generally in agreement, but both maps show a more detailed distribution for different areas. The classification also differs between the two products in some cases. This could be caused by the different types of approaches for derivation of degrading areas. The main reason for differences, however, can be assumed to arise from the different periods of observation. In summary the main regions in this district, the North and the South, are found to be degrading by both approaches. Other comparisons to regional studies in Namibia, Botswana, Lesotho and South Africa show similar coinciding results.

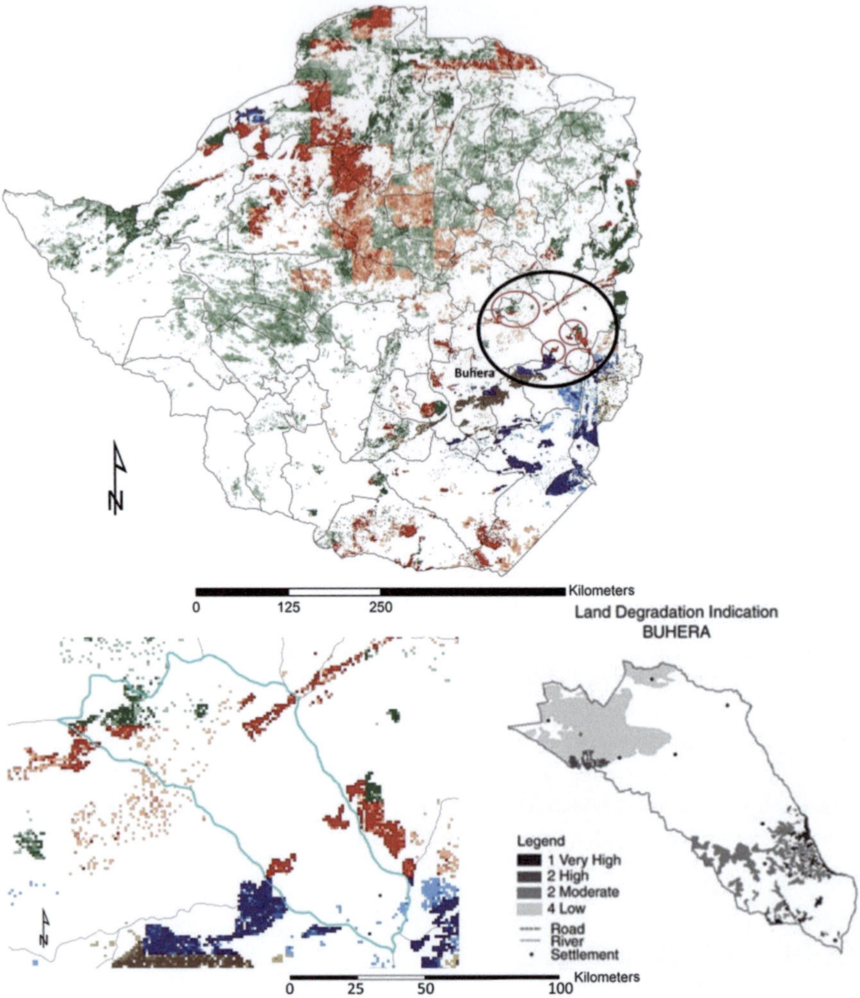

Fig. 12.9 Degradation index for Zimbabwe (**a**) and a zoom to the Buhera District (**b**) in comparison to the Landsat-NDVI derived land degradation indication (**c**) from Mambo and Archer (2007) (Color legend and categorization is according to Fig. 12.6 and Table 12.3)

12.6 Conclusion

The purpose of this study was to develop an approach to derive a Land Degradation index (LDI) for arid and semi-arid regions. For this we used calculated Net Primary Productivity (NPP) from our remote sensing data driven model BETHY/DLR. As area of interest the southern African countries of Namibia, Botswana, Zimbabwe, South Africa, Lesotho and Swaziland were chosen. From meteorological (ECMWF), soil (FAO/IIASA), land cover and phenological (SPOT) data, we calculated annual NPP sums at 1 km spatial resolution. With this data we could derive variations in vegetative productivity for the years 1999–2010 to cover a time period of 11 years. With these variations the grade of land degradation could be assessed.

Additionally, variations in climatological time-series for mean annual air temperature and annual precipitation have been computed from the ECMWF model input data. From these time-series we were able to estimate the influence of meteorological changes over the observed period of 11 years. These three variations have been combined to an index of land degradation for the sub-continental region of southern Africa. The degradation map of southern Africa not only shows area of degradation but additional parameters controlling degradation as meteorological factor (air temperature and/or precipitation) and non-climatic factors which we relate to human induced changes.

Wide ranges of non-climatic caused degradation for example can be found in Botswana and Namibia. For these countries intense land use by grazing and farming is reported by several studies. Comparisons with regional studies show spatial consistency on regional scale, but detailed agreement can hardly be achieved due to spatial and/or temporal discrepancies in the scope of these works. However, with this method a degradation assessment of sub-continental regions can be performed for a whole decade with a spatial resolution of 1 km. As the temporal and spatial coverage of remote sensing data is driven to higher resolutions, the presented calculation of the variation in NPP will gain more and more in significance. This time-series will be extended and optimized as remote sensing products will continuously be enhanced.

Acknowledgements This study was funded by the EOS-Network of the Helmholtz Centres in Germany. We thank ECMWF, Medias France and Vito Belgium for providing their data.

References

Abel NOJ, Blaikie PM (1989) Land degradation, stocking rates and conservation policies in the communal rangelands of Botswana and Zimbabwe. Land Degrad Rehabil 1(2):101–123

Beck C, Grieser J, Rudolf B (2005) A new monthly precipitation climatology for the global land areas for the period 1951 to 2000. Climate status report 2004. German Weather Service, Offenbach, pp 181–190

Bosch O, vanRensburg FJ (1987) Ecological status of species on grazing gradients on the shallow soils of the western grassland biome in South Africa. J Grassl Soc South Afr 4(4):143–147

Bosch O, van Rensburg FJ, Truter ST, Truter DUT (1987) Identification and selection of benchmark sites on litholitic soils of the western grassland biome of South Africa. J Grassl Soc South Afr 4(2):59–62

CGER (2000) Ecological indicators for the nation. The National Academies Press, Washington, DC

Collatz GJ, Ribas-Carbo M, Berry JA (1992) Coupled Photosynthesis – stomatal conductance model for leaves of C4 plants. Aust J Plant Physiol 19:519–538

Darkoh MBK (1999) Case studies of rangeland desertification, chapter desertification in Botswana. Agricultural Research Institute, Rekjavik, pp 61–74

De Queiroz JS (1993) Range degradation in Botswana myth or reality? Technical report. Pastoral Development Network, Overseas Development Institute, London

Dougill AJ, Thomas DSG, Heathwaite AL (1999) Environmental change in the Kalahari: integrated land degradation studies for nonequilibrium dryland environments. Ann Assoc Am Geogr 89(3):420–442

Eisfelder C, Kuenzer C, Dech S, Buchroithner M (2012) Comparison of two remote sensing based models for NPP estimation – a case study in Central Kazakhstan. IEEE J Select Top Appl Earth Obs Remote Sens 6(4):1843–1856

Eisfelder C, Klein I, Niklaus M, Kuenzer C (2013) Net primary productivity in Kazakhstan, its spatio-temporal patterns and relation to meteorological variables. J Arid Environ 103:17–30

FAO, IIASA, ISRIC, ISSCAS, JRC (2009) Harmonized world soil database (version 1.1). FAO/IIASA, Rome/Laxenburg

Farquhar GD, von Caemmerer S, Berry JA (1980) A biochemical model of photosynthesis in leaves of C3 species. Planta 149:58–90

Fox R, Rowntree K (2001) Redistribution, restitution and reform: prospects for the land in the Eastern Cape Province, South Africa. In: Conacher A (ed) Land degradation. Kluwer, London, pp 167–186

GCOS (2003) The second report on adequacy of global observation systems for climate in support of the unfccc – executive summary. Technical report, World Meteorological Organization, Geneva

Geiger R (1954) Landolt-Börnstein – Zahlenwerte und Funktionen aus Physik, Chemie, Astronomie, Geophysik und Technik, alte Serie vol 3, Chapter: Klassifikation der Klimate nach W. Köppen. Springer, Berlin, pp 603–607

Gessner U, Niklaus M, Kuenzer C, Dech S (2013) Intercomparison of leaf area index products for a gradient of sub-humid to arid environments in West Africa. Remote Sens 5(3):1235–1257

GTOS (2009) Biomass. Assessment of the status of the development of the standards for the terrestrial essential climate variables. Technical report, Food and Agriculture Organization of the United Nations (FAO), Rome

Hamandawana H, Eckardt F, Chanda R (2005) Linking archival and remotely sensed data for long-term environmental monitoring. Int J Appl Earth Obs Geoinform 7(4):284–298

Hamandawana H, Chanda R, Eckardt F (2007) Natural and human induced environmental changes in the semi-arid distal reaches of Botswana's Okavango delta. J Land Use Sci 2:57–78

Hoffman M, Todd S (2000) A national review of land degradation in South Africa: the influence of biophysical and socio-economic factors. J South Afr Stud 26:743–758

Knauer K, Gessner U, Dech S, Kuenzer C (2014) Remote sensing of vegetation dynamics in West Africa. Int J Remote Sens 35(17):6357–6396

Knorr W (1997) Satellite remote sensing and modelling of the global CO_2 exchange of land vegetation: a synthesis study. PhD thesis. Max-Planck-Institut für Meteorologie, Hamburg, Germany

Knorr W, Heimann M (2001a) Uncertainties in global terrestrial biosphere modeling, Part I: a comprehensive sensitivity analysis with a new photosynthesis and energy balance scheme. Glob Biogeochem Cycles 15(1):207–225

Knorr W, Heimann M (2001b) Uncertainties in global terrestrial biosphere modeling, Part II: global constraints for a process-based vegetation model. Glob Biogeochem Cycles 15 (1):227–246

Köppen W (1900) Versuch einer Klassifikation der Klimate, vorzugsweise nach ihren Beziehungen zur Pflanzenwelt. Geogr Z 12:657–679

Kottek M, Grieser J, Beck C, Rudolf B, Rubel F (2006) World map of the Köppen- Geiger climate classification updated. Meteorol Z 15(3):259–263

Le Houérou HN (1996) Climate change, drought and desertification. J Arid Environ 34 (2):133–185

Mambo J, Archer E (2007) An assessment of land degradation in the Save catchment of Zimbabwe. Area 39(3):380–391

Mitchel TD, Jones PD (2005) An improved method of constructing a database of monthly climate observations and associated high-resolution grids. Int J Climatol 25:693–712

Perkins JS, Thomas DSG (1993) Spreading deserts or spatially confined environmental impacts? Land degradation and cattle ranching in the Kalahari Desert of Botswana. Land Degrad Dev 4 (3):179–194

Ringrose S, Musisi-Nkambwe S, Coleman T, Nellis D, Bussing C (1999) Use of Landsat thematic mapper data to assess seasonal rangeland changes in the southeast Kalahari, Botswana. Environ Manag 23:125–138

Roeckner E, Bäuml G, Bonaventura L, Brokopf R, Esch M, Giorgetta M, Hagemann S, Kirchner I, Kornblueh L, Manzini E, Rhodin A, Schlese U, Schulzweida U, Tompkins A (2003) The atmospheric general circulation model ECHAM 5. Part I: model description. Max Planck Institute for Meteorology Report 349, Hamburg, Germany

Ross R (1999) A concise history of South Africa. Cambridge University Press, Cape Town

Stringer LC, Reed MS (2007) Land degradation assessment in Southern Africa: integrating local and scientific knowledge bases. Land Degrad Dev 18(1):99–116

Strohbach BJ (2000a) Vegetation degradation trends in the northern Oshikoto Region: I. The Hyphaene petersiana plains. Dinteria 26:45–62

Strohbach BJ (2000b) Vegetation degradation trends in the northern Oshikoto Region: II. The Colophospermum mopane shrublands. Dinteria 26:45–62

Strohbach B (2001) Vegetation degradation in Namibia. J Namibia Sci Soc 49:127–156

Van Vegten JA (1981) Man-made vegetation changes: an example from Botswana's Savanna. National Institute of Development and Cultural Research, Documentation Unit, University College of Botswana, University of Botswana and Swaziland, Gaborone

Wessels KJ, Prince SD, Frost PE, van Zyl D (2004) Assessing the effects of human-induced land degradation in the former homelands of northern South Africa with a 1 km AVHRR NDVI time-series. Remote Sens Environ 91(1):47–67

Wessels K, Prince S, Reshef I (2008) Mapping land degradation by comparison of vegetation production to spatially derived estimates of potential production. J Arid Environ 72 (10):1940–1949

Wisskirchen K (2005) Modellierung der regionalen CO_2-Aufnahme durch Vegetation. PhD thesis. Meteorologisches Institut der Rhein. Friedrich–Wilhelm–Universität, Bonn, Germany

Chapter 13
Investigating Fourteen Years of Net Primary Productivity Based on Remote Sensing Data for China

Christina Eisfelder and Claudia Kuenzer

Abstract Net primary productivity (NPP) is an important environmental indicator that provides information about vegetation productivity and carbon fluxes. Analyses of NPP time-series allow for understanding temporal patterns and changes in vegetation productivity. These are especially important in rapidly changing environments, such as China, the world's third largest country. In this study, we use the model BETHY/DLR (Biosphere Energy Transfer Hydrology Model) for derivation of NPP time-series for China for 14 years from 1999–2012. We analyse spatial and temporal NPP distributions. These include mean annual NPP distribution and mean productivities for different land cover classes. Monthly data provide information about temporal patterns of vegetation productivity for different regions in China and different vegetation types. Analyses of interannual NPP variability revealed considerable differences in the development of annual vegetation productivity within the analysed time period for different provinces. The decrease in NPP for the district Shanghai shows the strong influence of one of Asia's fastest growing megacities on the environment. The NPP time-series was additionally analysed for a forest region in North China, which has been affected by forest disturbances. Our results show that the NPP data are suitable for monitoring of forest disturbance and regrowth. The analyses and results presented in this study provide valuable information about spatial and temporal variation of vegetation productivity in the various regions within China.

C. Eisfelder (✉) • C. Kuenzer
German Remote Sensing Data Center, DFD, Earth Observation Center, EOC,
German Aerospace Center, DLR, Oberpfaffenhofen, Germany
e-mail: Christina.Eisfelder@dlr.de

C. Kuenzer et al. (eds.), *Remote Sensing Time Series*, Remote Sensing and Digital Image Processing, DOI 10.1007/978-3-319-15967-6_13

13.1 Introduction

Net primary productivity (NPP) is the dry matter production by vegetation. It quantifies the carbon uptake by plants per unit area and unit time and, thus, provides information about carbon emission or sequestration. NPP is a key variable for ecological and environmental monitoring and a sensitive indicator of climate change (Niemeijer 2002; Schimel 1995). It plays an important role in the global carbon cycle (Prentice et al. 2001).

Modelling of NPP is of special interest in a country that shows such immense and rapid development like China (United Nations Development Programme UNDP 2013). The People's Republic of China is the world's third largest country and also the most populous (Population Reference Bureau, PRB 2013). It experiences strong economic growth and migration trends, which put pressure on ecological resources (PRC 2012).

Monitoring of NPP and its spatio-temporal dynamics are important base information for understanding possible impacts of global change and land management (Eisfelder et al. 2014). In the past decade, several studies on NPP estimation for China have been published, but most focus on small study areas or short time-periods (e.g. Fu et al. 2013; Gao et al. 2013; Lu et al. 2013). National wide studies are fewer and do so far not present variations for the first decade of twenty-first century (Cao et al. 2003; Gao et al. 2004; Hou et al. 2013).

In this study, we model NPP for China for the 14-year period from 1999 to 2012. We apply the remote sensing data based Biosphere Energy Transfer Hydrology (BETHY/DLR) model. Our aims were to present annual NPP distribution for the period 1999–2012 for China, analyse monthly patterns of vegetation productivity, and investigate interannual variations in NPP for the time period covered. We also analyse whether the modelled NPP data can be used for monitoring of forest disturbance areas in North China. Thereby, we want to demonstrate the usability of the modelled NPP time-series for understanding patterns in NPP and for monitoring possible long-term changes in vegetation productivity.

13.2 Data and Methods

In this chapter, we introduce the study area and provide a short description of the NPP model used within this study.

13.2.1 Study Area

Our study area is the People's Republic of China. China covers an area of about 9.6 million km^2. It is one of the world's largest countries and encompasses diverse

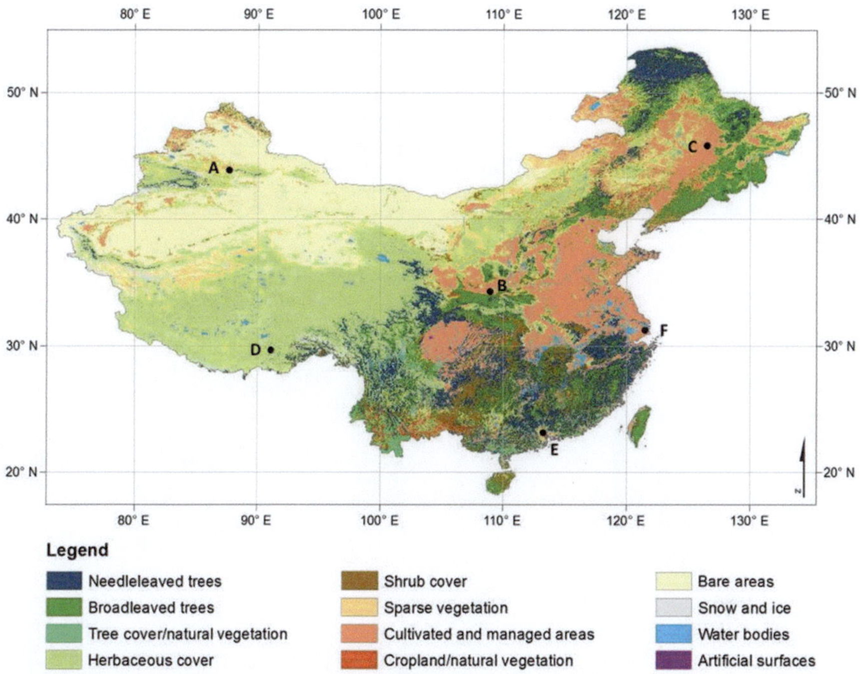

Fig. 13.1 Land cover for China from the GLC2000 (Bartholomé and Belward 2005; with simplified legend). Locations *A–F* refer to the precipitation and temperature diagrams from Fig. 12.2

climatic conditions. Figure 13.1 gives an overview on the land cover of China according to the land cover classification from the global land cover database for the year 2000 (GLC2000) (Bartholomé and Belward 2005).

In the western part of the country, grassland and sparsely vegetated areas are dominant. In the northeast boreal forest and agricultural areas can be found. The largest agricultural areas are located in the central eastern part of China. The regions in the southeast are mainly covered by a mixture of natural vegetation, mainly forest areas, shrub cover, and some agricultural areas (compare Fig. 13.1).

The southeastern part of China can be described as a warm temperate climate with fully humid to winter dry conditions and hot to warm summers (Kottek et al. 2006). In the northeastern part of the country snow climate with dry winters and hot to warm summers is dominant. The western part of China can be divided into a cold arid desert climate in the northern part and a polar tundra climate in the southern part including the Tibetan plateau (Kottek et al. 2006). Figure 13.2 shows selected precipitation and temperature diagrams for different regions of the country (locations given in Fig. 13.1).

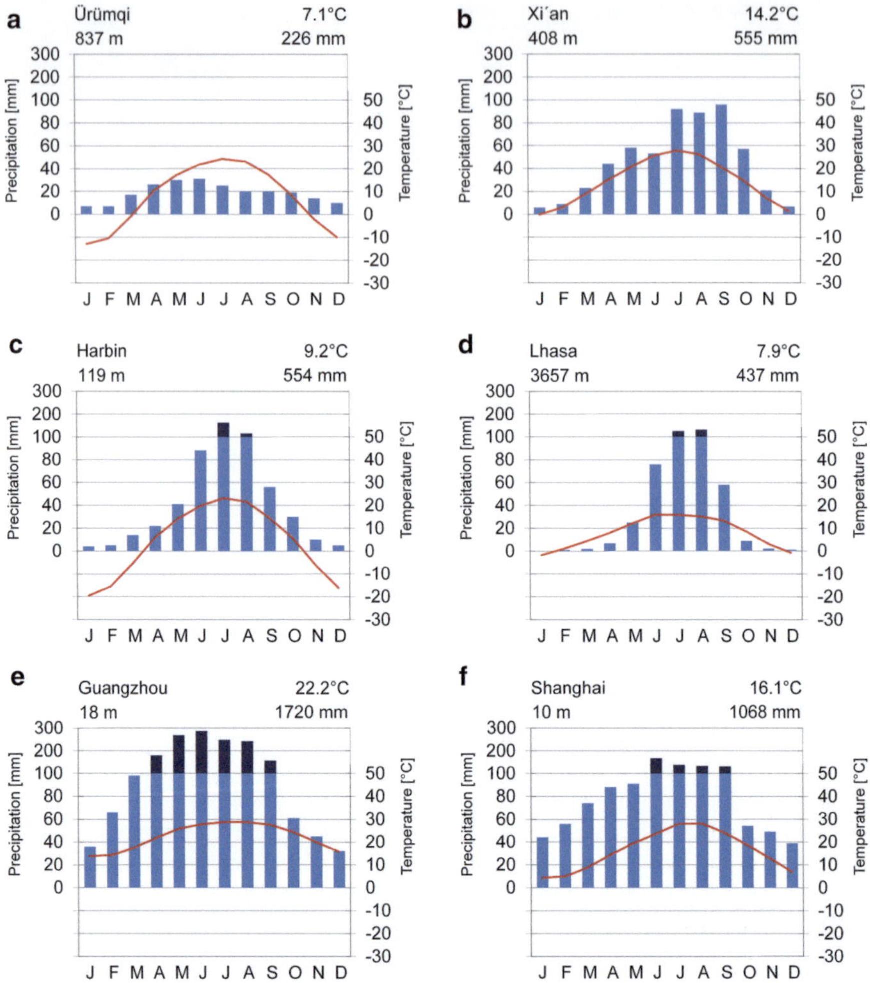

Fig. 13.2 Selected precipitation (average monthly precipitation) and temperature (average daily mean temperature) diagrams (extracted from Hijmans et al. 2005). Locations of the diagrams (**a**–**f**) are shown in Fig. 13.1: (**a**) Ürümqi, (**b**) Xi'an, (**c**) Harbin, (**d**) Lhasa, (**e**) Guangzhou, (**f**) Shanghai

13.2.2 The Model BETHY/DLR

In this study, we applied the model BETHY/DLR for NPP calculation for China. BETHY/DLR is a soil-vegetation-atmosphere-transfer (SVAT) model (Wißkirchen et al. 2013). It is based on meteorological and remote sensing derived input data. The CO_2 uptake by vegetation is simulated by BETHY/DLR as a process that is limited by light intensity, heat, and soil water availability (Knorr 1997). The parameterization of photosynthesis distinguishes between C_3 and C_4 plants (Farquhar et al. 1980; Collatz et al. 1992). The photosynthesis rate A is defined as the minimum of the carboxylation rate and the electron transport rate, minus dark respiration (Eisfelder et al. 2014).

BETHY/DLR currently differentiates 33 internal vegetation types, for which individual plant specific parameters are provided. These are needed for calculation of photosynthesis and include maximum carboxylation rate, maximum electron transport rate, maximum rooting depth, and maximum height. For each spatial unit (i.e. pixel), two vegetation types can be modelled. This allows modelling of mixed land cover classes as well as representation of fractional coverage of less than 100 %. Weighting factors define the fraction of primary and secondary vegetation type (Eisfelder et al. 2013; Wißkirchen et al. 2013).

BETHY/DLR is driven by meteorological and remote sensing based input data (Table 13.1). Meteorological data are available from the European Centre for Medium-Range Weather Forecasts (ECMWF) ERA-Interim reanalysis (Berrisford et al. 2011; Dee et al. 2011). These include data on precipitation, air temperature, cloud coverage, and wind speed. Further, a land cover map and leaf area index (LAI) data are used. The GLC 2000 (Bartholomé and Belward 2005) land cover map was used within this study. Remote sensing based LAI data were available from the geoland2 (Gio-GL 2013) data base. The LAI data are available globally and provided with 1 km^2 resolution as 10-day composites. Pre-processing of LAI time-series was performed before the data were used as input for BETHY/DLR. This included correction of gaps and outliers in the time-series with a harmonic analysis (Wißkirchen et al. 2013).

Additional input data for BETHY/DLR comprise a digital elevation model from the latest version of the NOAA/NGDC GTOPO30 product (USGS 1996) and soil types from the FAO soil map (FAO et al. 2009).

The spatial resolution of the NPP output from BETHY/DLR is 1 km^2. This equals the resolution of LAI and land cover input data. The continuous time-series of meteorological input data on at least daily basis (cf. Table 13.1) allows for a high temporal resolution. NPP outputs are calculated daily. Based on these daily data, monthly and annual sums and means are calculated. An overview on the processing scheme of BETHY/DLR is provided in Fig. 13.3.

13.3 Results and Discussion of NPP for China

Based on the monthly and annual NPP data, we analysed the spatial distribution of vegetation productivity in China and calculated productivities for individual land cover classes. Further, we discuss the monthly development of NPP and have a look at interannual variations in vegetation productivity.

Table 13.1 Input data for NPP modelling with BETHY/DLR for China

Input parameter	Spatial resolution	Temporal resolution	Source
Land cover map	~1 km	Once	GLC2000
LAI	~1 km	10-daily	geoland2
Soil map	~1 km	Once	FAO
Digital elevation model	~1 km	Once	GTOPO30
2 metre temperature	0.25°	>daily	ECMWF
Large-scale and convective precipitation	0.25°	>daily	ECMWF
Low, medium, and high cloud cover	0.25°	>daily	ECMWF
10 metre eastward and northward wind component	0.25°	>daily	ECMWF
Surface geopotential	0.25°	Once	ECMWF

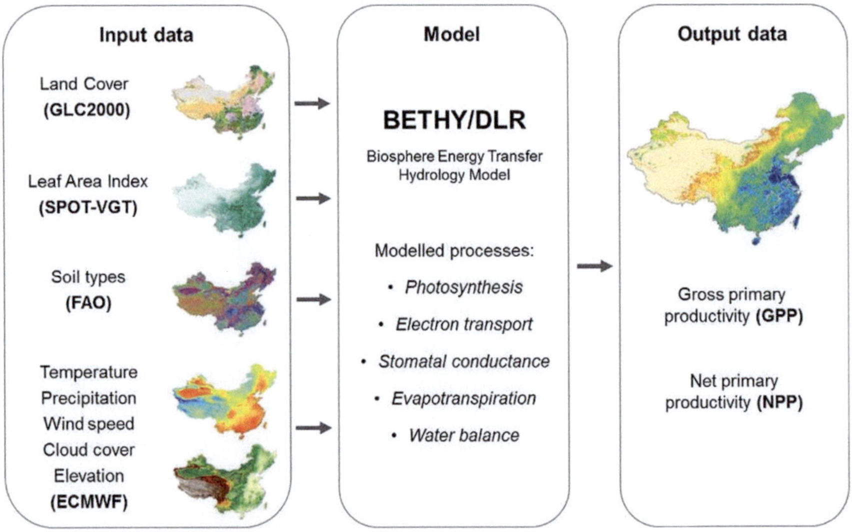

Fig. 13.3 Scheme of BETHY/DLR model

13.3.1 Spatial NPP Distribution

Figure 13.4 presents the calculated mean annual NPP for China for 1999–2012. Highest mean annual NPP values can be observed in areas covered by broadleaved forest or shrubs in the south-eastern part of China. In these areas, annual NPP sums range between 1,500 and 2,000 g C m^{-2}. Agricultural areas in the central eastern part of the country, roughly located between 32–36°N and 113–121°E, also show high NPP with productivity maxima reaching 1,500 g C m^{-2}.

The western part of China, a region covered by herbaceous vegetation, sparse vegetation, and bare areas (cf. Fig. 13.1), is characterized by very low vegetation

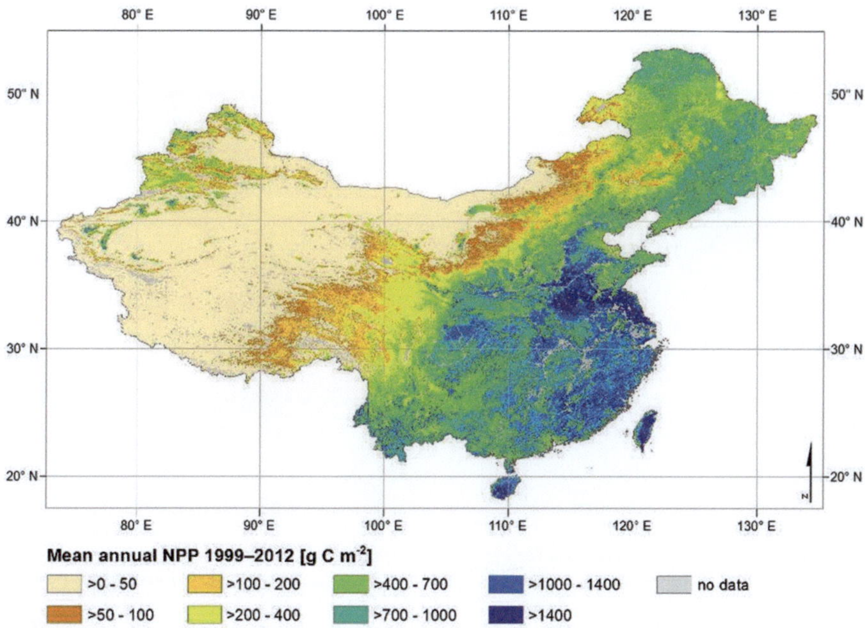

Fig. 13.4 Mean annual NPP for China for 1999–2012 calculated with BETHY/DLR

productivity. In the area between 37–42°N and 75–87°E some local spots with clearly visible higher productivity can be observed. These are cultivated and managed areas (cf. Fig. 13.1) and, further north, also some more grassland and managed areas can be identified. In addition, Fig. 13.5 provides an overview on average mean annual NPP values for individual provinces within China.

Average mean annual NPP values for individual land cover classes are presented in Table 13.2. Highest productivity can be observed for broadleaved evergreen and trees with NPP of 1,034 g C m^{-2}. This is followed by cultivated and managed areas (824 g C m^{-2}), broadleaved deciduous trees (718 g C m^{-2}), needleleaved trees (707 g C m^{-2}, 704 g C m^{-2}) and evergreen shrubs (686 g C m^{-2}). Lowest annual NPP in the period 1999–2012 have herbaceous cover (171 g C m^{-2}), sparse vegetation (57 g C m^{-2}), and bare areas (18 g C m^{-2}).

The NPP values from BETHY/DLR presented in Table 13.2 correspond to productivities published in other studies for China. Feng et al. (2007), for example, reported annual NPP values of 122.6 g C m^{-2} for grassland, 14.3 g C m^{-2} for barren areas, 342 g C m^{-2} for cropland, 363 g C m^{-2} for shrubland, and 552 g C m^{-2} at average for forests. Yan et al. (2006) found annual NPP for coniferous and broad-leaf mixed forest in the Dinghushan Biosphere Reserve of 1,148 g C m^{-2} for evergreen broad-leaf forest and 678 g C m^{-2}.

Matsushita and Tamura (2002) reported NPP of 1178 g C m^{-2} for broadleaved forests, 690 g C m^{-2} for broadleaved crops, 614 g C m^{-2} for needleleaved forest, and 460 g C m^{-2} for shrubs in East Asia. For grasslands in western China

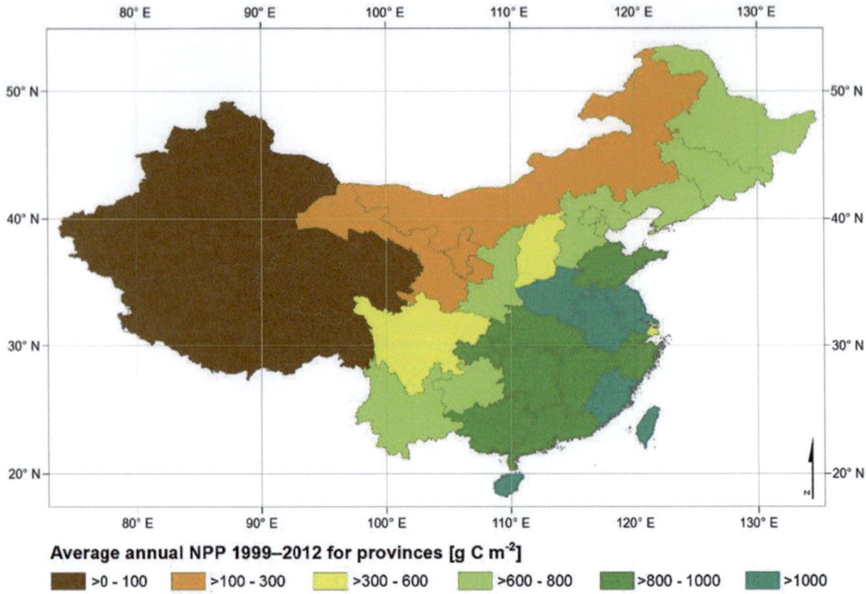

Fig. 13.5 Mean annual NPP for individual provinces within China for 1999–2012 calculated with BETHY/DLR

Table 13.2 Average mean annual NPP for different land cover types in China for 1999–2012

Land cover type	Mean annual NPP [g C m^{-2}]
Broadleaved trees, evergreen	1034.2
Cultivated and managed areas	824.1
Broadleaved trees, deciduous	718.1
Needleleaved trees, deciduous	707.3
Needleleaved trees, evergreen	703.5
Shrub cover, evergreen	685.8
Tree cover/natural vegetation	651.3
Cropland/natural vegetation	606.8
Mixed leaf trees	577.3
Regularly flooded area	482.3
Shrub cover, deciduous	429.2
Herbaceous cover	171.3
Sparse vegetation	56.9
Bare areas	17.5

productivities between 124 and 231 g C m^{-2} per year were presented by Lu et al. (2004). In the same study, rice was found to have annual productivity of 632 g C m^{-2} and other agricultural areas 492 g C m^{-2}.

Wang et al. (2008) analysed NPP of Chinese forest areas for the late 1990s and found NPP of 1,372 g C m^{-2} for evergreen broadleaved forest, 1,240 g C m^{-2} for evergreen needleleaved forest, 994 g C m^{-2} for deciduous broadleaved forest, and 852 g C m^{-2} for deciduous needleleaved forest. The available NPP data for similar vegetation types are in general consistent with the results obtained in our study (Table 13.2).

13.3.2 Monthly NPP Patterns

Figure 13.6 shows mean monthly NPP for China for the period 1999–2012. The monthly NPP values clearly identify variances in productivity for different parts of the country throughout the year.

The warm temperate climate in the southeast of China allows for vegetation productivity throughout the year. In the north and west pronounced winter dormancy can be observed, which lasts from November through March.

In April and May, high productivities of up to 330 g C m^{-2} are reached in the agricultural areas in the central eastern part of the country (32–36°N and 113–121°E). Forest areas, which have highest annual NPP in general (Table 13.2), reach maximum productivities a bit later, with about 210 g C m^{-2} per month in June/July/August. Vegetation growth starts later in the northeast of China. In this area, productivity can be observed from May on. The maximum NPP is reached in July and August in the northeastern part of the country. After the maximum, productivity quickly declines within 1 month. Highest productivity for grasslands and cultivated and managed areas in the very northwest of China can be observed in June and July.

Table 13.3 summarizes mean monthly NPP for vegetated land cover classes within China for the period 1999–2011. The results from BETHY/DLR clearly show the different temporal behaviour of evergreen and deciduous trees and shrubs (Table 13.3). Evergreen trees and shrubs show a longer growth period than deciduous trees/shrubs, but the maximum productivity, reached around July, is higher for deciduous trees/shrubs. Evergreen vegetation tends to show a slightly later maximum. This becomes also obvious in Fig. 13.7, which displays mean monthly NPP for the vegetated land cover classes with largest coverage within China.

Maximum vegetation productivity is reached in July or August for almost all analysed land cover classes. A late productivity peak in August can be observed for cultivated and managed areas (Table 13.3).

13.3.3 Interannual NPP Variability

Figure 13.8 shows annual NPP for China for individual years from 1999 to 2012 calculated with BETHY/DLR. Differences in NPP distribution between the years

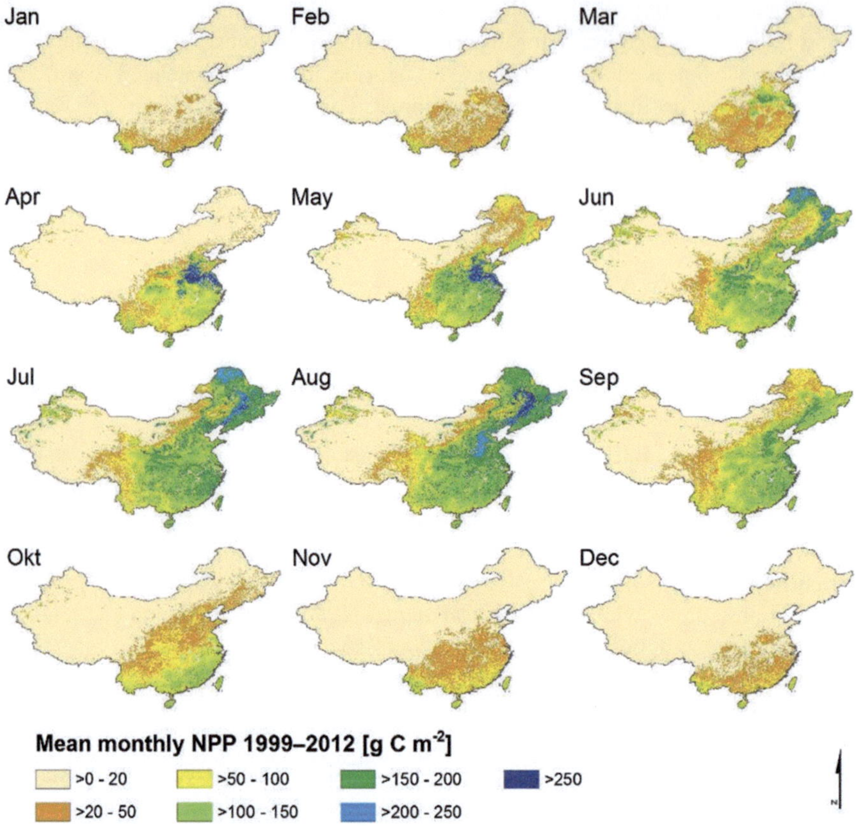

Fig. 13.6 Mean monthly NPP for China for the time period 1999–2012

are clearly visible. The agricultural areas in the central East, for example, show lower productivity in 2000, 2002, and 2003 than in the other years. Boreal forest areas in the northeast have comparatively lower productivities in 2003 and 2011. For the area near the coast in Southeast China higher productivities can be observed in 2003, 2004, 2011 and 2012, compared to the other years.

Interannual NPP variation for individual Chinese provinces is also displayed in Fig. 13.9. The provinces are grouped into regional clusters and mean values for the defined regions are also shown. The graphs allow analysing temporal NPP patterns in more detail. For the dry western part of China, with generally very low NPP, we can observe a slight trend towards higher productivity in the 13-year period analysed in this study (Fig. 13.9). This is not the case for most other regions, where interannual variation is more pronounced. Strong variations can be observed for some districts in the Central North and Southeast. In the northern districts, annual NPP is relatively stable or shows a slight increase. In the Northeast, higher values can be observed for the period 2004–2007. An increase in NPP, followed by a slight downward trend between 2005 and 2011 can be observed for most districts

Table 13.3 Mean monthly NPP for different land cover types in China for 1999–2012

Class name	Mean monthly NPP [g C m^{-2}]											
	Jan	Feb	Mar	Apr	May	Jun	Jul	Aug	Sep	Oct	Nov	Dec
Broadleaved trees, evergreen	29.9	36.4	51.1	87.8	123.1	126.7	140.4	141.7	114.8	90.2	54.3	38.1
Broadleaved trees, deciduous	0.4	1.2	4.5	18.1	79.8	166.4	172.4	156.5	93.3	21.9	3.1	0.6
Needleleaved trees, deciduous	14.3	18.1	28.2	54.5	82.6	93.3	109.8	110.0	83.2	58.0	31.4	20.1
Tree cover, needle-leaved, deciduous	0.0	0.0	1.5	9.5	69.3	195.1	194.8	162.2	67.0	7.6	0.2	0.0
Mixed leaf trees	0.0	0.0	1.9	10.3	60.2	154.4	151.0	127.9	61.4	9.6	0.7	0.0
Tree cover/natural vegetation	36.0	38.2	42.5	60.0	65.5	58.6	63.9	68.2	61.6	63.9	51.2	41.7
Shrub cover, evergreen	14.5	18.2	26.1	53.2	80.5	88.5	102.6	108.7	82.0	58.1	32.7	20.9
Shrub cover, deciduous	1.6	1.6	2.7	11.2	36.6	112.3	120.2	99.1	35.0	4.9	2.1	1.9
Herbaceous cover	0.9	1.2	2.5	6.6	13.8	27.4	42.4	43.2	22.3	7.4	2.5	1.2
Sparse vegetation	0.0	0.0	0.4	2.4	5.1	9.8	14.6	15.1	7.2	1.9	0.2	0.0
Regularly flooded area	8.9	10.5	18.8	34.2	49.2	67.5	90.1	90.4	58.8	28.5	14.9	10.4
Cultivated and managed areas	8.7	13.9	36.0	79.5	100.3	99.0	143.3	161.2	109.9	41.1	19.2	12.0
Cropland/natural vegetation	27.4	27.8	29.6	43.6	55.1	74.0	88.6	89.7	56.1	42.7	40.8	31.4
Bare areas	0.0	0.1	0.4	1.6	2.3	3.1	3.5	3.3	2.1	1.0	0.2	0.0

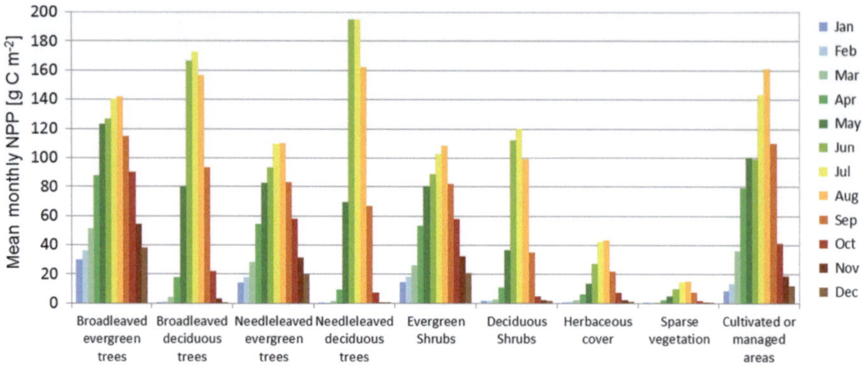

Fig. 13.7 Mean monthly NPP values for 1999–2012 for selected vegetated land cover classes in China calculated with BETHY/DLR

in the Central North and Central South regions. The development of annual NPP in the period 1999–2011 for the district Shanghai differs strongly from the development in the nearby districts. For Shanghai a clear downward trend can be observed in the period of investigation. In the Southwest, NPP values are quite stable, with an increase only for the last year. Annual NPP is more variable in the districts in the Southeast of China.

13.3.4 Impact of Urban Sprawl Around Shanghai

The development of NPP for the province of Shanghai within the observed time-period is remarkable (Fig. 13.9). A linear regression analysis reveals that the downward trend over the course of the observed time span is significant with $p < 0.001$. To quantify the loss of productivity, we calculated mean annual NPP within Shanghai and the two neighbouring prefectural-level cities Suzhou and Jiaxing. The difference between NPP for the years 2000 and 2010 for individual municipalities is displayed in Fig. 13.10. A clear downward trend in NPP can be observed for all municipalities in the area of investigation around Shanghai (Fig. 13.10). Absolute NPP loss is highest in the municipalities Fengxian, Jinshan, and Pinghu at the northern shore of Hangzhou Bay and Jiaxing, with a loss greater than 250 g C m^{-2} in the 10-year period. Relative loss is in general lowest with less than 20 % for the Jiaxing (Southwest), except for the area around the city of Jiaxing. Shanghai (East) and Suzhou (Northwest) have higher percentage NPP reduction. All municipalities except Shazhou in Suzhou have a loss of greater than 20 %. Five out of nine municipalities around Shanghai show a loss of more than 25 % in the period 2000–2010 (Fig. 13.10).

To ensure that years compared are representative and no extraordinary extreme years, we also calculated mean values over the 3-year periods of 1999–2001 and

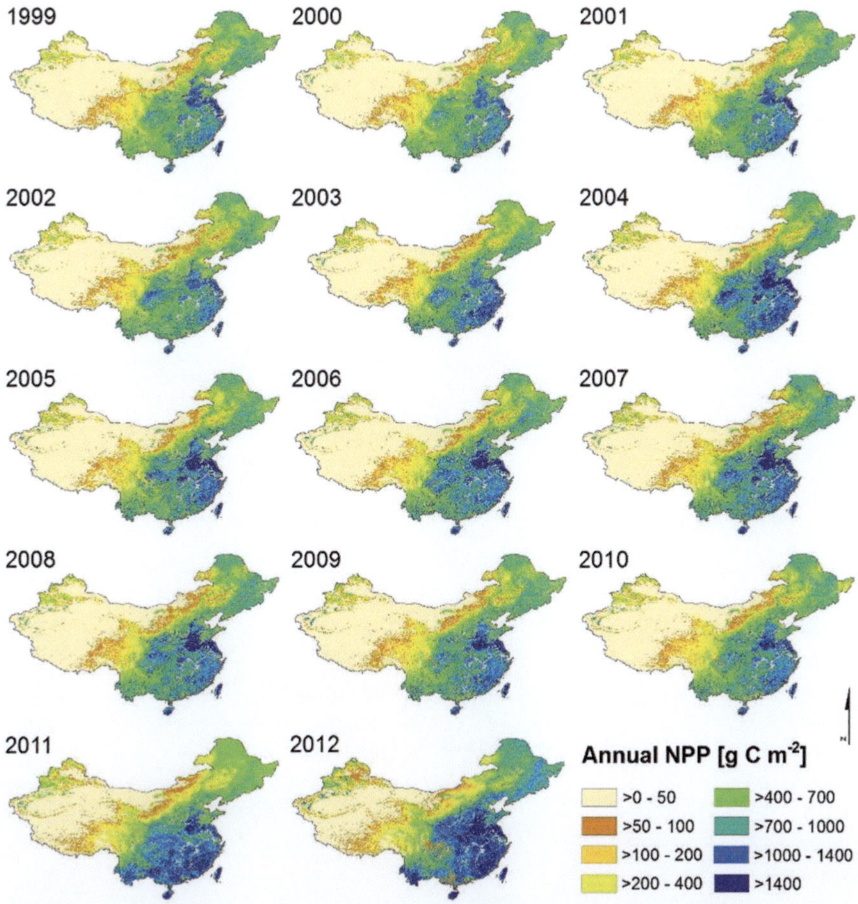

Fig. 13.8 Annual NPP for China for the years 1999–2012 calculated with BETHY/DLR

2010–2012. The results of this periodical comparison and resulting absolute and relative NPP loss values are given in Table 13.4.

The low absolute loss in NPP around the city centre of Shanghai can be explained by the already high urban sprawl in that area. Low absolute loss but high relative loss, such as observed for Baoshan and Minhang (Fig. 13.10), indicate that the area had already been closely built-up in 2000 and that until 2010 large parts of the remaining vegetation areas have been cleared. In general, we assume that a strong decrease in NPP has probably been caused by increasing urban growth. In the case of Wuxian, the high relative loss is due to the large water body of Taihu Lake. Comparatively low absolute NPP loss can be observed within the prefecture-level city of Jiaxing. This region is more rural and large areas are used for agriculture. NPP loss is not as severe as in other parts of the investigated area, but also a reduction of at least 10 % can be observed for the period 2000–2010.

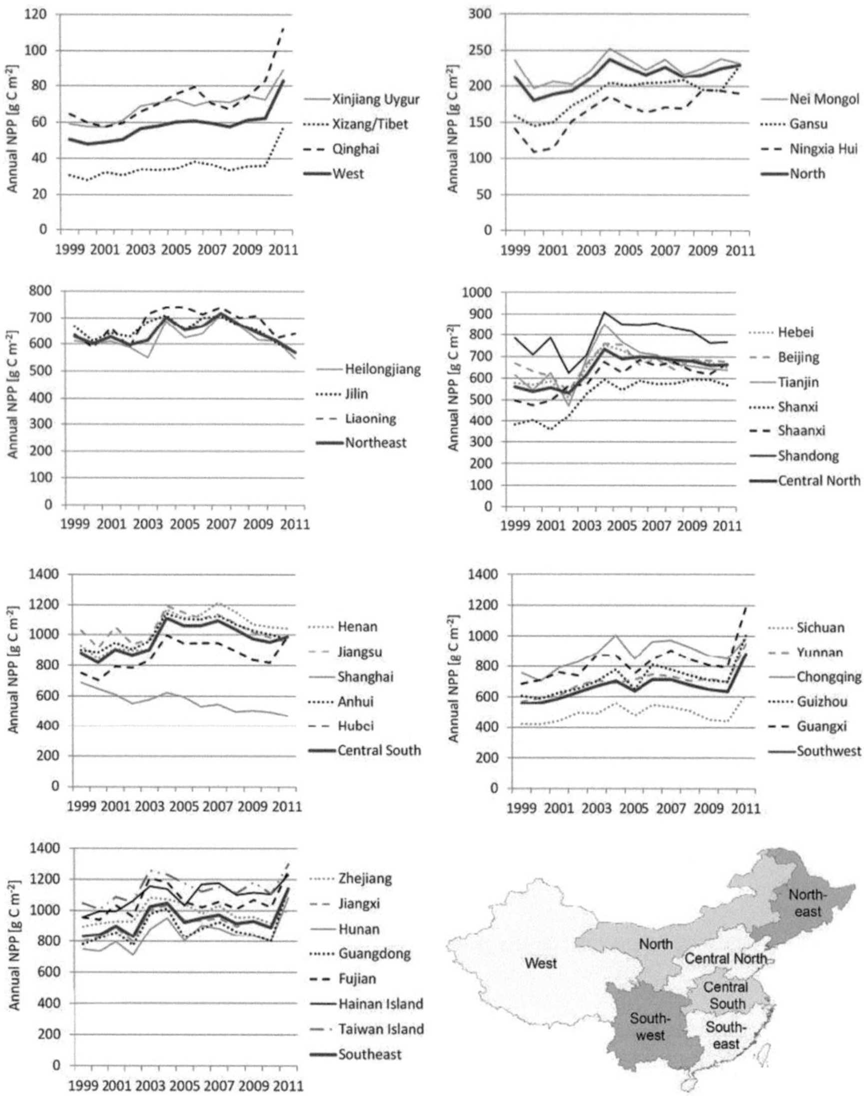

Fig. 13.9 Interannual variation in NPP for various regions in China within the time period 1999–2011. The *thick solid line* in each diagram gives the mean annual NPP variation for each region. The other lines show mean annual NPP for individual provinces. The figure in the lower right shows the location of the regions within China

13.3.5 Monitoring Forest Disturbance in North China

Figure 13.11 depicts a forested region in Northeast China at the border to Russia, which is mainly covered by needle-leaved forest. The area covers the northern part

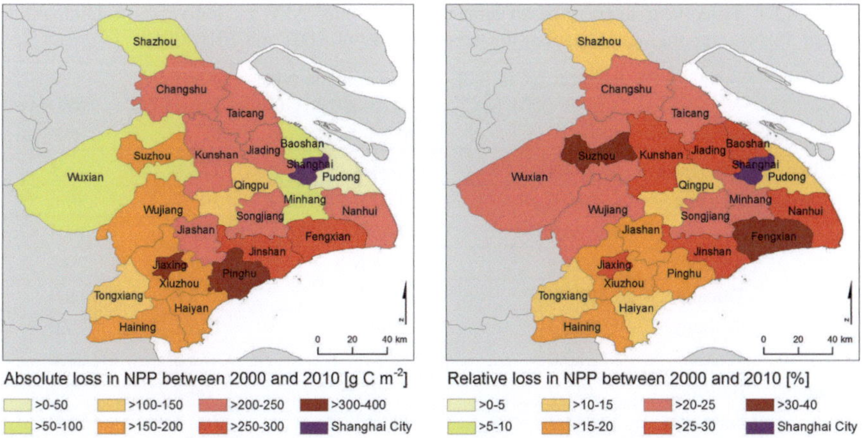

Fig. 13.10 Loss in NPP between 2000 and 2010 for individual municipalities within Suzhou, Jiaxing, and Shanghai. The city centre of Shanghai is shown in violet. *Left*: Absolute loss in annual NPP. *Right*: Relative loss in annual NPP with respect to the annual NPP in 2000

of the Greater Khingan mountains. The area belongs to the provinces Nei Mongol (West) and Heilongjiang (East). The three maps in Fig. 13.11 show annual NPP for the years 2002, 2003, 2006 and 2011. From the NPP data shown in Fig. 13.11, it can be seen that in the year 2003 three forest areas (red polygons) plus some smaller areas have been disturbed. In 2006, an additional area with recent disturbance can be observed (blue polygon in Fig. 13.11).

Figure 13.12 shows average annual NPP for 1999–2011 for the three polygons that were disturbed in 2003. From the annual variation in NPP, the disturbance impacts become also obvious. The vegetation productivity for these areas was high between 1999 and 2002 and suddenly dropped to low values in 2003.

The annual NPP data also clearly show the regrowth of vegetation within the three disturbed areas from 2003. After the rapid decrease in the year of the disturbance event, an increase of annual productivity can be observed. For all three disturbed areas, NPP went up again. The most quickly raise shows largest disturbed area close to the Russian border in the Northeast (forest area no. 1).

The forest disturbances we observe in this study based on the NPP data, have also been reported by Tao et al. 2013, who analysed causes of forest disturbance in Northeast China. The two large disturbance areas in 2003 (no. 1 and 2 from Fig. 13.11) as well as the forest disturbance area in 2006 have been caused by forest fires The small area in 2003 (no. 3 from Fig. 13.11) can be attributed to deforestation and land-use change (Tao et al. 2013).

Table 13.4 Three-year average mean annual and monthly NPP for individual municipalities within Suzhou, Jiaxing, and Shanghai for the periods 1999–2001 and 2010–2012

	Municipality	3-year mean NPP 1999–2001 [g C m^{-2}]	3-year mean NPP 2010–2012 [g C m^{-2}]	Absolute loss [g C m^{-2}]	Relative loss [%]
Suzhou	Changshu	982.4	804.4	178.0	18.1
	Kunshan	960.6	667.9	292.7	30.5
	Shazhou	720.8	645.4	75.4	10.5
	Suzhou	534.8	375.6	159.2	29.8
	Taicang	1088.0	882.2	205.7	18.9
	Wujiang	745.3	599.3	146.1	19.6
	Wuxian	360.0	284.5	75.5	21.0
Shanghai	Baoshan	187.5	132.7	54.8	29.2
	Fengxian	1048.7	786.3	262.4	25.0
	Jiading	863.6	646.9	216.7	25.1
	Jinshan	1464.7	1123.2	341.5	23.3
	Minhang	279.3	224.2	55.1	19.7
	Nanhui	911.0	694.9	216.1	23.7
	Pudong	267.0	238.0	29.0	10.9
	Qingpu	952.5	797.1	155.4	16.3
	Songjiang	1100.6	879.3	221.3	20.1
Jiaxing	Haining	1181.9	997.2	184.7	15.6
	Haiyan	1388.3	1218.8	169.5	12.2
	Jiashan	1210.3	938.3	272.0	22.5
	Jiaxing	1079.3	842.4	236.9	22.0
	Pinghu	1542.1	1272.8	269.3	17.5
	Tongxiang	1204.3	1080.5	123.8	10.3
	Xiuzhou	1334.4	1149.9	184.5	13.8
Total	ALL	930.8	751.4	165.1	19.8

13.4 Conclusions

We applied the model BETHY/DLR to calculate NPP time-series for China for 14 years from 1999 to 2012. The NPP results were analysed regarding spatial, monthly, and interannual variations. We first presented mean annual NPP distribution for 1999–2012 and derived mean productivities for different land cover classes. The results obtained with BETHY/DLR correspond well to previously published data.

The results of the monthly NPP analyses show, that the data are well suited to analyse differences in temporal patterns of vegetation productivity, both for different regions in China and for different vegetation types. The results are important base information for understanding growth efficiency of different plants in differing climatic regimes.

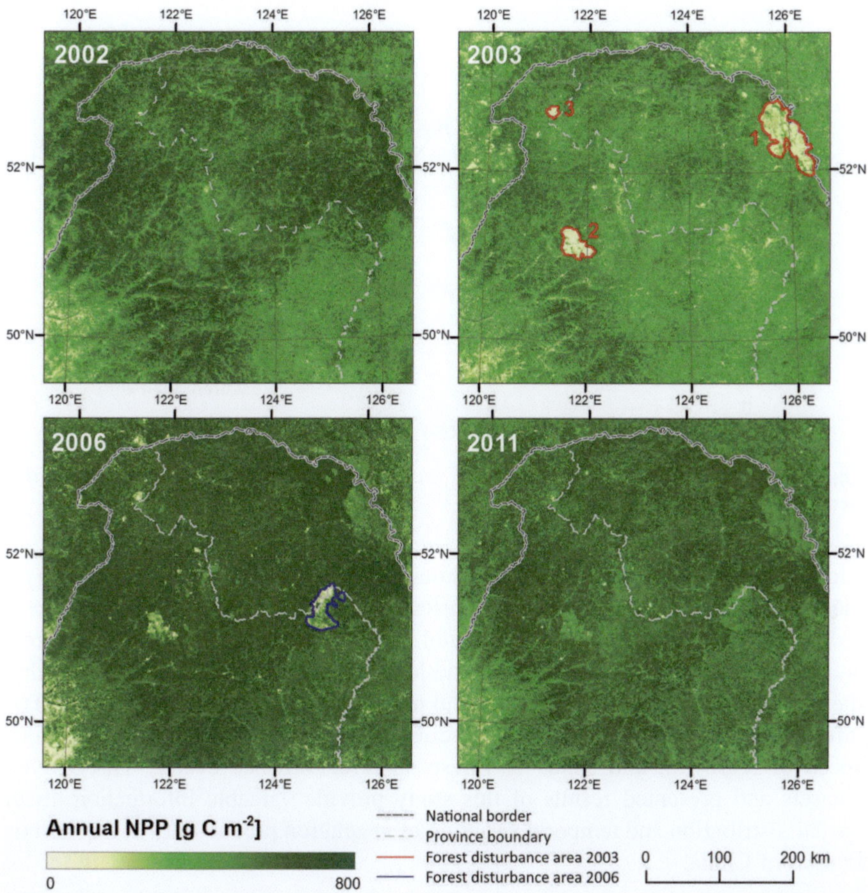

Fig. 13.11 Annual NPP for the years 2002, 2003, 2006, and 2011 for a forest region in North China. The displayed region borders Russia to the North (national border shown as *grey line*). The *red* and *blue* polygons highlight forest disturbance areas in 2003 and 2006 respectively. The numbering of disturbed forest areas in 2003 refers to Fig. 13.12

The analyses of interannual NPP variability provided an interesting insight to the development of annual vegetation productivity within the period under investigation. The annual productivity within different provinces varied considerably. For different regions of China, partly opposite trends could be observed. A decrease in NPP can be observed for the district Shanghai, which differs from the pattern in the surrounding region. This shows the strong influence of urbanization around one of Asia's fastest growing megacities on the environment.

A closer look at the region around Shanghai revealed that a strong impact on the environment can be observed based on the NPP time-series. For Suzhou, Jiaxing, and Shanghai, a decrease in vegetation productivity can be observed for the period 2000–2010. All municipalities in the area of investigation around Shanghai showed

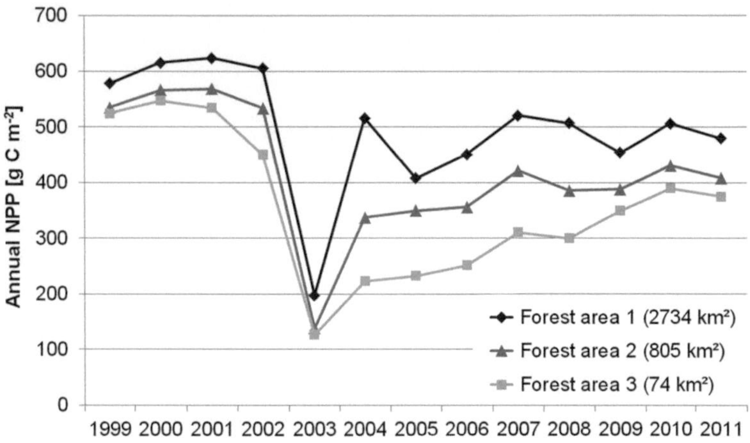

Fig. 13.12 Mean annual NPP for 1999–2011 within the three forest areas from Fig. 13.11 that were disturbed in 2003

a loss in NPP of at least 10 %. In our results, the strong influence of urbanization on the environment around one of Asia's fastest growing megacities becomes obvious.

The NPP time-series was also analysed for a region in North China. This forest area has been effected by forest fires and logging activity. Our analyses show that the NPP data can be used to identify and monitor areas of forest disturbance. The information derived from NPP time-series can support understanding impacts, speed of regrowth, and status of forests after disturbance events. The derived datasets and presented results of this study provide valuable information about spatial distribution and temporal variation of vegetation productivity in the various regions of China.

Acknowledgements We thank ECMWF, VITO, and IIASA for providing input data for NPP modelling.

References

Bartholomé E, Belward AS (2005) GLC2000: a new approach to global land cover mapping from Earth observation data. Int J Remote Sens 26(9):1959–1977

Berrisford P, Dee D, Poli P, Brugge R, Fielding K, Fuentes M, Kallberg P, Kobayasha S, Uppala S, Simmons A (2011) The ERA-Interim archive. In: ERA report series, no. 1, version 2.0. ECMWF, Reading

Cao M, Prince SD, Li K, Tao B, Small J, Shao X (2003) Response of terrestrial carbon uptake to climate interannual variability in China. Glob Chang Biol 9:536–546

Collatz GJ, Ribas-Carbo M, Berry JA (1992) Coupled Photosynthesis–stomatal conductance model for leaves of C_4 plants. Aust J Plant Physiol 19(5):519–538

Dee DP, Uppala SM, Simmons AJ, Berrisford P, Poli P, Kobayashi S, Andrae U, Balmaseda MA, Balsamo G, Bauer P, Bechtold P, Beljaars ACM, van de Berg L, Bidlot J, Bormann N,

Delsol C, Dragani R, Fuentes M, Geer AJ, Haimberger L, Healy SB, Hersbach H, Hólm EV, Isaksen L, Kållberg P, Köhler M, Matricardi M, McNally AP, Monge-Sanz BM, Morcrette JJ, Park BK, Peubey C, de Rosnay P, Tavolato C, Thépaut JN, Vitart F (2011) The ERA-Interim reanalysis: configuration and performance of the data assimilation system. Q J R Meteorol Soc 137(545):553–597

Eisfelder C, Kuenzer C, Dech S, Buchroithner MF (2013) Comparison of two remote sensing based models for regional net primary productivity estimation – a case study in semi-arid Central Kazakhstan. IEEE J Select Topics Appl Earth Obs Remote Sens 6(4):1843–1856

Eisfelder C, Klein I, Niklaus M, Kuenzer C (2014) Net primary productivity in Kazakhstan, its spatio-temporal patterns and relation to meteorological variables. J Arid Environ 103:17–30

FAO et al (2009) Harmonized world soil database (version 1.1). FAO/IIASA, Rome/Laxenburg

Farquhar GD, Caemmerer S, Berry JA (1980) A biochemical model of photosynthesis in leaves of C3 species. Planta 149(1):78–90

Feng X, Liu G, Chen JM, Chen M, Liu J, Ju WM, Sun R, Zhou W (2007) Net primary productivity of China's terrestrial ecosystems from a process model driven by remote sensing. J Environ Manag 85(3):563–573

Fu Y, Yu X, Zhao Y, Zeng X, Xia L (2013) Assessment impacts of weather and LandUse/Land Cover (LULC) change on Urban Vegetation Primary Productivity (NPP): a case study in Guangzhou, China. Remote Sens 5:4125–4144

Gao Q, Yu M, Wang J, Jia H, Wang K (2004) Relationship between regional primary production and vegetation patterns. Ecol Model 172:1–12

Gao Y, Zhou X, Wang Y, Wang C, Zhan Z, Chen L, Yan J, Qu R (2013) Vegetation net primary productivity and its response to climate change during 2001–2008 in the Tibetan Plateau. Sci Total Environ 444:356–362

Gio-GL (2013) Gio global land component – Lot 1 "Operation of the global land component", Product user manual, Leaf Area Index (LAI) – version 1. Issue I1.00. http://land.copernicus.eu/global/sites/default/files/products/GIO-GL1_PUM_LAIV1_I1.00.pdf. Accessed 18 Dec 2013

Hijmans RJ, Cameron SE, Parra JL, Jones PG, Jarvis A (2005) Very high resolution interpolated climate surfaces for global land areas. Int J Climatol 25:1965–1978

Hou Y, Wang S, Zhou Y, Yan F, Zhu J (2013) Analysis of the carbon dioxide concentration in the lowest atmospheric layers and the factors affecting China based on satellite observations. Int J Remote Sens 34(6):1981–1994

Knorr W (1997) Satellite remote sensing and modelling of the global CO_2 exchange of land vegetation: a synthesis study. PhD dissertation. Max-Planck-Institute for Meteorology, Hamburg

Kottek M, Grieser J, Beck C, Rudolf B, Rubel F (2006) World map of the Köppen-Geiger climate classification updated. Meteorol Z 15:259–263

Lu L, Li X, Veroustraete F, Dong QH (2004) Estimation of NPP in Western China using remote sensing and the C-Fix model. In Proceedings of IEEE international geoscience and remote sensing symposium, IGARSS 2004, 20–24 September, pp 12–15

Lu L, Li X, Veroustraete F, Kang E, Wang J (2013) Analysing the forcing mechanisms for net primary productivity changes in the Heihe River Basin, north-west China. Int J Remote Sens 30 (3):793–816

Matsushita B, Tamura M (2002) Integrating remotely sensed data with an ecosystem model to estimate net primary productivity in East Asia. Remote Sens Environ 18(1):58–66

Niemeijer C (2002) Developing indicators for environmental policy: data-driven and theory-driven approaches examined by example. Environ Sci Policy 5(2):91–103

PRB (2013) 2013 world population data sheet. http://www.prb.org/pdf13/2013-population-data-sheet_eng.pdf. Accessed 29 Jan 2014

PRC (2012) The People's Republic of China national report on sustainable development, 100 pp. http://www.china-un.org/eng/zt/sdreng/P020120608816288649663.pdf. Accessed 27 Nov 2013

Prentice IC, Farquhar GD, Fasham MJR, Goulden ML, Heimann M, Jaramillo VJ, Kheshgi HS, Le Quéré C, Scholes RJ, Wallace DWR (2001) The carbon cycle and atmospheric carbon dioxide. In: Houghton JT, Ding Y, Griggs DJ, Noguer M, van der Linden PJ, Dai X, Maskell K, Johnson CA (eds) Climate change 2001: the scientific basis. Contribution of working group I to the third assessment report of the Intergovernmental Panel on Climate Change. Cambridge University Press, Cambridge/New York

Schimel DS (1995) Terrestrial biogeochemical cycles: global estimates with remote sensing. Remote Sens Environ 51(1):49–56

Tao J, Zhang Y, Yuan X, Wang J, Zhang X (2013) Analysis of forest fires in Northeast China from 2003 to 2011. Int J Remote Sens 34(22):8235–8251

UNDP (2013) China national human development report 2013. Sustainable and liveable cities: towards ecological civilization, p 200. http://www.undp.org/content/dam/china/docs/Publications/UNDP-CH-HD-Publication-NHDR_2013_EN_final.pdf. Accessed 26 Nov 2013

USGS (1996) GTOPO30, Global 30-Arc-Second Elevation Data Set. U.S. Geological Survey, EROS Data Center, Sioux Falls, South Dakota. http://eros.usgs.gov/#/Find_Data/Products_and_Data_Available/GTOPO30. Accessed 9 Dec 2013

Wang Y, Zhao P, Ren H, Kakubar Y (2008) Spatiotemporal dynamics of forest net primary production in China over the past two decades. Glob Planet Chang 61(3–4):267–274

Wißkirchen K, Tum M, Günther KP, Niklaus M, Eisfelder C, Knorr W (2013) Quantifying the carbon uptake by vegetation for Europe on a 1 km^2 resolution using a remote sensing driven vegetation model. Geosci Model Dev Dis 6:2457–2489

Yan J, Wang Y, Zhou G, Zhang D (2006) Estimates of soil respiration and net primary production of three forests at different succession stages in South China. Glob Chang Biol 12(5):810–821

Chapter 14
The Utility of Landsat Data for Global Long Term Terrestrial Monitoring

David P. Roy, Valeriy Kovalskyy, Hankui Zhang, Lin Yan, and Indrani Kommareddy

Abstract The utility of satellite time series data for monitoring land surface change is well established. This chapter highlights recent Landsat research, product developments, and opportunities, for global long term Landsat monitoring, that are now evolving rapidly with the opening of the Landsat archive. Specifically, it introduces the NASA (National Aeronautics and Space Administration) funded global Web Enabled Landsat Data products, and overviews Landsat time series phenology and land cover change monitoring applications and research, and prospectives for Landsat time series monitoring.

14.1 Introduction

The Landsat series of satellites provide the longest temporal record of space-based global coverage observations of terrestrial emitted and reflected electromagnetic radiation (Roy et al. 2014a). A free Landsat data policy initiated in 2008 (Woodcock et al. 2008) has opened a new era for utilizing the now more than four million globally distributed Landsat acquisitions stored in the Landsat archive at the United States Geological Survey (USGS) Earth Resources Observation and Science (EROS) center. The Landsat sensors are well calibrated (Markham and Helder 2012; Schott et al. 2012) and over the conterminous United States are geolocated to sub-pixel accuracy (Lee et al. 2004). It is now possible to obtain at no cost Landsat data that were sensed decades apart over the same location. Global 30 m Landsat observations have been provided by the Landsat 4 and 5 Thematic Mapper (TM), Landsat 7 Enhanced Thematic Mapper Plus (ETM+), and Landsat 8 Operational Landsat Imager (OLI), spanning a period from 1982 to present, and at lower spatial resolution by the Multispectral Scanner System (MSS) starting in 1972 (Loveland and Dwyer 2012). The Landsat TM, ETM+ and OLI sensors provide 30 m

D.P. Roy (✉) • V. Kovalskyy • H. Zhang • L. Yan • I. Kommareddy
Geospatial Science Center of Excellence, South Dakota State University, Brookings, SD, USA
e-mail: David.Roy@sdstate.edu

C. Kuenzer et al. (eds.), *Remote Sensing Time Series*, Remote Sensing and Digital Image Processing, DOI 10.1007/978-3-319-15967-6_14

multispectral observations that can be used to identify and monitor landscape features for resource management and for climate and global change studies.

Prior to the opening of the Landsat archive, global coverage data sets were provided to the user community at no cost. The Global Land Survey (GLS) provides relatively cloud-free single date, manually selected, Landsat acquisitions of the land for the 1970s, 1990s and 2000s (Tucker et al. 2004). With the advent of free Landsat data it becomes feasible to apply temporal compositing approaches to multi-temporal Landsat acquisitions of the same location. Compositing procedures are applied independently on a per-pixel basis to gridded satellite time series and provide a practical way to reduce cloud and aerosol contamination, fill missing values, and reduce the data volume of moderate resolution global near-daily coverage satellite data (Holben 1986; Cihlar et al. 1994; Roy et al. 1997). Thus, instead of spatially mosaicing select relatively cloud-free Landsat acquisitions together (Zobrist et al. 1983), all the available multi-temporal acquisitions may be considered and at each gridded pixel the acquisition that satisfies some compositing criteria selected. In this way, the GLS 2005 Landsat ETM+ data set was generated by compositing up to three circa 2005 low cloud cover acquisitions (Gutman et al. 2008).

The opening of the Landsat archive has enabled the development of new land monitoring approaches that are based upon using all the available Landsat images for a given region and time period rather than just a select subset of cloud-free images. Previously, data costs precluded continental-scale studies, much less global ones. Studies were based on what Landsat data researchers could afford, not on what data they truly needed (Wulder et al. 2012). Arguably, the capability of any Landsat sensor for large area or long term monitoring has not yet been fully assessed. Moreover, although the Landsat data are provided as radiometrically and geometrically corrected images, processing them into spatially and temporally explicit information presents significant technical challenges.

The provision of consistently processed 'higher-level' products has been advocated by the Landsat Science Team as needed to meet outstanding information needs for climate and global change studies, to help national and international reporting linked to multilateral environmental agreements, and for regional and national resource management applications (Roy et al. 2014a). One step towards meeting these needs has been the NASA (National Aeronautics and Space Administration) funded Web Enabled Landsat Data (WELD) project that has demonstrated the capability to generate and distribute temporally composited and spatially mosaicked Landsat products of the conterminous United States and Alaska. The WELD products and distribution are an early example of the likely emergence of regional to national Landsat based mapping and monitoring services. Moreover, the acceptance and utility of the global Moderate Resolution Imaging Spectroradiometer (MODIS) Land products for science and applications (Justice et al. 2002; Masuoka et al. 2010) indicate the need for similar global coverage products but at Landsat resolution.

This chapter first introduces the current WELD products and showcases global WELD product prototyping, illustrating the potential for Landsat monitoring anywhere, and then reviews the Landsat time series applications and research that are developing rapidly in response to the opening of the Landsat archive. The chapter concludes with future research challenges and prospectives.

14.2 Web Enabled Landsat Data (WELD) Products

14.2.1 Current WELD Products

The WELD products provide Landsat 7 ETM+ 30 m temporally composited and spatially mosaicked products of the conterminous United States (CONUS) and Alaska with a weekly, monthly, seasonal and annual reporting frequency (Roy et al. 2010). The products are designed to provide consistent data that can be used to derive land cover and geo-physical and bio-physical products. For example, they have enabled the development of turnkey approaches to CONUS land cover and land cover change characterization (Hansen et al. 2011, 2014) due to the systematic Landsat processing, including conversion of digital numbers to calibrated top of atmosphere reflectance and brightness temperature, cloud masking, and reprojection into a gridded continental map projection (Roy et al. 2010).

The WELD products are generated from every available Landsat 7 ETM+ Level 1 T data in the Landsat archive with cloud cover $\leq$80 %, about 10,000 scenes per year for the CONUS and Alaska. The weekly, monthly, seasonal and annual products are defined by application of a compositing algorithm that attempts to select the best Landsat observation of each pixel in the reporting period (Roy et al. 2010). For each 30 m pixel location the six reflective Landsat 7 ETM+ bands, the two thermal bands, bit packed band saturation information, the Normalized Difference Vegetation Index (NDVI) derived as the near-infrared minus the red reflectance divided by their sum, two cloud masks, the day of year that the pixel was sensed on, and the number of Landsat observations considered in the product reporting period are stored. The products are defined in the Albers Equal Area conic projection in separate geolocated tiles of 5000 $\times$ 5000 30 m pixels.

The current WELD Version 1.5 products do not include radiometric correction for the effects of the atmosphere or topographic variations. A number of atmospheric correction methodologies have been developed but those using radiative transfer algorithms and atmospheric characterization data provide the most potential for automated large area application (Vermote et al. 2002; Masek et al. 2006; Ju et al. 2012). The impact of the atmosphere on Landsat data can be pronounced, for example, considering 53 million 30 m pixel locations sampled systematically across the CONUS over 12 months indicated that the mean CONUS absolute difference between surface and top of atmosphere NDVI expressed as a percentage of the surface NDVI was 28 % for "vegetated" surfaces (Roy et al. 2014b). Landsat atmospheric correction may not always be reliable, particularly in the shorter wavelength bands (Ju et al. 2012), and for this reason some large area land cover and land cover change mapping algorithms do not use atmospherically corrected data (Hansen et al. 2011, 2013, 2014). Although radiometric correction for topographic variations has been shown to improve land cover classification accuracy (Hantson and Chuvieco 2011; Vanonckelen et al. 2013), reliable systematic correction over large areas is challenging because this requires an adequate characterization of the atmosphere (especially aerosols) and modelling of the interaction of scattered radiation with adjacent surface components that is particularly complex when there is variable topography (Tanre et al. 1981; Vermote and Kotchenova 2008). None of the currently available large area Landsat or MODIS top of atmosphere or surface reflectance datasets include topographic correction.

The WELD products are made freely available in both HDF and GeoTIFF formats from portals hosted and maintained by the USGS EROS. Currently, the Version 1.5 weekly, monthly, seasonal and annual WELD products for the CONUS and Alaska are available for 10 years (2002–2012), a product volume of 4 TB per year with HDF internal compression. The WELD HDF tiled products are available via HTTP (http://e4ftl01.cr.usgs.gov/WELD) to support bulk data ordering. In addition, in response to user requests to browse and order the WELD products interactively, a What You See Is What You Get (WYSIWYG) Internet distribution interface was developed and ported to the USGS EROS (http://weld.cr.usgs.gov, Fig. 14.1). Users are able to interactively select and view any of the CONUS or Alaska WELD products, flick through the chronology of preceding and subsequent product time periods, and pan and zoom within selected product browse imagery. Registered users may order any arbitrary rectangular geographic area of interest, up to 2 GB, in a way that the product tiling file structure is transparent to the user. In addition, a single pixel product time series dump functionality is provided to allow users to extract WELD product time series for any 30 m pixel location, selected either interactively or by entering latitude and longitude coordinates. Figure 14.2 shows an example resulting weekly WELD product time series. Before the WELD project, to make this plot would require the separate ordering of 213 Landsat images, application of image-specific calibration parameters, conversion of the digital numbers to radiance, conversion of the radiance to brightness temperature, and then precise geographic pixel location and value extraction, and notation of each Landsat acquisition date.

14.2.2 Planned Global WELD Products

The global Landsat data volume is much greater than the global MODIS land product generation volume. For example, there are approximately 150 million 1 km global land pixels, whereas just for the CONUS there are approximately 11,000 million 30 m land pixels (Roy et al. 2010). To process global coverage Landsat data volume requires substantial computational resources. Figure 14.3 shows 12 global monthly WELD products generated on the NASA Earth Exchange (NEX) for climate year 2010. These 30 m products have a similar tiled format as the

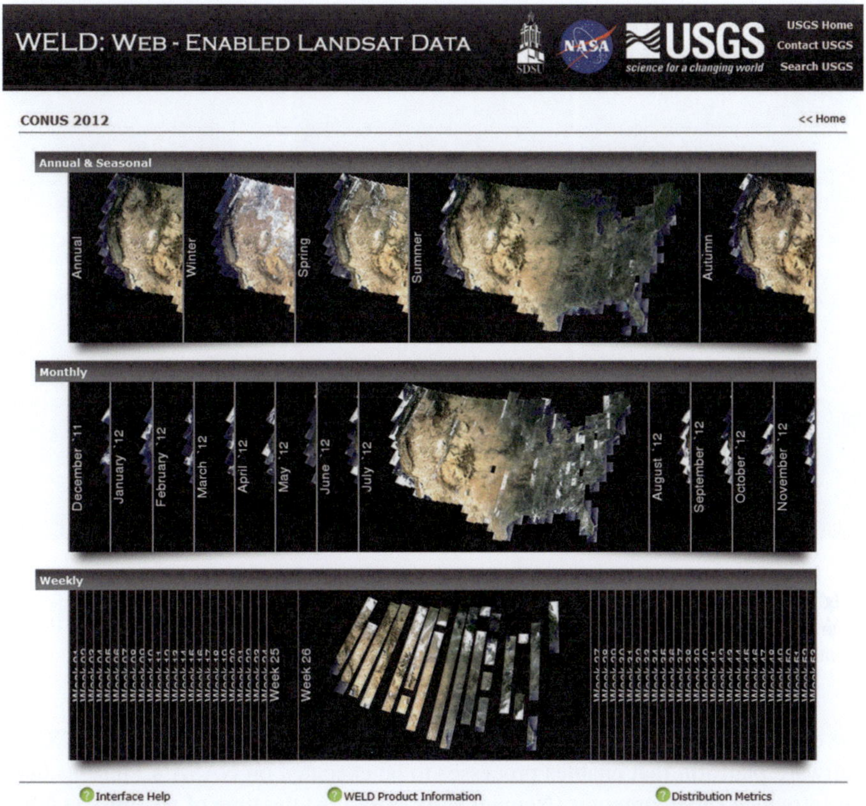

Fig. 14.1 Screen shot of the What You See Is What You Get (WYSIWYG) WELD Internet distribution interface at (http://weld.cr.usgs.gov) showing the product selection page for the conterminous United States (CONUS) 2012 WELD products: annual and four seasonal products (*top row*), 12 monthly products (*middle row*), and 52 weekly products (*bottom row*)

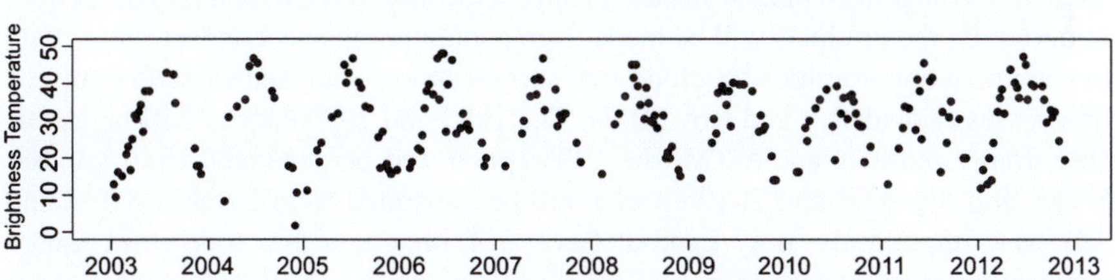

Fig 14.2 Ten year 30 m WELD Landsat 7 brightness temperature weekly time series ordered using the WELD WYSIWYG (http://weld.cr.usgs.gov) for a single desert pixel near Socorro, NM, USA. A total of 213 pixel values are shown

current WELD products but are defined in the MODIS land product sinusoidal projection (Wolfe et al. 1998). Each global monthly product took approximately 3 h to generate once the global WELD code was correctly configured to make optimal

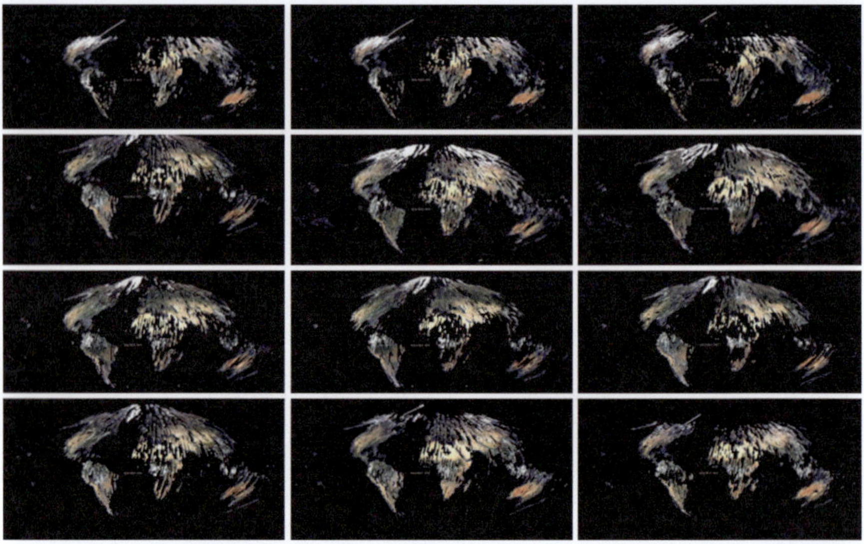

Fig 14.3 Twelve prototype NEX global monthly WELD products generated from ~7000 L1T Landsat 7 ETM+ acquisitions per month. Each 1.8 km true color browse pixel shown is generated from 60 × 60 30 m pixels. Black shows where there were no Landsat ETM+ L1T data in that month. MODIS Sinusoidal equal area projection

use of the NEX parallel processing capabilities. The key component of the NEX is a software platform that enables processes to be executed on NASA's Pleiades high performance supercomputer (Nemani et al. 2011). At the time of writing the NEX has 9 TB of online storage and 180,000 cores.

Planned global monthly WELD products are being generated under new NASA funding. The products will be generated for six three year epochs centered on 1985, 1990, 1995, 2000, 2005 and 2010, selected because they are spaced evenly 5 years apart and so enable 30 m land surface change to be monitored over a 25 year period. Importantly, the products will be made from contemporaneous Landsat sensor data to maximize opportunities for cloud free surface observation and so denser surface time series (Kovalskyy and Roy 2013). The 1985 and 1990 epochs will be generated from Landsat 4 and 5 TM, the 1995 epoch will be generated from Landsat 5 TM, and the 2000 and 2010 epochs will be generated from Landsat 5 TM and 7 ETM+. Importantly, only Landsat Level 1 T images with a root mean square geolocation error of less than a 30 m pixel dimension are used to enable more accurate mapping and monitoring applications, the generation of less smoothed temporally composited Landsat products (Roy 2000), and more accurate multi-temporal change detection applications (Townshend et al. 1992).

At the time of publication of this chapter, 3 years (2009–2011) of prototype global WELD products have been made available via HTTP (http://globalweld.cr.usgs.gov/collections) and via a WYSIWYG Internet distribution interface (http://globalweld.cr.usgs.gov).

14.3 Landsat Time Series Applications and Research That Are Developing Rapidly in Response to the Opening of the Landsat Archive

14.3.1 Phenology Monitoring

Phenology concerns the seasonal periodicity of plant and animal life cycles (Noormets 2009). At a recent U.S. National Phenology Network meeting the need to make phenological information more amenable to managers, decision makers, and the public was advocated (Enquist et al. 2012). Sparse collections of ground-based phenology records, such as the timing of flowering and budding of various plant species, are increasingly complemented by phenological studies that use satellite data (Morisette et al. 2008).

Most satellite based phenological studies apply curve fitting techniques to time series of spectral indexes that are sensitive to vegetation dynamics. The NDVI has been widely used (White et al. 2009). A variety of polar orbiting satellites that provide coarse spatial (100 m to 1 km) but high, near daily, temporal resolution NDVI data have been used, including from the Advanced Very High Resolution Radiometer (AVHRR) (Myneni et al. 1997; Schwartz et al. 2002; White and Nemani 2006; Maignan et al. 2008; Kross et al. 2011; Jeganathan et al. 2014), SPOT/Vegetation (Atzberger et al. 2014; Shen et al. 2014), and MODIS (Zhang and Goldberg 2011; Bolton and Friedl 2013; Shuai et al. 2013; Hmimina et al. 2013; Hilker et al. 2014).

The 30 m resolution of Landsat provides an appropriate scale for phenological monitoring. Studies such as Kovalskyy et al. (2011), Elmore et al. (2012), Hufkens et al. (2012), and Melaas et al. (2013), provide evidence that the use of moderate spatial resolution Landsat data increases the ability to capture phenological variations imposed by micro-climatic and topographic effects, reduces the mixing of different landscape components that may have different phenologies, and enables more precise scaling of ground-based phenological observations.

The potential for Landsat data to derive phenological information has to date been constrained primarily by the nominal 16-day Landsat temporal observation frequency (Fisher et al. 2006; Fisher and Mustard 2007). Integration of data from contemporaneous but different Landsat sensors provides more opportunities for cloud free surface observation and so denser time series (Kovalskyy and Roy 2013). For example, Fig. 14.4 shows 12 years of WELD weekly NDVI data for a single forest pixel in the Columbia River Valley, WA, USA. These data were atmospherically corrected and cloud-screened and importantly were derived by selecting in each week the best pixel observation from contemporaneous Landsat 5 TM and 7 ETM+ data. The Landsat 5 TM and 7 ETM+ sensors are almost identical and have 15° fields of view and each overpasses the same location every 16 days. Combined, the nominal repeat cycle they provide is 8 days but because satellite paths converge at higher latitudes, the repeat coverage may be more frequent (Kovalskyy and Roy 2013). A total of 306 observations are illustrated and were selected from

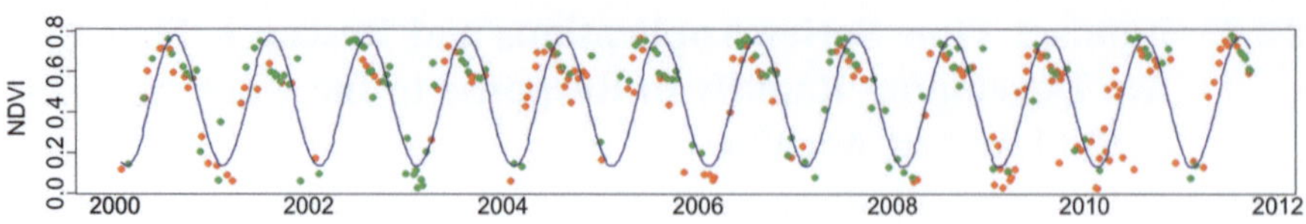

Fig. 14.4 Twelve years of weekly WELD processed Landsat 7 ETM+ (*green*) and Landsat 5 TM (*orange*) cloud-screened and atmospherically corrected 30 m NDVI for a single forest pixel in the Columbia River Valley, WA, USA. The *blue line* is a harmonic ordinary least squares regression fit ($R2 = 0.78$, 306 points) of the form: $NDVI = 0.46 - 0.001x - 0.252\cos(x) - 0.123\sin(x) + 0.011\cos(2x) - 0.091\sin(2x)$, where $x = ((DOY/365) + YEAR\text{-}2000)$, YEAR and DOY and are the year and the day of the year that each Landsat pixel was observed on respectively

173 Landsat 7 ETM+ and 212 Landsat 5 TM cloud-free observations of the pixel location. There were more Landsat 5 observations because in May 2003 the Landsat 7 ETM+ sensor had a failure and thereafter 22 % of the sensed pixels were dropped (Markham et al. 2004).

The potential utility of the data illustrated in Fig. 14.4 for tracking phenology is evident. A simple second order harmonic regression fit of the NDVI data is shown (blue line). A number of researchers have advocated harmonic functions to enable disturbance detection as new Landsat observations become available (Brooks et al. 2012; Zhu et al. 2012). More sophisticated non-linear harmonic functions that enable inter-annual variability in the phase and amplitude of the NDVI are recommended (Carrao et al. 2010). Phenological metrics of interest to the research and applications community include not only metrics that capture the inter-annual trend in the vegetation index but also that capture the date of the start and end of the growing season, the duration of the growing season, and the start of vegetation senescence. Considerable effort has been expended on developing vegetation index time series fitting functions, including fitting bell shaped curves (Badhwar 1984; Tucker et al. 2001), Gaussian functions (Jönsson and Eklundh 2004), quadratic fits of aggregated growing degree days (de Beurs and Henebry 2004), and piece-wise log curve fitting (Zhang et al. 2003, 2009). An outstanding challenge is to assess the most reliable and accurate methodology. Currently however there is no consensus best approach and different methodologies have been found to prove useful under different environmental conditions (White et al. 2009; de Beurs and Henebry 2010; Atkinson et al. 2012).

14.3.2 Land Cover Change Monitoring

Classification is regarded as a fundamental process in remote sensing used to relate pixel values to the land cover present at the corresponding location on the Earth's surface. Land cover classification and change detection approaches have evolved considerably in the last several decades (Franklin and Wulder 2002; Coppin et al. 2004; Lu and Weng 2007; Hansen and Loveland 2012). Recent Landsat based mapping and change detection approaches have focused on quantifying disturbance or land cover change over a range of time scales that were rarely feasible before no-cost access to the Landsat archive (Hansen and Loveland 2012). Large area Landsat land cover change algorithms applications have primarily focused on mapping forest land cover change and forest disturbance. Per-pixel time series approaches that identify significant changes by examination of the temporal trajectory of surface reflectance or vegetation indices have been adopted. The temporal trajectory is interpreted through sets of rules, which can be predefined, as in the Vegetation Change Tracker (Masek et al. 2013; Huang et al. 2010) and in the LandTrendr algorithms (Kennedy et al. 2010, 2012; Cohen et al. 2010), or statistically derived from training data (Hansen et al. 2013; Sexton et al. 2013). Landsat long time-series approaches have also been developed to document spatial patterns of gradual land cover condition (Hostert et al. 2003; Goodwin et al. 2008; Vogelmann et al. 2012).

New methods have been developed that allow modeling of the seasonal trajectory of spectral information to enable disturbance detection as new Landsat observations become available (Brooks et al. 2012; Zhu et al. 2012), including changes in both land use and land management within a given land use (Brooks et al. 2014). In particular, with the improved Landsat 8 acquisition frequency, about 60 % more global acquisitions per day than from Landsat 7 (Roy et al. 2014a), the opportunity exists to begin to monitor moderate resolution land cover disturbances and rapid land cover changes shortly after they occur.

14.4 Prospectives for Landsat Time Series Monitoring

The opening of the Landsat archive has fostered new analytical approaches for describing land surface condition and dynamics and these will likely continue to evolve. Landsat time series applications are fundamentally challenged however by cloud cover, particularly in tropical and boreal regions that are routinely cloudy at the time of Landsat overpass (Ju and Roy 2008; Kovalskyy and Roy 2013). Moreover, the amount of Landsat data in the U.S. Landsat archive has not been constant among Landsat sensors, from year to year, or geographically, outside of the conterminous United States, because of differing Landsat data acquisition strategies, data reception capabilities, and system health issues (Markham et al. 2004; Goward et al. 2006; Loveland and Dwyer 2012). This is evident in Fig. 14.3, where successive months with no Landsat coverage, for example in parts of Africa, are apparent. These two factors, missing data and cloud obscuration, significantly challenge the implementation of time series approaches. For example, temporal interpolation or curve fitting, like in Fig. 14.4, will have variable reliability depending on the temporal frequency of the clear observations and the dynamics of the land surface (Jackson et al. 2004; Verbesselt et al. 2010; Zhu et al. 2012; Yan and Roy 2014). Temporal compositing approaches have been developed to select a best Landsat observation from all the Landsat observations collected over some reporting period (White et al. 2014), for example, to generate gridded weekly, monthly, seasonal and annual composites, such as the WELD products (Roy et al. 2010) (Fig. 14.1). Similarly, automated cloud screening and mosaicking approaches have allowed wall-to-wall mapping of large areas of tropical forest and boreal regions (Helmer and Ruefenacht 2005; Broich et al. 2011; Potapov et al. 2011). Even with combined contemporaneous Landsat data, there may be insufficient observations to capture certain rapid surface changes, such as due to fire, pest outbreak, or flooding, particularly if the changes occur in persistently cloudy periods.

A potential solution to provide more frequent cloud-free surface observations is to fuse Landsat observations with data from other remote sensing systems. In this respect the Moderate Resolution Imaging Spectroradiometer (MODIS) is well suited and the MODIS Terra satellite was placed in the same morning orbit as Landsat 7 to provide opportunities for multi-scale global land surface change monitoring (Skole et al. 1997). Fusion can be undertaken empirically or by using physically based approaches. The spatial and temporal adaptive reflectance fusion model (STARFM) blends 16 day 30 m Landsat with daily 500 m MODIS data to generate synthetic daily Landsat-like 30 m reflectance data (Gao et al. 2006) and represents a significant step for empirical fusion research. Landsat fusion is not straightforward because of differences among sensor spectral band configurations, sensor and sun geometries, and surface changes. The STARFM technique, while providing useful information, requires scene-dependent tuning parameters and does not explicitly handle the directional dependence of reflectance as a function of the sun–target–sensor geometry, usually described by the Bi-directional Reflectance Distribution Function (BRDF). Roy et al. (2008) developed a semi-physical fusion approach that uses the MODIS BRDF 500 m product (Schaaf et al. 2002) to predict 30 m Landsat spectral reflectance for a desired date as the product of the observed Landsat reflectance and the ratio of the 500 m surface reflectance modeled using the MODIS BRDF spectral model parameters and the sun-sensor geometry on the predicted and observed Landsat dates. Recent research has modeled MODIS vegetation index time series using phenological modelling approaches to adjust and reconstruct sparse Landsat time series (Zhang et al. 2014). Other researchers have explored techniques based on STARFM that are somewhat adaptive to land surface change including the Spatial Temporal Adaptive Algorithm for mapping Reflectance Change (Hilker et al. 2009) and the Enhanced STARFM (Zhu et al. 2010). Fusion methods have gained popularity. This is especially true for STARFM, facilitated by its publically available software, which has been applied for applications including classification of conservation tillage land (Watts et al. 2011), analysis of dryland forest phenology (Walker et al. 2012), evapotranspiration retrieval (Anderson et al. 2011; Jia et al. 2012), and derivation of urban environmental variables in support of public health studies (Liu and Weng 2012).

Further research on fusion approaches is required, including fusion technique comparison studies (Emelyanova et al. 2013), and more detailed consideration of satellite geometric and sensor differences. The 2015 launch of Sentinel-2 by the European Space Agency (ESA) will offer additional unique and complementary observations to the data provided by Landsat. There are two planned Sentinel-2 satellites designed to provide multiple global acquisitions with similar spectral and spatial characteristics as Landsat (Drusch et al. 2012). Combined, the Landsat 8 and Sentinel-2 data streams will provide the opportunity for near daily land surface monitoring and may herald a new era for terrestrial change monitoring.

The sensor improvements of Landsat 8 compared to the previous Landsat sensors are well documented (Irons et al. 2012) and offer refined opportunities for quantitative approaches to Landsat information extraction (Roy et al. 2014a). It is likely that computationally intensive processing algorithms that have been developed but only demonstrated for local studies will be increasingly applied to large Landsat data sets. Physically-based information extraction approaches via model inversion and data assimilation, and automated computer vision algorithms, may become more prevalent. For example, recently, an automated Landsat agricultural crop field extraction methodology was demonstrated using WELD time series (Yan and Roy 2014).

Finally, it is established that satellite derived products and information should be validated, particularly if they are used for policy making or fiscal scrutiny, for example, Landsat deforestation and carbon assessments made in support of REDD (Reducing Emissions from Deforestation and Forest Degradation in Developing Countries) activities (UN-REDD 2008). Satellite product validation is typically undertaken by comparison with independent reference data (Justice et al. 2000) distributed using appropriate sampling schemes (Stehman 2013). However, validation of time series products is particularly challenging and adequate independent time series reference data rarely exist. Methods that couple human interpretation of time series data with higher spatial resolution aircraft and satellite imagery to allow characterization of land cover change at sample points have been developed (Cohen et al. 2010) and likely this will remain an active and needed research domain.

14.5 Summary

This chapter has demonstrated the key role that Landsat based remote sensing can play in providing spatially and temporally explicit data and derived data products for land surface monitoring. The opening of the Landsat archive has fostered new analytical approaches for describing land surface condition and dynamics and these will continue to evolve. Recent Landsat mapping and change detection approaches have focused on quantifying disturbance or land cover change over a range of time scales that were rarely feasible before no-cost access to the Landsat archive.

Future developments will likely include the development of global scale Landsat applications that take advantage of the historical Landsat archive, applications that fuse Landsat data with other remotely sensed data to reduce the impacts of missing Landsat data and surface cloud obscuration, and the development of applications that detect changes shortly after Landsat data are sensed. The challenge is to transfer the methodologies developed in the research domain into the operational domain to provide information with known accuracy and consistency. This is particularly important if Landsat based products are used in strategic ways and in support of policy making.

Acknowledgements This work was funded by NASA grant NNX13AJ24A and USGS contract G12PC00069.

References

Anderson MC, Kustas WP, Norman JM, Hain CR, Mecikalski JR, Schultz L, González-Dugo MP, Cammalleri C, D'Urso G, Pimstein A, Gao F (2011) Mapping daily evapotranspiration at field to continental scales using geostationary and polar orbiting satellite imagery. Hydrol Earth Syst Sci 15:223–239

Atkinson PM, Jeganathan C, Dash J, Atzberger C (2012) Inter-comparison of four models for smoothing satellite sensor time-series data to estimate vegetation phenology. Remote Sens Environ 123:400–417

Atzberger C, Formaggio AR, Shimabukuro YE, Udelhoven T, Mattiuzzi M, Sanchez GA, Arai E (2014) Obtaining crop-specific time profiles of NDVI: the use of unmixing approaches for serving the continuity between SPOT-VGT and PROBA-V time series. Int J Remote Sens 35 (7):2615–2638

Badhwar GD (1984) Use of Landsat-derived profile features for spring small-grains classification. Int J Remote Sens 5(5):783–797

Bolton DK, Friedl MA (2013) Forecasting crop yield using remotely sensed vegetation indices and crop phenology metrics. Agric For Meteorol 173:74–84

Broich M, Hansen MC, Potapov P, Adusei B, Lindquist E, Stehman SV (2011) Time-series analysis of multi-resolution optical imagery for quantifying forest cover loss in Sumatra and Kalimantan, Indonesia. Int J Appl Earth Obs Geoinform 13:277–291

Brooks EB, Thomas VA, Wynne RH, Coulston JW (2012) Fitting the multitemporal curve: a fourier series approach to the missing data problem in remote sensing analysis. IEEE Trans Geosci Remote Sens 50(9):3340–3353

Brooks EB, Wynne RH, Thomas VA, Blinn CE, Coulston JW (2014) On-the-fly massively multitemporal change detection using statistical quality control charts and Landsat data. IEEE Trans Geosci Remote Sens 52(6):3316–3332

Carrao H, Gonalves P, Caetano M (2010) A nonlinear harmonic model for fitting satellite image time series: analysis and prediction of land cover dynamics. IEEE Trans Geosci Remote Sens 48(4):1919–1930

Cihlar J, Manak D, D'Iorio M (1994) Evaluation of compositing algorithms for AVHRR data over land. IEEE Trans Geosci Remote Sens 32:427–437

Cohen WB, Yang Z, Kennedy R (2010) Detecting trends in forest disturbance and recovery using yearly Landsat time series: 2. TimeSync – tools for calibration and validation. Remote Sens Environ 114:2911–2924

Coppin P, Jonckheere I, Nackaerts K, Muys M (2004) Digital change detection methods in ecosystem monitoring: a review. Int J Remote Sens 25:1565–1596

de Beurs KM, Henebry GM (2004) Land surface phenology, climatic variation, and institutional change: analyzing agricultural land cover change in Kazakhstan. Remote Sens Environ 89 (4):497–509

de Beurs K, Henebry G (2010) Spatio-temporal statistical methods for modelling land surface phenology. In: Hudson IL, Keatley MR (eds) Phenological research. Springer, Dordrecht

Drusch M, Del Bello U, Carlier S, Colin O, Fernandez V, Gascon F, Hoersch B, Isola C, Laberinti P, Martimort P, Meygret A, Spoto F, Sy O, Marchese F, Bargellini P (2012)

Sentinel-2: ESA's optical high-resolution mission for GMES operational services. Remote Sens Environ 120:25–36

Elmore AJ, Guinn SM, Minsley BJ, Richardson AD (2012) Landscape controls on the timing of spring, autumn, and growing season length in mid-Atlantic forests. Glob Chang Biol 18 (2):656–674

Emelyanova IV, McVicar TR, Van Niel TG, Li LT, van Dijk AI (2013) Assessing the accuracy of blending Landsat–MODIS surface reflectances in two landscapes with contrasting spatial and temporal dynamics: a framework for algorithm selection. Remote Sens Environ 133:193–209

Enquist CAF, Rosemartin A, Schwartz MD (2012) Identifying and prioritizing phenological data products and tools, Eos. Trans Am Geophys Union 93(37):356

Fisher JI, Mustard JF (2007) Cross-scalar satellite phenology from ground, Landsat, and MODIS data. Remote Sens Environ 109(3):261–273

Fisher JI, Mustard JF, Vadeboncoeur MA (2006) Green leaf phenology at Landsat resolution: scaling from the field to the satellite. Remote Sens Environ 100(2):265–279

Franklin SE, Wulder MA (2002) Remote sensing methods in medium spatial resolution satellite data land cover classification of large areas. Prog Phys Geogr 26(2):173–205

Gao F, Masek J, Schwaller M, Hall F (2006) On the blending of the Landsat and MODIS surface reflectance: predicting daily Landsat surface reflectance. IEEE Trans Geosci Remote Sens 44:2207–2218

Goodwin NR, Coops NC, Wulder MA, Gillanders S (2008) Estimation of insect infestation dynamics using a temporal sequence of Landsat data. Remote Sens Environ 12:3680–3689

Goward S, Arvidson T, Williams D, Faundeen J, Irons J, Franks S (2006) Historical record of Landsat global coverage: mission operations, NSLRSDA, and international cooperator stations. Photogramm Eng Remote Sens 72:1155–1169

Gutman G, Byrnes R, Masek J, Covington S, Justice C, Franks S, Kurtz R (2008) Towards monitoring land-cover and land-use changes at a global scale: the global land survey 2005. Photogramm Eng Remote Sens 74:6–10

Hansen MC, Loveland TR (2012) A review of large area monitoring of land cover change using Landsat data. Remote Sens Environ 122:66–74

Hansen MC, Egorov A, Roy DP, Potapov P, Ju J, Turubanova S, Kommareddy I, Loveland TR (2011) Continuous fields of land cover for the conterminous United States using Landsat data: first results from the Web-Enabled Landsat Data (WELD) project. Remote Sens Lett 2:279–288

Hansen MC, Potapov PV, Moore R, Hancher M, Turubanova SA, Tyukavina A, Thau D, Stehman SV, Goetz SJ, Loveland TR, Kommareddy A, Egorov A, Chini L, Justice CO, Townshend JRG (2013) High-resolution global maps of 21st-century forest cover change. Science 342:850–853

Hansen MC, Egorov A, Potapov PV, Stehman SV, Tyukavina A, Turubanova SA, Roy DP, Goetz SJ, Loveland TR, Ju J, Kommareddy A, Kovalskyy V, Forsythe C, Bents T (2014) Monitoring conterminous United States (CONUS) land cover change with Web-Enabled Landsat Data (WELD). Remote Sens Environ 140:466–484

Hantson S, Chuvieco E (2011) Evaluation of different topographic correction methods for Landsat imagery. Int J Appl Earth Obs Geoinform 13:691–700

Helmer EH, Ruefenacht B (2005) Cloud-free satellite image mosaics with regression trees and histogram matching. Photogramm Eng Remote Sens 71:1079–1089

Hilker T, Wulder MA, Coops NC, Linke J, McDermid G, Masek JG, Gao F, White JC (2009) A new data fusion model for high spatial- and temporal-resolution mapping of forest disturbance based on Landsat and MODIS. Remote Sens Environ 113:1613–1627

Hilker T, Natsagdorj E, Waring RH, Lyapustin A, Wang Y (2014) Satellite observed widespread decline in Mongolian grasslands largely due to overgrazing. Glob Chang Biol 20(2):418–428

Hmimina G, Dufrêne E, Pontailler JY, Delpierre N, Aubinet M, Caquet B, de Grandcourt A, Burban B, Flechard C, Granier A, Gross P, Heinesch B, Longdoz B, Moureaux C, Ourcival JM, Rambal S, Saint André L, Soudani K (2013) Evaluation of the potential of MODIS satellite

data to predict vegetation phenology in different biomes: an investigation using ground-based NDVI measurements. Remote Sens Environ 132:145–158
Holben B (1986) Characteristics of maximum-value composite images from temporal AVHRR data. Int J Remote Sens 7:1417–1434
Hostert P, Roder A, Hill J (2003) Coupling spectral unmixing and trend analysis for monitoring of long-term vegetation dynamics in Mediterranean rangelands. Remote Sens Environ 87:183–197
Huang CQ, Coward SN, Masek JG, Thomas N, Zhu ZL, Vogelmann JE (2010) An automated approach for reconstructing recent forest disturbance history using dense Landsat time series stacks. Remote Sens Environ 114:183–198
Hufkens K, Friedl M, Sonnentag O, Braswell BH, Milliman T, Richardson AD (2012) Linking near-surface and satellite remote sensing measurements of deciduous broadleaf forest phenology. Remote Sens Environ 117:307–321
Irons JR, Dwyer JL, Barsi JA (2012) The next Landsat satellite: The Landsat data continuity mission. Remote Sens Environ 122:11–21
Jackson TJ, Chen DY, Cosh M, Li FQ, Anderson M, Walthall C, Doriaswamy P, Hunt ER (2004) Vegetation water content mapping using Landsat data derived normalized difference water index for corn and soybeans. Remote Sens Environ 92(4):475–482
Jeganathan C, Dash J, Atkinson PM (2014) Remotely sensed trends in the phenology of northern high latitude terrestrial vegetation, controlling for land cover change and vegetation type. Remote Sens Environ 143:154–170
Jia Z, Liu S, Xu Z, Chen Y, Zhu M (2012) Validation of remotely sensed evapotranspiration over the Hai River Basin, China. J Geophys Res Atmos 117:D13113
Jönsson P, Eklundh L (2004) TIMESAT – a program for analyzing time-series of satellite sensor data. Comput Geosci 30(8):833–845
Ju J, Roy DP (2008) The availability of cloud-free Landsat ETM+ data over the conterminous United States and globally. Remote Sens Environ 112:1196–1211
Ju J, Roy DP, Vermote E, Masek J, Kovalskyy V (2012) Continental-scale validation of MODIS-based and LEDAPS Landsat ETM+ atmospheric correction methods. Remote Sens Environ 122:175–184
Justice C, Belward A, Morisette J, Lewis P, Privette J, Baret F (2000) Developments in the validation of satellite products for the study of the land surface. Int J Remote Sens 21:3383–3390
Justice C, Townshend J, Vermote E, Masuoka E, Wolfe R, Saleous N, Roy DP, Morisette J (2002) An overview of MODIS Land data processing and product status. Remote Sens Environ 83:3–15
Kennedy RE, Yang Z, Cohen WB (2010) Detecting trends in forest disturbance and recovery using yearly Landsat time series: 1. LandTrendr – temporal segmentation algorithms. Remote Sens Environ 114:2897–2910
Kennedy RE, Yang Z, Cohen WB, Pfaff E, Braaten J, Nelson P (2012) Spatial and temporal patterns of forest disturbance and regrowth within the area of the Northwest Forest Plan. Remote Sens Environ 122:117–133
Kovalskyy V, Roy DP (2013) The global availability of Landsat 5 TM and Landsat 7 ETM+ data land surface observations and implications for global 30 m Landsat data product generation. Remote Sens Environ 130:280–293
Kovalskyy V, Roy DP, Zhang X, Ju J (2011) The suitability of multi-temporal Web-Enabled Landsat Data (WELD) NDVI for phenological monitoring – a comparison with flux tower and MODIS NDVI. Remote Sens Lett 3(4):325–334
Kross A, Fernandes R, Seaquist J, Beaubien E (2011) The effect of the temporal resolution of NDVI data on season onset dates and trends across Canadian broadleaf forests. Remote Sens Environ 115(6):1564–1575
Lee DS, Storey JC, Choate MJ, Hayes R (2004) Four years of Landsat-7 on-orbit geometric calibration and performance. IEEE Trans Geosci Remote Sens 42:2786–2795

Liu H, Weng QH (2012) Enhancing temporal resolution of satellite imagery for public health studies: a case study of West Nile Virus outbreak in Los Angeles in 2007. Remote Sens Environ 117:57–71

Loveland TR, Dwyer JL (2012) Landsat: building a strong future. Remote Sens Environ 122:22–29

Lu D, Weng Q (2007) A survey of image classification methods and techniques for improving classification performance. Int J Remote Sens 2(8):823–870

Maignan F, Bréon FM, Bacour C, Demarty J, Poirson A (2008) Interannual vegetation phenology estimates from global AVHRR measurements: comparison with in situ data and applications. Remote Sens Environ 112(2):496–505

Markham BL, Helder DL (2012) Forty-year calibrated record of earth-reflected radiance from Landsat: a review. Remote Sens Environ 122:30–40

Markham BL, Storey JC, Williams DL, Irons JR (2004) Landsat sensor performance: history and current status. IEEE Trans Geosci Remote Sens 42:2691–2694

Masek JG, Vermote EF, Saleous NE, Wolfe R, Hall FG, Huemmrich KF (2006) A Landsat surface reflectance dataset for North America, 1990–2000. IEEE Trans Geosci Remote Sens Lett 3 (1):68–72

Masek JG, Goward SN, Kennedy RE, Cohen WB, Moisen GG, Schleeweis K, Huang CQ (2013) United States forest disturbance trends observed using Landsat time series. Ecosystems 16:1087–1104

Masuoka E, Roy DP, Wolfe R, Morisette J, Sinno S, Teague M, Saleous N, Devadiga S, Justice C, Nickeson J (2010) MODIS land data products – generation, quality assurance and validation. In: Ramachandran B, Justice C, Abrams M (eds) Land remote sensing and global environmental change: NASA's Earth observing system and the science of ASTER and MODIS, vol 11, Remote sensing and digital image processing series. Springer, Dordrecht

Melaas EK, Friedl MA, Zhu Z (2013) Detecting interannual variation in deciduous broadleaf forest phenology using Landsat TM/ETM data. Remote Sens Environ 132:176–185

Morisette JT, Richardson AD, Knapp AK, Fisher JI, Graham EA, Abatzoglou J, Wilson BE, Breshears DD, Henebry GM, Hanes JM, Liang L (2008) Tracking the rhythm of the seasons in the face of global change: phenological research in the 21st century. Front Ecol Environ 7 (5):253–260

Myneni RB, Keeling CD, Tucker CJ, Asrar G, Nemani RR (1997) Increased plant growth in the northern high latitudes from 1981 to 1991. Nature 386:698–702

Nemani R, Votava P, Michaelis A, Melton F, Milesi C (2011) NASA Earth exchange: a collaborative supercomputing environment for global change science. Trans EOS 13:109–110

Noormets A (2009) Phenology of ecosystem processes: applications in global change research. Springer, Dordrecht

Potapov PV, Turubanova S, Hansen MC (2011) Regional-scale boreal forest cover and change mapping using Landsat data composites for European Russia. Remote Sens Environ 115:548–561

Roy DP (2000) The impact of misregistration upon composited wide field of view satellite data and implications for change detection. IEEE Trans Geosci Remote Sens 38:2017–2032

Roy DP, Kennedy P, Folving S (1997) Combination of the normalized difference vegetation index and surface temperature for regional scale European forest cover mapping using AVHRR data. Int J Remote Sens 18:1189–1195

Roy DP, Ju J, Lewis P, Schaaf C, Gao F, Hansen M, Lindquist E (2008) Multi-temporal MODIS–Landsat data fusion for relative radiometric normalization, gap filling, and prediction of Landsat data. Remote Sens Environ 112:3112–3130

Roy DP, Ju J, Kline K, Scaramuzza PL, Kovalskyy V, Hansen MC, Loveland TR, Vermote EF, Zhang C (2010) Web-enabled Landsat Data (WELD): Landsat ETM+ composited mosaics of the conterminous United States. Remote Sens Environ 114:35–49

Roy DP, Wulder MA, Loveland TR, Woodcock CE, Allen RG, Anderson MC, Helder D, Irons JR, Johnson DM, Kennedy R, Scambos TA, Schaaf CB, Schott JR, Sheng Y, Vermote EF, Belward

AS, Bindschadler R, Cohen WB, Gao F, Hipple JD, Hostert P, Huntington J, Justice CO, Kilic A, Kovalskyy V, Lee ZP, Lymburner L, Masek JG, McCorkel J, Shuai Y, Trezza R, Vogelmann J, Wynne RH, Zhu Z (2014a) Landsat-8: science and product vision for terrestrial global change research. Remote Sens Environ 145:154–172

Roy DP, Qin Y, Kovalskyy V, Vermote EF, Ju J, Egorov A, Hansen MC, Kommareddy I, Yan L (2014b) Conterminous United States demonstration and characterization of MODIS-based Landsat ETM+ atmospheric correction. Remote Sens Environ 140:433–449

Schaaf C, Gao F, Strahler A, Lucht W, Li X, Tsang T, Strugnell N, Zhang X, Jin Y, Muller JP, Lewis P, Barnsley M, Hobson P, Disney M, Roberts G, Dunderdale M, d'Entremont R, Hu B, Liang S, Privette J, Roy D (2002) First operational BRDF, albedo and nadir reflectance products from MODIS. Remote Sens Environ 83:135–148

Schott JR, Hook SJ, Barsi JA, Markham BL, Miller J, Padula FP, Raqueno NG (2012) Thermal infrared radiometric calibration of the entire Landsat 4, 5, and 7 archive (1982–2010). Remote Sens Environ 122:41–49

Schwartz MD, Reed BC, White MA (2002) Assessing satellite-derived start-of-season measures in the conterminous USA. Int J Climatol 22(14):1793–1805

Sexton JO, Urban DL, Donohue MJ, Song C (2013) Long-term landcover dynamics by multi-temporal classification across the Landsat-5 record. Remote Sens Environ 128:246–258

Shen M, Zhang G, Cong N, Wang S, Kong W, Piao S (2014) Increasing altitudinal gradient of spring vegetation phenology during the last decade on the Qinghai–Tibetan Plateau. Agric For Meteorol 189–190:71–80

Shuai Y, Schaaf C, Zhang X, Strahler A, Roy D, Morisette J, Wang Z, Nightingale J, Nickeson J, Richardson AD, Xie D, Wang J, Li X, Strabala K, Davies JE (2013) Daily MODIS 500 m reflectance anisotropy direct broadcast (DB) products for monitoring vegetation phenology dynamics. Int J Remote Sens 34(16):5997–6016

Skole DS, Justice CO, Janetos A, Townshend JRG (1997) A land cover change monitoring program: a strategy for international effort. In: Mitigation and adaptation strategies for global change. Kluwer, Amsterdam, pp 1–19

Stehman SV (2013) Estimating area from an accuracy assessment error matrix. Remote Sens Environ 132:202–211

Tanre D, Herman M, Deschamps PY (1981) Influence of the background contribution upon space measurements of ground reflectance. Appl Opt 20(20):3676–3684

Townshend JRG, Justice CO, Gurney C, McManus J (1992) The impact of misregistration on change detection. IEEE Trans Geosci Remote Sens 30:1054–1060

Tucker CJ, Slayback DA, Pinzon JE, Los SO, Myneni RB, Taylor MG (2001) Higher northern latitude normalized difference vegetation index and growing season trends from 1982 to 1999. Int J Biometeorol 45(4):184–190

Tucker CJ, Grant DM, Dykstra JD (2004) NASA's global orthorectified Landsat data set. Photogramm Eng Remote Sens 70:313–322

UN-REDD (2008) UN collaborative programme on REducing Emissions from Deforestation and Forest Degradation in the Developing Countries (UN-REDD), Framework document. http://www.un-redd.org/AboutunREDDProgramme/tabid/583/Default.aspx. Accessed 30 Nov 2014

Vanonckelen S, Lhermitte S, Van Rompaey A (2013) The effect of atmospheric and topographic correction methods on land cover classification accuracy. Int J Appl Earth Obs Geoinform 24:9–21

Verbesselt J, Hyndman R, Newnham G, Culvenor D (2010) Detecting trend and seasonal changes in satellite image time series. Remote Sens Environ 114:106–114

Vermote EF, Kotchenova S (2008) Atmospheric correction for the monitoring of land surfaces. J Geophys Res 113:D23S90

Vermote EF, El Saleous NZ, Justice CO (2002) Atmospheric correction of MODIS data in the visible to middle infrared: first results. Remote Sens Environ 83:97–111

Vogelmann JE, Xian G, Homer C, Tolk B (2012) Monitoring gradual ecosystem change using Landsat time series data analyses: case studies in selected forest and rangeland ecosystems. Remote Sens Environ 122:92–105

Walker JJ, de Beurs KM, Wynne RH, Gao F (2012) Evaluation of Landsat and MODIS data fusion products for analysis of dryland forest phenology. Remote Sens Environ 117:381–393

Watts JD, Powell SL, Lawrence RL, Hilker T (2011) Improved classification of conservation tillage adoption using high temporal and synthetic satellite imagery. Remote Sens Environ 115:66–75

White MA, Nemani RR (2006) Real-time monitoring and short-term forecasting of land surface phenology. Remote Sens Environ 104(1):43–49

White MA, De Beurs KM, Didan K, Inouye DW, Richardson AD, Jensen OP, O'Keefe J, Zhang G, Nemani RR, Van Leeuwen WJD, Brown JF, De Wit A, Schaepman M, Lin X, Dettinger M, Bailey AS, Kimball J, Schwartz MD, Baldocchi DD, Lee JT, Lauenroth WK (2009) Intercomparison, interpretation, and assessment of spring phenology in North America estimated from remote sensing for 1982–2006. Glob Chang Biol 15(10):2335–2359

White JC, Wulder MA, Hobart GW, Luther JE, Hermosilla T, Griffiths P, Coops NC, Hall RJ, Hostert P, Dyk A, Guindon L (2014) Pixel-based image compositing for large-area dense time series applications and science. Can J Remote Sens 30(40):192–212

Wolfe R, Roy D, Vermote E (1998) The MODIS land data storage, gridding and compositing methodology: L2 grid. IEEE Trans Geosci Remote Sens 36:1324–1338

Woodcock CE, Allen R, Anderson M, Belward A, Bindschadler R, Cohen WB, Gao F, Goward SN, Helder D, Helmer E, Nemani R, Oreopoulos L, Schott J, Thenkabail PS, Vermote EF, Vogelmann J, Wulder MA, Wynne R (2008) Free access to Landsat imagery. Science 320:1011

Wulder MA, Masek JG, Cohen WB, Loveland TR, Woodcock CE (2012) Opening the archive: how free data has enabled the science and monitoring promise of Landsat. Remote Sens Environ 122:2–10

Yan L, Roy DP (2014) Automated crop field extraction from multi-temporal Web Enabled Landsat Data. Remote Sens Environ 144:42–64

Zhang X, Goldberg MD (2011) Monitoring fall foliage coloration dynamics using time-series satellite data. Remote Sens Environ 115(2):382–391

Zhang X, Friedl MA, Schaaf CB, Strahler AH, Hodges JCF, Gao F, Reed BC, Huete A (2003) Monitoring vegetation phenology using MODIS. Remote Sens Environ 84(3):471–475

Zhang X, Friedl MA, Schaaf CB (2009) Sensitivity of vegetation phenology detection to the temporal resolution of satellite data. Int J Remote Sens 30(8):2061–2074

Zhang HK, Chen JM, Huang B, Song H, Li Y (2014) Reconstructing seasonal variation of Landsat vegetation index related to leaf area index by fusing with MODIS data. IEEE J Select Top Appl Earth Obs Remote Sens 7(3):950–960

Zhu XL, Chen J, Gao F, Chen XH, Masek JG (2010) An enhanced spatial and temporal adaptive reflectance fusion model for complex heterogeneous regions. Remote Sens Environ 114:2610–2623

Zhu Z, Woodcock CE, Olofsson P (2012) Continuous monitoring of forest disturbance using all available Landsat imagery. Remote Sens Environ 122:75–91

Zobrist AL, Bryant NA, McLeod RG (1983) Technology for large digital mosaics of Landsat data. Photogramm Eng Remote Sens 49:1325–1335

Chapter 15
Forest Cover Dynamics During Massive Ownership Changes – Annual Disturbance Mapping Using Annual Landsat Time-Series

Patrick Griffiths and Patrick Hostert

Abstract Remote sensing is a core tool for forest monitoring. Landsat data has been widely used in forest change detection studies but many approaches lack capabilities such as assessing changes for long temporal sequences. Moreover, most methods are not capable of detecting gradual long term processes, such as post disturbance recovery. Following the open Landsat data policy implemented in 2008, but also due to the improved level 1 processing standards, Landsat remote sensing experienced considerable innovation, with many novel algorithms for automated preprocessing and also for change detection. Among these, trajectory based change detection methods provide new means for assessing forest cover changes using Landsat data. For example, disturbances can be assessed on a yearly basis and residual noise in the time series is effectively reduced, enabling the previously impossible detection of gradual changes (e.g. recovery, degradation). We here demonstrate the analytic power of an annual time series approach (using the Landsat based detection of trends in disturbance and recovery (LandTrendr) algorithm) by assessing forest cover dynamics for an area in Romania, Eastern Europe. Our results illustrate that trajectory-based time series approaches can successfully be applied in relatively data scarce regions. Annual disturbance patterns allow for improved process understanding, and provide valuable inputs to a range of applications, including resource management, climate modelling or socio-ecological systems understanding, as in the case of Romania.

P. Griffiths (✉)
Geography Department, Humboldt-Universität zu Berlin, Berlin, Germany
e-mail: patrick.griffiths@geo.hu-berlin.de

P. Hostert
Geography Department, Humboldt-Universität zu Berlin, Berlin, Germany

IRI THESys, Humboldt-Universität zu Berlin, Berlin, Germany

C. Kuenzer et al. (eds.), *Remote Sensing Time Series*, Remote Sensing and Digital Image Processing, DOI 10.1007/978-3-319-15967-6_15

15.1 Introduction

Spatially and temporally explicit information of forest ecosystems is important for a broad range of applications in the context of forest management, carbon budget assessments or biodiversity studies (Foley et al. 2005; GLP 2005). Remote sensing has become a central tool for forest management and the monitoring of forest cover dynamics (Wulder and Franklin 2003). Alternative information sources, such as forest inventory data, commonly involve high labor costs, are very time consuming, and costly to obtain. While not all information needs of forest managers can be addressed by remote sensing based analyses, satellite observations can complement or even replace certain types of field based data.

Observations by remote sensing are spatially explicit, repeatable and objective. Thus, they allow assessing changes over time and provide information independent of administrative or political boundaries (Kuemmerle et al. 2006). Moreover, field based data on forest attributes might not be available for a targeted time period or not comparable if data from different management entities are compared. A large variety of remote sensing based monitoring and change detection methods have been developed over the last decades (Coppin and Bauer 1996; Coppin et al. 2004; Lu et al. 2004). These span a broad range of methodological complexity, from simple image differencing over integrated multi-temporal image classification to spectrally grounded change vector analysis. As a forest's spectral appearance commonly follows clearly defined temporal patterns over the course of a season, relatively simple change detection methods may already allow for a sound assessment of forest cover changes between two or few points in time (i.e. multi-temporal images).

Most change detection methods, however, are not well suited for assessing forest cover changes for a multitude of temporal instances, for example, when assessing changes on a yearly basis over a period of 10 years. Multi-temporal classification can facilitate change assessments for multiple points in time but is accompanied by considerable complexity, for example regarding training data requirements. Furthermore, forest canopy changes that spectrally result in more subtle changes, such as drought stress of damages related to infestation by insects, are much more difficult to detect based on the comparison of few individual images. Attributing the causes of forest cover changes however, is not straightforward and often related to high levels of uncertainty. Thus, many remote sensing studies of forest cover adopt the ecological forest disturbance concept, which comprises natural (e.g. fires, wind throw or snow break) as well as anthropogenic (e.g. logging, clearing for development) causes of forest changes.

Time-series approaches provide a methodological framework that potentially allows assessing forest cover changes with higher temporal resolution and that are equally suited to detect and quantify long term trends associated with gradual canopy changes. In the context of remote sensing, time series methods were predominantly used for large field-of-view sensors that provide observations with high temporal frequency. Using such datasets, continental to global assessments of vegetation trends and phenology became feasible (Reed et al. 1994; Friedl et al. 2002; Tucker et al. 2005). For many forest ecosystems however, the coarse spatial resolution of these sensors prohibits the successful detection of forest disturbances caused by forest management or natural process regimes. Moderate resolution imagery (20–100 m pixel sizes) provides a good compromise between the spatial coverage and spatial resolution, while simultaneously providing systematic repeat acquisitions, and is therefore better suited to map forest cover changes in different areas around the globe.

Time series approaches have only in few cases been adapted to moderate resolution imagery (Kennedy et al. 2007). With regard to Landsat imagery, many of the prohibitive restrictions have largely disappeared over the recent past (Wulder et al. 2012). Most important were changes in data policy after which formerly expensive data could be immediately downloaded over the internet. Almost equally important were improvements to the level 1 data quality, specifically the unprecedented accuracy of the geometric and radiometric calibration. These developments have driven considerable innovation into Landsat based change detection science. For example, time series analysis concepts that use fitting approaches rather than simple trend seeking are relatively new to Landsat data analyses, but have recently been published and evaluated (Huang et al. 2010; Kennedy et al. 2010). These annual time series approaches represent each year in the time series through a single observation rather than utilizing all observations made within a given year. This allows assessing forest cover changes at annual intervals and thereby facilitates the quantification of gradual and long term forest change processes.

The following case study intends to demonstrate the utility of annual time series approaches for characterizing forest changes over a time period of more than 25 years using Landsat data (Griffiths et al. 2012). A key motivation for conducting this research were the massive forest ownership changes that occurred in post-socialist Romania following the reintroduced of private property after 1990. Romania's restitution process started in 1991 and was implemented in three individual laws (1991, 2000 and 2005) and is still ongoing today. More than half of the countries forests has been transferred into private ownership, and more than 800,000 new forest owners have emerged (Abrudan et al. 2005; Ioras and Abrudan 2006). In order to better understand to what extent these ownership changes modified forest disturbance regimes, annual disturbance dynamics, post disturbance recovery as well as long term growth was identified from a Landsat time series using the LandTrendR (Landsat based detection of trends in disturbance and recovery) time series segmentation and fitting algorithms (Kennedy et al. 2010).

15.2 Data and Methods

The LandTrendR algorithms provide the means to fit simplified spectral-temporal trajectories to each pixel in a time series of radiometrically corrected and/or normalized index values through the process of temporal segmentation and fitting.

These simplified trajectories minimize residual noise in the time series signal, which might be related to phenological and atmospheric differences between images or illumination differences caused by e.g. topographic shading. The fitted trajectories can finally be categorized based for example on shape, duration and timing. The workflow involves data preprocessing, creation of the annual composites, segmentation and fitting, post processing and finally, mapping and validation.

15.2.1 Study Region

Our analysis focused on one Landsat footprint (path 183, row 023) in central Romania. It is located on the main arc of the Carpathian Mountains, where the north-south aligned eastern Carpathians transition into the east-west oriented southern Carpathians (Fig. 15.1). The area is characterized by mountainous terrain extending up to 1,600 m in the eastern- and over 2,500 m in the southern Carpathians. The intermontane basins and plateaus feature hilly and undulating terrain, with a mix of agriculture, pasture and settlements. Forests cover roughly half of the study region and comprise beech dominated deciduous forests, spruce and fir dominated coniferous forests, as well as mixed forests (accounting for approximately 37 %, 26 % and 37 % of the forest area, respectively).

15.2.2 Landsat Imagery

In order to assemble an annual time series stack, we obtained all precision terrain corrected imagery (L1T) that was acquired during the seasonal window of main photosynthetic activity (June – September) for the years from 1984 to 2010. While a narrower seasonal window would be preferred, a general data scarcity and prevailing (often topography related) cloud coverage hinders the assembly of such a time series in the study region. For years where peak growing season imagery was very cloudy we provided several scenes in order to provide observations for most pixels in the study area. As the United States Geological Survey (USGS) Landsat archive contained no imagery for the years 1996–1999, we additionally acquired three images from the European Landsat archive and co-registered these L1G images to the L1T geometry. Moreover, for years where all the available imagery was strongly cloud contaminated, we additionally incorporated imagery from the east and westward overlapping footprints (path 182 and 184, row 023). This resulted in a total of 52 Landsat Thematic Mapper (TM) and Enhanced Thematic Mapper Plus (ETM+) images (4 x Landsat-4, 31 x Landsat-5, 17 x Landsat-7). No suited imagery was available for 4 years (1985, 1990, 1992 and 1997).

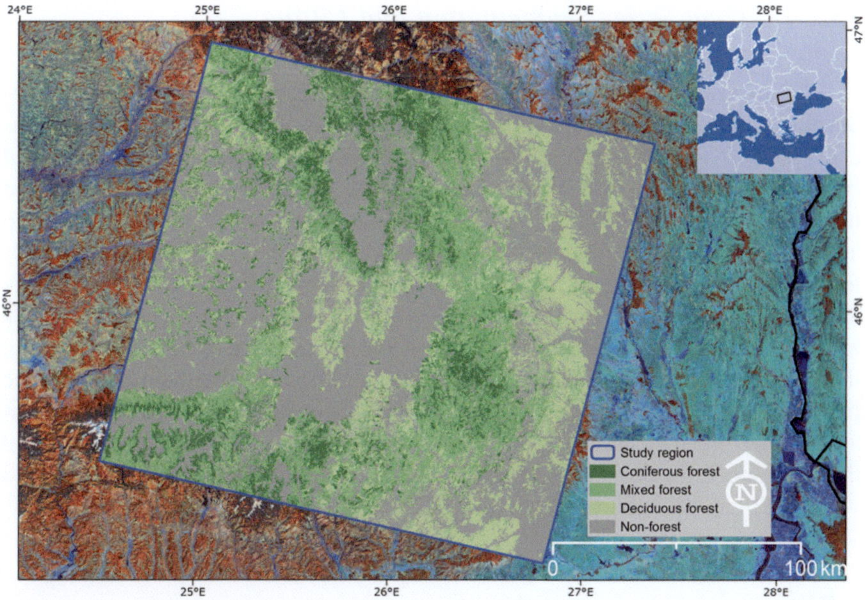

Fig. 15.1 The study region (*blue frame*) showing the distribution of the predominant forest types. The overview map (*top right*) shows the location within Europe. A pixel-based image composite for the 2010 leaf-on period is shown in the background (RGB = 4, 5, 3). For details on the forest classification and the image composite see Griffiths et al. (2014)

15.2.3 Preprocessing

In order to ensure spectral and radiometric comparability across acquisition dates and Landsat sensors, we combined atmospheric correction and radiometric normalization. First, we atmospherically corrected an image from August 27th 1998 using the COST approach (Chavez 1996). All other images were subsequently radiometrically normalized to this corrected image based on the Multivariate Alteration Detection algorithm (Canty and Nielsen 2008). The LandTrendR approach uses a single spectral index to represent the annual time series of observations. We tested different vegetation indices and spectral transformations, and decided for the Tasseled Cap Wetness (TCW) component as this index provided the best compromise between deciduous forests with strong near infrared reflectivity and extremely low reflective coniferous forests. Thus after radiometric normalization, all images were transformed into Tasseled Cap (TC) space (Crist and Cicone 1984). LandTrendR subsequently performs a simple best value compositing for those years that have more than one image, prioritizing acquisitions close to the all year acquisition median day-of-year (DOY), which is 223, equivalent to August 11th. The time series stack of these annual best value composites is in the following referred to as source values and these provide the basis for time series segmentation and fitting.

15.2.4 Time Series Segmentation and Fitting

In order to restrict time series segmentation and fitting to forested pixels only, we developed a forest mask following a supervised classification approach (Griffiths et al. 2012). The resulting mask included pixels in the forest category if a pixel had been forested in any point in time and assigned the non-forest category if a pixel had never been forested over the study period. Subsequently, time series segmentation and fitting is carried out on the time series source values. A detailed description of this process is provided by Kennedy et al. (2010). An essential element is the identification of vertices in the trajectory which can be thought of as turning points in the spectral-temporal development of a pixel. Vertices are identified by locating the largest deviation from a series of sequential regressions between points in the time series. Subsequently, different simplified versions of the trajectory are evaluated and the one with the best (according to the p-value) statistical fit is selected and the fitted spectral values are written into the fitted version of the time series (fitted values). Figure 15.2 illustrates a schematic time series for a single pixel.

15.2.5 Disturbance, Recovery and Growth Mapping

In order to reduce commission errors during the disturbance mapping, we filtered trajectories in the time series according to the relative percent cover change magnitude. We approximated a percent cover estimation for each observation in the time series by linearly scaling the observed TCW values of bare soil (i.e. zero percent cover) with those of dense canopies (i.e. 100 % cover), both of which could be identified in the imagery. Subsequently the relative disturbance magnitude was derived as the difference between pre- and post-disturbance percent cover values relative to the pre-disturbance cover estimate. It should be noted that these percent cover estimates only provide a crude cover estimate that allows for effective reduction of false detections. They do not directly compare to cover estimates derived through more quantitative (e.g. field or modelling) methods.

For each detected disturbance we extracted disturbance onset, relative disturbance magnitude and the duration of the disturbance event. We then extracted all disturbance trajectories (i.e. those that feature a segment with decreasing TCW values) that featured at least 30 % relative percent cover decrease over one year and at least 5 % cover decrease over a 20 year duration, using a sliding scale for years in between. These thresholds are more conservative than in other studies (Kennedy et al. 2010), in order to account for higher noise levels in our time series. Additionally, the pre-disturbance cover estimate had to be at least 10 % to be included in the disturbance mapping. This resulted in a total of 22 disturbance classes (1986–2010, 4 years without data). Finally, we applied a spatial filter that assigned groups of pixels less than 7 of a given disturbance class, to the surrounding class.

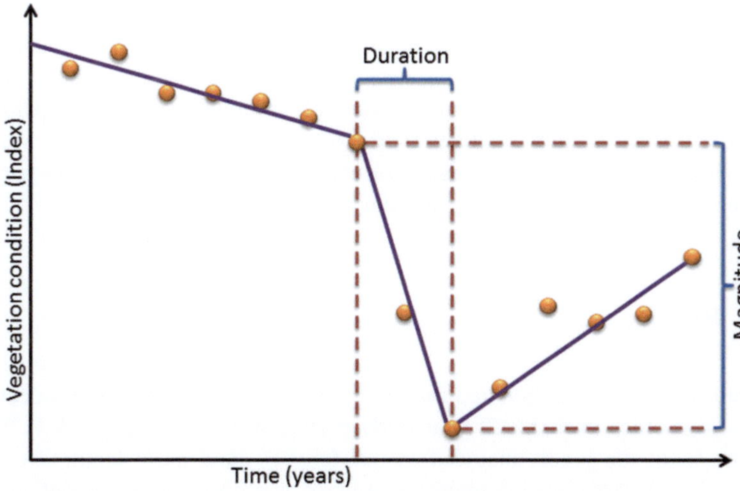

Fig. 15.2 Schematic illustration of an annual pixel time series. The observed values from the imagery are shown as orange points, the fitted time series trajectory is shown as a *purple line*. This example shows an idealized disturbance trajectory where a period of slight loss of vegetative cover is followed by an abrupt decrease (i.e. disturbance), which then again is followed by a recovery segment. Parameters such as disturbance duration and magnitude are illustrated

In order to quantify post-disturbance recovery and long term growth dynamics, we extracted all segments of vegetative increase where the change in percent cover was at least 15 % and that occurred over a 3–26 year period (26 years is the entire time span of the time series). Based on these segments, we mapped areas of long growth as well as areas recovering from previous disturbances. Post disturbance recovery was then assessed on the basis of individual disturbance patches. We assessed post-disturbance recovery state after five years by averaging the percent cover value of all pixels in an individual disturbance patch. Pixel trajectories increasing over at least 10 years or at least 20 years were summarized separately. These areas were not affected by any disturbances within the study period and we categorized two groups based on their absolute percent cover increase: (1) cover increase between 20 and 50 %, and (2) cover increase between 50 and 100 %.

15.2.6 Validation

The validation of the disturbance map was based on an independent point sample that we obtained through stratified random sampling. This yielded 30 points per disturbance class and 300 points for the stable forest and non-forest classes. Samples were interpreted regarding the presence/absence as well as the timing of disturbance events. For this we visualized the time series trajectories along with the image data in a dedicated visualization and interpretation tool (Cohen et al. 2010).

Additionally, high resolution imagery in Google Earth was available as well as field data from several field trips to the region. We constructed a confusion matrix and accounted for the stratified sampling design by calculating an error adjusted area estimates as well as the 95 % confidence intervals around these estimates (Olofsson et al. 2013). Unfortunately, validation of growth and recovery dynamics was not feasible based on the historic Landsat imagery itself and was therefore not pursued. These results therefore only reveal tendencies of vegetation dynamics.

15.3 Results

The time series segmentation and fitting algorithms depicted temporal trajectories of forest dynamics over the 27 year study period overall well (Fig. 15.3). While about 50 % of the pixels included in our forested land mask featured at least 21 cloud free observations, in many cases much less observations (about 12 % of pixels had 17 or less observations) were available due to topographically induced cloud cover or a lack of image acquisitions. Periods of relative stability were identified despite year to year variation in the observed signal (Fig. 15.3(1)). In many cases, intra annual variability in the time series signal was considerable. For example, Fig. 15.3(2) shows a disturbance event followed by recovery on a northward facing slope (aspect 40°). While the fitted time series underestimated the magnitude of the disturbance event, the fitted recovery segment compensated much of the inter-annual variability in the signal. Areas exhibiting long term growth of forest biomass were also successfully fitted in many cases. We provide one illustrative example of a full 27 year growth trajectory in Fig. 15.3(4). Here the inter-annual variability during the early years of the trajectory is considerable, due to variations in the herbaceous dominated land cover. For the period from 1995 to 2010, only six observations were available due to cloud coverage and scan line gaps (related to the scan line corrector failure on Landsat-7 in May 2003), but the estimated fit still depicted the overall trend in biomass increase well. Some cases of erroneous fits related to disturbances that occurred during the earlier years of the time series. These are often caused through too few or even only a single observation before a disturbance detection. While pixels that experienced multiple disturbance events were overall rare (<2 %), Fig. 15.3(5) depicts such a case. Here a first disturbance event occurred in 1995 and was then followed by a second disturbance in 2006. The parameterization of the segmentation and fitting routines resulted in a very much generalized trajectory suggesting a continuous decrease in biomass since 1984 while the trajectory and image chips in Fig. 15.3 clearly suggest two distinct disturbance events.

Regionally, disturbances were overall rare in the eastern part of the study region and predominantly occurred in the eastern Carpathians and to a lesser extent in the southern Carpathians (Fig. 15.4). Our disturbance map indicated that in many areas disturbances were basically absent until the later years within the study period (e.g. Fig. 15.4(5)). Several areas showed extensive disturbance patterns

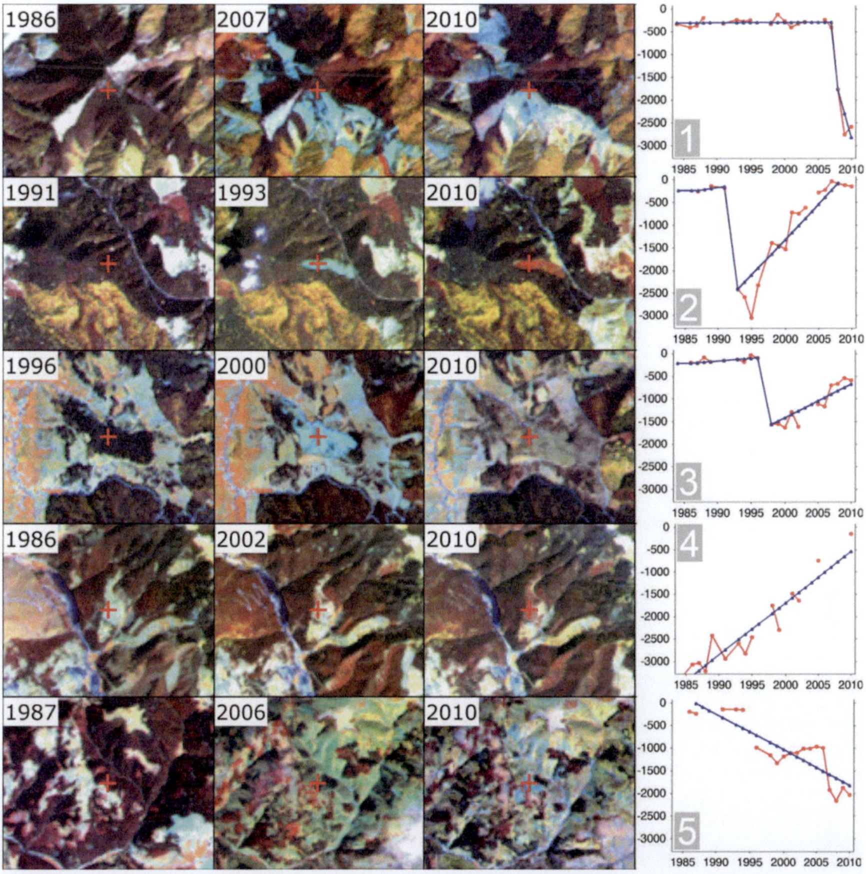

Fig. 15.3 Five examples of the time series data are provided in each row. Each example provides three image chips (RGB = Landsat bands 4/5/3) depicting key points in the development of the illustrated pixels. The trajectory plots (*right column*, x-axis is time in years, y-axis is Tasseled Cap Wetness index) show the time series for the center pixel indicated in the image chips. The observed time series data (source values) are shown in *red* where gaps indicate missing observations. The fitted time series trajectories are shown in *blue*. The scale of all image windows is 1:125,000

(e.g. Fig. 15.4(4) and Fig. 15.3(5)), commonly dominated by disturbances detected in 1996 which likely relate to a major wind throw that occurred in the region in November 1995 (Mihalciuc et al. 1999).

Approximately 50 % of the study region was forested at one point in time during the study period. We detected roughly 70,000 ha of forest area that experienced disturbance over the 27 year time period which relates to 4.2 % of the assessed forest area. Annually we detected between 604 ha in 1991 and 8,596 ha in 1996 as experiencing disturbance (Fig. 15.5). Generally, the yearly area mapped as disturbed declined towards the time of socialist collapse in Romania (i.e. 1989) when compared to the very early years. The annually disturbed area was relatively low

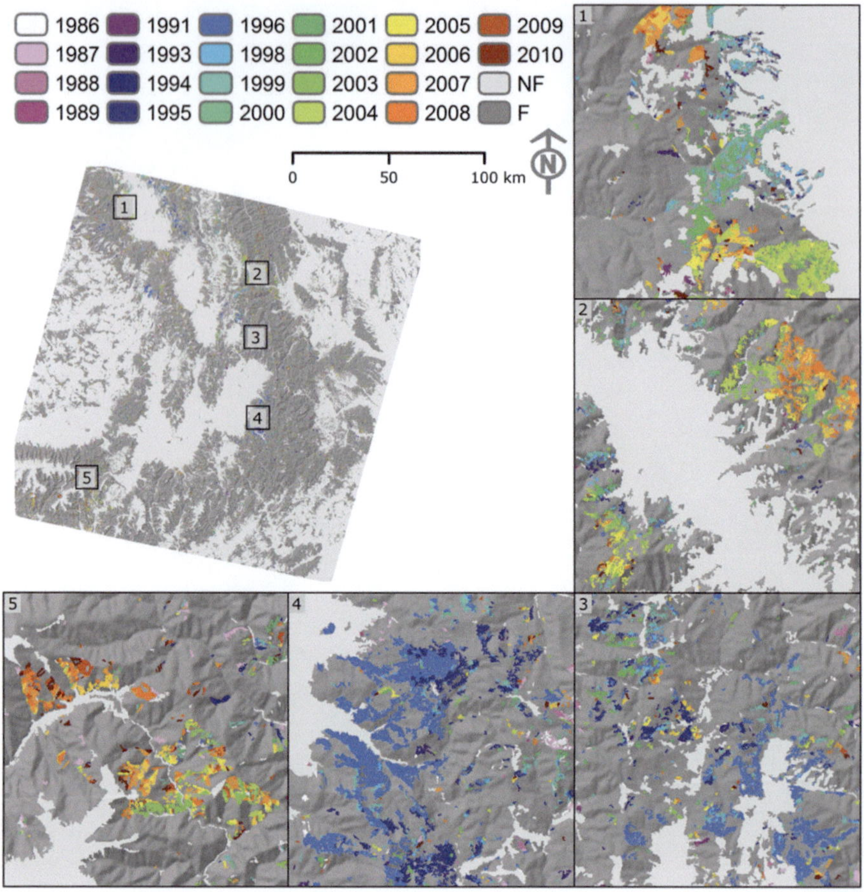

Fig. 15.4 The disturbance map with five detailed close-up. Scale of all close-up windows is 1:300,000, "NF" is the *Non Forest* and "F" is the stable *Forest* class

following the first restitution law in 1991, but then increased up to 1995 and 1996 where we detected more than 5,500 ha and 8,500 ha of disturbed forest area, respectively. Following the second restitution year in 2000, the annually disturbed area was relatively low but increased again towards 2004 when the third law was passed. During the years following the third restitution law in 2004, the annual detected disturbance area was above 4,000 ha with the exception of 2005 and 2009 (Fig. 15.5).

The assessment of post-disturbance recovery after 5 years showed that on average, disturbance patches had recovered 50 % of the relative disturbance magnitude. Exceptionally high recovery rates were observed for 1986 and 1994 (~55 % each), while lower recovery rates were observed for disturbances detected in 1996 (43 % of the relative disturbance magnitude). Areas exhibiting substantial increases in forest cover amounted to a total of 110,000 ha (6.6 % of the assessed

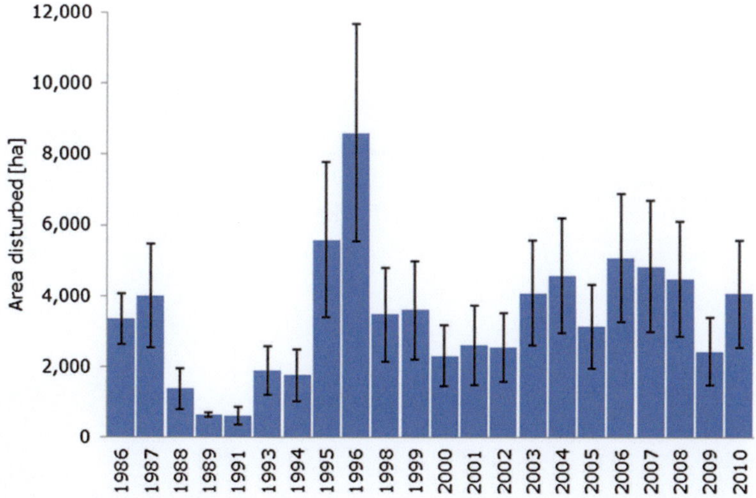

Fig. 15.5 Bar chart summarizing the annual disturbance dynamics (error adjusted area estimates)

forest pixels) for growth trajectories over 10 or more years. Regarding growth trajectories evolving over 20 or more years, about 3.2 % of the assessed forest pixels (52,000 ha) increased between 20 and 50 % relative cover while less than 1 % increased by more than 50 %. Growth trajectories evolving over 10–20 years amounted to 39,000 ha (2.4 %) for 20–50 % cover increase while cases increasing by more than 50 % over the same time period were about 4,500 ha (0.3 %). Figure 15.6 provides a spatial depiction of long term growth dynamics.

The overall accuracy for the disturbance map was 95.7 % after accounting for the stratified random sampling design for the validation point sample (Table 15.1). On the level of individual disturbance classes, the highest commission error was observed for 1986 (26 %) and the highest omission error for 1988 (40 %). The more recent disturbance classes generally featured lower errors when compared to the earliest years in the time series. Stable forest and non-forest areas where validated with omission and commission errors below 5 %.

15.4 Discussion and Conclusion

Time series approaches have recently become an attractive option for Landsat based remote sensing studies and were favored by data quality and policy issues as well as increased computational capabilities. Compared to conventional change detection methods, time series approaches allow the assessment of land changes over long periods and for many points in time and thus offer a cost effective alternative to other mapping methods. The annual time step that algorithms such as LandTrendR or the Vegetation Change Tracker are based on, is ideal for

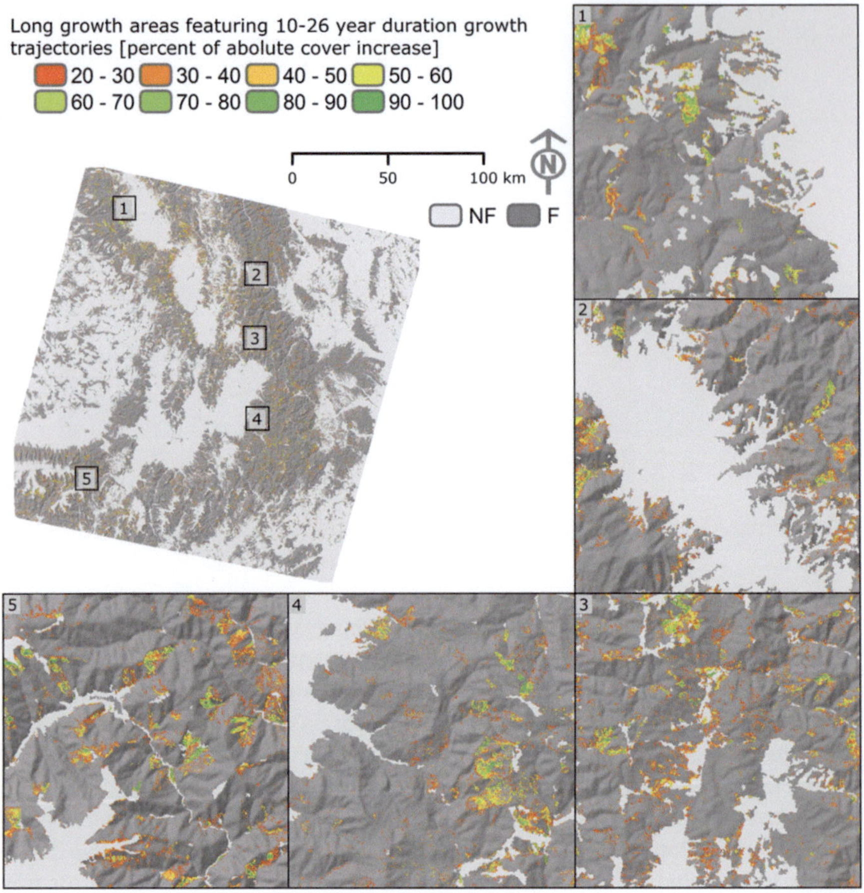

Fig. 15.6 Long term growth map depicting pixels with temporal trajectories suggesting substantial increase in percent cover over 10–26 years. "NF" is the *Non Forest* and "F" is the stable *Forest* class

monitoring forest change dynamics in a temperate setting (Huang et al. 2010; Kennedy et al. 2010).

Our case study was centered on major societal changes in the context of the post socialist transition in Romania. Re-privatization and restitution of forest property profoundly altered the ownership structure in Romania. In order to estimate to what extent these developments affected forest disturbance regimes, disturbance dynamics needed to be quantified on an annual basis. This offers unprecedented detail into year-to-year disturbance patterns which can be related to the key periods such as the individual restitution laws. However, disturbance maps as those developed in this study capture man-made as well as natural forest cover changes and need to be interpreted with caution even though both disturbance regimes are often intertwined (Griffiths et al. 2012). Moreover, other studies suggest that certain forest types

Table 15.1 Summary of the disturbance map validation providing errors of omission and commission for the disturbance classes as well as the stable forest and non-forest class

Year	Omission (%)	Commission (%)
1986	0.00	25.81
1987	12.15	16.13
1988	39.53	16.67
1989	5.32	0.00
1991	0.00	13.33
1993	11.45	10.00
1994	0.00	22.58
1995	6.17	22.58
1996	7.88	3.33
1998	9.83	9.68
1999	1.87	12.90
2000	21.39	10.00
2001	2.58	32.26
2002	4.75 %	10.00
2003	4.39	3.33
2004	8.37	3.23 %
2005	0.00	6.45
2006	6.10	6.25
2007	0.00	6.67
2008	5.86	0.00
2009	0.00	12.90
2010	2.21	12.12
NF	4.73	3.39
F	3.73	4.83
OAC =	95.72	

relate to higher disturbance levels due to higher susceptibility to natural disturbance agents and forest management considerations (Griffiths et al. 2014). Assessing the long term development of forested pixels through time series analyses allows quantifying post disturbance recovery and forest growth patterns. These are, however, spectral measures of recovery and thus not directly comparable to recovery in an ecological sense. Our results therefore indicate tendencies, for example, that the area that experienced disturbances was smaller than the areas undergoing substantial increase in forest cover. This indicates a high productivity of these forests. However, regional disturbance hotspots drastically changed within the study period.

In this case study, we demonstrated the utility of Landsat time series analyses in general and the LandTrendR approach specifically. The performance of the time series fitting algorithms is overall very satisfying, as abrupt changes (e.g. disturbances), as well as gradual long term processes (recovery, long term growth), were well captured. Work remains to be done in order to develop validation techniques and quantify the errors related to regrowth and recovery mapping. In some cases a different parameterization could have further improved the results of the time series fitting (e.g. Fig. 15.3(5)), yet optimizing the parameterization for few individual cases comes at the cost of losing general applicability. Overall, annual time series approaches, such as LandTrendR, represent recent and innovative approaches to land monitoring and change detection. While the availability of moderate resolution imagery will greatly improve in the coming years (Drusch et al. 2012; Roy et al. 2014), this study showed that trajectory-based change detection approaches are also feasible using historic imagery in relatively data scarce regions (Griffiths et al. 2012; Main-Knorn et al. 2013). Reliable disturbance products on an annual basis are highly valuable for a range of applications, e.g. for investigating the effects of forest management regimes or as input to climate models. These approaches hold great potential for improved process understanding through new conceptualization of land change processes in the context of optical remote sensing data analyses (Kennedy et al. 2014).

Acknowledgements This research was funded by the Belgian Science Policy, Research Program for Earth Observation Stereo II, contract SR/00/133, as part of the FOMO project (Remote sensing of the forest transition and its ecosystem impacts in mountain environments). Support is gratefully acknowledged by Humboldt-University Berlin.

References

Abrudan IV, Marinescu V, Ignea G, Codreanu C (2005) Present situation and trends in Romanian forestry. In: Abrudan IV, Schmitthüsen FJ, Herbst P (eds) Legal aspects of European forest sustainable development, Proceedings of the 6th international symposium, Poiana Brasov, Romania, pp 157–171

Canty MJ, Nielsen AA (2008) Automatic radiometric normalization of multitemporal satellite imagery with the iteratively re-weighted MAD transformation. Remote Sens Environ 112:1025–1036

Chavez PS (1996) Image-based atmospheric corrections revisited and improved. Photogramm Eng Remote Sens 62:1025–1036

Cohen WB, Yang Z, Kennedy R (2010) Detecting trends in forest disturbance and recovery using yearly Landsat time series: 2. TimeSync – tools for calibration and validation. Remote Sens Environ 114:2911–2924

Coppin P, Bauer ME (1996) Digital change detection in forest ecosystems with remote sensing imagery. Remote Sens Rev 13:207–234

Coppin P, Jonckheere I, Nackaerts K, Muys B, Lambin E (2004) Digital change detection methods in ecosystem monitoring: a review. Int J Remote Sens 25:1565–1596

Crist EP, Cicone RC (1984) A physically-based transformation of Thematic Mapper data – the TM Tasseled Cap. IEEE Trans Geosci Remote Sens 22:256–263

Drusch M, Del Bello U, Carlier S, Colin O, Fernandez V, Gascon F, Hoersch B, Isola C, Laberinti P, Martimort P, Meygret A, Spoto F, Sy O, Marchese F, Bargellini P (2012) Sentinel-2: ESA's optical high-resolution mission for GMES operational services. Remote Sens Environ 120:25–36

Foley JA, DeFries R, Asner GP, Barford C, Bonan G, Carpenter SR, Chapin FS, Coe MT, Daily GC, Gibbs HK, Helkowski JH, Holloway T, Howard EA, Kucharik CJ, Monfreda C, Patz JA, Prentice IC, Ramankutty N, Snyder PK (2005) Global consequences of land use. Science 309:570–574

Friedl MA, McIver DK, Hodges JCF, Zhang XY, Muchoney D, Strahler AH, Woodcock CE, Gopal S, Schneider A, Cooper A, Baccini A, Gao F, Schaaf C (2002) Global land cover mapping from MODIS: algorithms and early results. Remote Sens Environ 83:287–302

GLP (2005) Science plan and implementation strategy, IGBP report no. 53/IHDP report no. 19. IGBP Secretariat, Stockholm

Griffiths P, Kuemmerle T, Kennedy RE, Abrudan IV, Knorn J, Hostert P (2012) Using annual time-series of Landsat images to assess the effects of forest restitution in post-socialist Romania. Remote Sens Environ 118:199–214

Griffiths P, Kuemmerle T, Baumann M, Radeloff VC, Abrudan IV, Lieskovsky J, Munteanu C, Ostapowicz K, Hostert P (2014) Forest disturbances, forest recovery, and changes in forest types across the Carpathian ecoregion from 1985 to 2010 based on Landsat image composites. Remote Sens Environ 151:72–88

Huang C, Coward SN, Masek JG, Thomas N, Zhu Z, Vogelmann JE (2010) An automated approach for reconstructing recent forest disturbance history using dense Landsat time series stacks. Remote Sens Environ 114:183–198

Ioras F, Abrudan IV (2006) The Romanian forestry sector: privatisation facts. Int For Rev 8:361–367

Kennedy RE, Cohen WB, Schroeder TA (2007) Trajectory-based change detection for automated characterization of forest disturbance dynamics. Remote Sens Environ 110:370–386

Kennedy RE, Yang Z, Cohen WB (2010) Detecting trends in forest disturbance and recovery using yearly Landsat time series: 1. LandTrendr – temporal segmentation algorithms. Remote Sens Environ 114:2897–2910

Kennedy RE, Andrefouet S, Cohen WB, Gomez C, Griffiths P, Hais M, Healey SP, Helmer EH, Hostert P, Lyons MB, Meigs GW, Pflugmacher D, Phinn SR, Powell SL, Scarth P, Sen S, Schroeder TA, Schneider A, Sonnenschein R, Vogelmann JE, Wulder MA, Zhu Z (2014) Bringing an ecological view of change to Landsat-based remote sensing. Front Ecol Environ 12:339–346

Kuemmerle T, Hostert P, Perzanowski K, Radeloff VC (2006) Cross-border comparison of land cover and landscape pattern in Eastern Europe using a hybrid classification technique. Remote Sens Environ 103:449–464

Lu D, Mausel P, Brondizio E, Moran E (2004) Change detection techniques. Int J Remote Sens 25:2365–2407

Main-Knorn M, Cohen WB, Kennedy RE, Grodzki W, Pflugmacher D, Griffiths P, Hostert P (2013) Monitoring coniferous forest biomass change using a Landsat trajectory-based approach. Remote Sens Environ 139:277–290

Mihalciuc V, Simionescu A, Mircioiu L (1999) Sanitary state of coniferous stands calamited by wind and snow on 5./6. November 1995 in Eastern carpathians from Romania. In: Forster B, Knízek M, Grodzki W (eds) Methodology of forest insect and disease survey in Central Europe. Second workshop of the IUFRO WP 7.03.10. Swiss Federal Institute for Forest, Snow and Landscape Research (WSL), Sion-Châteauneuf, pp 238–241

Olofsson P, Foody GM, Stehman SV, Woodcock C (2013) Making better use of accuracy data in land change studies: estimating accuracy and area and quantifying uncertainty using stratified estimation. Remote Sens Environ 129:122–131

Reed BC, Brown JF, Vanderzee D, Loveland TR, Merchant JW, Ohlen DO (1994) Measuring phenological variability from satellite imagery. J Veg Sci 5:703–714

Roy DP, Wulder MA, Loveland TR, Woodcock CE, Allen RG, Anderson MC, Helder D, Irons JR, Johnson DM, Kennedy R, Scambos TA, Schaaf CB, Schott JR, Sheng Y, Vermote EF, Belward AS, Bindschadler R, Cohen WB, Gao F, Hipple JD, Hostert P, Huntington J, Justice CO, Kilic A, Kovalskyy V, Lee ZP, Lymburner L, Masek JG, McCorkel J, Shuai Y, Trezza R, Vogelmann J, Wynne RH, Zhu Z (2014) Landsat-8: science and product vision for terrestrial global change research. Remote Sens Environ 145:154–172

Tucker CJ, Pinzon JE, Brown ME, Slayback DA, Pak EW, Mahoney R, Vermote EF, El Saleous N (2005) An extended AVHRR 8-km NDVI dataset compatible with MODIS and SPOT vegetation NDVI data. Int J Remote Sens 26:4485–4498

Wulder MA, Franklin SE (2003) Remote sensing of forest environments: concepts and case studies. Kluwer, Boston

Wulder MA, Masek JG, Cohen WB, Loveland TR, Woodcock CE (2012) Opening the archive: how free data has enabled the science and monitoring promise of Landsat. Remote Sens Environ 122:2–10

Chapter 16
Radar Time Series for Land Cover and Forest Mapping

Christiane Schmullius, Christian Thiel, Carsten Pathe, and Maurizio Santoro

Abstract Radar time series are powerful means to improve retrieval algorithms about land surface characteristics in the following ways: (i) as information for identification of land surface conditions, (ii) as source of multivariate statistics for mapping methodologies, (iii) to select the right scene(s) for dedicated retrieval procedures, or (iv) to train model parameters in physical retrievals. Albeit radar data from air- and spaceborne platforms have been investigated since 40 years, operational applications are limited – partly due to the non-intuitive handling of complex microwave backscatter signals, and partly due to restricted geometric and temporal resolutions or frequency and polarization constraints. This chapter gives an overview of 20 years of pilot projects performed by the authors and their collaborators with the goal of large-area radar data exploration. All studies lead to innovative pre-operational applications, several with promising discoveries that can now be realized with a new and expanding fleet of radar satellites. Four case studies for land cover, forest mapping, forest cover change and savannah monitoring conclude this chapter.

16.1 Introduction

Radar remote sensing applications for land surfaces have been investigated extensively since the 1980s, when NASA deployed their airborne system AIRSAR over various test sites to prepare and accompany a series of spaceborne Shuttle Imaging Radar (SIR) experiments (1981: SIR-A, 1984: SIR-B, April and October 1994: SIR-C/X-SAR) jointly with the German Aerospace Centre (DLR) and the Italian

C. Schmullius (✉) • C. Thiel • C. Pathe
Department for Earth Observation, Institute of Geography, Faculty for Chemistry and Geosciences, Friedrich-Schiller University, Jena, Germany
e-mail: c.schmullius@uni-jena.de

M. Santoro
GAMMA Remote Sensing Research and Consulting AG, Gümligen, Switzerland

C. Kuenzer et al. (eds.), *Remote Sensing Time Series*, Remote Sensing and Digital Image Processing, DOI 10.1007/978-3-319-15967-6_16

Space Agency (ASI). These Multi-sensor Airborne Campaigns (MACs) and SIR-missions represent the first proof-of-concept of retrievals which have been developed with ground scatterometer systems in the decades before. A summary by Ulaby et al. 1986 of these in situ backscatter time series under various environmental conditions with explanations of the underlying physics still serves as a major reference for applications with spaceborne data. This transferability of concepts is based on the unique radar calibration procedure which allows comparisons over time and between sensors as long as the same wavelength and polarisation is being used. The fundamental relation between the characteristics of the radar system, the transmitted signal, the target, and the received signal is called the radar equation. Since radar systems actively emit microwave radiation, the illuminating geometry and technical characteristics are known and can thus be accounted for in the retrieval algorithm for a specific surface phenomenon (such as phenological state, biomass, or soil moisture).

Albeit the information content in radar images corresponds to well-defined surface parameters (i.e. volumetric moisture, surface roughness and structure of the scattering medium), this remote sensing technique has only a limited user community due to the complexity of the necessary pre-processing steps. These procedures originate from the principle of using backscattered intensities of actively emitted microwave pulses, thus causing speckle, corner reflections or geometric obstacles such as layover, foreshortening and rectangular pixels – these are all non-intuitive phenomena for the optically trained remote sensing expert. Hence, radar image understanding needs specific training, especially for non-engineers. This gap has for example been tackled in the Radar remote sensing education initiative (SAR-EDU) of the German Space Administration (DLR Raumfahrtmanagement) which provides educational material to interested university teachers, federal agency employees and PhD students through a dedicated DLR web-portal since December 2014: https://saredu.dlr.de. The reader is referred to this portal for background material, since an introduction to radar theory would go beyond the scope of this section.

The following paragraphs contain an overview of three concepts of how to use radar time series as tools for land cover and forestry applications: (1) information about what + where + state of land surface characteristics, (2) source of multivariate statistics as mapping characteristics, (3) choosing the right scene for the product retrieval. Four examples of operational applications for land cover, forest cover and -change and savannah monitoring are given in the third section.

16.2 Radar Time Series as a Tool for Land Monitoring

Land surface properties unfold their characteristics in radar data sets – as for optical images – when exploiting temporal information. Time series are of special importance for radar applications because of the limited spectral space (usually only one wavelength and two polarisations) and because of the radar-specific suitability for multi-temporal exploitation due to the sensor's high radiometric accuracies. Additionally, due to the capability of radar signals to penetrate the atmosphere even in the presence of clouds and rain, radar time series allow to focus on the changing surface constituents over months, years, and even decades as will be demonstrated.

16.2.1 Time Series for Information Retrieval About Land Cover State

Operational land cover mapping with spaceborne radar data became possible with the launch of the European satellite ERS-1 in 1991. For the first time, SAR images became available globally and repeatedly. A first summary publication by the European Space Agency (ESA) (Wooding et al. 1995) showed the remarkable behaviour of C-band VV-polarized time series for crop mapping (Fig. 16.1). Backscatter and attenuation features, already described in Ulaby et al. 1986 (ibid. Fig. 17.19.32, p. 1566), were now observable for large areas and thus their spatial manifestation was proven.

Global C-Band time series were successfully continued through the launches of ESA's follow-up satellites ERS-2 (European Remote Sensing) in 1995 and ASAR (Advanced Synthetic Aperture Radar) on ENVISAT (Environmental Satellite) in 2002. Providing over a decade of C-band data allowed the development of automated routines for land cover mapping. The German ENVILAND-1 and -2 projects, supported by DLR's Space Administration from 2004 to 2012, investigated the potential of radar-optical fusion techniques. The goal was the development of an automated retrieval of land cover and biophysical characteristics through exploitation of both sensor types. The proposed classification procedure consisted of three main stages. The first processing step comprises the segmentation of the optical Earth observation (EO) data. Next, potential training sites are being selected automatically by applying a decision tree with flexible, scene-specific thresholds that are calculated based on expert knowledge and histogram analyses. Finally, as the third step, training samples are being used as input to a supervised classification. The overall accuracy of the final land cover map increased by 6 % when C-band texture and backscatter information was included during classification. The results are illustrated in Fig. 16.2. A significant improvement was achieved for the classes Urban and Grassland as well as for Forest. The absolute number of crop classes could be increased (Riedel et al. 2008).

Radar time series not only support identification of phenological states and improve classification accuracies; they can also be used innovatively in cloud removal. Eckardt et al. (2013) reconstructed cloud-contaminated pixels in Landsat images using the Synthetic Aperture Radar (SAR) data. Cloud contamination was simulated with masks of varying size to systematically investigate the developed technique. The authors used three radar images during 1 month of the growing season to establish a small temporal signature. Each image, though, originated from

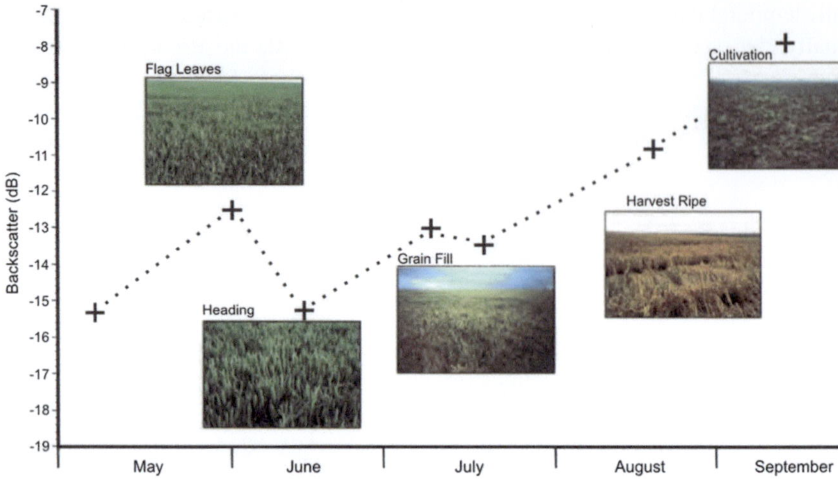

Fig. 16.1 Multi-temporal signature of ERS-1 C-band backscatter signals (Wooding et al. 1995, modified): the decrease between end of May and mid-June is characteristic for the increasing attenuation of VV-polarized radar pulses during the lengthening of cereal stalks or, similarly, grass canopies. With ripening and yellowing, attenuation is minimized and backscatter increases again. This process is especially observable at C-band because of the correspondence in size between the wavelength and geometry of the scattering objects

a different radar satellite (ERS-2, TerraSAR-X and ALOS-1 (Advanced Land Observing Satellite)). They developed a Closest Feature Vector (CFV) based on the assumption of a physical relation between reflectances in the VIS/NIR (Visible/near-infrared) part of the electromagnetic spectrum, the temporal microwave back-scatter signature, and the respective plant conditions at the acquisition dates. SAR data thus guided the selection of the needed reflectances as an integral part of the processing procedure to fill cloudy image gaps. Figure 16.3 illustrates the information content of radar false-colour composites compared to a Landsat VIS-band colour composite.

16.2.2 *Time Series as Source for Statistical Land Surface Indicators*

Comparable to coarse resolution optical image acquisitions (e.g. Medium Resolution Imaging Spectrometer (MERIS), Satellite Pour l'Observation de la Terre-Vegetation (SPOT-VGT), National Oceanic and Atmospheric Administration (NOAA) Advanced Very High Resolution Radiometer (AVHRR)), new ScanSAR modes became available having a reduced geometric resolution but an improved temporal repetition rate. The first ScanSAR-mode from a satellite was provided from the Canadian RADARSAT-1 launched in 1995, followed by the ENVISAT ASAR system in 2002. These new types of data sets enabled continental, wall-to-

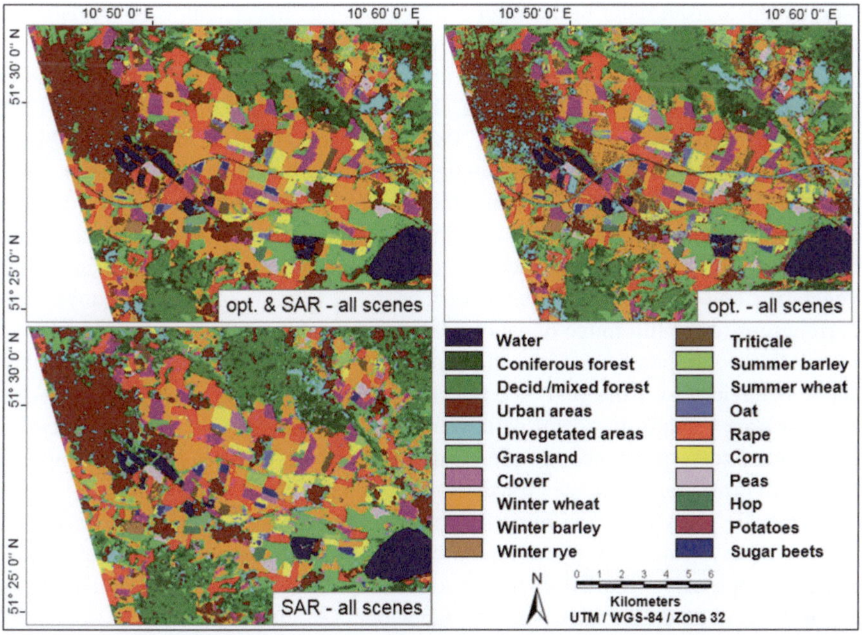

Fig. 16.2 Land cover maps from multi-temporal Landsat, ASAR and ERS-2 scenes from 2005, test site Nordhausen/Thuringia, Germany. The achieved overall accuracies are: optical & SAR: 83,7 %; optical only: 77,9 %; SAR only: 80,2 % (© Uni Jena, 2007). Only two Landsat scenes from 21. April and 10. July were available due to typical cloudy weather conditions during the growing season, but nine SAR images could be used

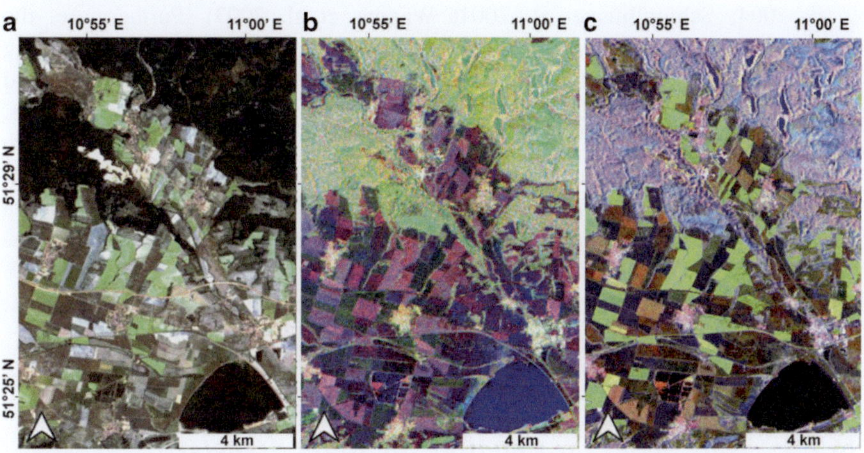

Fig 16.3 Comparison of colour composites to illustrate the image information content: (**a**) Landsat TM5 B3|B2|B1; (**b**) X-HH|L-HH|C-VV; (**c**) X-HH|X-HV|L-HH (Eckardt et al. 2013). The radar scenes are being used as an additional source of spectral information to overcome the problem of missing data due to cloud cover

wall products from radar backscatter retrievals using "hyper-temporal" statistics, a term introduced for radar applications by Schmullius and Santoro in 2007.

One of the first hyper-temporal retrievals for land cover applications had been performed with a non-imaging system, the wind scatterometer on-board ERS-1 for northern Eurasia (Schmullius 1997). Two indices had been introduced: the Slope Index (representing the incidence angle dependence of daily scatterometer measurements collected over 3 years), and the Radar Backscatter Index (RBSI) (defined as the ratio of the mean backscatter from large incidence angles over the absolute Slope Index). The RBSI is a measure of the amount of volume scattering from vegetation and showed – despite the very coarse resolution of 50 km – convincing correlations to the full range of canopy densities (best $r = 0.93$ for July). The RBSI correlation coefficients were investigated month by month and showed a relatively stable behaviour over the year (see Fig. 16.4). The dense time series revealed – for the first time – spatio-temporal pattern of freeze/thaw-processes, and secondly, an unexpected correspondence between 5-day interval backscatter time series anomalies with crop yield estimates for the former Soviet wheat belt.

About 10 years later, the hyper-temporal approach was extended to ScanSAR radar imagery: stacks of hyper-temporal ASAR Wide-Swath data led to a new retrieval approach – the BIOMASAR algorithm (Santoro et al. 2011). The algorithm is based on the famous Water Cloud Model (Attema and Ulaby 1978) and its extension to interferometry (Askne et al. 1997; Santoro et al. 2002). Collecting a minimum of 60 individual backscatter acquisitions per location, the typical radar noise is strongly reduced and a systematic sensitivity with even high forest densities and thus forest growing stock volume (GSV) was found. This finding was of special interest because C-band backscatter and coherence data were regarded to saturate at low biomass levels (Balzter et al. 2002; Balzter and Schmullius 2001; Luckmann et al. 2004; Schmullius et al. 2001; Wagner et al. 2003). Furthermore, the BIOMASAR algorithm does not require in-situ reference data for model training and completely relies on statistical estimates from backscatter data of unvegetated areas versus dense forest plots. These can be found by using other canopy products, e.g. MODIS Vegetation Continuous Fields (VCF). Validation showed that the approach is performing very well and is comparable to conventional approaches involving in-situ data for model training. No signal saturation is observed up to a GSV level of 300 m^3/ha and the relative root mean square error (RMSE) was between 34.2 and 48.1 % at a 1 km pixel size and consistently between 20 and 30 % at an 0.5° resolution as used by global vegetation models (Santoro et al. 2011, 2013a).

16.2.3 Time Series for Choosing the Right Scene(s)

Radar time series can also be used to select images of the most appropriate day-of-year, season or temporal combination for a specific application. For example, ASAR C-band time series were analysed as part of the ESA GMES (Global

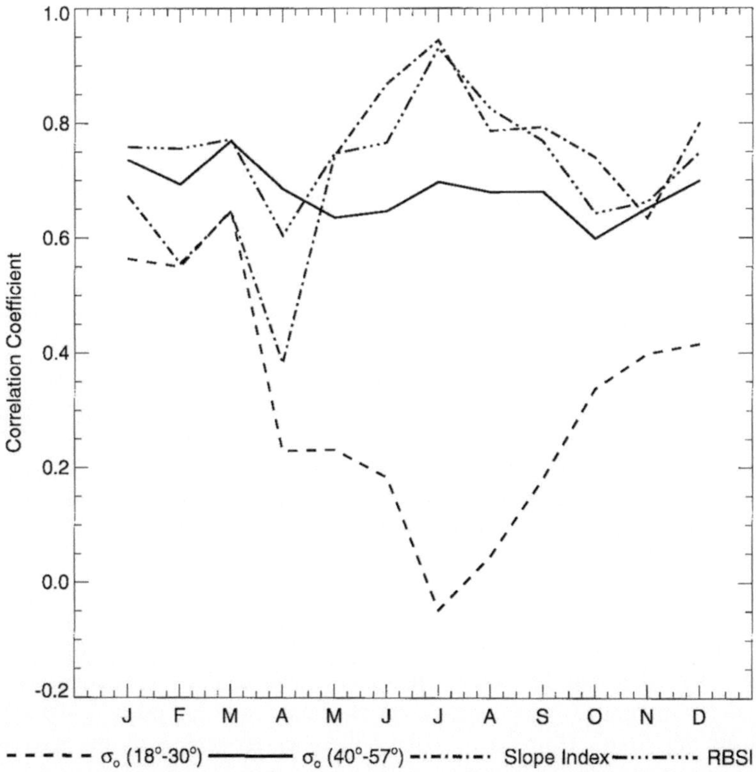

Fig. 16.4 Time series of correlation coefficients for four ERS-1 wind scatterometer parameters with canopy densities at 20 meteorological stations in Siberia (Schmullius 1997): normalised backscatter intensities at steep and large incidence angles, the slope of the incidence angle dependencies per month, and the Radar Backscatter Index (explained in the text). Values are monthly, three year averages (1992–1994). Canopy densities range from 5 to 90 %, the RBSI has a dynamic range from 4 to 12

Monitoring for Environment and Security) Service Element (GSE) Project "Forest Monitoring" with the unexpected result that April-scenes outperformed any other month to map forest disturbances in the boreal zone due to the specific scattering processes during the thawing period (Thiel et al. 2007, 2008). Ackermann (2015) investigated a total of 222 radar satellite scenes from TerraSAR-X, Cosmo-SkyMed (Constellation of small Satellites for Mediterranean basin Observation) and PALSAR (Phased Array type L-band Synthetic Aperture Radar) of a Thuringian forest site with an extensive set of ground data to quantify the impact of forest structure and temporal influences. Figure 16.5 shows one of the evaluated X-band time series, which can be summarized as temporally stable except major precipitation events leading to an increase. The time series also reveal an unexpected stability of the difference between spruce and beech forests – regardless of the chosen co-polarization HH or VV- even in winter times under leafless conditions.

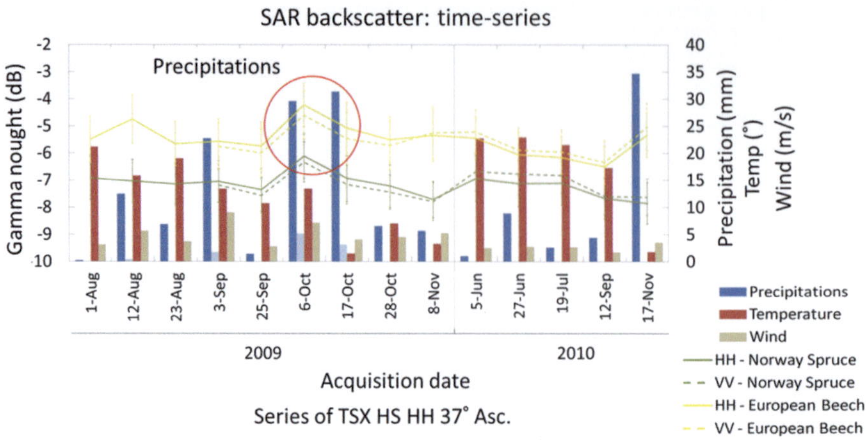

Fig. 16.5 TerraSAR-X backscatter time series over Thuringian forest test site. The signal stability demonstrates the all-year capabilities of X-band data for species discrimination between spruce needleleaf and beech broadleaf canopies (Ackermann et al. 2010). Major precipitation events increase overall backscatter (*red circle*), but separability is maintained

L-band backscatter intensities are strongly impacted by meteorological influences on the land surface conditions due to its much longer wavelength and thus deeper penetration into the canopy and, partly, into the upper soil layer. Moisture changes and particularly freezing events alter the scattering process significantly (compare Fig. 16.6). However, L-band coherence (coherence is the measure of phase correlation between two radar acquisitions) had revealed a large sensitivity and dynamic range to forest biomass (see Fig. 16.7). Coherence from winter showed a clear correlation with forest GSV. For summer scenes, the spread in the values was too large to give reliable results (Eriksson et al. 2003). Therefore, Eriksson et al. (2005) evaluated coherence from eight 44-day image pairs acquired with the Japanese Earth Resources Satellite 1 (JERS-1) during frozen winter conditions over several test sites in Siberia. A simple empirical exponential model was used for the retrieval. The results showed that under frozen winter conditions L-band repeat-pass coherence is a useful data source for stem volume retrieval, although accuracy decreases for higher stem volumes. The lowest retrieval error was 59 m^3/ha and with one exception the relative RMSE stayed within the range 42–76 %.

With increasing availability of L-band coherence time series from JERS-1 (1992–1998) and PALSAR on-board ALOS-1 (2006–2011) and -2 (launch 2014) a systematic investigation of the signal stability and influencing factors became feasible to support selection of best scenes for forest retrievals. Thiel and Schmullius extensively published on the retrieval performance under varying environmental conditions, which is described in the following illustrations:

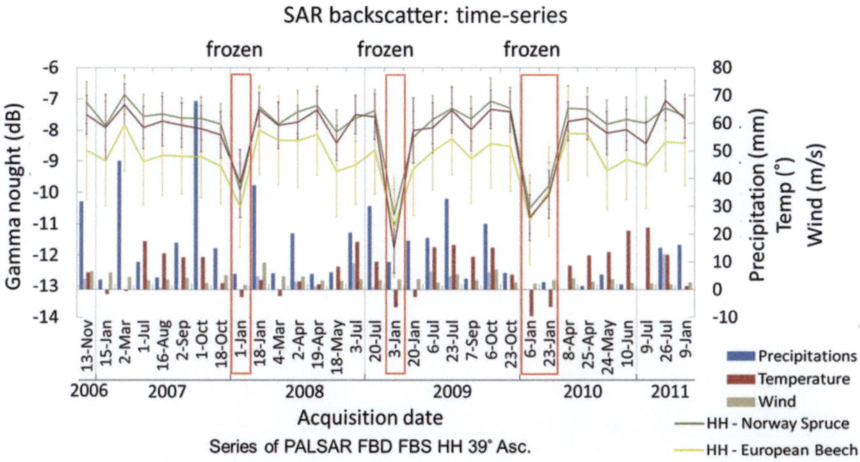

Fig. 16.6 PALSAR L-band time series from Thuringian forest test site exhibit very strong dependency on frozen conditions due to attenuation (Ackermann 2015)

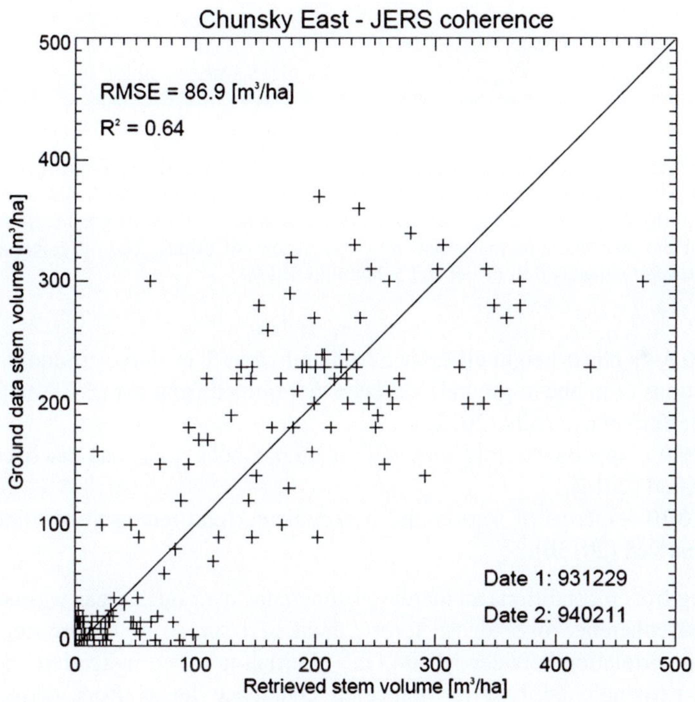

Fig. 16.7 Stem volume retrieval for the Siberian forest test site Chunsky East, Krasnoyarsk Kray. This result belongs to the first large-area application of an inversion model applied to L-band winter coherence values over a 1 mio km^2 forest area between Lake Baikal and the river Yenisei (Eriksson et al. 2005). This region represents commercially valuable forest stands that experience increased logging activities (compare Sect. 16.3.2, ZAPÁS Study)

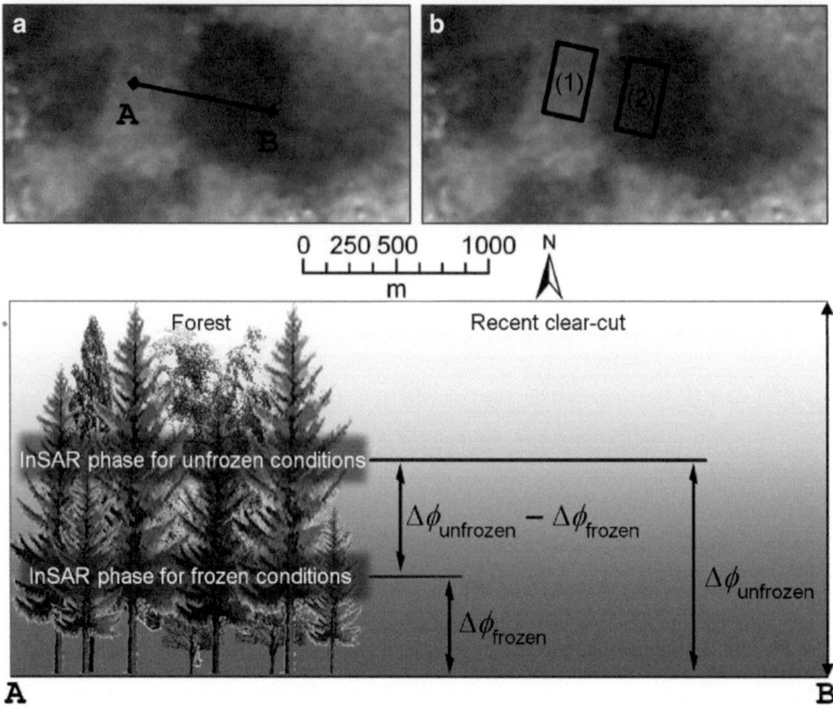

Fig. 16.8 PALSAR interferometric phase image from Chunsky test site in Siberia (57°49′12″ N, 97°2′48″ E). Bright areas are forest-covered, dark areas feature clear-cuts. (**a**) *A*–*B* defines the used transect and (**b**) shows the selected areas for phase calculation. The *bottom* graphic illustrates the location of the phase heights under unfrozen and frozen conditions. The difference in height amounts to approximately 4 m (Thiel and Schmullius 2013a)

Figure 16.8 – a phase height difference from unfrozen (8 m above ground) to frozen conditions (4 m above ground) has been determined from over 370 samples and 36 interferometric pairs (2013a),

Figure 16.9 – species-specific seasonal coherence behaviour was discovered and quantified (2014),

Figure 16.10 – effects of season and perpendicular and temporal baselines were summarized (2013b).

During frozen conditions an increased coherence over open areas was observed, decreased coherence over dense forest, decreased spread of coherence, and an improved correlation between |γ| and GSV. Furthermore, no indication was found that the perpendicular baseline impacted coherence levels from dense forest, whereas a difference was observed at unfrozen conditions. Consequently, at frozen conditions temporal decorrelation is the major source for diminishing coherence values of up to 0.3 points. Hence, |γ| acquired under frozen conditions has good and consistent potential for GSV mapping. The relationship between GSV and |γ| was

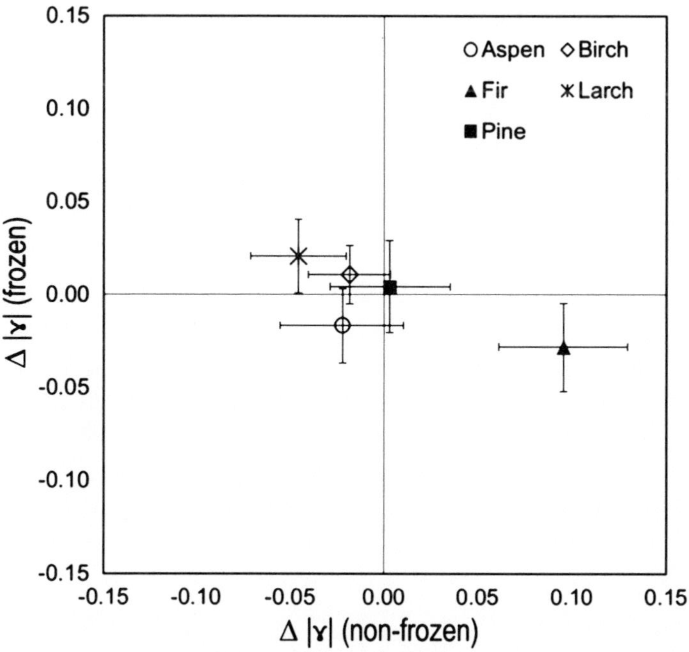

Fig. 16.9 Deviation Δ of species-specific coherences γ from average coherence values for dense forests under frozen and non-frozen conditions for all sites (11 forest enterprises with a total of 12,243 stands covering 3,097 km^2). Frozen conditions exhibit species-independent stable coherence values, whereas intra-species variance and species-dependency increases for non-frozen states (Thiel and Schmullius 2014)

found to have an average coefficient of determination R^2 of 0.6. Saturation occurs at about 250 m^3/ha.

16.3 Case Studies

The results presented in Sect. 16.2 refer to discoveries and manifestations that have already proven their operational character for land or forest cover monitoring for large-area mapping – either through the sheer size of the investigated region or the extremely large number of temporal data stacks. In this section, we present state-of-the-art case studies which have not been tested in space or time for their operationality, but that are ready for a proof-of-concept.

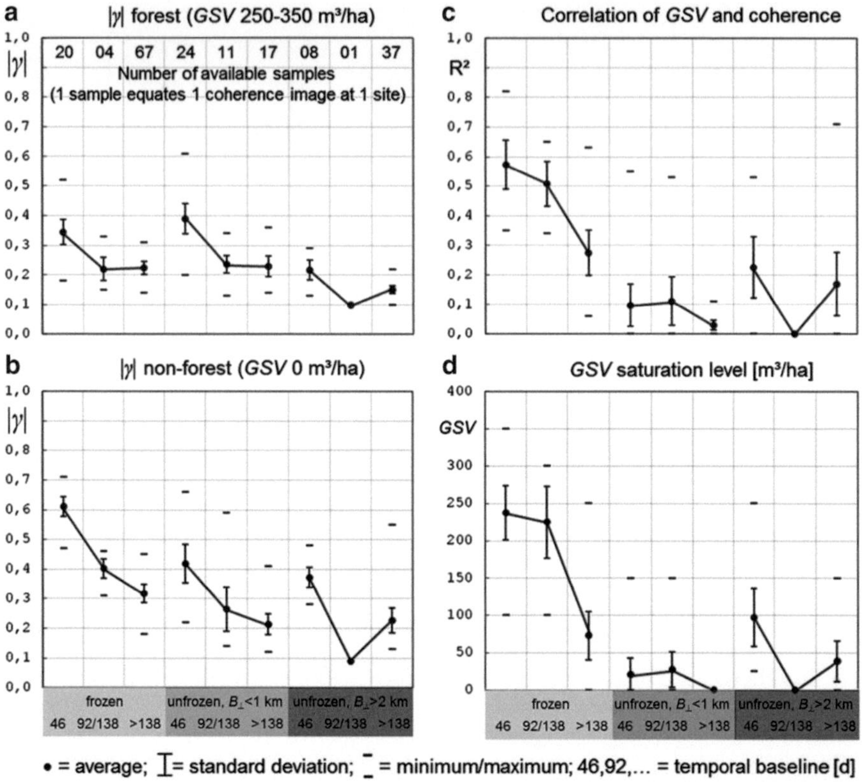

Fig. 16.10 Summary of coherence characteristics based on 300 interferograms from ALOS-1 PALSAR data and 12,243 forest stands in central Siberia (Thiel and Schmullius 2013b): (**a**) |γ| for forest, (**b**) |γ| for non-forest, (**c**) correlation coefficient for GSV and |γ|, and (**d**) saturation level for GSV. All graphs are separated into three columns for frozen and six columns for unfrozen conditions; unfrozen conditions are further divided into perpendicular baselines <1 km and >2 km. Each of the three groups has one column representing a temporal Baseline of 46 days, one for either 92 or 138 days, and one for more than 138 days. See the *first row* in (**a**) for the number of available samples for each of the nine columns

16.3.1 Land Cover Classification Using Multi-temporal C-Band Data

Sentinel-1, the new European radar satellite launched on 3rd April 2014, represents the start of ESA's third decade of continuous C-band sensors in space: ERS-1 and-2 since 1991 and 1995 respectively, and ENVISAT ASAR from 2002 to 2012. Compared to its predecessors, Sentinel-1 provides much shorter revisit times of only 12 days and an increased geometric resolution of 10 m in its Interferometric Wide-Swath Mode. These improved specifications represent an increased potential for operational land cover products. Two ESA studies which specifically focussed on the benefits of radar time series for land cover mapping to prepare the operational use of Sentinel-1 are presented in this section: AMOC-I (Acoustic Monitoring of the Ocean Climate) (Cartus et al. 2008), and AMOC-II (Thiel 2010). Mapping approaches based on radar data need to consider the issue of speckle respectively radiometric accuracy. For mono-temporal intensity images, speckle reduction can only be achieved by using filters that reconstruct the local radar cross section by means of spatial averaging causing a loss of geometric resolution. For time series, Quegan and Yu (2001) suggested a multi-temporal filtering approach, eventually leading to multi-temporal statistics. For the classification of five basic (Level 1) land cover classes, i.e. Water, Forest, Settlement, Grassland and Agriculture, Cartus et al. (2008) analysed four multi-temporal metrics characterising the temporal variation of SAR backscatter:

- minimum / maximum / mean backscatter for each pixel in all images,
- mean annual variation (MVA).

MVA can be calculated as follows (Quegan et al. 2000):

$$mva = 10 \cdot \log\left[\frac{2}{N(N-1)}\sum_{i=1}^{N-1}\sum_{j>i} R_{ji}\right] \quad R_{ij} = \max\left({}^{I_i}/_{I_j}, {}^{I_j}/_{I_i}\right) \tag{16.1}$$

where N represents the number of images and R the normalized ratio of intensities. When calculating the MVA for the speckle filtered VV, HH and HV polarised data, a clear difference between agricultural areas and the land cover classes forest, settlement and grassland was found. This difference was most pronounced for the HV data (Fig. 16.11).

Figure 16.12 shows histograms of the four multi-temporal metrics. The MVA generally reflects the high temporal stability of backscatter over forest, settlements and grassland. The best contrast between these classes and agricultural areas is found using the HV polarization. Forest has an MVA <1 dB in the co-polarised and <2 dB in the cross-polarised images. Settlements show very stable backscatter, but for some dense built-up areas the MVA is high because of different ascending and descending viewing directions (Henderson and Lewis 1998: p. 741). The very low HV backscatter allows a simple discrimination of water from other classes. Differentiation of forest and grassland is not possible using the MVA, but using the annual mean, maximum and minimum backscatter in all three polarisations. Differentiation of forest and settlement seems not to be possible based on multi-temporal metrics. Thus, textural measures are needed.

The analysis of the multi-temporal statistics resulted in a procedure, which classifies C-band data reliably into five Level-1 land cover classes by innovatively using multi-temporal metrics instead of using only backscatter intensities (see Fig. 16.13). Thinning studies about the amount of input data showed that the requested overall accuracy of 85 % and individual class accuracies of at least 70 % can be reached when a minimum of four C-band acquisitions during the growing season are available with HH/HV- or VV/VH-polarisations (Thiel et al. 2009c). ESA's Sentinel-1 satellite now consistently provides the required

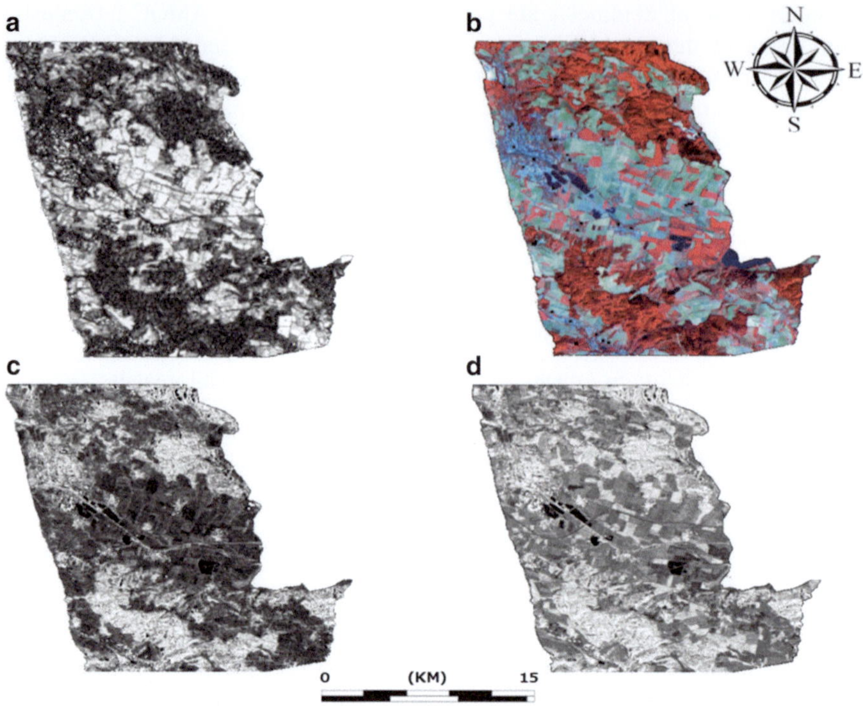

Fig. 16.11 Multi-temporal metrics of 14 ASAR AP HV polarimetric intensity images with (**a**) MVA (explanation in text), (**c**) annual minimum σ^0, (**d**) annual mean σ^0. (**b**) shows for comparison a Landsat ETM+ NIR-Red-Blue RGB-colour composite acquired Sep. 4th 1999 (Cartus et al. 2008). The geographical area is the same as in Figs. 16.2 and 16.3 with UL coordinates 53°33′N/10°44′E, and LR 51°22′N/11°02′E

dual polarisation C-band measurements with short revisit times. These data sets could be used stand-alone for operational land cover mapping as shown in Fig. 16.14.

The results presented in Sect. 16.2 refer to discoveries and manifestations that have already proven their operational character for land or forest cover monitoring for large-area mapping – either through the sheer size of the investigated region or the extremely large number of temporal data stacks. In this section, we present state-of-the-art case studies which have not been tested in space or time for their operationality, but that are ready for a proof-of-concept.

16.3.2 Forest Mapping Using Radar Time Series

Until multi-temporal C-band backscatter intensity and coherence acquisitions became available, forest mapping capabilities were assigned to L-band data or

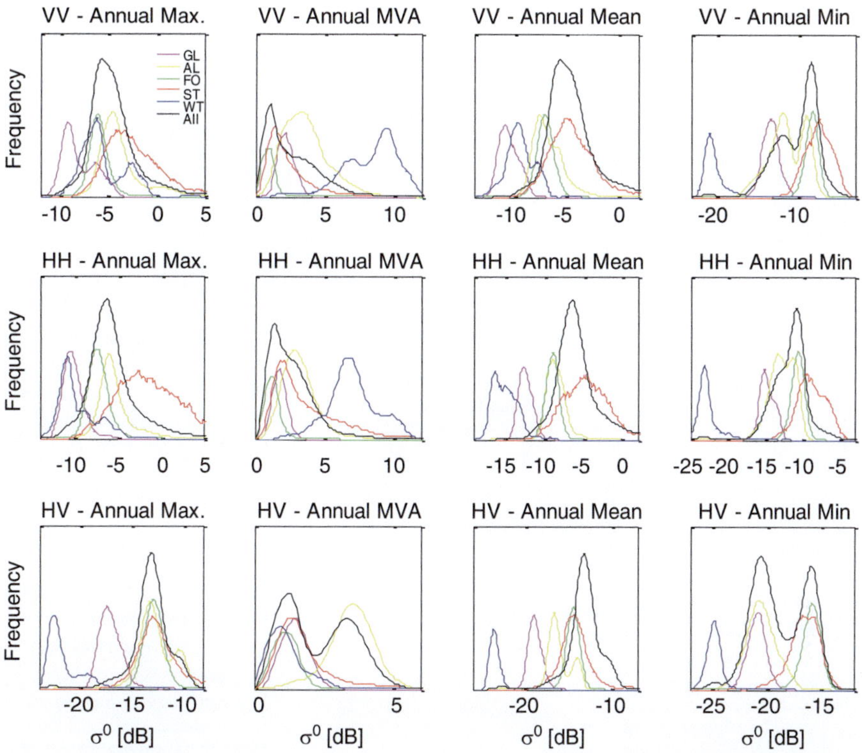

Fig. 16.12 Histograms of multi-temporal metrics for land cover classes Grassland 'GL' (magenta), Agriculture 'AL' (*yellow*), Forest 'FO' (*green*), Settlement 'ST' (*red*), and Water 'WT' (*blue*) (Cartus et al. 2008)

even longer wavelengths (e.g. Dobson et al. 1992). Therefore, interpretation of 1-day repeat-pass "Tandem" coherence images from ESA's ERS-1 and -2 satellites in the late 1990s showed surprising results with stable correlations to forest density – a feature which was then extensively applied for the first 1 million km^2 radar-retrieved biomass map of Siberian forests (Schmullius and Rosenqvist 1997b; Schmullius et al. 2001). A further discovery followed about 10 years later with the availability of "hyper-temporal" C-band Wide-Swath and Global ScanSAR Mode time series from ESA's ASAR sensor on-board ENVISAT: the sensitivity to GSV could be increased to over 300 m^3/ha (Santoro and Cartus 2010).

Two case studies are hence described in this section: the application of the former SIBERIA algorithm to forest mapping in China in the framework of the ESA DRAGON-1 program (http://earth.esa.int/dragon/); and secondly, the production of the first radar-retrieved continuous GSV map of the Northern Hemisphere based on the BIOMASAR-algorithm (http://biomasar.org/).

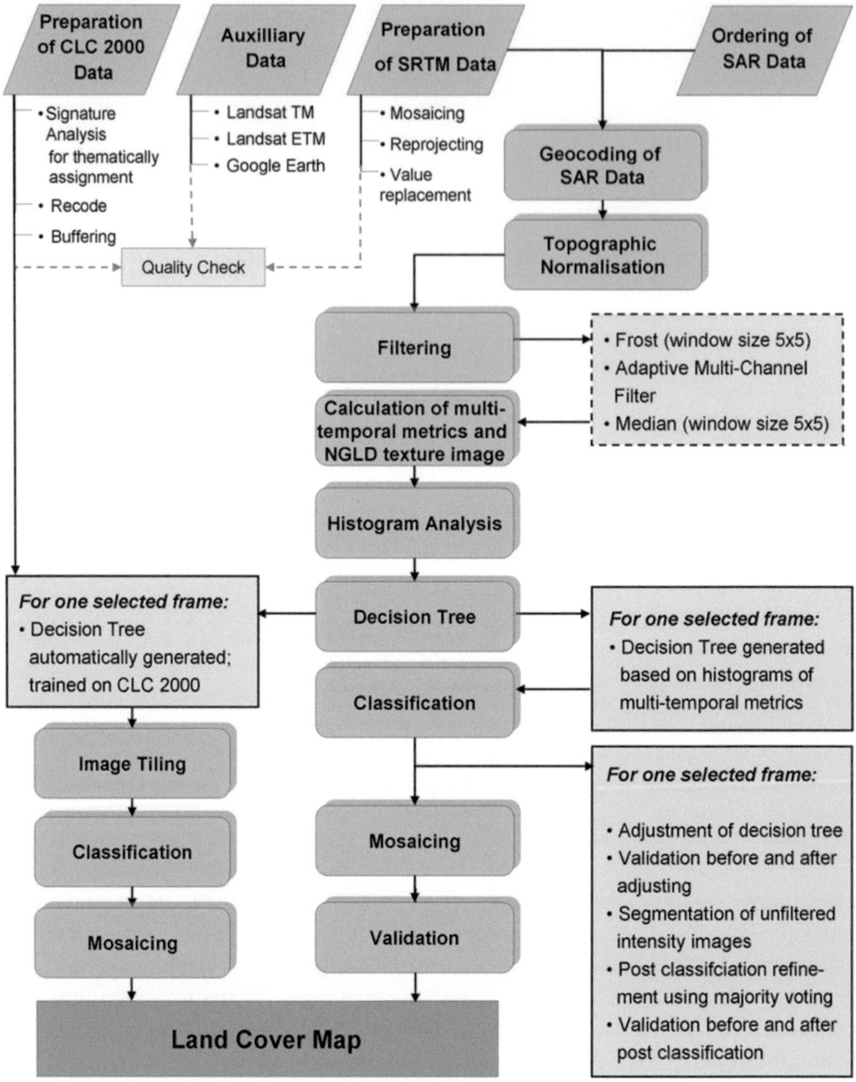

Fig. 16.13 SAR pre-processing and classification chain in ESA AMOC II Study to prepare operational use of Sentinel-1 data for land cover mapping (Thiel 2010). The resulting land cover map of this procedure is illustrated in Fig. 16.14

16.3.2.1 DRAGON-1 Case Study

ERS-1 and -2 Tandem coherence has high potential for mapping boreal forest stem volume (e.g. Askne and Santoro 2005). Large-scale application, however, is hindered by the variability of coherence with meteorological and environmental conditions. Retrieval procedures therefore need to be based on model training

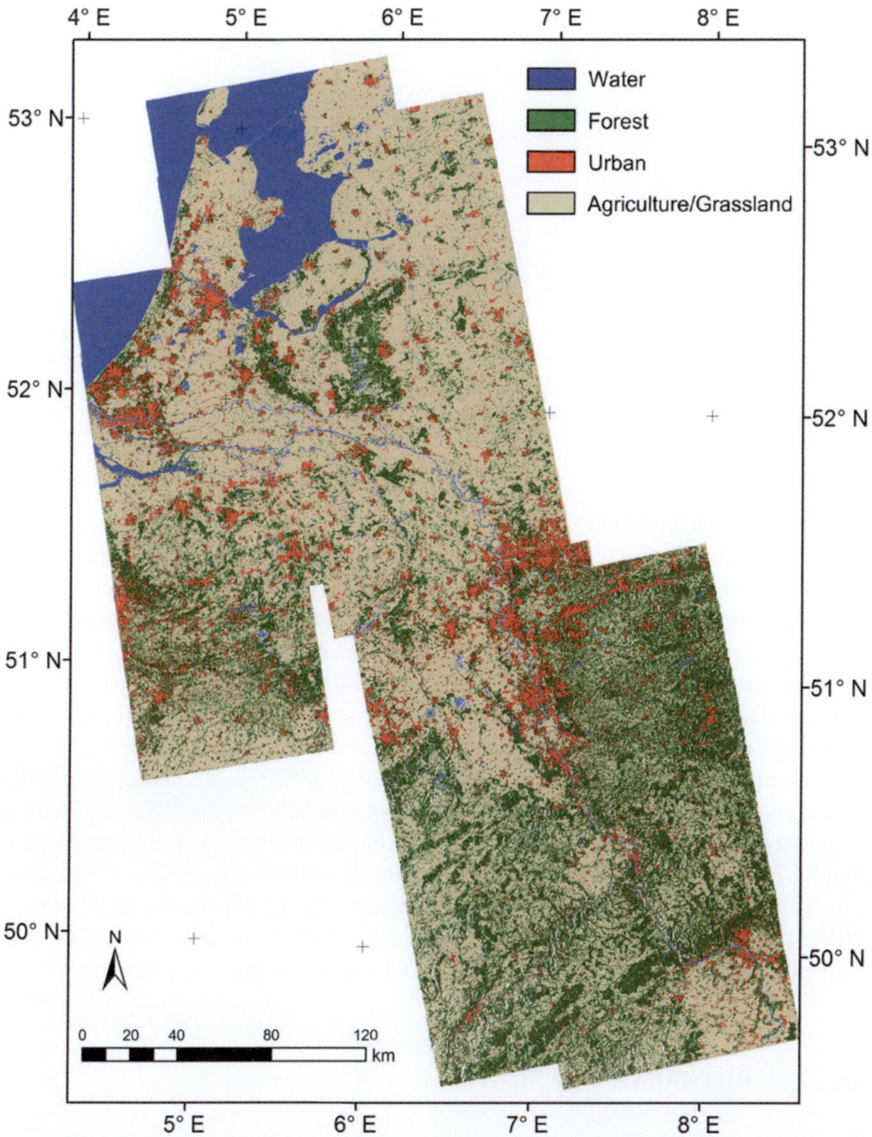

Fig. 16.14 In order to generate a large land cover map of ca. 75,000 km^2 from SAR data (covering parts of the Netherlands, Belgium and Germany), a knowledge-based decision tree has been applied to 11 Envisat ASAR frames without any local adjustment. The incidence angles vary between modes IS 1 and IS 3. The classification procedure is based on backscatter intensity, multi-temporal metrics and texture features. The map was validated using 50 geo-referenced Quick Bird snapshots with 496 reference points. The overall accuracy of the complete land cover map (with a merged class containing grassland and agriculture, as originally intended by ESA) was found to be 89.72 % (Thiel 2010)

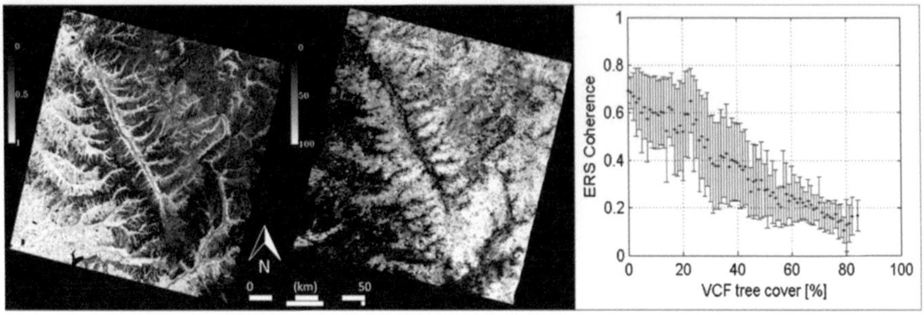

Fig. 16.15 ERS-1/-2 Tandem coherence (*left*) and VCF tree cover (*middle*) for a forested area in NE China. The coherence image is based on acquisitions from 3./4. Oct 1997. The *right* plot shows the decrease of coherence with increasing VCF tree cover ±1 standard deviation. VCF values end at 80 % cover, hence higher density and biomass classes cannot be trained without inventory information (Cartus et al. 2011)

relating coherence to stem volume by using forest inventory data, which is generally only available for a few small test sites. In the Forest-DRAGON project, a new approach was developed that allows model training based on the MODIS Vegetation Continuous Fields canopy cover product (Hansen et al. 2003) without further need for in situ information. A comparison is shown in Fig. 16.15.

As a test for wall-to-wall nation-wide applications, the retrieval method was applied to a multi-seasonal Tandem dataset consisting of 223 ERS-1/-2 image pairs covering northeast China (~1.5 million km^2). The coverage is shown in Fig. 16.16. Four stem volume classes were produced (0–20, 20–50, 50–80, and >80 m^3/ha). The agreement in terms of the kappa coefficient was between $\kappa = 0.52$–0.87 with a standard deviation of 20 m^3/ha. For images acquired in winter, κ was between 0.71 and 0.87; for images acquired in fall and spring it was between 0.52 and 0.78. The producer and user accuracies of the intermediate volume classes reached >80 %. Hence, a regression-based retrieval for forest stock discrimination with ERS-1/-2 Tandem coherence and the VCF-based model training approach appears justified.

16.3.2.2 BIOMASAR Case Study

Promising results from a hyper-temporal retrieval experiment concerning biomass mapping for Sweden, Quebec and central Siberia by Santoro and Cartus (2010) led to an unprecedented endeavor employing more than 647,000 ASAR ScanSAR C-band backscatter data sets acquired between October 2009 and February 2011 over the North American and Eurasian continent: the production of the BIOMASAR forest GSV map of the Northern Hemisphere (Santoro et al. 2013b). Figure 16.17 gives an impression of the hyper-temporal data quantity. The innovative aspect of the algorithm is its independence from in situ measurements for model training. Model parameter estimates are obtained from tendency statistics of

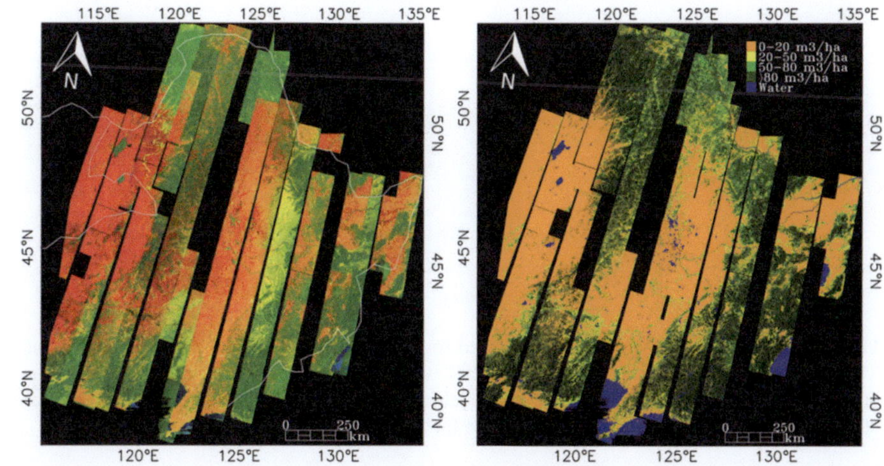

Fig. 16.16 *Left*: Mosaicked ERS-1/-2 false-color composite of northeast China: R-Tandem coherence, G-ERS-1 intensity, B-ERS-1/-2 intensity difference. The mosaic consists of 223 frames with 50×50 m^2 pixel size. Color differences between different tracks are caused by environmental effects. *Right*: Mosaic of the classified GSV maps from ERS-1/-2 Tandem data. The adaptive retrieval model has effectively removed the differences in the input data (Cartus et al. 2011)

the backscatter measurements for unvegetated and dense forest areas, which can be selected using a continuous tree canopy cover product, such as MODIS VCF.

Data processing included multi-looking to 1 km pixel size, terrain geocoding to a pixel size of 0.01°, speckle filtering and correction for slope-induced effects on the backscatter. Validation of the GSV maps was carried out at the full resolution of 0.01° as well as at the aggregated level by comparing against in situ and optically-based official inventory information. Detailed validation at the full 1 km resolution with the Swedish, Quebecan and central Siberian forest inventories proved consistent GSV estimates with a good agreement up to 300 m^3/ha and an underestimation of GSV for biomass levels above 300 m^3/ha. Overall, the RMSE is between 28.7±10.0 % at 1 km pixel size (Santoro et al. 2015). Larger errors were obtained at 100 m spatial resolution because of local errors in the reference datasets. Averaging GSV estimates over neighboring pixels improved the retrieval statistics substantially. For an aggregation factor of 10×10 pixels, the relative RMSE was below 25 %. In general, the spatial patterns of the estimated SAR-based GSV showed good agreement with those of existing reference data sets at a similar spatial resolution. The resulting BIOMASAR map is shown in Fig. 16.18.

16.3.3 Forest Cover Change with Radar Time Series

Forest cover monitoring requires regular observations throughout the year, depending on the biome and the respective fire and/or logging activities. The

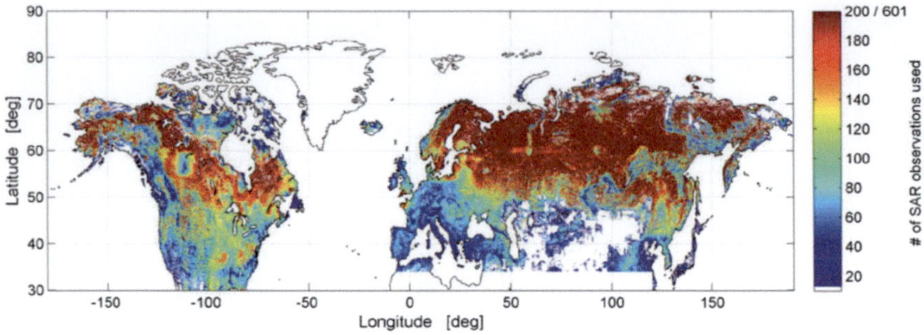

Fig. 16.17 Number of SAR observations used for the Growing Stock Volume (GSV) retrieval. The color bar is constrained between 10 and 200, while the number of observations reached up to 601. Zero is assigned to unmapped pixel and pixels for which the number of SAR observations was less than 10, i.e. represent unreliable estimates for the retrieval model (Santoro et al. 2013b)

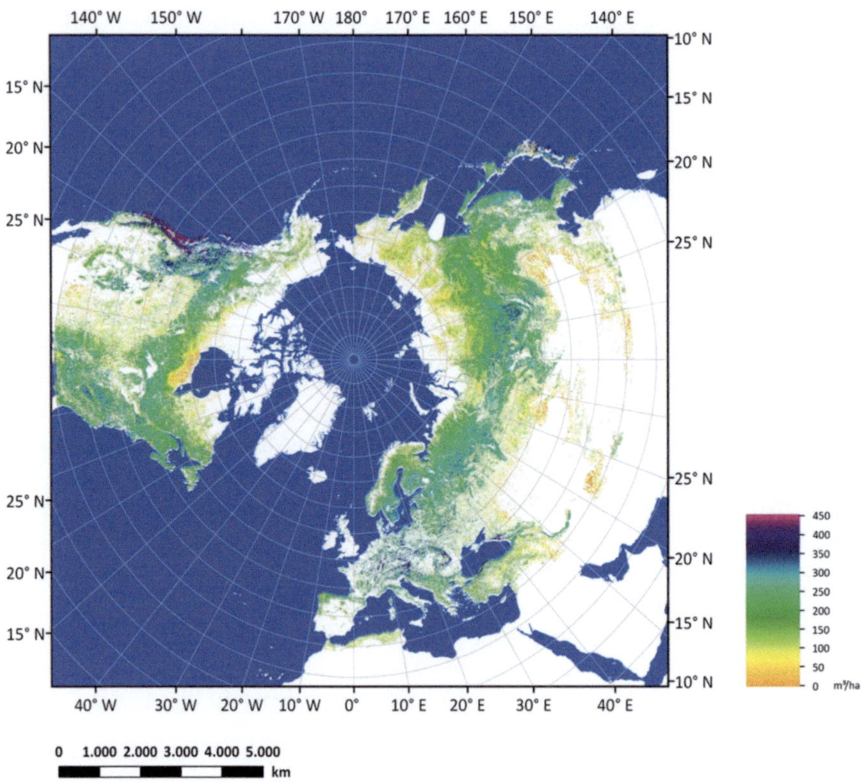

Fig. 16.18 The BIOMASAR Growing Stock Volume (GSV) Map from ENVISAT, www.esa.int/Our_Activities/Observing_the_Earth. The color bar is constrained between 0.1 and 450 m^3/ha, while the retrieved GSV reached up to 1,019 m^3/ha (Santoro et al. 2013a)

Japanese space agency Japan Aerospace Exploration Agency (JAXA) has been the only SAR data provider offering a consistent observation strategy based on the recommendations of its Kyoto and Carbon Panel for more than two decades since its first radar sensor JERS-1 (e.g. Rosenqvist et al. 2007). These L-band data sets reveal less temporal decorrelation when calculating interferometric coherence due to the longer wavelength and thus larger and more stable scattering elements. Hence, multi-temporal coherence and multi-temporal backscatter intensity can both be exploited to build an operational change procedure, as demonstrated by Thiel et al. (2009a, b) where ALOS-1 PALSAR summer intensities and winter coherences led to segment-based classification accuracies of 93 % for a 100,000 km^2 study area for seven classes: water, urban, arable, burnt, recent and old clear-cuts.

The following two case studies extend the application of radar remote sensing to multiple satellites. The first study describes how three radar time series products have been jointly exploited to map forest cover change – independently of the fact that they have been generated from different sensors (ERS-1/-2 vs. ASAR) and different algorithms (SIBERIA vs. BIOMASAR). The two respective methods were described in Sect. 16.3.2. The second study comprises a joint usage of the PALSAR mosaic time series with the multi-temporal MODIS Enhanced Vegetation Index to increase the reliability of disturbance mapping.

16.3.3.1 DRAGON-2 Case Study: Merging Space and Time

The forest project contributing to the second ESA cooperation program with China, DRAGON-2 (https://saredu.dlr.de), focused on the evaluation of multi-temporal, multi-sensor and multi-scale Earth Observation products of northeastern China. First, the GSV map produced with ERS-1/-2 coherence images for 1995–1998 using the SIBERA-algorithm (Fig. 16.16) and the GSV map produced from Envisat ASAR ScanSAR data from 2007 to 2008 with the BIOMASAR-algorithm (Fig. 16.19) were compared with several land cover/forest cover products of optical origin. These comparisons were used to assess the plausibility of the respective GSV estimates to undertake the change study. The change analysis was carried out in the regions of Daxinganling (~200 × 200 km) and Xiaoxinganling (~300 × 300 km) in northeast China.

Leiterer et al. (2010) developed a multi-scale cross-comparison assessment design that uses few in situ measurements and data quality flags and applies existing land cover products such as GlobCover, MODIS VCF, GLC2000 (Global Land Cover (GLC2000)) and the AVHRR LCC(Lambert conic conformal). The sampling design for the comparison is based on the FAO (Food and Agriculture Organization) Forest Resource Assessment 2010, which uses a 1° sampling grid with 10 × 10 km sample plots. The results are shown in Table 16.1. Additionally, forest GSV inventory data including >1,100 point measurements were available for a small test site in northeast China showing good overall accuracy of 76 % for the ERS-retrieved map.

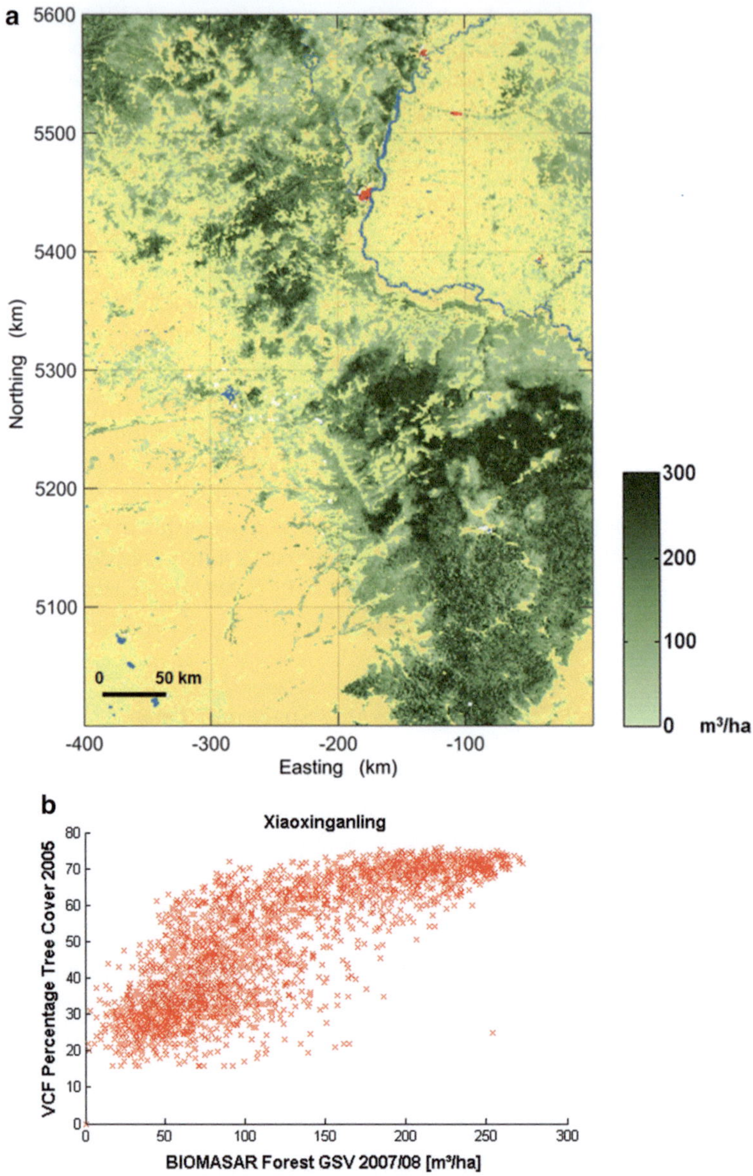

Fig. 16.19 (**a**) ASAR-retrieved GSV map 2007/2008 (pixel size 1 km) illustrating great amount of heterogeneity; green tones: forest GSV, other colours non-forest land cover types. (**b**) scattergram of ASAR GSV versus Vegetation Continuous Field percentages showing the known VCF-threshold at 80 % cover and that the ASAR-product has not reached saturation yet (Reiche et al. 2010)

Table 16.1 Overall agreement (OA) between ERS-1/-2 GSV map (50 m resolution, Fig. 16.16) and four optical land cover products for test areas 1-Daxinganling and 2-Xiaoxinganling in NE China based on aggregated forest/non-forest classes (Leiterer et al. 2010)

LC-product	Site 1	Site 2
NLCD	0.73	0.80
GlobCover	0.78	0.85
VCF (15 % CC)	0.91	0.89
GLC2000	0.89	0.83
AVHRR LCC	0.79	0.67

The Chinese national land cover product NLCD (National Land Cover Database) (based on Landsat images) identified that misclassification of the ERS-maps primarily took place at croplands. Thus, land cover information should be used to support the radar-retrievals particularly for transition zones between forest and shrub land. Cross-comparison for the ASAR forest maps revealed, that the spatial variability was more reliably captured (using NLCD as reference again) than with other land cover products. Figure 16.19a illustrates the heterogeneity of the forest cover in the Xiaoxingaling test area as identified by the ASAR-retrieved GSV map and Fig. 16.19b) shows the scatterplot comparison with VCF tree cover percentages for the same area.

To undertake the forest change exercise, the ERS data were reprocessed to the coarser ASAR Global Monitoring Mode resolution of 1 km. The resulting change map is given in Fig. 16.20. It is the first 10-year forest stock change representation from multiple SAR-sensor time series and thus a milestone in radar remote sensing. Missing tracks are a result of missing ERS-Tandem acquisitions. Solely ASAR-based GSV estimates for the time period 2005–2010 are shown in Fig. 16.21 and have been checked for plausibility during field campaigns and using related Earth observation products (compare Fig. 16.22). Patterns of increase and decrease in that 5-year period can be explained with growth models and fire events respectively. However, a distributed geo-spatial error quantification could not be performed due to missing in situ information.

16.3.3.2 ZAPÁS Case Study – Operational Radar-Optical Synergy

Forest cover disturbance rates of Siberian forests are increasing due to intensification of human activities and changing climate conditions. Hüttich et al. (2014a) used a radar-optical data concept to develop an automated forest cover change detection: pre-classification change-detection techniques were applied to annual ALOS PALSAR backscatter mosaics (2007–2010, see Fig. 16.23) to assess yearly forest biomass loss; and time series of the MODIS Enhanced Vegetation Index product (EVI, 2000–2014) were integrated in a web-based middleware system to establish near-real time detection of forest disturbances using the Breaks For Additive Season and Trend (BFAST) method. Using the Earth Observation Monitor

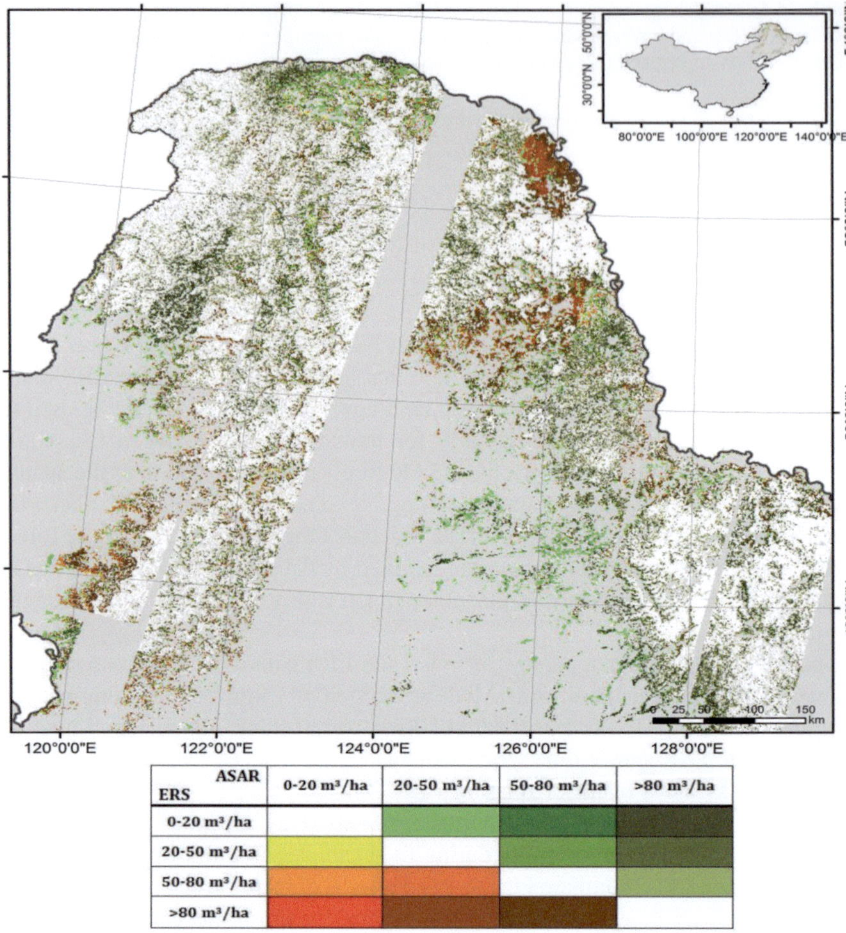

Fig. 16.20 Forest stock change of a ten year period is mapped in four Growing Stock Volume classes using an ERS Tandem dataset and a hyper-temporal ASAR image stack. The tandem product is based on the SIBERIA algorithm, the hyper-temporal method on the BIOMASAR retrieval method. The twelve change classes correctly identify in green colors areas of extensive re-forestation after very large forest fires in 1987; the brown regions indicate burnt zones as a result of a 2006 fire (Reiche et al. 2010). This map represents the first published growing stock volume change product from multiple radar sensors using different retrieval algorithms that combine interferometric with hyper-temporal techniques. The Tandem GSV product with a geometric resolution of 50 m was resampled for this purpose to the 1 km resolution of the ASAR hyper-temporal GSV product. (© FSU Jena, GAMMA RS)

(EOM, www.earth-observation-monitor.net) an operational monitoring system was assessed for the capabilities to detect biomass loss and to analyze temporal patterns of forest cover loss related to logging activities, fire events or other disturbances.

A pre-classification approach using a decision-tree classifier was implemented. Only changes from forest to non-forest were considered. In order to identify

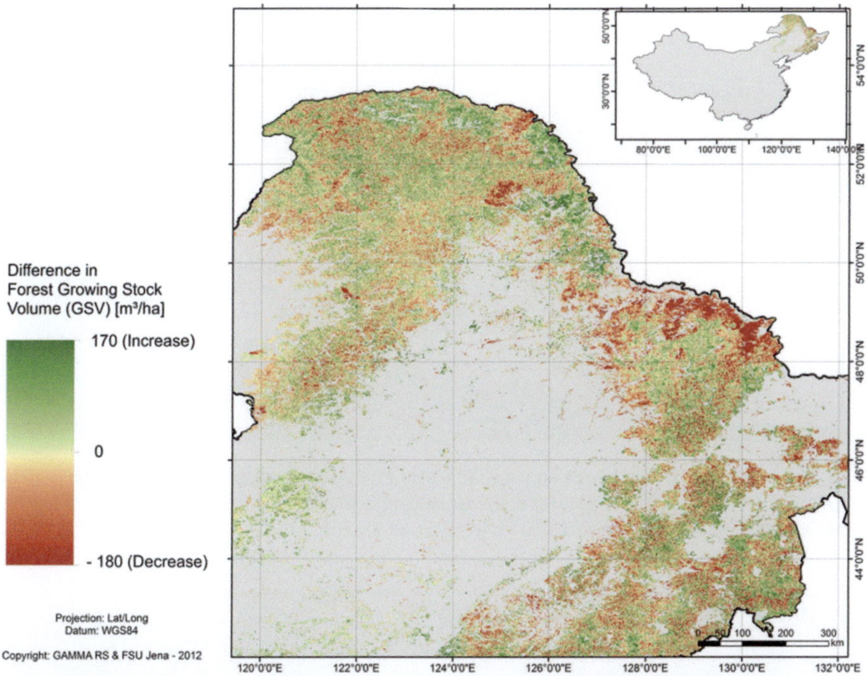

Fig. 16.21 Growing Stock Volume (GSV) change map from estimates obtained with the BIOMASAR algorithm using Envisat ASAR ScanSAR images acquired in 2005 and 2010 (Schmullius et al. 2012). The study area has a total coverage of about 540,000 km^2 including forest and shrubland areas according to the GLC2000 land cover information. The map is characterized by several changes through fire, logging and re-forestation. The GSV estimated with the BIOMASAR algorithm presents an uncertainty quantified on the order of 10 % regardless of the GSV level as proven with comparable inventory data from Siberia

common classification thresholds eight different multi-temporal metrics were analyzed (compare Fig. 16.24). The forest/non-forest change thresholds were detected using non-parametric support vector machines (SVM).

The PALSAR-based average accuracy of forest loss detection was 70 %, whereas the MODIS-based change assessment using breakpoint detection achieved average accuracies of 50 % for trend-based breakpoints and 43.4 % for season-based breakpoints. Time series tracking of phenological activities was realized using the MODIS EVI product with a 16-day temporal resolution for additional information on forest disturbances. Two years (2009 and 2010) showed higher user accuracies of 80.0 %. It is assumed that the PALSAR input mosaics for those years had been acquired under more stable radiometric conditions, which demonstrates the need for free access to raw data rather than processed data products. The sensor's 46-day revisit time is not a limitation due to the stable environmental conditions during winter (even for coherence estimates). This is of particular importance for the development of an operational large-scale forest-monitoring system for the boreal region: the mentioned accuracies are well in the range of

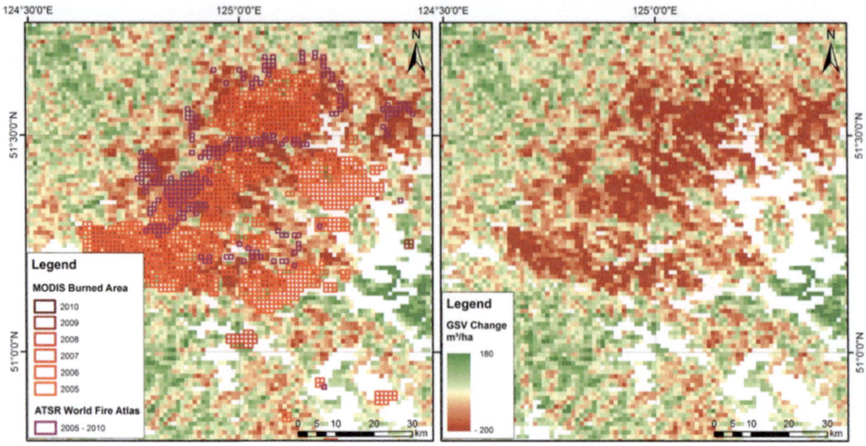

Fig. 16.22 MODIS Burned Area and ATSR World Fire Atlas overlaid on the GSV change map (*left*) and GSV change map (*right*) for a fire detected in 2006 showing very good agreement for the areas affected by stock volume losses. (Schmullius et al. 2012)

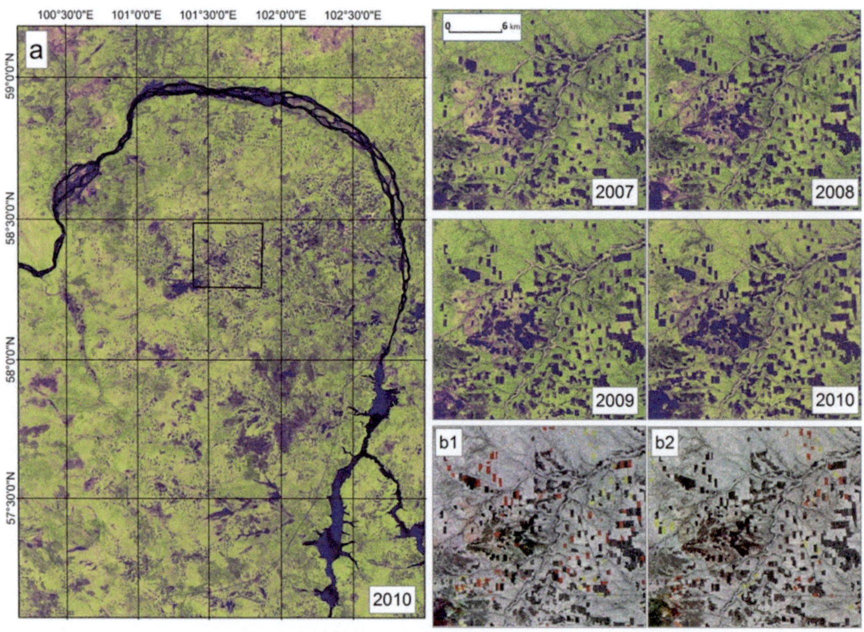

Fig. 16.23 False-color RGB composites of ALOS PALSAR 25 m mosaic images, acquisition periods each year May-October, and incidence angle 34.3°. (*a*) *Red*: HH-intensity, *Green*: HV-intensity, *Blue*: ratio HH/HV; same color scheme applies for zoomed area in 2007, 2008, 2009, and 2010 (Hüttich et al. 2014a, b). Change is shown for the zoom region as color composites with (*b1*) Red-2007, Green-2008, Blue-2009; and (*b2*) Red-2008, Green-2009, Blue-2010. The data were made available through the ALOS Kyoto and Carbon Initiative of JAXA

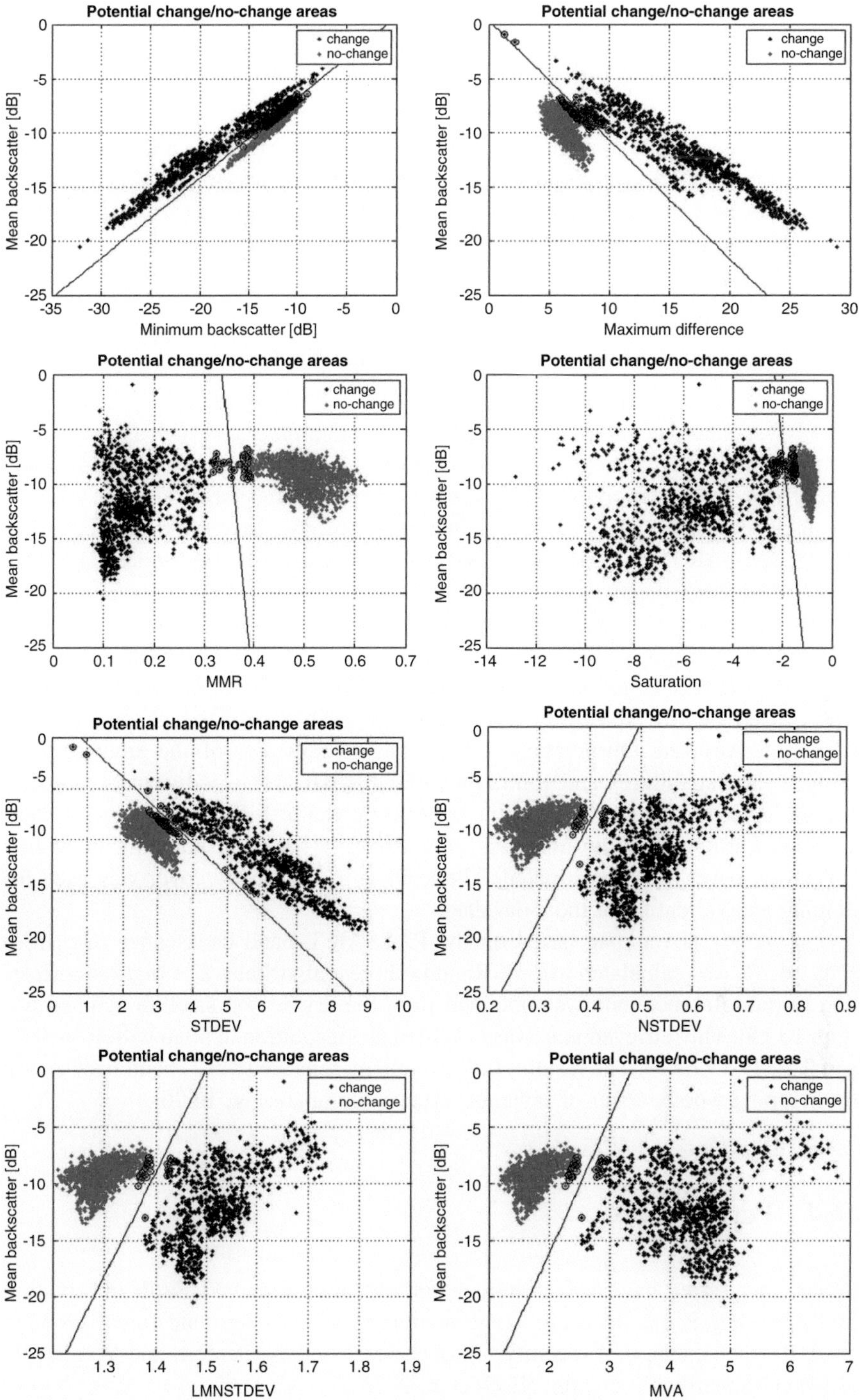

Fig. 16.24 Multi-temporal metrics to identify thresholds for forest change/no-change classification with PALSAR data: maximum-minimum ratio (MMR), normalized standard deviation (NSTDEV), mean average variability (MVA), and logarithmic measure based on normalized standard deviation (LMNSTDEV). The largest class separability was calculated for HV-polarization (Hüttich et al. 2014a, b)

inventory errors of 34–39 %, as has been investigated stand-wise in Hüttich et al. (2014b).

It can be concluded, that the freely available PALSAR mosaics are an applicable tool for up-to-date forest monitoring – even considering varying accuracies. Monitoring can be enhanced through web-based middleware exploiting MODIS time series and the BFAST-method to suggest the type of disturbance.

16.3.4 Special Seasonal Case: Savanna Systems

Woody vegetation cover affects a range of ecosystem processes such as carbon and water cycling, energy fluxes, and fire regimes. Information on the spatial distribution of woody vegetation over large areas is needed to understand the dynamics of savanna ecosystems. In this study, fractional woody cover was mapped using ALOS PALSAR L-band HH- and HV-polarised backscatter time series. The SAR backscatter intensity and two polarimetric decompositions were compared with woody cover obtained from high-resolution airborne LiDAR (Light detection and ranging) data using a semi-empirical exponential model. The SAR data were acquired at different seasonal cycles between 2007 and 2010. The LiDAR survey was carried out in April/May 2008 with the LiDAR component of the CAO (Carnegie Airborne Observatory, USA). The overall aim of the study was to analyze the capabilities and limitations of SAR data for woody cover mapping and the investigation of the potential synergistic use of LiDAR jointly with radar systems. Furthermore, the influence of seasonality for radar mapping of woody vegetation cover was investigated. The LiDAR-based woody cover was used for training and validation of the radar data.

The woody cover map based on the PALSAR L-band backscatter intensities (Fig. 16.27) was calculated using a Random Forest algorithm. The highest correlation to the reference data was obtained from the dry season satellite acquisitions (Fig. 16.25). The retrieved map was validated at a resolution of 50 m with $R^2 = 0.73$ and RMSE = 7.62 %. The results show promising sensitivity of L-band backscatter for mapping woody cover of savanna surfaces (compare Fig. 16.26).

16.4 Perspectives

Radar time series have led to innovative and unprecedented Earth observation products such as extended crop type mapping and above-ground forest growing stock (this Chapter), or – recently – global water bodies (Santoro and Wegmüller 2014) and the urban foot-print (Esch et al. 2013).

In the past, radar data were not available in the temporal or geometric resolution needed – or too expensive. Therefore, the knowledge created since the 1980s could not be transformed into operational routines – until now. With the new fleet of

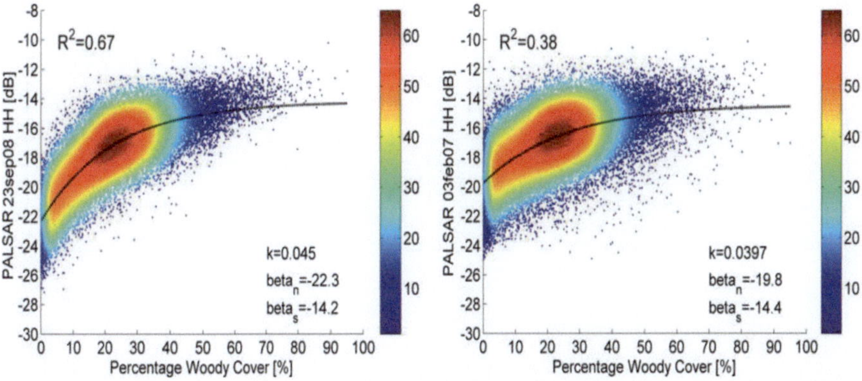

Fig. 16.25 Regression analysis between PALSAR L-band backscatter and airborne LiDAR-retrieved woody cover: .(*left*) winter dry season, (*right*) summer wet season (Urbazaev et al. 2013). Due to an increase of low vegetation and soil moisture during the wet season the backscatter intensity from low woody cover regions is increased and sensitivity to the woody components decreased. The dry season is therefore recommended for savanna woody cover mapping with L-band sensors

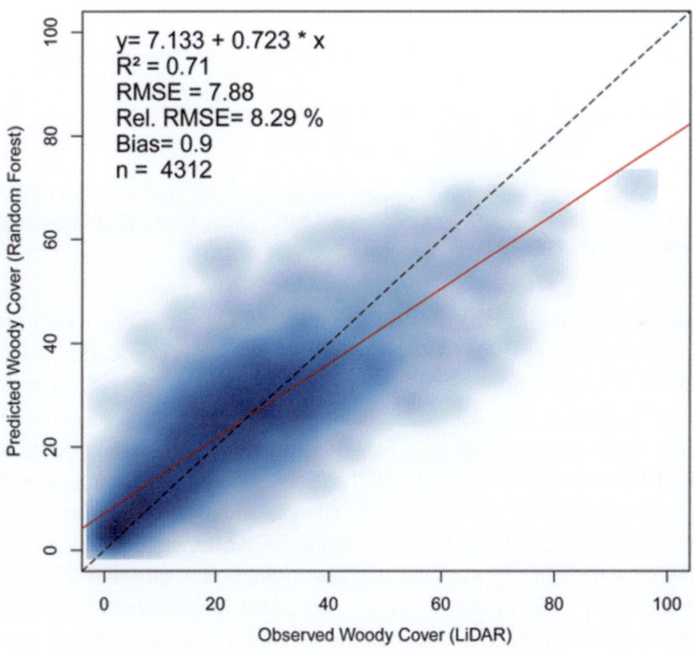

Fig. 16.26 PALSAR L-band predicted vs. airborne LiDAR-observed woody cover of undisturbed savanna vegetation in Kruger National Park, South Africa (Urbazaev et al. 2013). The red line indicates the regression line. The typical exponential behavior can be observed with a slight saturation effect above 60 % woody cover

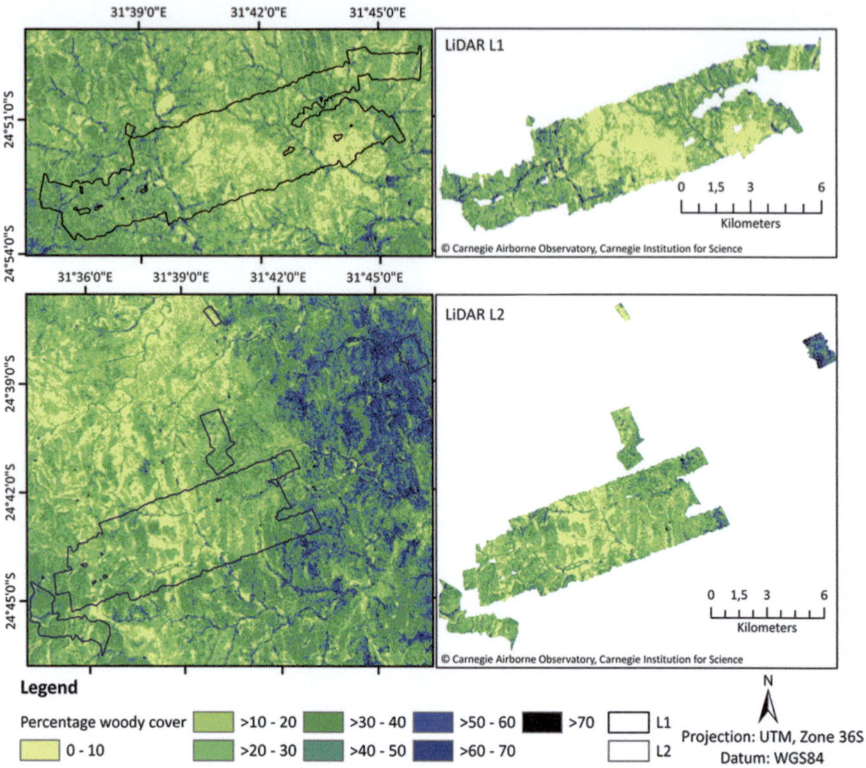

Fig. 16.27 Comparison between PALSAR-retrieved woody cover (*left*) and CAO LiDAR-based woody cover (*right*) for two sites in Kruger National Park, South Africa. Both spatial patterns are in very good agreement (Urbazaev et al. 2013)

active radar sensors being launched during this 2nd decade of the new millennium (ALOS-2 PALSAR, BIOMASS, COSMO-SkyMED Second Generation, RADARSAT Constellation, Sentinel-1a/-1b, TerraSAR Next Generation, and more planned), further very advanced analysis techniques such as differential and polarimetric SAR interferometry (DInSAR, PolInSAR), persistent scatterer interferometry (PSI) or radar tomography will become operational from space. Now, all advantages of microwave remote sensing such as independence from solar illumination, penetration capabilities through clouds and volumetric media, sensitivity to water content, capability to measure surface movements, etc. will lead to reliable Earth observation contributions. Such consistent observables are desperately needed for environmental models and model-data fusion methods to develop benchmarks (Thurner et al. 2013) or to constrain model calculations (Quegan et al. 2011) in an increasingly fast changing and complex world. Hence, the famous radar equation, which was mentioned in the introduction, will persist to be one of the most fundamental descriptions for Earth observation.

Acknowledgements The authors want to thank the funding agencies German Federal Minister of Education and Research (BMBF), Federal Ministry for Economic Affairs and Energy (BMWI), Alliance of Science Organisations in Germany (DFG), German Aerospace Centre (DLR), European Commission (EC), European Space Agency (ESA) and Japan Aerospace Exploration Agency's (JAXA) Kyoto & Carbon Initiative, who have facilitated through their support in funding and data supply our continuous progress in SAR applications for large-area land surface mapping.

References

Ackermann N (2015) Growing stock volume estimation in temperate forested areas using a fusion approach with SAR satellites imagery series: Springer theses XXV, 307 p 141, illus., 121 illus. in color. ISBN 978-3-319-13138-2

Ackermann N, Thiel C, Borgeaud M, Schmullius C (2010) Potential of fusion of SAR and Optical satellite imagery for biomass estimation in temperate forested areas. In: Proceedings of the ESA living planet symposium, 28 June–2 July 2010, Bergen, Norway. http://earth.eo.esa.int/workshops/livingplanetsymposium2010/sessions/CXNL_10a04_870140.htm. Accessed 30 Nov 2014

Askne J, Santoro M (2005) Multitemporal repeat pass SAR interferometry of boreal forests. IEEE Trans Geosci Remote Sens 43(6):1219–1228

Askne J, Dammert P, Ulander LMH, Smith G (1997) C-band repeat-pass interferometric SAR observations of forest. IEEE Trans Geosci Remote Sens 35:25–35

Attema EPW, Ulaby FT (1978) Vegetation modeled as a water cloud. Radio Sci 13(2):357–364

Balzter HC, Schmullius C (2001) Mapping Siberian forests. First boreal forest map of Siberia from spaceborne radar. GIM Int 15:40–43

Balzter HC, Talmon E, Wagner W, Gaveau D, Plummer S, Yu J, Quegan S, Davidson M, Le Toan T, Gluck M, Shvidenko A, Nilsson S, Tansey K, Luckmann A, Schmullius C (2002) Accuracy assessment of a large-scale forest cover map of Central Siberia from synthetic aperture radar. Can J Remote Sens 28(6):719–737

Cartus O, Eckardt R, Richter N (2008) GMES/Sentinel-1 analysis of multi-temporal land observation at C-band (AMOC ESTEC contract no. 21302/07/NL/CB). Final report, Department for Earth Observation, University Jena, Germany

Cartus O, Santoro M, Schmullius C, Li Z (2011) Large area forest stem volume mapping in the boreal zone using synergy of ERS-1/2 tandem coherence and MODIS vegetation continuous fields. Remote Sens Environ 115:931–943

Dobson MC, Ulaby FT, Le Toan T, Beaudoin A, Kasichke ES, Christensen N (1992) Dependence of radar backscatter on coniferous forest biomass. IEEE Trans Geosci Remote Sens 30:412–416

Eckardt R, Berger C, Thiel C, Schmullius C (2013) Removal of optically thick clouds from multi-spectral satellite images using multi-frequency SAR data. Remote Sens 5(6):2973–3006

Eriksson LEB, Santoro M, Wiesmann A, Schmullius C (2003) Multi-temporal JERS repeat-pass coherence for growing stock volume estimation of Siberian forest. IEEE Trans Geosci Remote Sens 41(7):1561–1570

Eriksson L, Askne J. Santoro M, Schmullius C, Wiesmann A (2005) Stem volume retrieval with spaceborne L-band repeat-pass coherence. Multi-temporal combination for boreal forest. In: Proceedings of IGARSS'05. Seoul, 25–29 July 2005, pp 3591–3594

Esch T, Marconcini M, Felbier A, Roth A, Heldens W, Huber M, Schwinger M, Taubenbock H, Muller A, Dech S (2013) Urban footprint processor—fully automated processing chain generating settlement masks from global data of the TanDEM-X mission. Geosci Remote Sens Lett 10(6):1617–1621

Hansen MC, DeFries RS, Townshend JRG, Carroll M, Dimiceli C, Sohlberg RA (2003) Global percent tree cover at a spatial resolution of 500 meters: first results of the MODIS vegetation continuous fields algorithm. Earth Interact 7(10):1–15

Henderson FM, Lewis AJ (1998) Principles and applications of imaging radar – manual of remote sensing, vol 2. Wiley, New York

Hüttich C, Stelmaszczuk-Górska MA, Eberle J, Kotzerke P, Schmullius C (2014a) Operational forest monitoring in Siberia using multi-source earth observation data. Sib J Forest Sci 5:38–53

Hüttich C, Korets M, Bartalev S, Zharko V, Schepaschenko D, Shvidenko A, Schmullius C (2014b) Exploiting growing stock volume maps for large scale forest resource assessment: cross-comparisons of ASAR- and PALSAR-based GSV estimates with forest inventory in central Siberia. Forests 5(7):1753–1776. doi:10.3390/f5071753

Leiterer R, Reiche J, Cartus O, Santoro M, Schmullius C, Li Z (2010): Multiscale Comparison of forest maps based on forest stem volume classification and existing land cover products. In: Proceedings of the 2010 dragon 2 symposium, 17–21 May 2010, Guilin, China

Luckmann A, Tansey K, Schmullius C (2004) The construction of ERS-tandem and JERS SAR image mosaics for large-scale mapping of boreal forest in Siberia. Int J Remote Sens 25:751–768

Quegan S, Yu JJ (2001) Recursive multi-channel filtering of SAR images. IEEE Trans Geosci Remote Sens 39(11):2373–2379

Quegan S, Le Toan T, Yu JJ, Ribbes F, Floury N (2000) Multitemporal ERS SAR analysis applied to forest monitoring. IEEE Trans Geosci Remote Sens 38(2):741–753

Quegan S, Beer C, Shvidenko A, McCallum I, Handoh I, Peylin P, Rödenbeck C, Lucht W, Nilsson S, Schmullius C (2011) Estimating the carbon balance of central Siberia using a landscape-ecosystem approach, atmospheric inversion and dynamic global vegetation models. Glob Chang Biol 17:351–365

Reiche J, Leiterer R, Cartus O, Santoro M, Schmullius C, Zengyuan L (2010) FOREST DRAGON 2: large-area forest stem volume mapping in China, using ERS-1/2 tandem coherence. In: Proceedings CD of ESA living planet symposium, 27 June–2 July 2010, Bergen, Norway

Riedel T, Thiel C, Schmullius C (2008) Fusion of multispectral optical and SAR images towards operational land cover mapping in Central Europe. In: Blaschke T, Lang S, Hay GJ (eds) Object-based image analysis – spatial concepts for knowledge-driven remote sensing applications. Springer, Berlin, pp 493–512

Rosenqvist A, Shimada M, Ito N, Watanbe M (2007) ALOS PALSAR: a pathfinder mission for global-scale monitoring of the environment. IEEE Trans Geosci Remote Sens 45 (11):3307–3316

Santoro M, Cartus O (2010) STSE-BIOMASAR: validating a novel biomass retrieval algorithm based on hyper-temporal wide-swath and global monitoring Envisat ASAR datasets. Final report, ESA ESRIN contract no. 21892/08/I-EC

Santoro M, Wegmüller U (2014) Multi-temporal synthetic aperture radar metrics applied to map open water bodies. J Select Top Appl Earth Obs Remote Sens 7(8):3225–3238. doi:10.1109/JSTARS.2013.2289301

Santoro M, Askne J, Smith G, Fransson JES (2002) Stem volume retrieval in boreal forests from ERS-1/2 interferometry. Remote Sens Environ 81:19–35

Santoro M, Beer C, Cartus O, Schmullius C, Shvidenko A, McCallum I, Wermüller U, Wiesmann A (2011) Retrieval of growing stock volume in boreal forest using hyper-temporal series of Envisat ASAR ScanSAR backscatter measurements. Remote Sens Environ 115(2):490–507

Santoro M, Cartus O, Fransson J, Shvidenko A, McCallum I, Hall R, Beaudoin A, Beer C, Schmullius C (2013a) Estimates of forest growing stock volume for Sweden, Central Siberia, and Québec using Envisat advanced synthetic aperture radar backscatter data. Remote Sens 5:4503–4532

Santoro M, Schmullius C, Pathe C Schwilk J, Beer C, Thurner M, Fransson JES, Shvidenko A, Schepaschenko D, McCallum I, Beaudoin A, Hall R (2013b) Estimates of forest growing stock volume of the Northern Hemisphere from Envisat ASAR. In: Proceedings of the living planet symposium 1. http://biomasar.org/fileadmin/biomasar/papers/2849916santoro.pdf. Accessed 30 Nov 2014

Santoro M, Beaudoin A, Beer C, Cartus O, Fransson J, Hall R, Pathe C, Schmullius C, Shvidenko A, Thurner M, Wegmuller U (2015) Forest growing stock volume of the northern hemisphere: spatially explicit estimates for 2010 derived from Envisat ASAR data. Remote Sens Environ (in review)

Schmullius C (1997) Monitoring Siberian forests and agriculture with the ERS-1 Windscatterometer. IEEE Trans Geosci Remote Sens 35(5):1363–1366

Schmullius C, Rosenqvist A (1997) Closing the gap – a Siberian boreal forest map with ERS-1/2 and JERS-1. 3rd ERS symposium on space at the service of the environment, Florence, 17–21 March, 1997. http://earth.esa.int/workshops/ers97/papers/schmullius/. Accessed 30 Nov 2014

Schmullius C, Baker JS, Balzter H, Davidson M, Eriksson L, Gaveau D, Gluck M, Holz A, Luckman A, Marschalk U, McCallum I, Nilsson S, Oeskog A, Quegan S, Rauste Y, Roth A, Shvidenko A, Skuding V, Strozzi T, Tansey K, Vietmeier J, Voloshuk L, Wagner W, Wegmüller U, Wiesmann A (2001) SIBERIA – SAR imaging for boreal ecology and radar interferometry applications. Final report. EC Center for Earth Observation, project reports, contract no. ENV4-CT97-0743-SIBERIA. http://www.siberia1.uni-jena.de. Accessed 30 Nov 2014

Schmullius C, Leiterer R, Burjack I, Traut K, Santoro M, Li ZY, Ling FL (2012) FOREST DRAGON-2 – final results of the european partners. In: Dragon-2 final results & dragon 3 Kick-off symposium, Beijing, China. http://www.zora.uzh.ch/77312/. Accessed 30 Nov 2014

Thiel C (2010) GMES/Sentinel-1: software and prototyping of level 2 products for land-use-mapping analysis of multi-temporal land observation at C-band (AMOC II – ESA CONTRACT C21302/07/NL/CB). Final report, Department for Earth Observation, University Jena, Germany

Thiel C, Schmullius C (2013a) Investigating the impact of freezing on the ALOS PALSAR InSAR phase over Siberian forests. Remote Sens Lett 4(9):900–909

Thiel C, Schmullius C (2013b) Investigating ALOS PALSAR interferometric coherence in central Siberia at unfrozen and frozen conditions. Can J Remote Sens 39(3):232–250

Thiel C, Schmullius C (2014) Impact of tree species on magnitude of PALSAR interferometric coherence over Siberian forest at frozen and unfrozen conditions. Remote Sens 6:1124–1136. doi:10.3390/rs6021124

Thiel C, Thiel C, Riedel T, Schmullius C (2007) Analysis of ASAR APP time series over Siberia for optimising forest cover mapping – a GSE forest monitoring study. In: Proceedings CD of 6th ESA Envisat symposium, 23–27 April 2007, Montreux, Switzerland. https://earth.esa.int/workshops/envisatsymposium/proceedings/sessions/3D4/451296th.pdf. Accessed 30 Nov 2014

Thiel C, Thiel C, Riedel T, Schmullius C (2008) Object based classification of SAR data for the delineation of forest cover maps and the detection of deforestation – a viable procedure and its application in GSE forest monitoring. In: Blaschke T, Lang S, Hay GJ (eds) Object-based image analysis – spatial concepts for knowledge-driven remote sensing applications. Springer, Berlin, pp 327–344

Thiel C, Santoro M, Cartus O, Thiel C, Riedel T, Schmullius C (2009a) Perspectives of SAR based forest cover, forest cover change and biomass mapping. In: Vasse CP (ed) The Kyoto protocol: economic assessments, implementation mechanisms, and policy implications. Nova Science Publishers, New York, pp 13–56

Thiel C, Thiel C, Schmullius C (2009b) Operational large area forest monitoring in Siberia using ALOS PALSAR summer intensities and winter coherence. IEEE Trans Geosci Remote Sens 47 (12):3993–4000

Thiel Ca, Cartus O, Eckardt R, Richter N, Thiel Ch, Schmullius C (2009c) Analysis of multi-temporal land observation at C-band. IEEE geoscience and remote sensing symposium (IGARSS) III, pp 318–321. http://dx.doi.org/10.1109/IGARSS.2009.5417764

Thurner M, Beer C, Santoro M, Carvalhais N, Wutzler T, Schepaschenko D, Shvidenko A, Kompter E, Ahrens B, Levick S, Schmullius C (2013) Carbon stock and density of northern boreal and temperate forests. Glob Ecol Biogeogr 23(3):297–310

Ulaby FT, Moore RK, Fung AK (1986) Microwave remote sensing: active and passive, vol. III – volume scattering and emission theory, advanced systems and applications. Artech House, Inc., Dedham

Urbazaev M, Schmullius C, Thiel C, Renaud M, Naidoo L, Levick S, Smit I, Asner G, Leblon B (2013) Delineation of woody cover by using full and dual polarimetric L- and C-band datasets in the Kruger National Park, SA. In: ESA living planet symposium, Edinburgh, 9–13 September 2013. http://seom.esa.int/LPS13/5867d0b0/LP_2013_Urbazaev.pdf. Accessed 30 Nov 2014

Wagner W, Luckman A, Vietmeier J, Tansey K, Baltzer H, Schmullius C, Davidson M, Gaveau D, Gluck M, Letoan T, Quegan S, Shvidenko A, Wiesman A, Yu JJ (2003) Large-scale mapping of boreal forest in SIBERIA using ERS tandem coherence and JERS backscatter data. Remote Sens Environ 85:125–144

Wooding M, Attema E, Aschbacher J, Borgeaud M, Cordey RA, De Groot H, Harms J, Lichtenegger J, Nieuwenhuis G, Schmullius C, Zmuda A (1995) Satellite radar in agriculture. Experience with ERS-1. ESA Scientific Publications, Noordwijk

Chapter 17
Investigating Radar Time Series for Hydrological Characterisation in the Lower Mekong Basin

Daniel Sabel, Vahid Naeimi, Felix Greifeneder, and Wolfgang Wagner

Abstract Radar remote sensing is beneficial for retrieval of hydrological information such as soil moisture and flood extents due to the strong influence of water on the radar signal. The proper monitoring and analysis of such temporally dynamic phenomena requires dense time series data. Radar time series data is also useful for mitigating uncertainties in individual images, e.g. for the mapping of permanent water bodies. This chapter reviews capabilities, potentials and challenges of spaceborne radar time series data for the mapping of permanent water bodies, the monitoring of floods, and the retrieval of soil moisture content. The focus is put on the Lower Mekong Basin (LMB) in Southeast Asia. Two thirds of the LMB's population of 60 million people live directly from agriculture and fisheries. The Mekong River's resources are under pressure among others from an increasing population, intensified agriculture, and the expansion of hydropower. A thorough understanding of water resources in the LMB is therefore crucial to the sustainable development in the region. The chapter provides an outline of radar remote sensing for retrieval of hydrological information as well as an overview of the relevant operational capabilities of radar missions. A map of permanent water bodies of the entire Lower Mekong Basin derived from a time series of ENVISAT Advanced Synthetic Aperture Radar (ASAR) data is presented. Potentials and challenges of flood monitoring with SAR are illustrated with ASAR imagery showing the evolution of the floods that occurred around Tonle Sap Lake in Cambodia in 2011. Finally, the spatial and temporal dynamics of soil moisture across the LMB are analysed with the use of 14 years of scatterometer time series data acquired by the ERS-1, ERS-2, Metop-A and Metop-B satellites. The average seasonal soil moisture cycle was computed at the sub-catchment level. An anomaly analysis of the temporal soil moisture dynamics revealed large inter-annual variability across the Lower Mekong Basin.

D. Sabel (✉) • V. Naeimi • W. Wagner
Department of Geodesy and Geoinformation, Vienna University of Technology, Vienna, Austria
e-mail: Daniel.Sabel@geo.tuwien.ac.at

F. Greifeneder
European Academy of Bozen/Bolzano, EURAC, Bolzano-Bozen, Italy

C. Kuenzer et al. (eds.), *Remote Sensing Time Series*, Remote Sensing and Digital Image Processing, DOI 10.1007/978-3-319-15967-6_17

17.1 Introduction

Water is essential to life. Despite two thirds of the Earth's surface being covered with oceans, the amount of accessible fresh water is limited and for many nations across the globe, water scarcity represents a fundamental challenge to their economic and social development. Up-to-date knowledge of the distribution of water in its various forms is required in order to manage this valuable resource in a sustainable, fair and secure manner. Such knowledge is also fundamental to the understanding of the hydrologic cycle and the climate system (Dent 2012; OECD 2008).

Remote sensing with Earth orbiting satellites is invaluable for monitoring water resources as it can provide frequent observations on regional to global scales. With the use of radars, measurements can be performed independent of solar conditions and cloud coverage, thereby enabling day-and-night, all weather observations. This is particularly useful in frequently cloud covered areas. In the Mekong Basin in Southeast Asia, Leinenkugel et al. (2013) reported an average of 85–95 % cloud coverage during the rainy season, making the use of optical imagery e.g. for flood mapping very challenging. Radars are also attractive for retrieval of hydrological information due to the strong interaction of microwaves with water resulting from the high relative dielectric constant of liquid water compared to most other naturally occurring materials.

Most aspects of the use of radar time series data reviewed in this chapter are generally applicable. The focus is however put on the Lower Mekong Basin (LMB), which covers an area of more than 600,000 km^2 in Southeast Asia. The diverse ecosystems in the LMB are crucial to the livelihood of over 60 million people. Two thirds of its largely rural population lives directly from agriculture and fisheries. The sustainable development in the LMB is threatened by challenges such as population growth and increased pressure on the Mekong River's resources, in particular from intensified agriculture and the expansion of hydropower (Kuenzer et al. 2013a). Information that can support water resource management and improve the understanding of the hydrology in the region is therefore highly valuable.

17.1.1 Hydrological Information

Knowledge of the distribution of permanent water bodies such as lakes, wetlands, reservoirs and rivers is essential to the assessment of current and future water resources. Such water bodies serve to store, clean and distribute water and represent therefore core components of the hydrological and biogeochemical water cycles (Lehner and Döll 2004). Floods represent one of the most frequent and at the same time most devastating type of disasters. In 2011, more than 100 million people were affected by flooding worldwide. The LMB was also affected, with vast floods in Cambodia. However, regularly occurring floods with average amplitudes can bring benefits in some regions. This is the case in the Mekong Delta and the Tonle Sap Basin in the LMB, where annual floods are essential to agriculture as suppliers of sediments providing nutrition to the soil (Renaud and Kuenzer 2012). Another important type of hydrological information is soil moisture, i.e. the water contained in the pores of the soil. It is in fact a central component in hydrology and an essential variable in the climate system. Soil moisture directly influences the fluxes of energy and moisture transferred between land and atmosphere as well as the partitioning of precipitation into infiltration and runoff. It describes the temporal condition of water available to plants and provides an integrated assessment of the relative state of water supply versus water demand (Legates et al. 2011). While there are hydrological stations in the LMB measuring water levels and river discharges, soil moisture in situ measurements are very scarce.

All of the mentioned aspects of hydrological information benefit from dense time series data. It will be demonstrated that the mapping of permanent water bodies can benefit in several ways from the use of time series data. The mapping of floods requires imagery delivered in a timely fashion. As floods can develop rapidly, frequent updates are also highly desirable. Soil moisture is highly variable in space and time, in particular in the top layer of the soil. The monitoring and analysis of soil moisture dynamics therefore require dense time series data.

17.1.2 Radar Remote Sensing

Radars transmit electromagnetic pulses at microwave frequencies and record the echo scattered back to the instrument from the target or area of observation. The ratio between transmitted and backscattered energy is usually expressed in terms of the normalized radar cross-section (NRSC) which is a dimensionless property (m^2m^{-2}) often referred to as sigma nought (σ^0) and expressed in decibel (dB). The value of sigma nought depends both on target attributes, such as dielectric properties, geometric shape and roughness, and on the radar system's frequency, polarisation configuration and observation geometry. Radars have a staggering potential for measuring physical properties of ground features, providing information quite different to that of instruments working in the optical and infrared portion of the electromagnetic spectrum. An excellent introduction to radar remote sensing can be found in Woodhouse (2006).

17.1.3 Operational Capabilities

This chapter will deal with two types of radar systems, namely Synthetic Aperture Radars (SARs) and scatterometers. A key selling point of SAR missions is high spatial detail. Current civilian spaceborne systems are able to achieve resolution down to about 1 m. The ground coverage of such high resolution acquisitions are however relatively limited. For example, the swath widths of the highest resolution modes on TerraSAR-X and COSMO-SkyMED are 10 km while RADARSAT-2 and Sentinel-1 obtain swath widths of 20 km and 80 km, respectively. Wider swaths can be achieved with scanning techniques (ScanSAR) which uses electronic beam-steering to image several parallel swaths along the flight direction in rapid succession. During post-processing, the individual swaths are combined to create a single, wide swath. In such modes the aforementioned instruments achieve swath widths of 100 km, 200 km, 500 km and 410 km, respectively, with spatial resolutions reduced to 16 m for TerraSAR-X, 40 m for Sentinel-1 and 100 m in the cases of COSMO-SkyMED and RADARSAT-2 (Covello et al. 2009; Drusch et al. 2012; Morena et al. 2004; Pitz and Miller 2010; Sentinel-1 Team 2013). SAR instruments are characterized by high power consumption, thermal heating and high data throughput. They are therefore able to acquire high data rate imagery during only a part of each orbit.

The strong focus on high spatial resolution rather than geographical coverage and frequent acquisitions have resulted in spaceborne SAR missions typically achieving observation intervals ranging from weeks to months on average. While such sparse time series' may be sufficient for observing slowly varying phenomena such as permanent water bodies, they are in usually insufficient for the monitoring of soil moisture and the mapping of flood developments. With dedicated acquisition tasking, it is possible to monitor selected regions more frequently. Revisit times can also be reduced with satellite constellation or through opportunistic combination of data from different missions. In order to satisfy a wide range of user requirements, SAR instruments often provide several modes of operation with different sets of compromises between spatial resolution, geographical coverage, radiometric resolution and polarisation configurations.

Scatterometers are real aperture radars providing measurements with a spatial resolution in the order of several tens of kilometres. Their main advantages over SARs are their high radiometric accuracy and their capability of frequently mapping large areas. In contrast to SAR instrument, they can acquire data during the entire orbit. With the use of wide ground swaths, astonishing geographical coverage can be achieved. The ASCAT (Advanced Scatterometer) scatterometers onboard the Metop-A and Metop-B satellites each provide daily coverage of about 82 % of the Earth's surface. Furthermore, scatterometer data have been provided free of charge. Scatterometer data are therefore attractive to applications for which spatial coverage, frequent observations and long time series are more important than high spatial resolution.

After a description of the study site, the focus is turned to the mapping of permanent water bodies and floods using SAR time series data. A map of permanent water bodies covering the entire LMB derived from several years of SAR time series data is presented. Potentials and challenges of flood monitoring with SAR are illustrated with ASAR imagery showing the evolution of the floods that occurred around Tonle Sap Lake in Cambodia in 2011. The last section is concerned with the retrieval of soil moisture using scatterometer time series data and an analysis of the spatial and temporal dynamics of soil moisture across the basin. A summary and an outlook can be found in the end of the chapter.

17.2 The Lower Mekong Basin

The Mekong River rises in the Tanggula mountain range in the Qinghai province in China and runs more than 4,300 km through the Tibet Autonomous region, the Yannan province of China, Myanmar, Laos, Thailand, Cambodia and Vietnam before reaching the South China Sea. It drains the Mekong River Basin, which comprises a large network of tributaries and watersheds covering an area greater than 795,000 km^2. The south part of the Mekong River Basin, shown with a red outline in Fig. 17.1, is known as the Lower Mekong Basin (LMB).

The climate of the LMB is strongly influenced by the rainy southwest monsoon occurring between May and October and dry northeast monsoon in the period of October to March. Rainfall is strongly seasonal with about 90 % of precipitation between the May and October. The mean annual precipitation ranges from 1,000 mm in northeast Thailand to more than 3,200 mm in the mountainous regions of Laos (Leinenkugel et al. 2013). The annual range of mean temperature between the hottest and coldest months throughout the region is just five degrees, with somewhat lower mean temperatures towards the east and in the Central Highlands than elsewhere in the LMB. Mean annual evaporation is 1,500 mm, varying from 1,000 mm in the Central Highlands to 2,000 mm in the Khorat Plateau. Due to the high relative air humidity, annual evaporation exhibits little variability from year to year (MRC 2010).

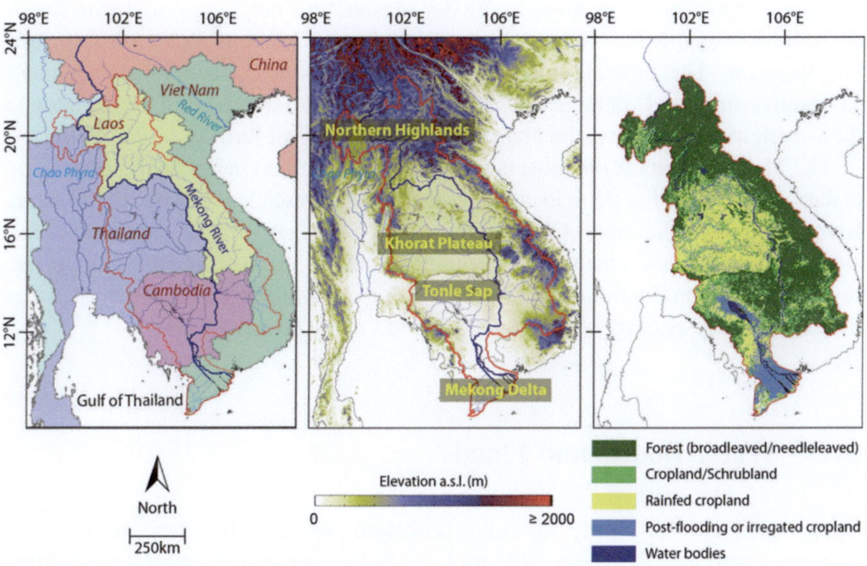

Fig. 17.1 Overview of the geography of the Lower Mekong Basin, shown with a red outline in the three maps. *Left*: the flow of the Mekong River through the basin. *Middle*: topography and the LMB's major sub-basins. *Right*: land cover according to the GlobCover 2009 dataset (Bicheron et al. 2008)

The following paragraphs give an overview of the geography and climatology of the LMB in terms of its four major sub-basins shown in the middle graphic in Fig. 17.1.

The Northern Highlands are mountainous with peaks above 2,000 m and valley floors more than 600 m below the mountain crests. It is the region with the highest amount of rainfall in the LMB. It comprises relatively sparsely populated regions in the northern parts of Thailand and Laos and the upland region of northeastern Myanmar.

The Khorat Plateau lies largely within northeastern Thailand. Its vast, low-lying terrain consists mainly of sediments and eroded bedrock. It is the driest region in the LMB with the highest evapotranspiration rate despite the annual rainfall between 1,000 and 1,600 mm (MRC 2010).

The Tonle Sap Basin extends across the majority of Cambodia. It contains the largest freshwater lake in Southeast Asia, called Tonle Sap Lake or simply the Great Lake (see right graphic in Fig. 17.1). During the dry season, the large but shallow Great Lake drains slowly via the Tonle Sap River into the Mekong River at their confluence in Phnom Pehn. The flow of the Mekong River follows a distinct annual cycle with a single flood pulse during the wet season. It usually starts rising in May and peaks in September or October. As the Mekong River rises above the level of the Great Lake, the flow of the Tonle Sap River is reversed, filling up the lake together with its surrounding floodplains. The area of the lake typically expands from about 3,000 km^2 to more than 15,000 km^2 (Kite 2001; Kuenzer 2013). The lake serves as a natural reservoir storing flood water from the surrounding watersheds and regulating river flows in the dry season, thus helping to moderate floods and relieve droughts in southern Cambodia and in the Mekong Delta in southwestern Vietnam. The annual inundation of the Cambodian floodplains can incur substantial structural damage but is at the same time an essential contributor to the wealth of biodiversity, the abundance of fish and soil fertility in the region.

The Mekong Delta is the most densely populated region in the LMB. The region is flat with 90 % of the delta located below 3 m above sea level. Here, the Mekong River drains into the South China Sea through a vast network of natural distributaries and man-made channels. The delta is one of the world's most important regions for rice production and at the same time one of the most endangered places on Earth with respect to sea level rise (Renaud and Kuenzer 2012).

17.3 Water Bodies and Floods

Microwaves exhibit nearly specular reflection on smooth water surfaces. As scatterometers and SARs are side-looking instruments, such specular reflection results in a large portion of the incident radiation being directed away from the radar, with a resulting low backscatter intensity. Other land covers such as bare soil, grassland, agricultural land and forests commonly exhibit higher backscatter as they are characterized by a certain degree of isotropic volume scattering or diffuse surface scattering. Water surfaces can therefore often be distinguished in SAR images as features of low backscatter. This fact is used as a fundamental assumption in SAR based flood mapping and water body mapping algorithms. However, local environmental conditions such as rain and wind may roughen water surfaces resulting in high backscatter from such features. In shallow waters, vegetation emerging above the water surface may result in so-called double bounce, whereby the radiation scattered off the water surface is re-directed towards the radar by the emerging vegetation, resulting in high backscatter values (Henderson and Lewis 1998). Microwaves tend to interact most strongly with objects having physical sizes similar to the wavelength of the radiation. In the case of vegetation, C-band microwaves ($\lambda = 3.75$–7.5 cm) will tend to interact most strongly with leaves and small branches while at L-band ($\lambda = 15$–30 cm), they will interact most strongly with larger branches and trunks. L-band is therefore capable of greater penetration through the vegetation and has been used to map flooded areas under forests (Hess et al. 2003; Frappart et al. 2005; Martinez and Le Toan 2007; Rosenqvist et al. 2007; Alsdorf et al. 2007). Zhang et al. (2014) reported high contrast between water bodies and other targets in both HH and VV polarisations in the area of Tonle Sap Lake with the use of high spatial resolution RADARSAT-2 imagery (C-band). They used polarimetric information to classify water bodies, flooded shrub, flooded forest and several other land cover types.

17.3.1 Permanent Water Bodies

The mapping of permanent water bodies can benefit from the use of time series data, whereby uncertainties related to highly dynamic local conditions such as wind and rain can be reduced through temporal filtering. Such filtering can also reduce random fluctuations stemming from thermal noise and speckle. The influence of local environmental conditions on SAR imagery is exemplified in Fig. 17.2 with two 150 m resolution images acquired by the ENVISAT Advanced Synthetic Aperture Radar (ASAR, C-band) instrument in Wide Swath (WS) around Great Lake in Cambodia. Such data have been used for water mapping purposes in several studies, e.g. (Bartsch et al. 2012; Kuenzer et al. 2013b; Matgen et al. 2011). While the right image in the figure shows an acquisition on June 8 2009, the left image shows the result of averaging a stack of 17 images acquired during the dry seasons (1**st** December to 31**st** March) in the years 2007–2011. Only dry season imagery was used in order to reduce the influence of large scale floods which occur mainly during the wet season. The mean backscatter image on the left in Fig. 17.2 represents near ideal conditions for water body mapping with a high contrast between the vast majority of water and terrain pixels. The lake as well as the Tonle Sap River

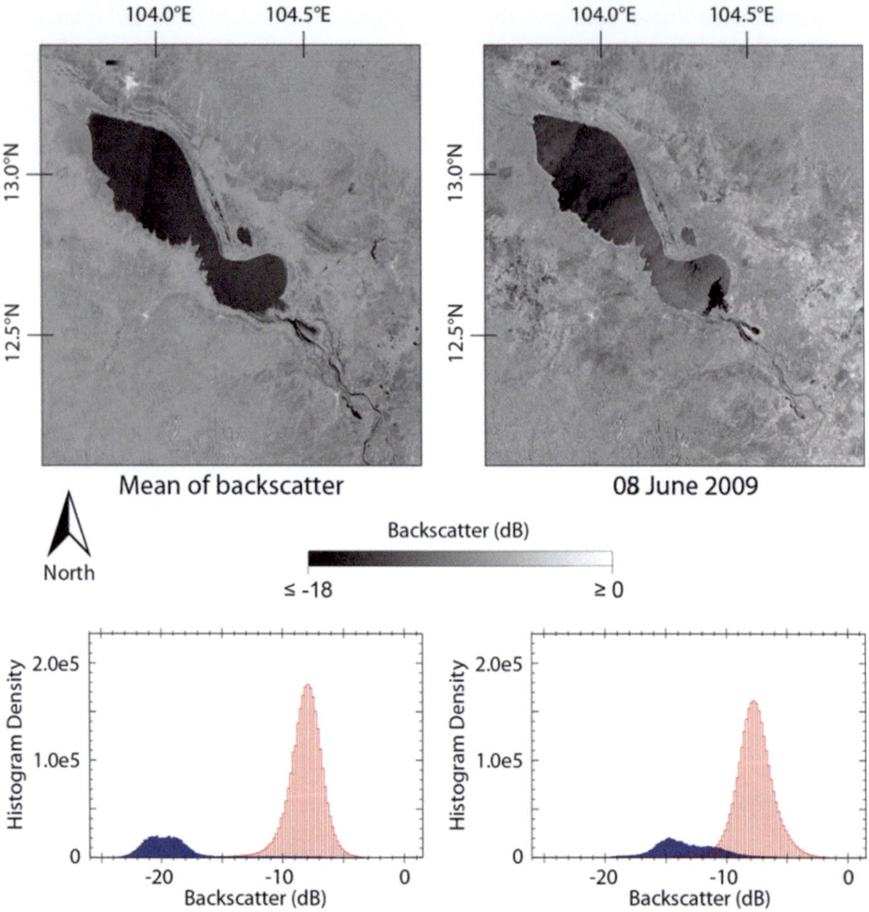

Fig. 17.2 The upper left graphic shows an image of backscatter averaged over time, allowing straightforward separation of water bodies (dark pixels) from land. The corresponding backscatter distributions, with water pixels in *blue* and terrain pixels in *red*, are shown in the *left* histogram. The *upper right* graphic illustrates the effects of wind and rain roughened water surfaces with an image acquired on June 8 2009. The resulting water and terrain backscatter distributions overlap, complicating accurate classification of water bodies (*left* histogram)

emanating from the lake's south-east shore can be well distinguished. The acquisition on 8th June 2009 is affected by wind and/or rain which increased the surface roughness over large parts of the lake, resulting in higher backscatter.

Also the river exhibits higher backscatter than in the mean backscatter image. The corresponding backscatter distributions of water and terrain overlap, making the classification of water bodies less straightforward.

Radiometric thresholding is a simple but efficient and widely used method for water body classification with SAR. Here, the classification thresholds for the images in Fig. 17.2 were determined with Otsu's method (Otsu 1979) and found at −11.5 and −10.3 dB for the left and right image respectively. With the 300 m resolution GlobCover 2009 dataset as reference, the omission error for classification of water was 10.1 % with the single ASAR image and 1.9 % with the mean backscatter image. The corresponding Kappa coefficient increased from 0.71 for the single image to 0.94 in case of the mean image. Due to the mismatch in spatial resolution as well as uncertainties in the GlobCover 2009 dataset, these results should be interpreted in relative rather than absolute terms. They clearly demonstrate the benefit of temporal aggregation for classification of permanent water bodies.

Based on the time series coverage that was build up over several years by the ASAR instrument, it was possible to extend the exercise in Fig. 17.2 to the entire LMB. Acquisitions were available on average every two to 4 weeks, depending on location. A total of 160 ASAR WS datasets acquired during the dry seasons in the period from 2007 to 2011 were processed. The datasets were radiometrically calibrated and geocoded with the SAR processing software Next ESA SAR Toolbox (NEST) and subsequently resampled to a regular grid. In order to combine data acquired at different incidence angles, each measurement was normalized to a local incidence angle of 30° using a linear model fitted to the time series of backscatter and local incidence angles at each location. Such linear models have been used for C-band data in the past, e.g. by Frison and Mougin (1996), Gauthier et al. (1998), Loew et al. (2006) and Sabel et al. (2012). The entire processing was integrated and managed with the use of the SAR Geophysical Retrieval Toolbox (SGRT) developed by Vienna University of Technology (TU Wien). The resulting 150 m resolution water body map is shown in Fig. 17.3, overlaid on the mean backscatter image. The dominating features at basin scale are the course of the Mekong River, the Great Lake in Cambodia and several large lakes in Thailand and Laos. Urban areas can be seen as features of high backscatter resulting from double bounce of the microwaves against surfaces and buildings.

While the water body map was produced mainly to demonstrate the capability of C-band SAR time series data, it provides a valuable source of information that complement currently available water body products. For instance, it provides an improvement in resolution relative the 250 m MODIS (Moderate Resolution Imaging Spectroradiometer) Global Raster Water Mask product (Carroll et al. 2009) and represents a more recent epoch than the SRTM Water Body Data product which was based mainly on SAR data acquired by the Shuttle Radar Topography Mission (SRTM) in February 2000.

17.3.2 Flood Mapping

No single SAR flood mapping technique can be considered appropriate for all SAR images or all types of environments. Various mapping techniques have been explored, including simple visual interpretation, supervised classification

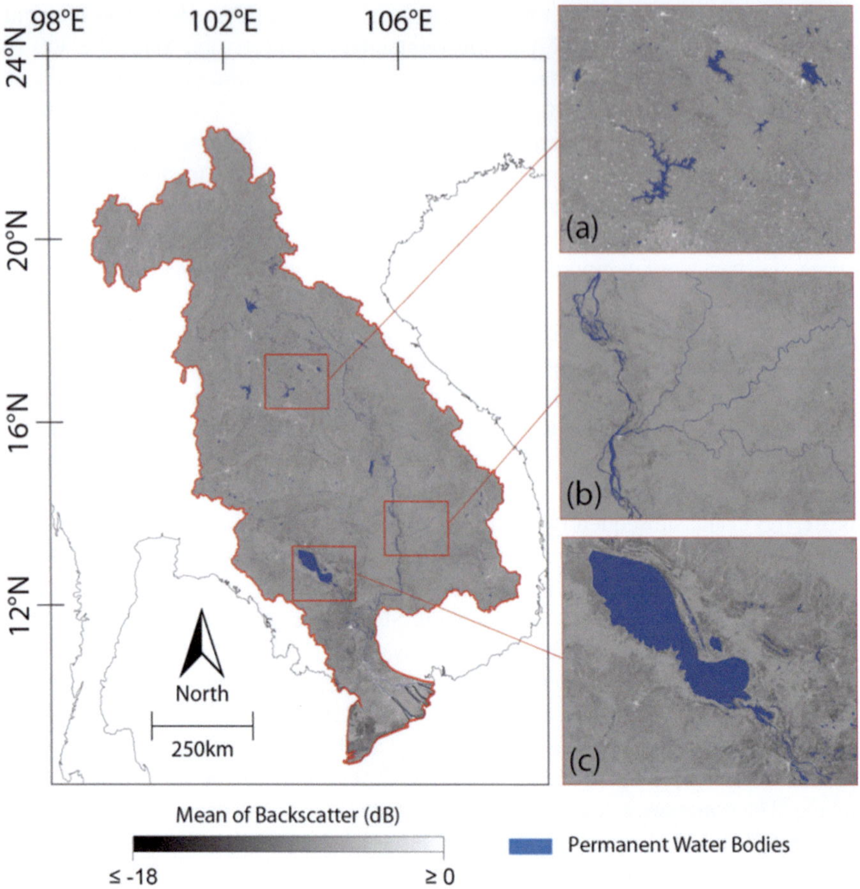

Fig. 17.3 Map of permanent water bodies in the Lower Mekong Basin based on 150 m resolution ENVISAT ASAR time series data. Water bodies (in *blue*) are superimposed on the mean backscatter image. Three features are shown in more detail: (**a**) Lakes in northeast Thailand and a multitude of towns visible as spots of high backscatter. (**b**) Confluences between (from north to south) Tonle Kong, Tonle San and Tonle Srepok rivers before joining the Mekong River in northern Cambodia. (**c**) The great Tonle Sap Lake in Cambodia

(Townsend 2002), image segmentation and texture algorithms (Han et al. 2005; Pulvirenti et al. 2011) as well as radiometric thresholding, change detection and region growing approaches (Gstaiger et al. 2012; Kuenzer et al. 2013b; Matgen et al. 2011).

In very dry regions such as deserts, bare ground may exhibit backscatter levels similar to those of water surfaces. O'Grady et al. (2011) showed that misclassifications of non-flooded pixels due to low backscatter over terrain can be reduced with the use of image differencing approaches by relating each SAR image to a pre-flood reference. State of the art review on SAR flood mapping can be found e.g. in Schumann et al. (2009) and Di Baldassarre et al. (2011).

The purpose of this section is to illustrate some of the challenges and potentials of flood mapping with SAR in the LMB. Examples of SAR imagery are provided around the Great Lake in Cambodia, which experienced particularly large floods in 2011. The floods affected 1.64 million Cambodians and caused 247 deaths. More than 50,000 households had to be evacuated and 220,000 ha of rice crops were destroyed. The reasons for the large floods were strong southwest monsoon activity starting in early June and the appearance of an Inter-Tropical Convergence Zone during whole of September. This resulted in intense rainfall, in particular in the middle and lower reaches of the LMB. The onset of a series of tropical storms early in the wet season also contributed to Mekong mainstream water levels exceeding long-term averages (MRC 2011). ASAR WS imagery capturing the evolution of the inundation around the Great Lake in 2011 is shown in Fig. 17.4. The permanent water body mask presented in the previous section has been overlaid to ease the identification of flooded areas. With the use of the ScanSAR technique, it was possible for the ASAR instrument to achieve a wide swath of 405 km, allowing the lake, the surrounding floodplains and flooded areas south of Phnom Penh to be mapped in a single acquisition. Near the end of the dry season the Great Lake occupied its nominal extent, as can be seen in the upper left graphic in Fig. 17.4. With increased flow of the Mekong River and subsequent reversal of the Tonle Sap River, the floodplains around the Great Lake gradually became inundated with vast areas flooded by the end of September (lower left graphic). During the second week of October, water levels in the Cambodian floodplains reached critical levels, especially at Phnom Penh.

In 2008, Cambodia did not experience any unusual floods (MRC 2009) and the inundation around Great Lake reached average proportions. A comparison of the flooding situation in the end of November in 2008 and 2011 is shown in Fig. 17.5. As before, dark pixels corresponding to low backscatter represents predominantly open water surfaces. Flood patterns are visible in both images (compare with upper left graphic in Fig. 17.4) and the vast extent of the floods in 2011 become evident, in particular to the east and southeast of the Great Lake. However, the 2011 image exhibits lower contrast than the 2008 image in the southeast part of the mapped area. The reason is different observation geometries. Both images were acquired while the satellite was moving along southward orbital tracks. As the ASAR instrument was right-looking relative to its flight direction, the eastern parts of the images were acquired at the steepest incidence angles. The 2011 acquisition was however acquired from a satellite track further west than in 2008. The eastern parts of the images were therefore observed at steeper incidence angles in 2011, with the region around Phnom Penh observed at an incidence angle of 19.3° in 2011, compared to 32.8° in 2008. Backscatter from water surfaces increases with decreasing incidence angle as more radiation is scattered back to the radar. Hence the higher contrast in the 2008 image. This example illustrates the importance of taking into account the observation geometry for interpretation of SAR images and

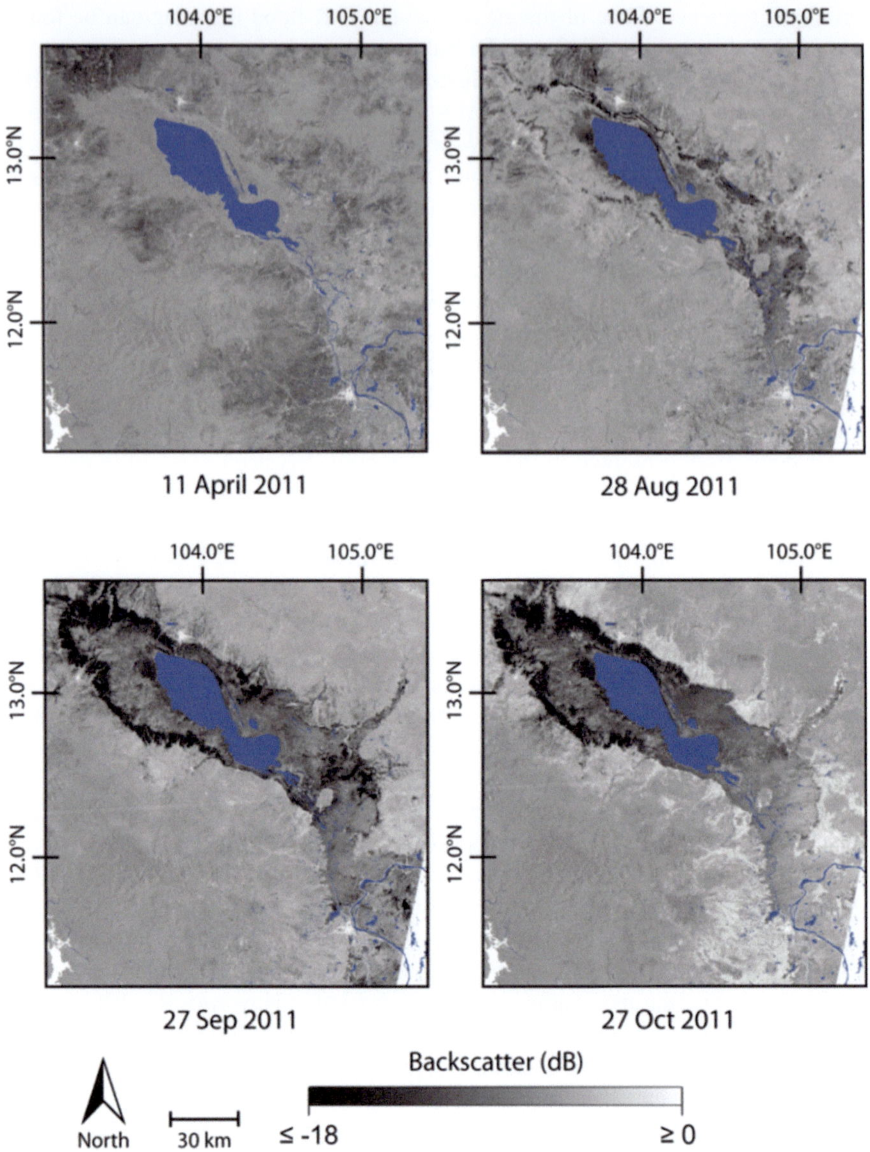

Fig. 17.4 Examples of C-band SAR imagery capturing the evolution of the large scale inundation around Tonle Sap Lake in 2011. Permanent water bodies are shown in *blue*

together with Fig. 17.4 demonstrates the high potential of C-band SAR for mapping large scale floods in the LMB.

As floods can evolve rapidly, imagery should be provided as often as possible and with the least possible time latency after acquisition. As mentioned previously, long revisit times have been a weakness of SAR missions. Observation strategies

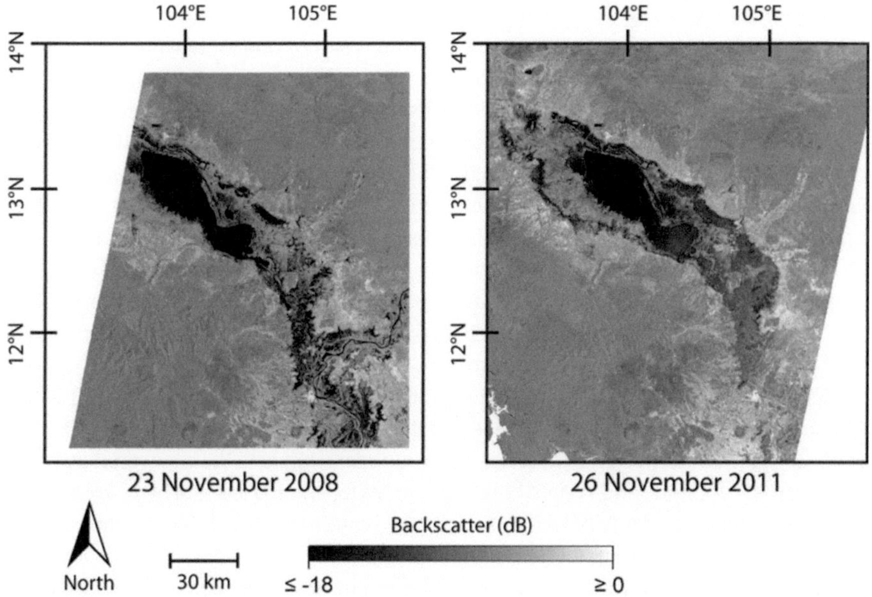

Fig. 17.5 Flood situation around the Great Lake by the end of November in 2008 and in 2011

have to varying degrees been determined by user requests. This often resulted in conflicting request as well as heterogeneous data coverage with missing or sparse time series data in some regions.

The ASAR instrument provided measurements of the region around Great Lake every 18 days on average in 2011, which can be considered as relatively frequent. Targeted acquisition planning can shorten observation intervals at specific locations but opportunities for acquisition are still subject to the satellite's position in the orbit as well as potential conflicts with other user requests.

17.4 Soil Moisture Monitoring

Soil moisture retrieval from spaceborne radar has been a field of active research since the 1970s. Comprehensive state-of-the-art reviews can be found in Barrett et al. (2009) and Kornelsen and Coulibaly (2013). Soil moisture retrieval in the microwave domain is attractive due to the high sensitivity to soil and plant water content, resulting from the high relative dielectric constant of liquid water compared to most other naturally occurring materials. The fact that microwaves can penetrate a short distance (centimetres) into the soil is beneficial for soil moisture retrieval.

Soil moisture is highly dynamic in space and time. Small scale variations are related to soil, vegetation and topography. As a result of atmospheric forcing in the form of precipitation and evaporation processes, soil moisture also exhibits a large scale component. Vinnikov et al. (1996) observed spatial correlation lengths of soil moisture in the order of 400–800 km in the upper 1 m of the soil in Russia. Entin et al. (2000) reported correlation lengths of several 100 km for test sites in USA, China, Russia and Mongolia. Traditional in situ measurements are point measurements, which often limits their spatial representativeness. Furthermore, in the LMB as in many other parts of the world, in situ soil moisture data are scarce or nonexistent. Remote sensing techniques on the other hand inherently provide a real measurements and extensive geographical coverage albeit with lower temporal sampling than in situ methods can provide.

The main challenges of soil moisture retrieval with radar are to account for the influence from vegetation and surface roughness on the backscatter signal (Barrett et al. 2009; Verhoest et al. 2008). Surface roughness impacts the measured average radar cross section and contributes to speckle. Increased surface roughness increases the proportion of incident energy scattered back towards the sensor. This effect is dependent on the surface roughness, the radar's frequency and polarisation configuration as well as on the angle of incidence, with the greatest sensitivity to surface roughness at shallow incidence angles. The variability of roughness, e.g. due to tilling characteristics in agricultural fields, and the difficult to accurately parameterize it for use in radar backscatter models means that surface roughness often poses a major problem for soil moisture retrieval with high resolution SAR systems (Ulaby et al. 1986; Verhoest et al. 2008; Wagner et al. 2007).

Microwaves penetrate through vegetation to some extent depending mainly on the plant water content and biomass and the radar carrier frequency. Dense and moist forests are usually opaque to C-band radar, while sparse forests, grassland and agricultural crops are partly transparent (Wagner et al. 2013). Vegetation therefore reduces the radar's sensitivity to soil moisture, while at the same time contributing to total backscatter intensity.

The ASCAT scatterometers onboard the Metop-A and Metop-B satellites, launched in 2006 and 2012 respectively, provide daily measurements globally with a spatial resolution of 25 km. At this resolution, fine scale variability is largely averaged out, making changes in surface roughness much less of an issue than in the case of SAR. Furthermore, the simultaneous triple-incidence angle measurements acquired by the instrument's for-, mid- and aft beams can be related to vegetation density, thereby providing a possibility to correct for the influence of vegetation. These features have allowed the development of global soil moisture products representing degree of soil moisture saturation in the pores of the soil in the surface layer (Surface Soil Moisture, SSM) and in the soil profile (Soil Water Index, SWI) (Naeimi et al. 2009; Wagner et al. 1999, 2013). The SSM and SWI products have been used and evaluated in numerous studies, e.g. (Albergel et al. 2009; Brocca et al. 2012; Brocca et al. 2010; Ceballos et al. 2005; Dorigo et al. 2010; Parajka et al. 2006; Scipal et al. 2008). The ASCAT soil moisture products are distributed in near-real-time by European Centre for Medium-Range Weather Forecasts (EUMETSAT) since 2008 and are used in an operational fashion by some Numerical Weather Prediction centres (de Rosnay et al. 2012; Dharssi et al. 2011).

In the following section, the spatial and temporal variability of soil moisture in the Lower Mekong Basin is analysed with the SWI approach applied to scatterometer data acquired by the ASCAT instrument acquired between 2007 and 2011 and by the ERS-1 and ERS-2 satellites in the period from 1991 to 2001.

17.4.1 Scatterometer Data Processing

The Scatterometer time series data were processed with the TU Wien soil moisture retrieval algorithm (Naeimi et al. 2009; Wagner et al. 1999, 2013). The algorithm is based on a change detection method that requires knowledge of backscatter levels representative of dry and wet conditions at each location. The method assumes a linear relation between surface soil moisture content and backscatter expressed in decibels. Variations in surface roughness are assumed to average out at the 25 km scale and therefore not to have a significant influence on temporal changes of backscatter. Backscatter levels for dry ($\sigma^0{}_{dry}$) and wet ($\sigma^0{}_{wet}$) soil conditions are derived at each location with a statistical analysis of the scatterometer time series. The difference between $\sigma^0{}_{dry}$ and $\sigma^0{}_{wet}$ represents the sensitivity to soil moisture changes.

The backscatter triplets acquired by the scatterometers' three beams are used to deduce the incidence angle behaviour of backscatter as a seasonal function with a daily time step. This function is used to normalize all backscatter measurements in the time series to a reference incidence angle of 40° and to remove the contribution from vegetation.

The SSM is computed by subtracting $\sigma^0{}_{dry}$ from the normalized backscatter values and dividing the difference by the sensitivity to soil moisture changes. The SSM is given in percent and is interpreted as the degree of saturation in the soil pores of the first few centimetres of the soil surface layer. One of the main challenges with the method is to obtain accurate estimates of $\sigma^0{}_{dry}$ and $\sigma^0{}_{wet}$ in regions where the soil rarely dries up or becomes completely saturated.

The soil moisture content in the deeper layers cannot be directly measured, but is closely related to the history of moisture in the surface layer via the process of infiltration. This allows the SSM time series to be used to estimate the moisture content in the deeper layers. Soil moisture varies more quickly over time near the surface than deeper down in the soil. Assuming that the water content in the lower layers is solely attributable to the history of moisture conditions in the surface layer,

Wagner et al. (1999) applied a simple infiltration model to the irregularly sampled time series of SSM measurements to compute the Soil Water Index (SWI):

$$SWI(t_n) = \frac{\sum_i^n SSM(t_i)e^{-\frac{t_n - t_i}{T}}}{\sum_i^n e^{-\frac{t_n - t_i}{T}}} \quad for\, t_i \leq t_n \tag{17.1}$$

Here, t is the time of the SSM measurement and T is the characteristic time length linking the deeper soil layer with the surface layer. T can be computed for a specific depth L according to $T = L/C$, where C is an area-representative pseudo-diffusivity constant. Depending on C, which differs between soil types, the value of T relates to different soil depths. Wagner et al. (1999) showed with in situ data over Ukraine that the model represents the first metre of soil best with a T-value of 20. Even though the T-value is expected to be somewhat different in the LMB, it was used as a reasonable approximation also in this study.

Finally the SWI measurements were spatially aggregated to the Basin Water Index (BWI) for each of 104 sub-divisions (sub-catchments), according to

$$BWI = \frac{\sum_i^N SWI_i}{N} \tag{17.2}$$

where N is the number of SWI measurements within the sub-catchment. The sub-catchments varied in size from 132 km^2 to more than 70,000 km^2, with an average size of about 6,000 km^2. Some of the sub-catchments were smaller than the resolution cell of the ASCAT scatterometer data, which raises the question if these measurements are representative for all sub-catchments. Several studies have shown that the ASCAT derived soil moisture often correlate well even with situ point scale measurements, e.g. Albergel et al. (2012), Brocca et al. (2012) and Matgen et al. (2012b). Therefore, it is reasonable to assume that in terms of spatial scale representativeness, the ASCAT derived measurements can be used for all the sub-catchments in the LMB.

17.4.2 Spatial and Temporal Soil Moisture Variability

The characteristic seasonal cycle of soil moisture condition across the basin is shown in Fig. 17.6. It was computed at the sub-catchment level by averaging the BWI for each calendar month cross the 14 years of scatterometer time series data. The widespread inundation occurring in the Tonle Sap Basin and the Delta during September through December prohibited the retrieval of soil moisture. Data

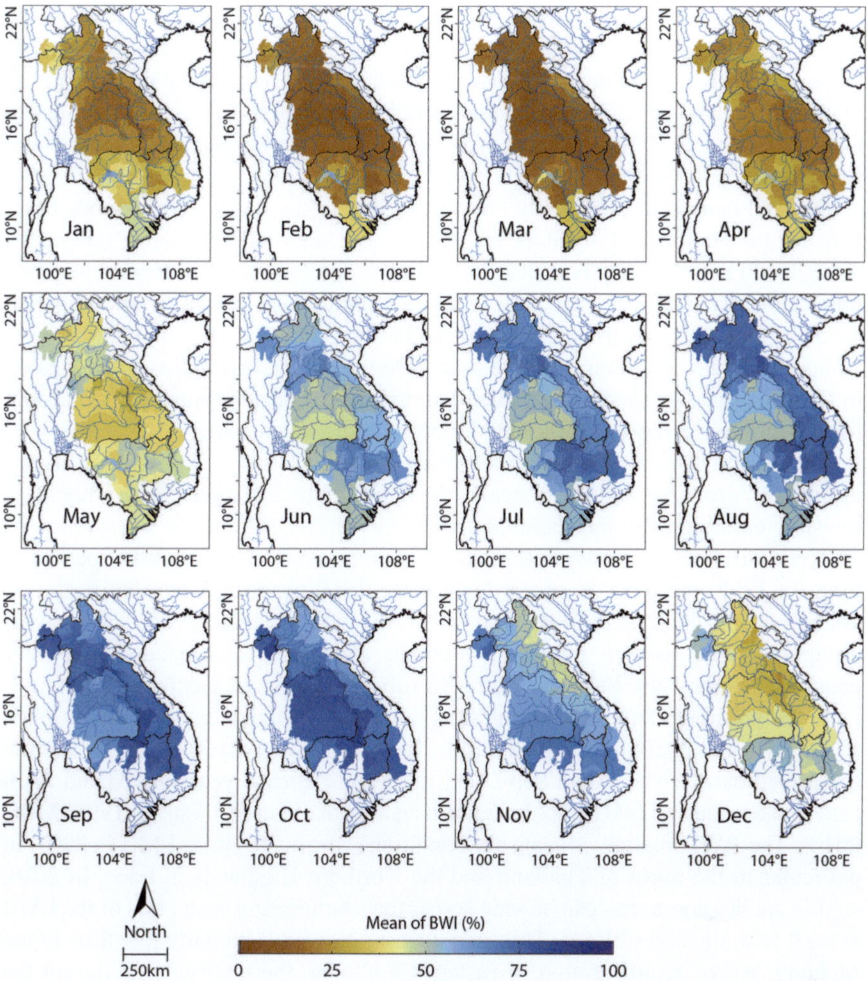

Fig. 17.6 Characteristic annual cycle of soil moisture conditions derived from Scatterometer time series observations acquired in the periods 1991–2000 and 2007–2011. Note that data acquired from September through December were excluded in the Tonle Sap Basin and the Delta due to widespread inundation

acquired over those catchments during those months of each year were therefore excluded.

As can be seen in Fig. 17.6, soil moisture conditions follow a clear seasonal cycle while exhibiting distinct spatial variability between the sub-catchments. January to April represents the driest period across the LMB while the wettest period occurs during June to October. The overall driest months are February and March. This seasonal cycle is strongly linked to the rainy southwest monsoon bringing precipitation between May and October and the dry northeast monsoon in the period of October to March.

During the wet season from May to September, the Khorat Plateau (see Fig. 17.1) exhibit on average dryer conditions than other regions. In October however, the Khorat Plateau is on average wetter than most other regions. In the first half of November soil moisture contents begins to fall across the basin. Cambodia and the Delta region typically remain wetter than the central and northern parts of the basin during December and January.

Drought can occur during any part of the year in the LMB. While floods bring benefits in the form of habitats for aquatic life and soil fertilizing sediments, droughts do not offer any benefits. Prolonged droughts have considerable negative impacts on fisheries and agriculture. The inter-annual soil moisture variability can be analysed with the use of anomalies. In this study, annual anomalies were computed as the deviation from the long term mean. The anomalies are expressed in terms of standard deviations rather than in units of degree saturation to account for the different temporal variability among the sub-catchments.

The annual BWI anomalies for the years 1991–2000 and 2007–2011 are shown in Fig. 17.7. In 1999 and 2000, nearly the entire LMB experienced very high soil moisture conditions. In the Cambodian lowlands and the Delta, the wet conditions in 2000 correlates with severe flooding that occurred over an extended period of time. In 1992, the most severe drought since 1960 occurred when the peak and volume of the flood were more than 40 % below the average (MRC 2005). This drought can be seen in the BWI anomaly which indicates exceptionally dry conditions in the Nam Mun and Nam Chi sub-catchments in Thailand.

In the Mekong Delta, the strongest negative BWI anomalies occurred in 1997, 1998, 2007 and 2010. In 1998, conditions were especially dry in the Delta and the Tonle Sap basins. The Tonle Sap Lake, which in a normal year can expand to an area of more than 15,000 m^2, only reached an area of about 7,000 m^2 in 1998 (MRC 2010). The BWI anomaly reveals dry conditions across the entire LMB in 2007, in particular in the north of Thailand and the Northern Highlands in Laos. In 2010, significant dry anomalies can be seen across the southern and west parts of the LMB as well as in the east of Laos. This coincides with an exceptionally low flow of the Mekong. At the Kratie station in eastern Cambodia, the total volume during the 2010 flood season was even lower than the corresponding measurements for 1992, which is generally regarded as the most severe drought year on record (MRC 2010). The BWI anomalies show that these dry conditions remained in 2011 in the east of Laos and the northern parts of the Tonle Sap basin. Further analysis as well as validation results of the scatterometer derived soil moisture data can be found in Naeimi et al. (2013).

17.5 Summary and Outlook

This chapter investigated the use of radar time series for the mapping of permanent water bodies, floods and the retrieval of soil moisture content in the Lower Mekong Basin (LMB) in Southeast Asia.

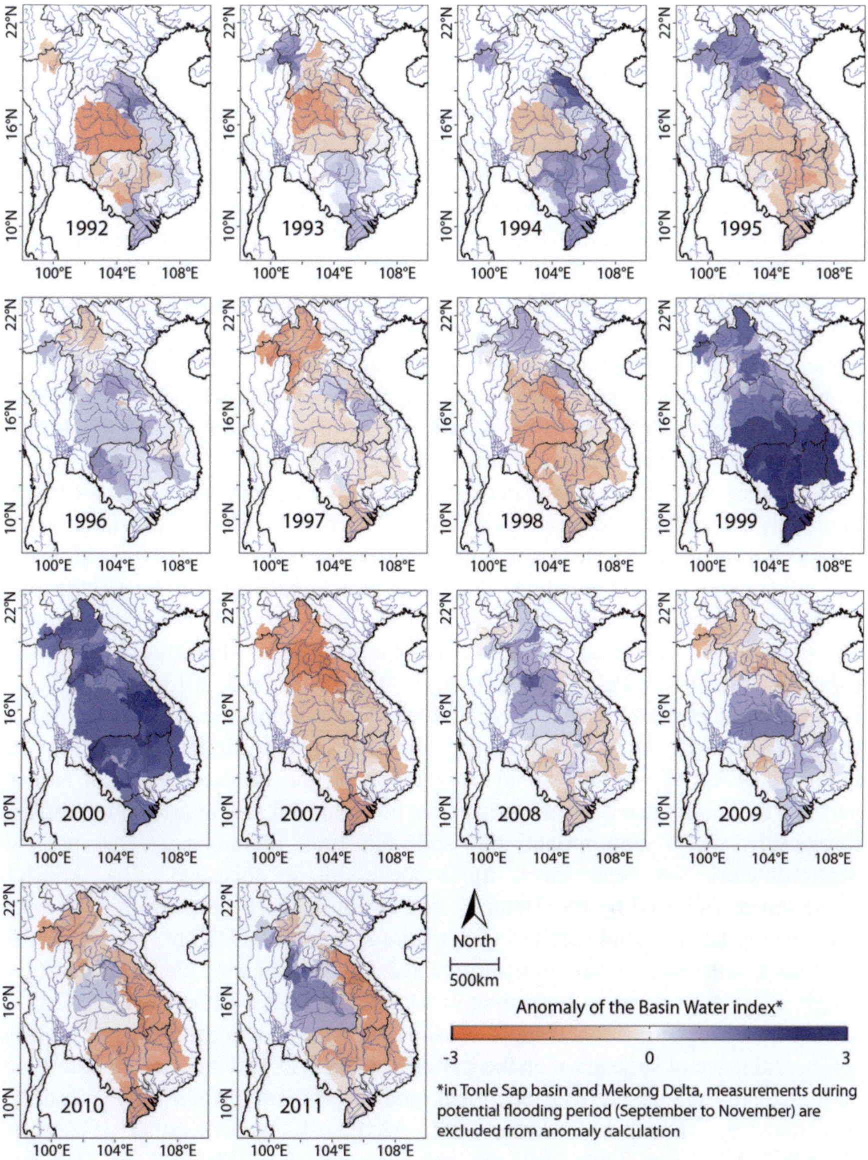

Fig. 17.7 Annual soil moisture anomalies relative the long term average given in units of standard deviations

The strong interaction of microwaves with water makes radars suitable for the mapping of water bodies and retrieval of soil moisture content. Key selling points for radar instruments are their capability of acquiring imagery independent of sun light and weather conditions. The latter is particularly valuable in areas frequently covered by clouds such as the Mekong Basin.

The high temporal variability of flood extents and soil moisture requires frequent observations. While SAR instruments can provide high spatial resolution down to about 1 m, they are associated with long revisit times. SAR instruments commonly offer modes of operation with increased geographical coverage at the expense of spatial resolution, resulting in denser time series data. Despite this, observation intervals for SAR instruments have been ranging from weeks to months on average.

The benefit of SAR time series for mapping permanent water bodies was demonstrated with the use of 160 ENVISAT ASAR Wide Swath images acquired in the dry seasons in the period between 2007 and 2011, providing a new 150 m resolution water mask of the entire LMB. The mask was shown to correlate well with the GlobCover 2009 dataset and clearly demonstrated the benefit of temporal aggregation for classification of permanent water bodies.

Some of the challenges and potentials of flood mapping with SAR were illustrated with a series of ENVISAT ASAR images of the region around Tonle Sap Lake, showing the evolution of the particularly large floods in 2011.

A number of commercial and non-commercial organisations have the capability to quickly create flood extent maps based on SAR data in response to a flood event (Westerhoff et al. 2013). However, automatic approaches are needed for systematic and fully objective flood mapping. Automatic approaches will also shorten the time from acquisition to delivery, which is critical in the context of crisis management. The development of robust automatic SAR flood mapping algorithms is on-going (Matgen et al. 2012a; Pulvirenti et al. 2011; Westerhoff et al. 2013).

The operational use of SAR for high resolution soil moisture retrieval has yet to become a reality, in part due to the complexity of accurately modelling the influence of surface roughness and vegetation on the backscatter signal and in part due to the fact that SAR missions have not achieved the frequent observations (preferably daily) appropriate for soil moisture monitoring. The use of Scatterometers has been much more successful to this end. The ASCAT scatterometers carried by the Metop-A and Metop-B satellites provide daily coverage of the Earth's land surfaces at a spatial resolution of 25 km. At this spatial resolution, fine scale variability is largely averaged out, making changes in surface roughness much less of an issue than in the case of SAR. Furthermore, ASCAT's simultaneous triple-incidence angle measurements provide a possibility to correct for the influence of vegetation on the signal. The ASCAT soil moisture products are systematically processed over the global land surface and made available in near-real-time by European Centre for Medium-Range Weather Forecasts (EUMETSAT) since 2008. With the expected launch of Metop-C in 2018 and plans for ASCAT's successor instrument, which will be flown onboard of one of the Second Generation satellites of the EUMETSAT Polar System well underway (Lin et al. 2012), continuity of scatterometer data for soil moisture retrieval is foreseen well into the 2020s. The spatial and temporal variability of soil moisture in the Lower Mekong Basin was analysed with Soil Water Index (SWI) measurements based on ASCAT time series data acquired between 2007 and 2011 as well as compatible scatterometer data acquired by the ERS-1 and ERS-2 satellites in the period from 1991 to 2001. The 14 year time series was used to produce monthly composites of the mean annual cycle of soil moisture across the LMB. Furthermore, annual anomalies of soil moisture relative the long term mean revealed large inter-annual variability, some of which could be linked to flooding events, droughts and anomalies in the flow of the Mekong River.

On 3rd April 2014, the first of the Sentinel-1 satellites was launched, carrying a C-band SAR instrument that provides continuity of measurements with ENVISAT ASAR. The Sentinel-1 mission will acquire imagery in a systematic fashion with the use of pre-programmed acquisition plans where user requests are accepted only in extraordinary cases. Each of the two Sentinel-1 satellites will potentially provide global data coverage over land with a resolution of 20 m at least every 12 days (Snoeij et al. 2011), which will allow the build-up of a more homogeneous data coverage and denser time series than previously possible with SAR. This will open up new possibilities such as more frequent and more detailed updates of permanent water bodies than previously possible with SAR instruments, allowing more detailed studies e.g. of the spatial and temporal dynamics of surface freshwater discharge. The Sentinel-1 data also significantly improve the possibility for high resolution, global monitoring of surface soil moisture (Hornacek et al. 2012) and more detailed soil moisture studies at regional to global scales. Systematic and regular acquisitions are also beneficial for flood monitoring as they provide pre-flood reference images required for change detection algorithms.

Acknowledgements The work presented in this chapter was partially funded by the WISDOM project (Water related Information Management System for the Sustainable Development of the Mekong Delta), funded by the German Ministry of Education and Research, BMBF. ENVISAT ASAR data were provided by European Space Agency.

References

Albergel C, Rüdiger C, Carrer D, Calvet JC, Fritz N, Naeimi V, Hasenauer S (2009) An evaluation of ASCAT surface soil moisture products with in-situ observations in Southwestern France. Hydrol Earth Syst Sci 13(2):115–124

Albergel C, de Rosnay P, Gruhier C, Muñoz-Sabater J, Hasenauer S, Isaksen L, Wagner W (2012) Evaluation of remotely sensed and modelled soil moisture products using global ground-based in situ observations. Remote Sens Environ 118:215–226. doi:10.1016/j.rse.2011.11.017

Alsdorf DE, Rodriguez E, Lettenmaier DP (2007) Measuring surface water from space. Rev Geophys 45(RG2002):1–24. doi:10.1029/2006RG000197

Barrett BW, Dwyer E, Whelan P (2009) Soil moisture retrieval from active spaceborne microwave observations: an evaluation of current techniques. Remote Sens 1(3):210–242. doi:10.3390/rs1030210

Bartsch A, Trofaier AM, Hayman G, Sabel D, Schlaffer S, Clark DB, Blyth E (2012) Detection of open water dynamics with ENVISAT ASAR in support of land surface modelling at high latitudes. Biogeosciences 9(2):703–714. doi:10.5194/bg-9-703-2012

Bicheron P, Defourny P, Brockmann C, Schouten L, Vancutsem C, Huc M, Bontemps S, Leroy M, Achard F, Herol M, Ranera F, Arino O (2008) GLOBCOVER – products description and validation report. MEDIAS-France. Retrieved from http://publications.jrc.ec.europa.eu/repository/handle/JRC49240

Brocca L, Melone F, Moramarco T, Wagner W, Hasenauer S (2010) ASCAT soil wetness index validation through in situ and modeled soil moisture data in central Italy. Remote Sens Environ 114(11):2745–2755. doi:10.1016/j.rse.2010.06.009

Brocca L, Moramarco T, Melone F, Wagner W, Hasenauer S, Hahn S (2012) Assimilation of surface- and root-zone ASCAT soil moisture products into rainfall-runoff modeling. IEEE Trans Geosci Remote Sens 50(7):2542–2555. doi:10.1109/TGRS.2011.2177468

Carroll M, Townshend J, DiMiceli C, Noojipady P, Sohlberg R (2009) A New global raster water mask at 250 meter resolution. Int J Dig Earth 2(4)

Ceballos A, Scipal K, Wagner W, Martinez-Fernandez J (2005) Validation of ERS scatterometer-derived soil moisture data in the central part of the Duero Basin, Spain. Hydrol Process 19 (8):1549–1566

Covello F, Battazza F, Coletta A, Manoni G, Valentini G. (2009) COSMO-SkyMed mission status: three out of four satellites in orbit. In: International Geoscience and Remote Sensing Symposium (IGARSS), vol 2. ASI – Agenzia Spaziale Italiana, Viale Liegi 26, 00198, Rome, Italy. pp II773–II776

De Rosnay P, Balsamo G, Albergel C, Muñoz-Sabater J, Isaksen L (2012) Initialisation of land surface variables for numerical weather prediction. Surv Geophys 35:607–621

Dent JE (2012) Climate and meteorological information requirements for water management: a review of issues. World Meterological Organization (WMO). ISBN 978-92-63-11094-7

Dharssi I, Bovis KJ, Macpherson B, Jones CP (2011) Operational assimilation of ASCAT surface soil wetness at the Met Office. Hydrol Earth Syst Sci 15(8):2729–2746

Di Baldassarre G, Schumann G, Brandimarte L, Bates P (2011) Timely low resolution SAR imagery to support floodplain modelling: a case study review. Surv Geophys 32(3):255–269. doi:10.1007/s10712-011-9111-9

Dorigo WA, Scipal K, Parinussa RM, Liu YY, Wagner W, De Jeu RAM, Naeimi V (2010) Error characterisation of global active and passive microwave soil moisture datasets. Hydrol Earth Syst Sci 14(12):2605–2616

Drusch M, Del Bello U, Carlier S, Colin O, Fernandez V, Gascon F, Bargellini P (2012) Sentinel-2: ESA's optical high-resolution mission for GMES operational services. Remote Sens Environ 120:25–36

Entin JK, Robock A, Vinnikov KY, Hollinger SE, Liu SX, Namkhai A (2000) Temporal and spatial scales of observed soil moisture variations in the extratropics. J Geophys Res-Atmos 105(D9):11865–11877

Frappart F, Seyler F, Martinez J-M, Leon JG, Cazenave A (2005) Floodplain water storage in the Negro River basin estimated from microwave remote sensing of inundation area and water levels. Remote Sens Environ 99:387–399. doi:10.1016/j.rse.2005.08.016

Frison PL, Mougin E (1996) Use of ERS-1 wind scatterometer data over land surfaces. IEEE Trans Geosci Remote Sens 34(2):550–560

Gauthier Y, Bernier M, Fortin JP (1998) Aspect and incidence angle sensitivity in ERS-1 SAR data. Int J Remote Sens 19(10):2001–2006

Gstaiger V, Huth J, Gebhardt S, Wehrmann T, Kuenzer C (2012) Multi-sensoral and automated derivation of inundated areas using TerraSAR-X and ENVISAT ASAR data. Int J Remote Sens 33(22):7291–7304

Han C, Guo H, Shao Y, Liao J (2005) Detection of the flood boundary in SAR image using texture. In: International Geoscience and Remote Sensing Symposium (IGARSS), vol 5. State Key Laboratory of Remote Sensing Science, Institute of Remote Sensing Applications, Chinese Academy of Sciences, Beijing, 100101, China, pp 3697–3699

Henderson FM, Lewis A (1998) Principles and applications of imaging radar, vol 2, 3rd edn, Manual of remote sensing. Wiley, New York

Hess LL, Melack JM, Novo EMLM, Barbosa CCF, Gastil M (2003) Dual-season mapping of wetland inundation and vegetation for the central Amazon basin. Remote Sens Environ 87(4):404–428

Hornacek M, Wagner W, Sabel D, Truong HL, Snoeij P, Hahmann T, Doubkova M (2012) Potential for high resolution systematic global surface soil moisture retrieval via change detection using sentinel-1. IEEE J Select Top Appl Earth Obs Remote Sens 5(4):1303–1311

Kite G (2001) Modelling the Mekong: hydrological simulation for environmental impact studies. J Hydrol 253(1–4):1–13. doi:http://dx.doi.org/10.1016/S0022-1694(01)00396-1

Kornelsen KC, Coulibaly P (2013) Advances in soil moisture retrieval from synthetic aperture radar and hydrological applications. J Hydrol 476(0):460–489. doi:http://dx.doi.org/10.1016/j.jhydrol.2012.10.044

Kuenzer C (2013) Threatening Tonle Sap: challenges for Southeast Asia's largest freshwater lake. Pac Geogr 40:29–31

Kuenzer C, Campbell I, Roch M, Leinenkugel P, Tuan V, Dech S (2013a) Understanding the impact of hydropower developments in the context of upstream–downstream relations in the Mekong river basin. Sustain Sci 8(4):565–584. doi:10.1007/s11625-012-0195-z

Kuenzer C, Guo H, Huth J, Leinenkugel P, Li X, Dech S (2013b) Flood mapping and flood dynamics of the Mekong delta: ENVISAT-ASAR-WSM based time series analyses. Remote Sens 5(2):687–715

Legates DR, Mahmood R, Levia DF, DeLiberty TL, Quiring SM, Houser C, Nelson FE (2011) Soil moisture: a central and unifying theme in physical geography. Prog Phys Geogr 35(1):65–86. doi:10.1177/0309133310386514

Lehner B, Döll P (2004) Development and validation of a global database of lakes, reservoirs and wetlands. J Hydrol 296(1–4):1

Leinenkugel P, Kuenzer C, Dech S (2013) Comparison and enhancement of MODIS cloud mask products for Southeast Asia. Int J Remote Sens 34(8):2730–2748

Lin CC, Betto M, Rivas MB, Stoffelen A, de Kloe J (2012) EPS-SG windscatterometer concept tradeoffs and wind retrieval performance assessment. IEEE Trans Geosci Remote Sens 50 (7):2458–2472. doi:10.1109/TGRS.2011.2180393

Loew A, Ludwig R, Mauser W (2006) Derivation of surface soil moisture from ENVISAT ASAR wide swath and image mode data in agricultural areas. IEEE Trans Geosci Remote Sens 44 (4):889–899

Martinez J-M, Le Toan T (2007) Mapping of flood dynamics and spatial distribution of vegetation in the Amazon floodplain using multitemporal SAR data. Remote Sens Environ 108:209–223. doi:10.1016/j.rse.2006.11.012

Matgen P, Hostache R, Schumann G, Pfister L, Hoffmann L, Savenije HHG (2011) Towards an automated SAR-based flood monitoring system: lessons learned from two case studies. Phys Chem Earth 36(7–8):241–252

Matgen P, Giustarini L, Hostache R (2012a) A new automatic synthetic aperture radar-based flood mapping application hosted on the European Space Agency's grid processing on demand fast access to imagery environment. Proc SPIE Int Soc Opt Eng 8538

Matgen P, Heitz S, Hasenauer S, Hissler C, Brocca L, Hoffmann L, Savenije HHG (2012b) On the potential of MetOp ASCAT-derived soil wetness indices as a new aperture for hydrological monitoring and prediction: a field evaluation over Luxembourg. Hydrol Process 26 (15):2346–2359

Mekong River Commission (MRC) (2005) Overview of the hydrology of the Mekong Basin. Mekong River Commission, Vientiane, p 73

Mekong River Commission (MRC) (2009) Flood situation report 2008. Mekong River Commission, Vientiane

Mekong River Commission (MRC) (2010) State of the basin report 2010. Mekong River Commission, Vientiane

Mekong River Commission (MRC) (2011) Flood situation report 2011. Mekong River Commission, Phnom Penh

Morena LC, James KV, Beck J (2004) An introduction to the RADARSAT-2 mission. Can J Remote Sens 30(3):221–234

Naeimi V, Scipal K, Bartalis Z, Hasenauer S, Wagner W (2009) An improved soil moisture retrieval algorithm for ERS and METOP scatterometer observations. IEEE Trans Geosci Remote Sens 47(7):1999–2013. doi:10.1109/TGRS.2008.2011617

Naeimi V, Leinenkugel P, Sabel D, Wagner W, Apel H, Kuenzer C (2013) Evaluation of soil moisture retrieval from the ERS and Metop scatterometers in the lower Mekong Basin. Remote Sens 5(4):1603–1623. doi:10.3390/rs5041603

O'Grady D, Leblanc M, Gillieson D (2011) Use of ENVISAT ASAR global monitoring mode to complement optical data in the mapping of rapid broad-scale flooding in Pakistan. Hydrol Earth Syst Sci 15(11):3475–3494. doi:10.5194/hess-15-3475-2011

OECD (2008) Space technologies and climate change: implications for water management, marine resources and maritime transport. OECD Publishing, Paris. doi:10.1787/9789264054196-en

Otsu N (1979) Threshold selection method from gray-level histograms. IEEE Trans Syst Man Cyber SMC-9(1):62–66

Parajka J, Naeimi V, Blöschl G, Wagner W (2006) Assimilating scatterometer soil moisture data into conceptual hydrologic models at the regional scale, hydrology and earth system sciences. Hydrol Earth Syst Sci 10(3):353–368

Pitz W, Miller D (2010) The TerraSAR-X satellite. IEEE Trans Geosci Remote Sens 48 (2):615–622

Pulvirenti L, Pierdicca N, Chini M, Guerriero L (2011) An algorithm for operational flood mapping from Synthetic Aperture Radar (SAR) data using fuzzy logic. Nat Hazards Earth Syst Sci 11(2):529–540. doi:10.5194/nhess-11-529-2011

Renaud F, Kuenzer C (2012) The water-development nexus: importance of knowledge, information and cooperation in the Mekong Delta. In: Renaud F, Kuenzer C (eds) The Mekong Delta system – interdisciplinary analyses of a river delta. Springer, Dordrecht, pp 445–458. doi:10.1007/978-94-007-3962-8

Rosenqvist A, Finlayson CM, Lowry J, Taylor D (2007) The potential of long-wavelength satellite-borne radar to support implementation of the Ramsar Wetlands Convention. Aquat Conserv – Mar Freshwat Ecosyst 17(3):229–244

Sabel D, Bartalis Z, Wagner W, Doubkova M, Klein JP (2012) Development of a Global Backscatter Model in support to the sentinel-1 mission design. Remote Sens Environ 120 (The Sentinel Missions – New Opportunities for Science):102–112. doi:10.1016/j.rse.2011.09.028

Schumann G, Bates PD, Horritt MS, Matgen P, Pappenberger F (2009) Progress in integration of remote sensing-derived flood extent and stage data and hydraulic models. Rev Geophys 47 (4):1–20. doi:10.1029/2008RG000274

Scipal K, Drusch M, Wagner W (2008) Assimilation of a ERS scatterometer derived soil moisture index in the ECMWF numerical weather prediction system. Adv Water Resour 31 (8):1101–1112

Sentinel-1 Team (2013) Sentinel-1 user handbook. https://sentinel.esa.int. Accessed 30 Dec 2013

Snoeij P, Brown M, Davidson M, Rommen B, Floury N, Geudtner D, Torres R (2011) Sentinel-1A and sentinel-1B CSAR status. Proc SPIE 8179. doi:10.1117/12.898646

Townsend PA (2002) Estimating forest structure in wetlands using multitemporal SAR. Remote Sens Environ 79(2–3):288–304

Ulaby FT, Moore RK, Fung AK (1986) Microwave remote sensing: active and passive. Volume III: from theory to applications. Microwave remote sensing: active and passive. Artech House, Norwood

Verhoest NEC, Lievens H, Wagner W, Álvarez-Mozos J, Moran MS (2008) On the soil roughness parameterization problem in soil moisture retrieval of bare surfaces from synthetic aperture radar. Sensors 8:4213–4248. doi:10.3390/s8074213

Vinnikov KY, Robock A, Speranskaya NA, Schlosser CA (1996) Scales of temporal and spatial variability of midlatitude soil moisture. J Geophys Res 101(D3):7163–7174

Wagner W, Lemoine G, Rott H (1999) A method for estimating soil moisture from ERS scatterometer and soil data. Remote Sens Environ 70(2):191–207. doi:10.1016/S0034-4257(99)00036-X

Wagner W, Blöschl G, Pampaloni P, Calvet J-C, Bizzarri B, Wigneron J-P, Kerr Y (2007) Operational readiness of microwave remote sensing of soil moisture for hydrologic applications. Nord Hydrol 38(1):1–20

Wagner W, Hahn S, Kidd R, Melzer T, Bartalis Z, Hasenauer S, Rubel F (2013) The ASCAT soil moisture product: a review of its specifications, validation results, and emerging applications. Meteorol Z 22(1):5–33

Westerhoff RS, Kleuskens MPH, Winsemius HC, Huizinga HJ, Brakenridge GR, Bishop C (2013) Automated global water mapping based on wide-swath orbital synthetic-aperture radar. Hydrol Earth Syst Sci 17(2):651–663

Woodhouse IH (2006) Introduction to microwave remote sensing. Taylor and Francis, Boca Raton

Zhang L, Guo H, Li X, Wang L (2014) Ecosystem assessment in the Tonle Sap Lake region of Cambodia using RADARSAT-2 wide fine-mode SAR data. Int J Remote Sens 35(8):2875–2892. doi:10.1080/01431161.2014.890301

Chapter 18
Land Surface Phenology Monitoring with SeaWinds Scatterometer Time Series in Eastern Asia

Linlin Lu, Huadong Guo, and Cuizhen Wang

Abstract Vegetation phenology tracks plants' lifecycle events and reveals the response of vegetation to global climate change. Microwave backscatter is insensitive to signal degradation from solar illumination and atmospheric effects and thus provides a useful tool for phenology monitoring. In this chapter, we analyzed a time series of Ku-band radar backscatter measurements from the SeaWinds scatterometer on board the Quick Scatterometer (QuickSCAT) to examine its effectiveness for land surface phenology monitoring across eastern Asia. The spatial pattern of annual mean backscatter follows regional vegetation type distributions. The Start Of Season (SOS) and End Of Season (EOS) were derived from the backscatter time series and compared with MODIS (Moderate Resolution Imaging Spectroradiometer) phenology products from 2003 to 2007. The failure of phenology metric detection for backscatter time series is caused by snow coverage and limited vegetation activity in arid areas. For tropical and semi-arid areas where optical observation is unavailable, backscatter data can provide valid phenological information. Due to their sensitivity to different factors, temporal discrepancies were observed between phenology products from backscatter and MODIS time series. Overall, the results indicate that SeaWinds backscatter provides an alternative view of vegetation phenology that is independent of optical sensors and can be applied to global phenology studies.

L. Lu (✉) • H. Guo
Key Laboratory of Digital Earth Sciences, Institute of Remote Sensing and Digital Earth, RADI, Chinese Academy of Sciences, CAS, Beijing, China
e-mail: lull@radi.ac.cn

C. Wang
Department of Geography, University of South Carolina, Columbia, SC, USA

C. Kuenzer et al. (eds.), *Remote Sensing Time Series*, Remote Sensing and Digital Image Processing, DOI 10.1007/978-3-319-15967-6_18

18.1 Introduction

Vegetation phenology tracks plants' lifecycle events such as bud break, flowering, and leaf senescence as the cumulative effects of daily weather at different developmental stages (Lieth 1974). Over the past century, phenological shifts have been observed in association with warming climate across a diverse range of plant taxonomy (IPCC 2007; Richardson et al. 2013). Land Surface Phenology (LSP) may be defined as seasonal patterns of variation in vegetated land surfaces observed from remote sensing (de Beurs and Henebry 2004). Critical phenological dates or metrics, such as the onset of greening, onset of senescence, timing of the maximum of the growing season and growing season length can be derived from time series of remote sensing data. These metrics are sensitive to the timing and duration of vegetation activity, which significantly affect fluxes of carbon, water, energy, and other trace gases and therefore are important to biosphere–atmosphere interactions (Morisette et al. 2009).

Satellite data at optical-infrared wavelengths from operational satellite sensors such as AVHRR (Advanced Very High Resolution Radiometer), MODIS (Moderate Resolution Imaging Spectroradiometer) and SPOT-VGT (Satellite Pour l'Observation de la Terre-Vegetation) and derived spectral vegetation indices (VI) have been widely used to LSP monitoring at spatial resolutions of 500–8,000 m (Ganguly et al. 2010). Cloud contamination and soil background in sparsely vegetated areas, however, often cast high uncertainties on these phenology products. Satellite microwave remote sensing records data at lower frequency wavelengths which are sensitive to changes of water content, canopy structure and biomass (Ulaby et al. 1982), and provides a useful alternative for phenology assessment. Microwave backscatter is insensitive to signal degradation from solar illumination and atmospheric effects. The wind scatterometer on ERS-l was compared with global VI data for monitoring vegetation dynamics (Wagner et al. 1999) and examining the seasonal vegetation development (Frison and Mougin 1996). The potential of passive and active microwave measurements in vegetation monitoring was jointly investigated at the global scale with the special sensor microwave/imager (SSM/I) and ERS scatterometer (Macelloni et al. 2003). Several microwave Vis (MVIs) have been retrieved from daily time-series brightness temperature of passive microwave radiometers. Comparing with NDVI (Normalized Difference Vegetation Index), the results proved that MVIs could provide significant new information of vegetation development (Min and Lin 2006; Shi et al. 2008).

Though its primary mission is to observe ocean winds, the SeaWinds on Quick Scatterometer (QuickSCAT) provides an opportunity for an active microwave scatterometer to assess phenological features associated with seasonal changes of vegetated landscapes. Hardin and Jackson (2003) modeled the monthly composites of SeaWinds Ku-band backscatter as a function of savanna grass biomass and leaf area, soil moisture, and other soil characteristics in South America, finding it promising in monitoring the dynamics of savanna grasslands. In boreal and subalpine evergreen coniferous forests, Kimball et al. (2004) employed a temporal change classification to detect the initiation and termination of the growing season from daily radar backscatter measurements from the SeaWinds scatterometer.

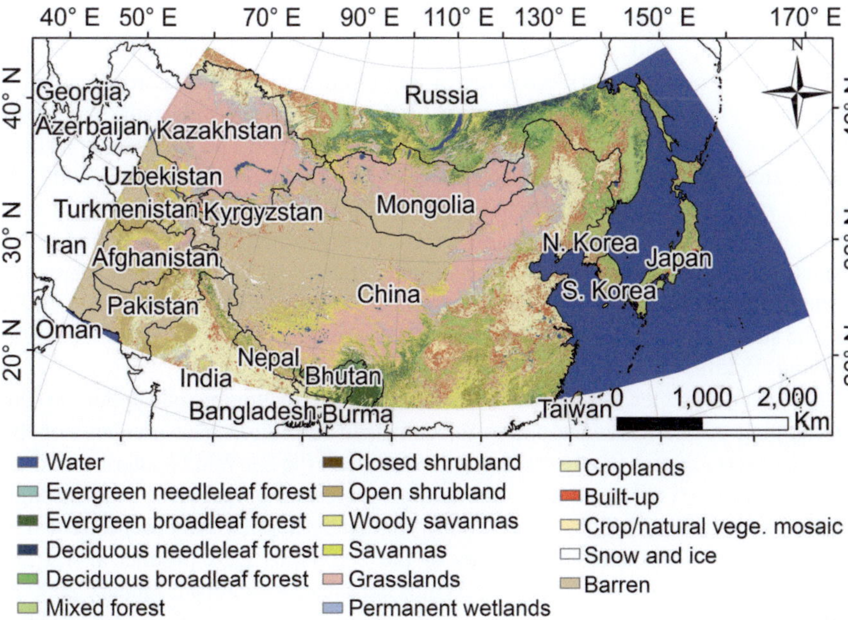

Fig. 18.1 Land cover types in eastern Asia

Frolking et al. (2005, 2006) found that the SeaWinds Ku-band backscatter and MODIS leaf area index (LAI) were strongly correlated in North American grasslands in the growing season, and backscatter was sensitive to seasonal variability in grassland biomass and productivity. Lu et al. (2013a) applied a three-year backscatter time series to vegetation phenology monitoring from 2003 to 2005 across China. Eastern Asia is a large region encompassing a wide range of climate and land cover types (Fig. 18.1). In this study, we extended the study by Lu et al. (2013a) and evaluated the effectiveness of SeaWinds Ku-band backscatter in vegetation phenology monitoring across eastern Asia. The spatial pattern and variability of backscatter was mapped and analyzed from 2000 to 2009. The phenology detection results from a 5-year backscatter time series were compared with the MODIS phenology products at regional level.

18.2 Data and Methods

18.2.1 SeaWinds Backscatter Data

Launched in June 19, 1999, SeaWinds scatterometer on-board the QuickSCAT satellite was originally intended to be a 'quick recovery' mission to fill the gap created by the unexpected failure of the NASA scatterometer (NSCAT) (King and

Greenstone 1999). The SeaWinds instrument consists of a rotating pencil-beam antenna, which provides contiguous measurement swaths of 1,400 km (inner-beam) and 1,800 km (outer-beam), coverage of approximately 70 % of Earth on a daily basis and 90 % global coverage every 2 days. The SeaWinds radar transmits microwave radiation at a 13.4 GHz (2.1 cm) wavelength and receives a surface backscatter signal with a 0.25 dB relative accuracy (King and Greenstone 1999). It records radar backscatter with dual polarization (vertical and horizontal) at two nominal incidence angles, 46.0° and 54.1° corresponding to the inner and outer beams. The inner beam has horizontal polarization while the outer beam has vertical polarization. The data records span from July 1999 to November 2009, and the standard processing of backscatter measurements yields a spatial resolution of about 25 km.

The Scatterometer Image Reconstruction (SIR) technique is applied to the overlapping passes to reduce noises for enhanced resolution measurements of the surface characteristics (Early and Long 1996). Since the SeaWinds outer beam data provide improved temporal coverage and an increased propagation path through vegetation volume (Frolking et al. 2006), this study applied the composite 4 day V-pol average backscatter time series of the SeaWinds L2A product at 4.5 km resolution (Early and Long 2001). SeaWinds backscatter data from January 2000 to November 2009 covering eastern Asia were downloaded from the NASA Scatterometer Climate Record Pathfinder database at Brigham Young University (http://www.scp.byu.edu) (Fig. 18.1). The Marine Geospatial Ecology Tools (MGET) were used to convert SIR files to raster format (Roberts et al. 2010).

18.2.2 MODIS Phenology and Land Cover Products

The V005 MODIS Land Cover Dynamics (MCD12Q2) product (informally called the MODIS Global Vegetation Phenology product) provides estimates of the timing of vegetation phenology at global scales (Ganguly et al. 2010). MCD12Q2 primarily uses the MODIS Enhanced Vegetation Index (EVI), which is computed from the MODIS Nadir Bidirectional Reflectance Distribution Function (BRDF)-Adjusted Reflectance (MODIS NBAR) product. The snow and ice flag included in the MODIS NBAR product is used to filter out data points associated with snow-covered surfaces from the input time series. In particular, a time series of EVI is extracted for each pixel. Periods of sustained EVI increase or decrease are identified after a gap-filling and smoothing process. Logistic models are fit to the time series and transition dates including the onset of EVI increase, the onset of EVI maximum, the onset of EVI decrease, and the onset of EVI minimum are identified as local maxima and minima in the rate of change of curvature of the fitted logistic function (Ganguly et al. 2010). In our study, the onset of EVI increase and onset of EVI decrease dates from 2003 to 2007 covering the study area are retrieved for the comparative analysis.

The MODIS 1 km land-cover product (MOD12Q1) was used to identify primary land covers in the study area. With the International Geosphere-Biosphere Programme (IGBP) land-cover legend (Friedl et al. 2002), the MODIS product of 2009 was downloaded and a total of 18 cover types were mapped and served as base information in this study (Fig. 18.1).

The MODIS Re-projection Tool was used to mosaic and convert Hierarchical Data Format (HDF) files to raster format. They were re-projected and resampled to the same resolution as the SeaWinds backscatter products at a local radius equal-area Lambert Projection.

18.2.3 Climate Data

In radar backscatter data, dielectric constants of land surfaces are strongly affected by frozen and thaw conditions. To differentiate data in the frozen and nonfrozen seasons and assign them different weights in the phenology model from the annual time series, a temperature mask was generated using temperature data. The daily air temperature grid data in a 2.5° spatial resolution was downloaded from the NCEP Daily Global Analyses data provided by the NOAA/OAR/ESRL PSD, Boulder, Colorado, USA, from their Web site at (http://www.esrl.noaa.gov/psd) (Trenberth and Olson 1988a, b). The onset of the fully thawed season was determined to be the last day of 6 consecutive days with mean air temperatures greater than 5 °C, and the end of the thaw season as the first day of 6 consecutive days with mean air temperature less than 50C (Karl et al. 1999; Peterson 2005). Gridded temperature masks defining the start and end of the thaw season were produced for each year. They were re-projected and spatially disaggregated to the same grid as the SeaWinds data.

18.2.4 Means and Variability of Backscatter

In this study, the backscatter data was used to calculate derivative datasets on annual basis, a common process for analyses of time series (Gessner et al. 2013; Kuenzer et al. 2009). The mean monthly backscatter was calculated from the monthly data based on the 2000–2009 time series. For a given pixel, the monthly 10-year arithmetic mean ($\overline{\text{Backscatter}_{x,y}}$) was calculated using Eq. (18.1), with *n* being the number of years and $\text{Backscatter}_{x,y}$ the monthly backscatter value for pixel x, y. The mean annual backscatter was calculated accordingly from the annual sums.

$$\overline{\text{Backscatter}_{x,y}} = \frac{1}{n}\sum\nolimits_1^n \text{Backscatter}_{x,y} \tag{18.1}$$

The deviation ($DevB_{x,\ y}$) from the 2000–2009 mean annual backscatter at a given spatial location for a certain year was calculated according to Eq. (18.2).

$$DevB_{x,y} = Backscatter_{x,y} - \overline{Backscatter}_{x,y} \tag{18.2}$$

In addition, the relative annual backscatter deviation ($rDevB_{x,\ y}$) was derived. This describes the deviation as a percentage from the mean annual backscatter, as given by Eq. (18.3).

$$rDevB_{x,y} = \frac{DevB_{x,y} \cdot 100}{\overline{Backscatter}_{x,y}} \tag{18.3}$$

The mean annual variability ($\overline{VB}_{x,y}$) was derived according to Eq. (18.4) with the relative annual backscatter deviation ($rDevB_{x,\ y}$) and n being the number of years.

$$\overline{VB}_{x,y} = \frac{1}{n}\sum\nolimits_{1}^{n} |rDevB_{x,y}| \tag{18.4}$$

18.2.5 Phenology Metric Detection

According to Frolking et al. (2006) and Lu et al.(2013a), radar backscatter has a significant relationship with LAI based on site-level linear regression analysis. Therefore, a weighted phenological model was applied to the backscatter time series to detect the timing of leaf flush and senescence in this study. The curve-fitting method, Asymmetric Gaussian implemented in the TIMESAT software is a common tool for time series analysis (Gao et al. 2008; Wang et al. 2011; Jones et al. 2012). It was applied to fit local model functions to the backscatter data of a vegetation growth cycle. The single growth cycle of vegetation typically consists of a growth and senescence phase. The Gaussian type of function was used as the following (Jönsson and Eklundh 2002, 2004):

$$f(t; x_1, x_2, \ldots x_5) = \left\{ \begin{array}{ll} e^{-\left(\frac{t-x_1}{x_2}\right)^{x_3}} & if\, t > x_1 \\ e^{-\left(\frac{x_1-t}{x_4}\right)^{x_5}} & if\, t < x_1 \end{array} \right\} \tag{18.5}$$

In this model, the coefficient x_1 determines the position of the maximum or minimum with respect to the independent time variable t. Coefficients x_2 and x_3 determine the width and flatness of the right half of the function, while x_4 and x_5 determine the width and flatness of the left half. In order to ensure smooth shapes of the model functions and consistency with data observations, these coefficients were restricted in certain conditions and were calculated using a separable Levenberg–Marquardt method (Madsen et al. 2004).

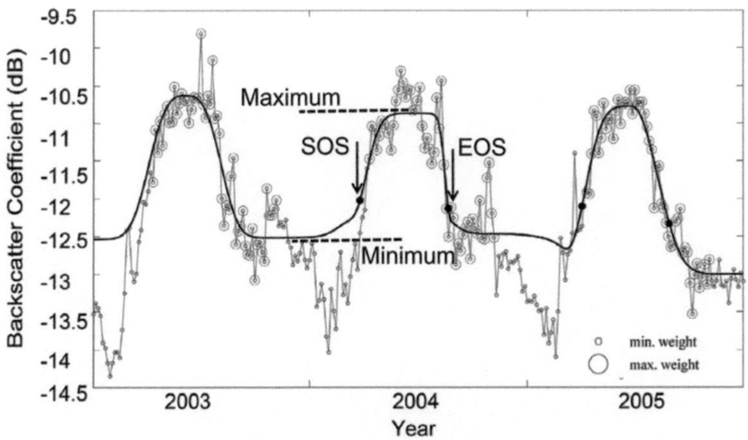

Fig. 18.2 An example of the 3-year SeaWinds backscatter times series and its curve-fitting results

With the asymmetric Gaussian fitting model, two seasonality metrics, namely the Start Of Season (SOS) and the End Of Season (EOS), were calculated. Assuming the seasonal amplitude as the difference between the maximum value and base level, the SOS was defined as the time when the left edge increased to 30 % of the seasonal amplitude measured from the left minimum level, and the EOS was defined as the time when the right edge has decreased to 30 % of the seasonal amplitude from the right minimum level (Jönsson and Eklundh 2004). The phenology detection of backscatter time series at a sample site is shown in Fig. 18.2. The sample site is a homogenous area (3 × 3 pixels) with main land cover as deciduous forest. The backscatter time series from 2003 to 2005 was weighted and fitted with the asymmetric Gaussian model. With this model, phenological metrics (SOS and EOS) of all pixels were extracted from the backscatter time series to examine their spatial patterns across the eastern Asia.

18.3 Results

18.3.1 Backscatter Mean and Variability

Figure 18.3 represents the spatial variation of the annual mean backscatter across eastern Asia for the 2000–2009 period. Comparing with the land cover type (Fig. 18.1), the spatial pattern of backscatter follows regional vegetation type distributions. For grasslands and barren lands with sparse vegetation in Kazakhstan, Mongolia, and northwestern China the backscatter value ranges from −20 to −15 dB. For areas dominated by forests in the south of China, it ranges from −15 to −11 dB. Agricultural lands in northern China exhibited medium backscatter

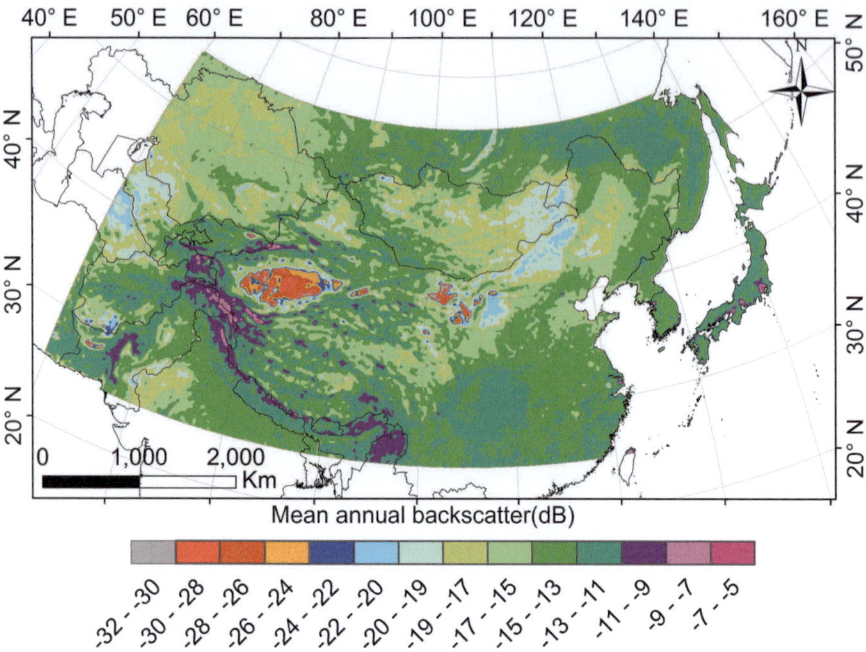

Fig. 18.3 Mean annual backscatter for the 2000–2009 period across eastern Asia

from −17 to −15 dB, whereas it shows higher value (−15 to −13 dB) in India and southern China. The desert in western China shows extremely low value (−30 to −26 dB). The mountain areas in the south western China are often covered with snow and show highest values (−11 to −5 dB).

The Fig. 18.4 illustrates how the mean annual backscatter varies from year to year for the 2000–2009 period. Areas with high variability show less stable backscatter values than areas with low variability. In general, the mean annual backscatter variability reveals moderate dynamic levels over the land surface (<10 %, and mostly <5 % in Fig. 18.4). The northern area of the study region shows higher variability than the southern. High variability can be observed in northeastern China. Similar variability was detected in the arid areas of Kazakhstan and the desert area of India. Fluctuation of backscatter is also observable for the Tibetan Plateau.

18.3.2 Regional Comparison of Backscatter- and MODIS-Derived Phenology Metrics

In order to reduce the uncertainties in phenological detection results, we used detected phenology dates from 2003 to 2007 for the comparative analysis. The

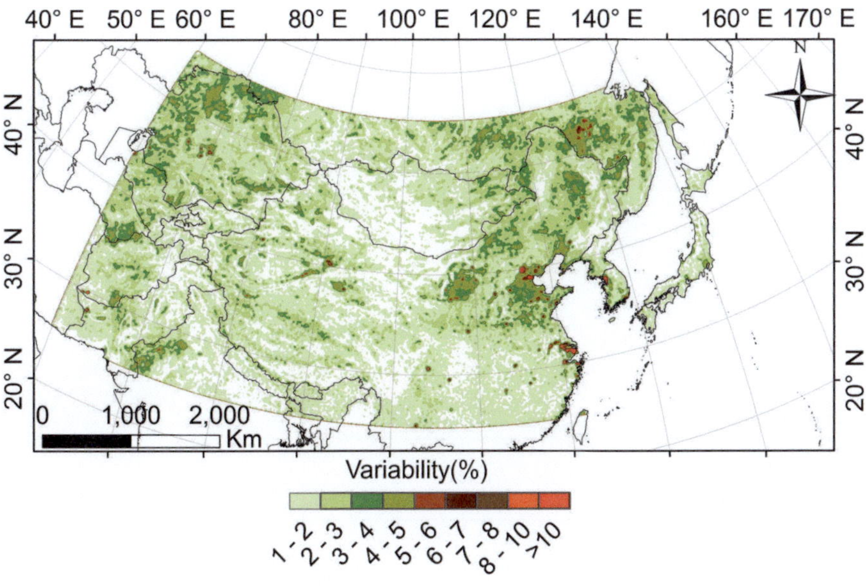

Fig. 18.4 Mean annual backscatter variability (%) for 2000–2009 across eastern Asia

numbers of successful annual retrievals for phenological metrics from backscatter and MODIS data are shown in Fig. 18.5. For backscatter data, missing values were distributed mostly in high latitude areas and the Tibetan Plateau. The failure of detection in MODIS data was distributed in the southern tropical and subtropical area, the arid and semiarid area in northwestern China and Sichuan province in southern China. For the multiple-cropping areas in eastern China, the phenological metrics can be detected from MODIS time series in only 3 or 4 years.

The mean SOS dates derived from backscatter time series and MODIS products for 2003–2007 are shown in Fig. 18.6a, b. The SOS dates ranged widely from approximately DOY 70 in the south to DOY 190 in the north. The grasslands extending from Kazakhstan to Mongolia (Fig. 18.1) showed an earlier SOS than areas at the same latitude. It shows a gradient from east to west, with southeastern Mongolia having a much later SOS from both datasets. Other areas with much later SOS from backscatter time series are northern China and northern India. The northern China is dominated by grassland and agriculture lands. The northern India is dominated by vast irrigated drylands and has a tropical desert climate.

Figure 18.7 shows the bias between backscatter SOS and MODIS greenup dates. For the agricultural lands in central China (Fig. 18.1), the backscatter data shows later greenup dates than MODIS data. For grasslands in the north, the backscatter data detects earlier SOS dates. In the south of China where there are mixed land covers of forest, shrub lands and croplands with a dominant subtropical and tropical monsoon climate, the backscatter data also detects earlier SOS.

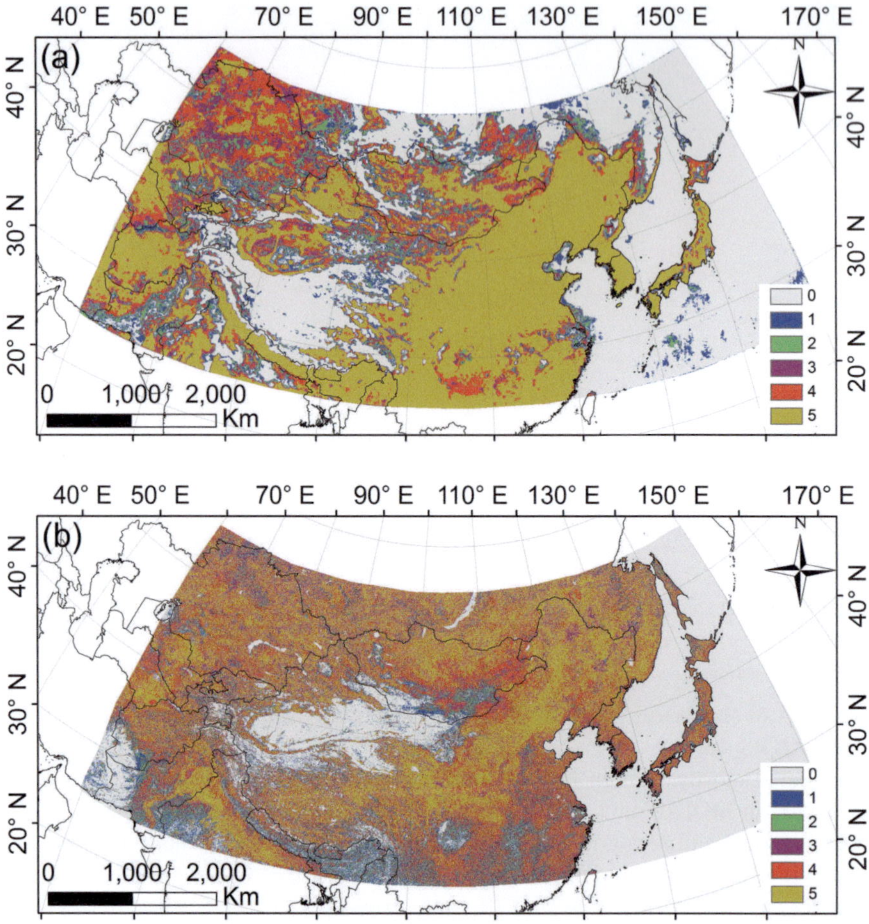

Fig. 18.5 Number of successful annual retrievals for phenology metrics derived from (**a**) backscatter time series and (**b**) MODIS phenology products at pixel level for 2003–2007 in eastern Asia

The end of season derived from backscatter time series and MODIS senescence date for 2003–2007 are shown in Fig. 18.8. The MODIS EOS dates range from DOY 100 to DOY 280, and backscatter EOS range from DOY 100 to DOY 360. Though the EOS detected by backscatter data is later than MODIS, it shows a clear spatial pattern from north to south regionally. A similar earlier EOS was detected in Kazakhstan from both backscatter and MODIS datasets. The EOS gradient detected by backscatter data across China shows an earlier growing season end in the north.

Figure 18.9 shows the bias between backscatter EOS and MODIS senescence dates. For most areas, the backscatter data detected later EOS dates. In the grasslands in Kazakhstan, Mongolia and China, the EOS detected by MODIS occurs later than that derived from backscatter. Another discrepancy occurred in the agricultural lands in central China and northern India.

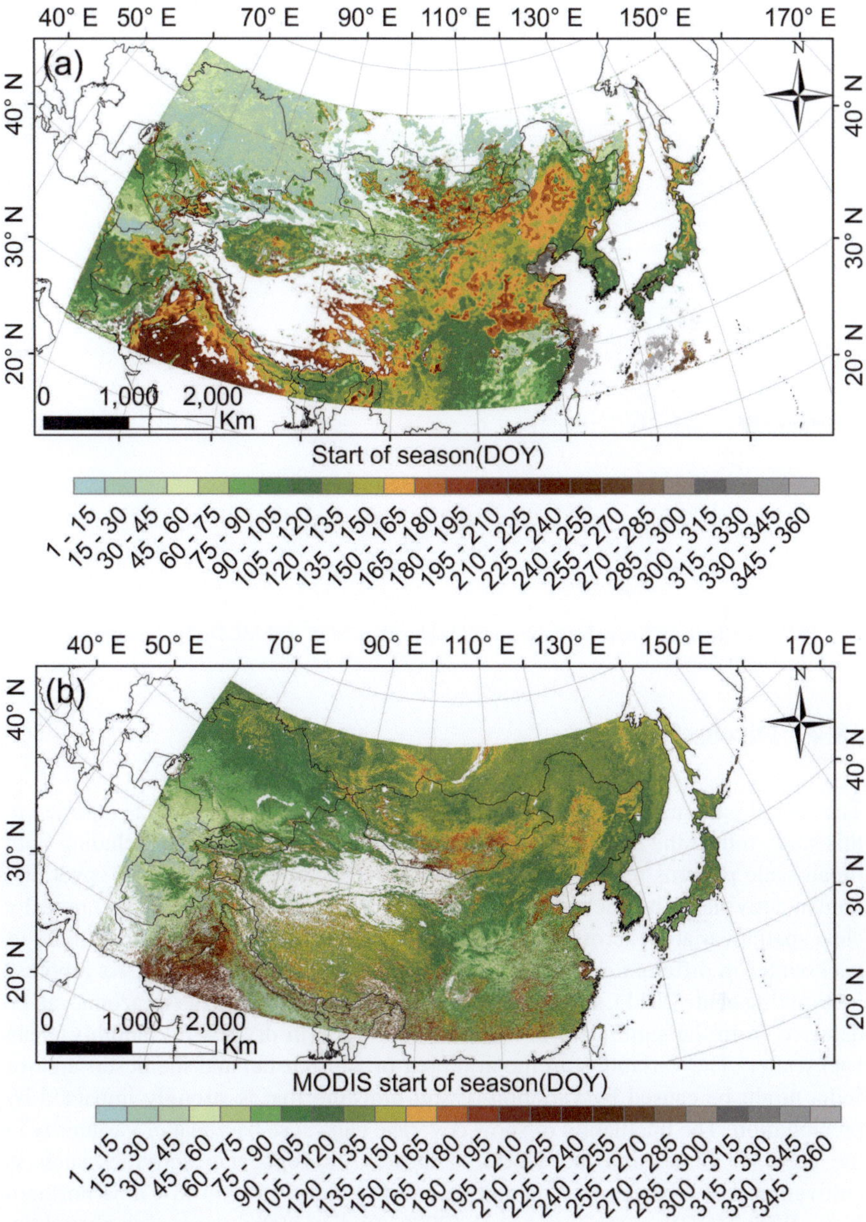

Fig. 18.6 The mean start of season derived from (**a**) backscatter time series for 2003–2007 and (**b**) MODIS greenup date for 2003–2007 in eastern Asia

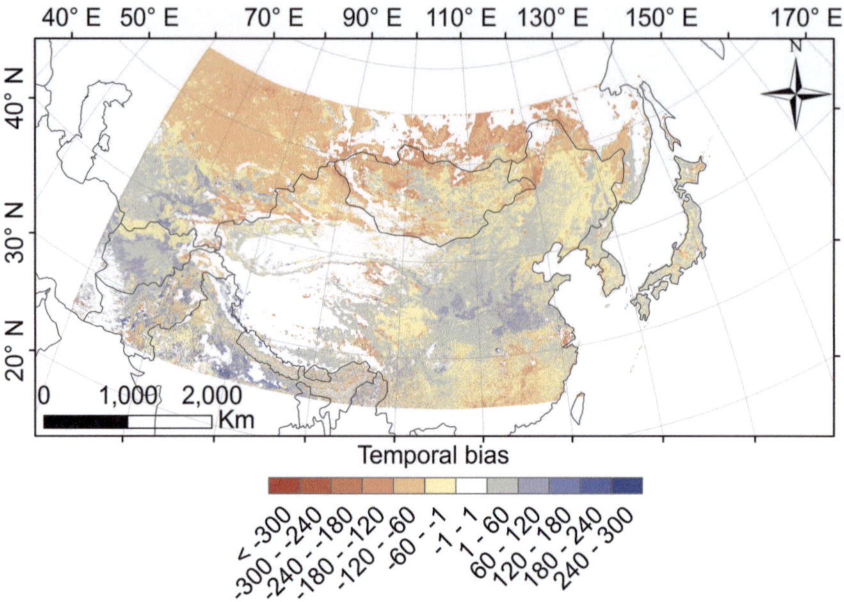

Fig. 18.7 The temporal bias between mean backscatter SOS and MODIS greenup date 2003–2007 in eastern Asia

18.4 Discussion

The spatial patterns of backscatter and derived phenological metrics as reported in this study reflect the controlling mechanisms of vegetation activity including both broad-scale patterns related to climate and local factors related to land cover and human activities. The backscatter and its variability (Figs. 18.3 and 18.4) reveal a clear spatial variation across the eastern Asia region. The inter-annual variability of backscatter in different areas might be attributed to different reasons. As reported by Frolking et al. (2011), severe drought occurred in southwestern Amazonia can be detected from the significant interannual variability in dry season monthly mean backscatter. The variability in the arid area of Kazakhstan and the desert area of India might be caused by variation of soil moisture that is strongly impacted by precipitation. The fluctuation of snow coverage can cause backscatter variability in the north of China and the Tibetan Plateau. Anthropogenic activities such as cultivation and urban expansion can lead to the high variation in China northern plain. With MODIS data, Zhang et al. (2006) studied global vegetation phenology and found that the timing of crop phenology can be quite variable, depending strongly on crop type and agricultural management. Frolking et al. (2013) reported that the significant backscatter increase of SeaWinds backscatter in major cities around the world was caused by changes of built-up infrastructure.

The numbers of successful annual retrievals for phenological metrics from time series (Fig. 18.5) reveal the algorithm sensitivity to data noise and gaps. For

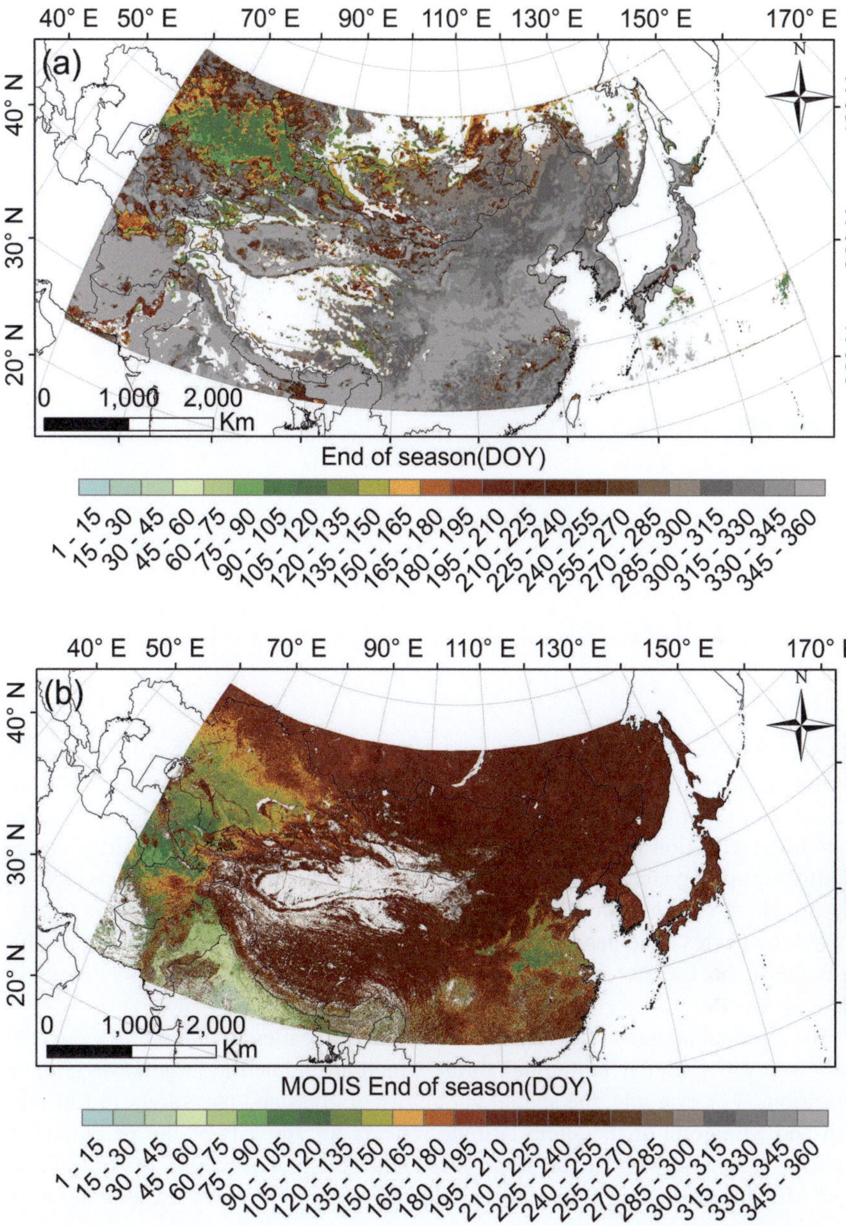

Fig. 18.8 The mean end of season derived from (**a**) backscatter time series for 2003–2007 and (**b**) MODIS senescence date for 2003–2007 in eastern Asia

SeaWinds backscatter time series, snow in the high latitude area and the Tibetan Plateau results in the failure of phenology detection algorithms (Fig. 18.5a). For MODIS phenology products, the persistent cloud cover, high levels of atmospheric

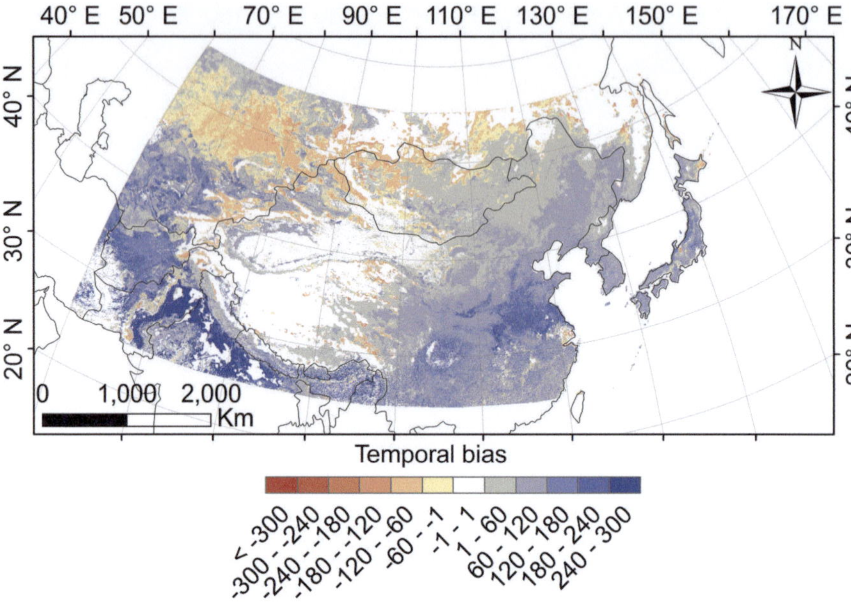

Fig. 18.9 The temporal bias between backscatter EOS and MODIS senescence date for 2003–2007 in eastern Asia

aerosols and weak seasonality present substantial challenges for land surface phenology algorithms (Ganguly et al. 2010). Missing values are observable in tropical and subtropical area (Fig. 18.5b). In arid and semiarid area of China, phenology detection failures are caused by the limited vegetation activities. In the multiple-cropping areas in eastern China, multiple vegetation growth cycles exit in 1 year, which can also cause the failure of phenology detection. Only three or four phenology cycles have been successfully detected in MODIS product (Fig. 18.5b). In contrast, backscatter imagery provides effective phenological observations in these areas.

The regional patterns of SOS and EOS from backscatter and MODIS (Figs. 18.6 and 18.8) show similar latitudinal shifts, which are related to large-scale climate transitions. For example, the spatial variability of grassland phenology in Kazakhstan and Mongolia at the same latitude reflects the impacts of the precipitation regimes. Agreeing with Begue et al (2014) and Zhang et al (2005), the SOS shifts from MODIS phenological products are linked to the timing of seasonal rainfall in the arid and semi-arid lands.

The backscatter-derived phenology metrics show different temporal bias from MODIS products in different land cover types. The bias of backscatter SOS dates and MODIS greenup dates is caused by the temporal shifts between backscatter increase and canopy greenup. For boreal forests in Russia, an earlier SOS and EOS were detected prior to greenup and senescence onset respectively. Kimball et al. (2004) studied boreal and subalpine evergreen forests with SeaWinds scatterometer, and revealed that radar remote sensing measurements of the initiation of the growing season corresponded strongly with both site measurements and ecosystem process model simulations. Jönsson et al. (2010) studied tree phenology with MODIS vegetation index, and revealed that without a sharp increase in greenness during spring, the seasonal changes in vegetation indices of evergreen trees were more related to snow dynamics than to changes in needle biomass. As seasonal thawing occurs, transpiration rates increase and water is allocated to canopy branches or existing leaves prior to new leaf construction (Waring et al. 1979). This may lead to backscatter seasonal increase and SOS prior to MODIS greenup date. For tropical and subtropical forests in southern China, an earlier greenup onset and later senescence onset were detected from backscatter data. The structural changes of the plants such as leaf fall and leaf flushing influence the seasonal variation of backscatter signals. In the steppes in Kazakhstan and Mongolia, the backscatter SOS and EOS generally precede the VI greenup. With low vegetation cover and biomass, the backscatter is more related to the fluctuation of soil moisture which is earlier than the increase of leaf greenness. In cropland-dominated areas, planting and harvest times drive the vegetation phenology cycles and the backscatter SOS generally follows NDVI greenup. Croplands are primarily barren or covered with non-photosynthetic residues prior to tilling and planting. Greenup occurs at seed germination or the initiation of visible above-ground photosynthetic vegetation growth, which can occur prior to significant biomass growth (Jones et al. 2011; Lu et al. 2014). The backscatter follows the initial VI greenup by several weeks or more following delayed increase in vegetation water content and development of above-ground biomass.

Despite these findings, there are significant limitations to our research. The data noise and gap in the time series may lead to large uncertainties of the phenology detection results. Different methods were used to detect phenological dates from backscatter and MODIS time series, which may cause their discrepancy (White et al. 2009). Moreover, time series of the eddy covariance measurements from tower sites can be integrated in further studies, which would improve understanding of our results (Jones et al. 2012; Melaas et al. 2013).

18.5 Conclusions

This study tested the feasibility and effectiveness of microwave backscatter data in regional vegetation phenology monitoring. The mean and variability of backscatter were derived from the SeaWinds Ku-band backscatter time series for the 2000–2009 period across eastern Asia. The vegetation phenology metrics (SOS and EOS) were extracted for the 2003–2007 period and compared with contemporary MODIS phenology products. The results can be summarized as follows.

The regional spatial patterns of annual mean backscatter in eastern Asia follow the spatial distributions of vegetation types. The mean annual variability observed in SeaWinds backscatter can be related to the dynamics of meteorological conditions such as drought and snowfall. Human activities, for example the intensified agricultural practices and expanded urban developments, also lead to its high variability.

In areas where optical time series is unavailable due to persistent cloud cover, high levels of atmospheric aerosols and low VI values, backscatter imagery provides effective observations. Different temporal biases were found in different land cover areas between the backscatter-extracted and the published MODIS phenology metrics. For boreal forests and steppes, the backscatter-extracted SOS and EOS preceded the MODIS greenup and senescence onsets, respectively. For tropical and subtropical forests, an earlier SOS and later EOS dates were detected with the backscatter data, while the SOS for croplands was found later than the MODIS greenup.

Overall, the application of backscatter time series for LSP monitoring can expand the breadth of the current optical satellites, thereby enhance our understanding of global vegetation dynamics and carbon cycle processes. However, given the complexity of Ku-band backscatter behaviors on vegetated landscapes, a critical challenge of our study is to understand the bio-ecological meaning of the temporal variation of backscatter time series. The inconsistency of phenology metric detection methodology should also be aware of. Our future research direction will focus on linking the variation of backscatter-extracted phenology metrics to the dynamics of ecosystem and its response to climate change. The integration of ground observation from tower sites will allow us to better interpret the remote sensing results.

Acknowledgments The resolution-enhanced SeaWinds backscatter data were obtained from the NASA Scatterometer Climate Record Pathfinder project (http://www.scp.byu.edu). This work was supported by the National Natural Science Foundation of China under grant No. 41471369 and the Major International Cooperation and Exchange Project 'Comparative study on global environmental change using remote sensing technology' under grant No. 41120114001.

References

Begue A, Vintrou E, Saad A, Hiernaux P (2014) Differences between cropland and rangeland MODIS phenology (start-of-season) in Mali. Int J Appl Earth Obs Geoinform 31:167–170

De Beurs KM, Henebry GM (2004) Land surface phenology, climatic variation, and institutional change: analyzing agricultural land cover change in Kazakhstan. Remote Sens Environ 89:497–509

Early DS, Long DG (1996) Error characteristics of the SIR resolution enhancement algorithm. In: Proceeding of IGARSS'96. IGARSS, Lincoln, pp 24–126

Early DS, Long DG (2001) Image reconstruction and enhanced resolution imaging from irregular samples. IEEE Trans Geosci Remote Sens 39:291–302

Friedl MA, Mciver DK, Hodges JCF, Zhang XY, Muchoney D, Strahler AH, Woodcock CE, Gopal S, Schneider A, Cooper A, Baccini A, Gao F, Schaaf C (2002) Global land cover mapping from MODIS: algorithms and early results. Remote Sens Environ 83:287–302

Frison PL, Mougin E (1996) Monitoring global vegetation dynamics with ERS-1 wind scatterometer data. Int J Remote Sens 17:3201–3218

Frolking S, Fahnestock M, Milliman T, Mcdonald K, Kimball J (2005) Interannual variability in North American grassland biomass/productivity detected by SeaWinds scatterometer backscatter. Geophys Res Lett 32:L21409

Frolking ST, Mcdonald MK, Kimball J, Zhao M, Fahnestock M (2006) Evaluation of the SeaWinds scatterometer for regional monitoring of vegetation phenology. J Geophys Res 111:D17302

Frolking S, Milliman T, Palace M, Wisser D, Lammers R, Fahnestock M (2011) Tropical forest backscatter anomaly evident in seawinds scatterometer morning overpass data during 2005 drought in Amazonia. Remote Sens Environ 115(3):897–907

Frolking S, Milliman T, Seto K, Friedl MA (2013) A global fingerprint of macro-scale changes in urban structure from 1999-2009. Environ Res Lett 8(2):10pp

Ganguly S, Friedl MA, Tan B, Zhang X, Verma M (2010) Land surface phenology from MODIS: characterization of the collection 5 global land cover dynamics product. Remote Sens Environ 114(8):1805–1816

Gao F, Morisette JT, Wolfe RE, Ederer G, Pedelty J, Masuoka E, Myneni R, Tan B, Nightingale J (2008) An algorithm to produce temporally and spatially continuous MODIS-LAI time series. IEEE Geosci Remote Sens Lett 5:60–64

Gessner U, Naeimi V, Klein I, Kuenzer C, Klein D, Dech S (2013) The relationship between precipitation anomalies and satellite-derived vegetation activity in Central Asia. Glob Planet Chang 110:74–87

Hardin PJ, Jackson MW (2003) Investigating SeaWinds terrestrial backscatter: equatorial savannas of south america. Photogramm Eng Remote Sens 69:1243–1254

IPCC (2007) Climate change 2007: synthesis report. Contribution of working groups I, II and III to the fourth assessment report of the Intergovernmental Panel on Climate Change. IPCC, Geneva, 104 p

Jones MO, Jones LA, Kimball JS, McDonald KC (2011) Satellite passive microwave remote sensing for monitoring global land surface phenology. Remote Sens Environ 115:1102–1114

Jones MO, Kimball JS, Jones LA, McDonald KC (2012) Satellite passive microwave detection of North America start of season. Remote Sens Environ 123:324–333

Jönsson P, Eklundh L (2002) Seasonality extraction by function fitting to time-series of satellite sensor data. IEEE Trans Geosci Remote Sens 40:1824–1831

Jönsson P, Eklundh L (2004) Timesat – a program for analyzing time-series of satellite sensor data. Comput Geosci 30:833–845

Jönsson AM, Eklundh L, Hellström M, Bärring L, Jönsson P (2010) Annual changes in MODIS vegetation indices of Swedish coniferous forests in relation to snow dynamics and tree phenology. Remote Sens Environ 114(11):2719–2730

Karl TR, Nicholls N, Ghazi A (1999) CLIVAR/GCOS/WMO workshop on indices and indicators for climate extremes: workshop summary. Clim Chang 42:3–7

Kimball JS, Mcdonald KC, Running SW, Frolking S (2004) Satellite radar remote sensing of seasonal growing seasons for boreal and subalpine evergreen forests. Remote Sens Environ 90:243–258

King MD, Greenstone R (eds) (1999) EOS reference handbook: a guide to NASA's earth science enterprise and the earth observing system. NASA document NP-1999-08-134-GSFC. NASA/ Goddasd Space Flight Centre, Greenbelt, 361 pp

Kuenzer C, Zhao D, Scipal K, Sabel D, Naeimi V, Bartalis Z, Hasenauer S, Mehl H, Dech S, Wagner W (2009) El Niño southern oscillation influences represented in ERS scatterometer-derived soil moisture data. Appl Geogr 29(4):463–477

Lieth H (ed) (1974) Phenology and seasonal modeling. Springer, New York

Lu L, Guo H, Wang C, Li Q (2013) Assessment of the SeaWinds scatterometer for vegetation phenology monitoring across China. Int J Remote Sens 34(15):5551–5568

Lu L, Wang C, Guo H, Li Q (2014) Detecting winter wheat phenology with SPOT-VEGETATION data in the North China Plain. Geocarto Int 29(3):244–255. doi:10.1080/10106049.2012.760004

Macelloni G, Paloscia S, Pampaloni P, Santi E (2003) Global scale monitoring of soil and vegetation using SSM/I and ERS wind scatterometer. Int J Remote Sens 24:2409–2425

Madsen K, Nielsen HB, Tingleff O (eds) (2004) Methods for non-linear least squares problems. Technical University of Denmark, Copenhagen

Melaas EK, Richardson AD, Friedl MA, Dragoni D, Gough CM, Herbst M, Montagnani L, Moors E (2013) Using FLUXNET data to improve models of springtime vegetation activity onset in forest ecosystems. Agric For Meteorol 171–172:46–56

Min QL, Lin B (2006) Determination of spring onset and growing season leaf development using satellite measurements. Remote Sens Environ 104:96–102

Morisette JT, Richardson AD, Knapp AK, Fisher JI, Graham EA, Abatzoglou J (2009) Tracking the rhythm of the seasons in the face of global change: phenological research in the 21st century. Front Ecol Environ 7:253–260

Peterson TC (2005) Climate change indices. WMO Bull 54(2):83–86

Richardson AD, Keenan TF, Migliavacca M, Ryu Y, Sonnentag O, Toomey M (2013) Climate change, phenology, and phenological control of vegetation feedbacks to the climate system. Agric For Meteorol 169:156–173

Roberts JJ, Best BD, Dunn DC, Treml EA, Halpin PN (2010) Marine geospatial ecology tools: an integrated framework for ecological geoprocessing with ArcGIS, Python, R, MATLAB, and C++. Environ Model Softw 25:1197–1207

Shi J, Jackson T, Tao J, Du J, Bindlish R, Lu L, Chen KS (2008) Microwave vegetation indices for short vegetation covers from satellite passive microwave sensor AMSRE. Remote Sens Environ 112:4285–4300

Trenberth KE, Olson JG (1988a) Evaluation of NMC global analyses: 1979-87. NCAR Technical note TN-299+STR. National Center for Atmospheric Research, Boulder, 82 pp

Trenberth KE, Olson JG (1988b) Intercomparison of NMC and ECMWF global analyses: 1980–1986. NCAR Technical note TN-301+STR. National Center for Atmospheric Research, Boulder, 81 pp

Ulaby FT, Moore RK, Fung AK (eds) (1982) Microwave remote sensing: active and passive, vol. II: radar. Addison-Wesley, Reading

Wagner W, Lemoine G, Borgeaud M, Rott H (1999) A study of vegetation cover effects on ERS scatterometer data. IEEE Trans Geosci Remote Sens 37:938–948

Wang C, Fritschi FB, Stacey G, Yang Z (2011) Phenology-based assessment of energy crops in north american tallgrass prairie. Ann Assoc Am Geogr 101:742–751

Waring RH, Whitehead D, Jarvis PG (1979) The contribution of stored water to transpiration in Scots pine. Plant Cell Environ 2:309

White MA, De Beurs KM, Didan K, Inouye DW, Richardson AD, Jensen OP, O'keefe J, Zhang G, Nemani RR, Van Leeuwen WJD, Brown JF, De Wit A, Schaepman M, Lin X, Dettinger M, Bailey AS, Kimball J, Schwartz MD, Baldocchi DD, Lee JT, Lauenroth WK (2009) Intercomparison, interpretation, and assessment of spring phenology in North America estimated from remote sensing for 1982–2006. Glob Chang Biol 15:2335–2359

Zhang XY, Friedl MA, Schaaf CB, Strahler AH, Liu Z (2005) Monitoring the response of vegetation phenology to precipitation in Africa by coupling MODIS and TRMM instruments. J Geophys Res Atmos 110:D12103, 14 pp

Zhang XY, Friedl MA, Schaaf CB (2006) Global vegetation phenology from moderate resolution imaging spectroradiometer (MODIS): evaluation of global patterns and comparison with in situ measurements. J Geophys Res 111:G04017

Chapter 19
Monitoring Recent Urban Expansion and Urban Subsidence of Beijing Using ENVISAT/ASAR Time Series Datasets

Xinwu Li, Huadong Guo, Huaining Yang, Zhongchang Sun, Lu Zhang, Shiyong Yan, Guozhuang Shen, Wenjin Wu, Lei Liang, and Meng Wang

Abstract With worldwide economic development and population increases, urban areas create significant stresses on the local, regional and global environment. Information about the spatial and temporal dynamics of the characteristics of urban areas is therefore needed to support sustainable urban development. Time series earth observation data obtained using radar satellites have provided effective data sources for monitoring urban areas. This chapter first describes the development of synthetic aperture radar as well as its important role in the detection and monitoring of urban areas. Then, the fundamental principle of time series radar data in monitoring urban areas is introduced and discussed. Next, to demonstrate the capacity of time series SAR (Synthetic Aperture Radar) imagery for monitoring urban areas using ENVISAT/ASAR (Environmental Satellite /Advanced Synthetic Aperture Radar) time series radar data, Beijing city in China was selected as a test site. Beijing has all of the typical problems of a megacity such as resource, environment and population problems arising from rapid urban expansion during recent decades. A C5.0 rulesets classifier and the Multi Temporal Interferometric Synthetic Aperture Radar (MTInSAR) method were used to map the urban expansion and the millimeter level urban subsidence, respectively and the results were validated via high resolution WorldView optical datasets and leveling benchmark measurement, respectively. The results demonstrate the effectiveness and high accuracy of the time series radar data for monitoring urban areas. Furthermore, the spatial-temporal characteristic of urban expansion and urban subsidence of Beijing city were analyzed. Finally, the mechanisms or driving factors for urban expansion and subsidence are addressed based on economic development, population growth and the impacts of recent Beijing government policy.

X. Li (✉) • H. Guo • Z. Sun • L. Zhang • G. Shen • W. Wu • L. Liang • M. Wang
Key Laboratory of Digital Earth Sciences, Institute of Remote Sensing and Digital Earth, RADI, Chinese Academy of Sciences, CAS, Beijing, China
e-mail: lixw@radi.ac.cn

H. Yang • S. Yan
National Earthquake Response Support Service, No. 1 Yuquan West Street, Shijingshan District, Beijing 100049, China

C. Kuenzer et al. (eds.), *Remote Sensing Time Series*, Remote Sensing and Digital Image Processing, DOI 10.1007/978-3-319-15967-6_19

19.1 Introduction

Over the previous decades, serious environmental issues such as deforestation, desertification, wetland loss, agricultural intensification in vulnerable areas, continuous urban development, and the disappearance of species deprived of their biotopes have significantly increased. Urban areas especially create significant stresses on the local, regional and global environment (Millennium Ecosystem Assessment 2005; Ban et al. 2014). Time series change information is one of most important tools to better monitor and manage environmental issues in urban areas. Specifically, the use of time series Synthetic Aperture Radar (SAR) observations will be critical to provide important information for monitoring urban areas (Guo 2001; Taubenböck et al. 2012).

Since its inception in the 1960s, SAR has become a global leading-edge technology and the core technology for active microwave earth observation due to its unique all-weather 24-h partial penetration capabilities (Guo 2001).

Radar satellites (see Table 19.1) that have been launched to date primarily include Seasat, ERS-1/2 (Earth Resources Satellite), ENVISAT/ASAR (Environmental Satellite /Advanced Synthetic Aperture Radar), JERS-1 (Japanese Earth Resources Satellite), RADARSAT-1/2, ALOS/PALSAR (Advanced Land Observing Satellite/ Phased Array type L-band Synthetic Aperture Radar), COSMO/ SKYMED (Constellation of small Satellites for Mediterranean basin Observation), TerraSAR-X, Sentinel-1 of the European Space Agency (ESA), ALOS-2/ PALSAR-2 (Lee and Pottier 2009; Shimada 2009; Schubert et al. 2014; Okada et al. 2014). These satellites played an important role in global environmental change (Guo et al. 2014), resource prospecting (Guo 2001), disaster evaluation and mitigation (Guo et al. 2009, 2010), urban environment (Zhang et al. 2010; Li et al. 2010) and lunar exploration (Guo 2014). Future launches include the P-band-SAR of the BIOMASS mission (Heliere et al. 2009), TerraSAR-L add-on for Digital Elevation Measurement (TanDEM-L) SAR of Germany (Moreira et al. 2009), and the Radar Satellite Constellation (RCM) of Canada (Séguin 2010). Most of the future radar satellites will not only have observational capabilities such as single band, multi-polarization, polarimetric, interferometric, high-resolution, and wide swath, similar to the radar satellite previously launched but also be enhanced with bi-static or constellation observation, polarimetric interferometry, high-resolution wide swath mapping, and three-dimensional information acquisition (Guo and Li 2011). Most of these SAR missions focus on understanding key scientific problems, such as the global carbon cycle or the water cycle (Heliere et al. 2009; Moreira et al. 2009; Guo et al. 2014). These advanced SAR satellites will likely play an increasingly important role in monitoring future urban expansion and urban subsidence on a regional and global scale, and make significant contributions to global urban development.

Time series optical datasets such as Landsat TM (Thematic Mapper) and LDCM (Landsat Data Continuity Mission) are often used to map urban expansion and capture the spatial-temporal characteristics of urban areas. Over a 30 year time

Table 19.1 SAR sensors and their characteristics

Sensors[a]	Launch/Operating	Nation/Agency	Band	Target revisit	Spatial resolution (m)
Seasat	1978–1978	USA/NASA&JPL	L	13 days	25
ERS-1/2	-1 1991–2000, -2 1995–2001	Italy/ESA	C	3 days	30
ENVISAT/ASAR	2002–2012	Italy/ESA	C	35	10
JERS-1	1992–1998	Japan /JAXA	L	44 days	18
RADARSAT-1	1995–	Canada/ CSA	C	24 days	30
RADARSAT-2	2007–	Canada/ CSA	C	24 days	3
TSX-1 (TerraSAR-X)	2007–2013	Germany/ DLR	X	11 days	1.48–3.49
TDX-1 (TanDEM-X)	2010–2014	Germany/ DLR	X	11 days	12 (DEM) (absolute vertical accuracy <10 m, relative accuracy <2 m)
Cosmo-SkyMed 1–4	-1 2007,	Italy/ITA	X	Daily-16 days	1
	-2 2007,				
	-3 2008,				
	-4 2009				
ALOS PALSAR-2 (Phase Array L-band SAR)	2013–2017	Japan/JAXA	L	14 days	3 × 1
Sentinel-1 A, B, C	A 2012–2020,	Italy/ESA	C	12 days	5
	B 2014–2022,				
	C 2019–2026				
RADARSAT Constelltion Mission (RCM) C-1, C-2, C-3	C1, C2 2014–2021, C3 2015–2022	Canada/ CSA	C	12 days	3
TanDEM-L	2017–2022	Germany/ DLR	L	8 days	20
BIOMASS	2016–2021	Italy/ESA	P	25–45 days	50 × 65

[a]Reference: Robust Imaging from Space. Adjunct Reference Document #1 (www.crcsi.com.au)

NASA National Aeronautics and Space Administration, *JPL* Jet Propulsion Laboratory, *ESA* European Space Agency, *CSA* Canadian Space Agency, *DLR* German Aerospace Centre, *ASI* Italian Space Agency, *JAXA* Japan Aerospace Exploration Agency span, a few studies used SAR time series imagery for regional or global urban expansion monitoring because of the complexity of their interactions with diverse urban features (Ban et al. 2014; Pesaresi et al. 2013; Taubenböck et al. 2012). Many studies are focused on SAR urban mapping methodology (Gamba and Herold 2009) and urban subsidence mapping using time series SAR datasets (Ferretti et al. 2001), for which optical datasets do not work well. However, with its all-weather and all-time imaging capability and its unique information content, SAR time series datasets have recently been increasingly used for regional and global urban extent extraction with promising results (Gamba et al. 2011). With the launch of TerraSAR-X, TanDEM-X and Sentinel-1 SAR, SAR urban mapping on a regional and global scale is becoming much easier, but it is still important to use ENVISAT/ASAR or other earlier SAR sensor data for urban area mapping from earlier times. Furthermore, besides all-weather and all-time imaging capability, SAR can be used for not only urban expansion mapping but also for urban subsidence mapping, an advantage with which optical time series imagery cannot compete.

In this chapter, to demonstrate the capacity of time series SAR imagery for monitoring urban expansion and urban subsidence simultaneously, ENVISAT/ASAR datasets were used to create a recent urban expansion and subsidence map and to reveal recent changes in spatial and temporal characteristics of an urban area and the mechanism or factors that drove the changes. The structure of this chapter is as follows: Sect. 19.1 is the introduction, Sect. 19.2 describes the basic principles of SAR time series observation of urban monitoring, Sect. 19.3 introduces the study area and datasets, Sect. 19.4 contains the SAR time series monitoring of the urban expansion and subsidence of Beijing city, and the final Sect. 19.5 provides a discussion and conclusions.

19.2 Study Area and Datasets

Beijing is located at 39.82°N ~ 40.12°N and 116.25°E ~ 116.63° E, and covers approximately 1,300 km^2 (Fig. 19.1). The sixth census survey in 2010 showed that the registered population of the city reached 19.612 million (Guo et al. 2014). The research group acquired 23 scenes of ENVISAT/ASAR repeat-pass single-look complex data (Track: 218, Frame: 2803) covering Beijing from 2006 to 2010, all of which are in the VV polarization mode, IS2 swath. The mean incidence angle of ASAR images is approximately 23.7°. Ten of the 23 scenes of ENVISAT/ASAR repeat-pass single-look complex data were used to detect urban area changes in the Beijing area. Because the urban expansion changes were not obvious from 2006 to 2010, to validate the results, 12 parts of one WorldView image (acquired on 11 February 2009) with a spatial resolution of 0.5 m were collected as reference data for an accuracy assessment of the urban area changes; and four benchmark measurements were collected in several areas of Beijing for validation of subsidence measurement results.

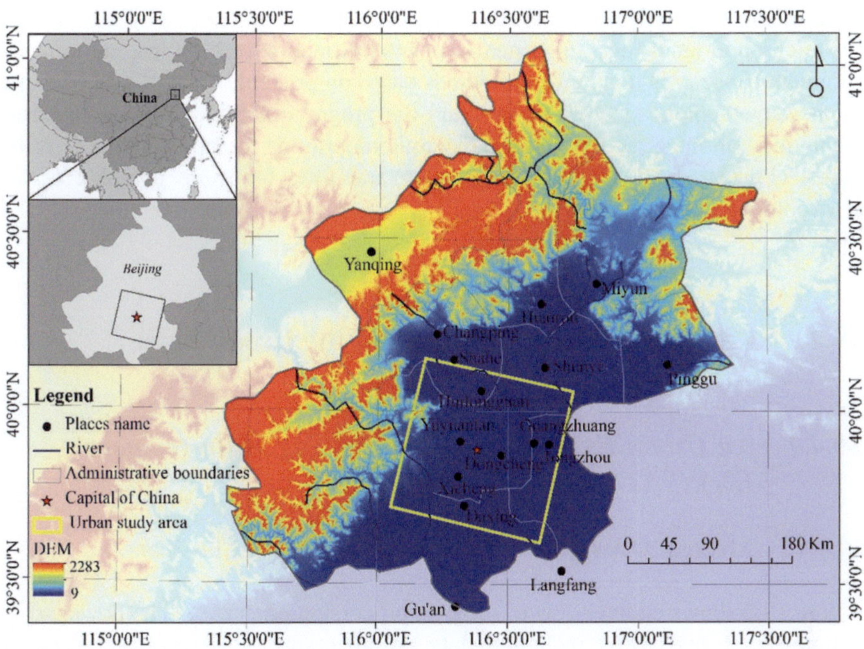

Fig. 19.1 The study area in Beijing, China

19.3 Basic Principles of SAR Time Series Observation of Urban Areas

There are three fundamental principles for using time series radar data to monitor the dynamics of urban areas. (1) Urban areas may be monitored based on the backscattering coefficient or intensity of ground objects or land cover (Rignot and van Zyl 1993). Most of the aforementioned satellite data can be used based on this principle. (2) Urban areas may be monitored based on the phase of the radar echo (Ferretti et al. 2000, 2001; Berardino et al. 2002). The urban area surface deformation caused by subsidence, seismicity or landslides might all contribute to the phase change of the radar echo. The dynamics of urban subsidence, seismic displacement or landslide deformation can be effectively monitored using the radar satellite interferometry technique, which is possessed by most of the radar satellites mentioned above. (3) Urban areas may be monitored based on the polarization, coherence, texture and other features of the SAR's echo (Conradsen et al. 2003; Lee and Pottier 2009; Moser and Serpico 2009; Li et al. 2012). Changes of ground objects will also change the polarization, coherence, texture and other features of the radar waves. Based on such changes, it is possible to monitor the dynamics of ground objects. Generally speaking, an urban area monitoring method based on principle (1) is simple and easy to implement. Methods based on principle (2) require SAR satellites that have interferometric measurement capacity; furthermore, most of these methods are only used to measure land surface changes caused by subsidence, seismicity and landslides. Polarimetric or interferometric or high resolution measurement capacity is required for methods based on principle (3). Overall, because of the development of SAR sensors and the progress of SAR image processing techniques, an integrated urban area dynamic monitoring method combining principles (1), (2) and (3) will be developed and used for more complex urban environments (Li et al. 2012).

19.4 SAR Time Series Monitoring of the Urban Area of Beijing City

19.4.1 An Urban Expansion Analysis Using Multi-temporal ENVISAT/ASAR Imagery

19.4.1.1 Method: Classification with the C5.0 Rulesets Classifier

Figure 19.2 shows an overview of the data processing procedure. The flowchart is composed of five parts, including data preprocessing, C5.0 classifier rulesets construction, a classification/accuracy assessment, an urban change detection analysis, and the interpretation of urban changes. For data preprocessing, all ASAR datasets were calibrated and then co-registered with a registration accuracy of approximately 1/8 pixel using DORIS (Delft object-oriented radar interferometric software). To reduce speckle noise in the SAR imagery, a multi-look processing algorithm was applied to reach a 5×1 look in azimuth and range, following a 3×3 enhanced Lee filter (Lee et al. 1999). Master and slave images were selected based on PSInSAR technology. After calculating a coherence map based on the master and slave images, three channels of data, including maps of master image magnitude, slave image magnitude and coherence, were selected and used to classify the data. As shown in Fig. 19.2, the construction of the C5.0 classification tree pixel-based method is an iterative process of sample data collection, model training, and validation. The samples were manually selected by cross-validating high resolution WorldView images, Google Earth images and SAR data. Subsequently, the three channels of SAR data were classified based on the constructed C5.0 rulesets and validated using an independent, randomly selected test data set. Finally, an urban change detection map was derived and dynamically analyzed based on multi-temporal classification maps using ArcGIS software. In addition, in a mountainous area, the bare rocks on the top of the mountain have a very high scattering coefficient similar to the back scattering value of urban buildings. In this study we masked the non-urban mountainous area using SRTM DEM (Shuttle Radar Topography Mission Digital Elevation Model) data. A detailed description of the classification with the C5.0 algorithm as well as accuracy assessment is given in the following sections.

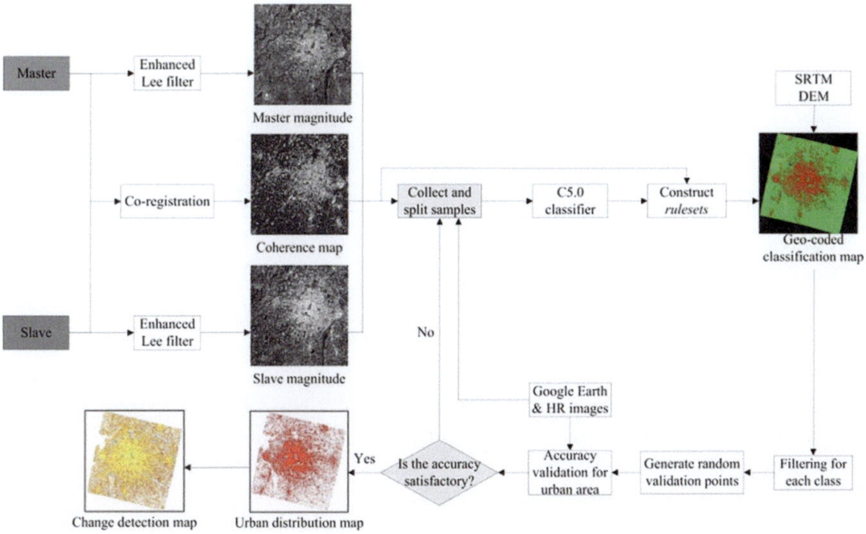

Fig. 19.2 Flowchart of the classification procedure. The *dark gray* boxes represent dataset inputs; the *white* boxes represent automatic computational processes; and the *light gray* boxes represent manual processes

Training samples were collected based on ENVISAT/ASAR data; and the collected training samples were selected by cross-checking the ENVISAT/ASAR data with high resolution Google Earth images and WorldView images in the study area. The target variables contained four land use land cover (LULC) classes: urban, vegetation, water bodies, and bare soil. Polygon regions of interest (ROIs) for each LULC class were manually drawn based on the three-channel SAR imagery, which were later divided into training and testing samples (see Table 19.2). In addition, the ROI polygon sizes for each LULC class were considered and kept approximately equal during the process of ROI creation. Two-thirds of the polygon ROIs were randomly selected as training samples for each class separately, and the remainder was used for assessing the accuracy of the classifier.

The classification was based on the commercially available decision tree classifier C5.0 (Quinlan 1993; Brown de Colstoun and Walthall 2006; Evrendilek and Gulbeyaz 2011). It uses the gain ratio criterion to determine the best attribute to separate different classes as well as the best possible threshold to make this separation (Quinlan 1993). Compared with other algorithms, the C5.0 method incorporates new methods in machine learning such as adaptive boosting, an ensemble method that has been widely shown to enhance classification accuracy and to minimize noise sensitivity (Freund and Schapire 1996; Evrendilek and Gulbeyaz 2011; Guo et al. 2014).

This study applied the See5/C5.0 program as an adaptive, boosted ruleset classifier to perform the classification. *Ruleset* classifiers are generally easier to understand than trees because each rule describes a specific context associated with

Table 19.2 Number of the polygons and pixels selected for each class in the training and testing samples

Types/samples	2006				2007			
	Training		Testing		Training		Testing	
	Polygons	Pixels	Polygons	Pixels	Polygons	Pixels	Polygons	Pixels
Urban	109	2,680	54	1,358	106	2,596	53	1,243
Vegetation	67	1,636	33	819	76	1,829	38	917
Water	71	1,617	35	808	67	1,616	33	701
Bare soil	63	1,292	32	645	64	1,367	33	685
Total	310	7,225	154	3,630	313	7,408	157	3,546
Types/samples	2008				2009			
	Training		Testing		Training		Testing	
	Polygons	Pixels	Polygons	Pixels	Polygons	Pixels	Polygons	Pixels
Urban	102	2,250	51	1,108	108	2,564	54	1,276
Vegetation	85	1,944	42	980	75	1,639	37	1,006
Water	81	1,801	40	934	78	1,742	39	917
Bare soil	62	1,263	32	576	62	1,305	32	679
Total	330	7,258	165	3,598	323	7,250	162	3,878

Types/samples	2010			
	Training		Testing	
	Polygons	Pixels	Polygons	Pixels
Urban	114	2,713	57	1,356
Vegetation	83	2,138	42	1,090
Water	87	2,055	43	983
Bare soil	64	1,351	33	791
Total	348	8,257	175	4,220

a class; additionally, they are often more accurate predictors than trees (Guo et al. 2014). In our research, the *ruleset* option was selected and used to construct classifiers. Three channels of data for each year, including maps of master image magnitude, slave image magnitude and coherence were used to construct *ruleset* classifiers and implement the classification based on the C5.0 adaptive boosting algorithm. The parameterization for the pruning rate and boosting were defined according to the recommendation of Evrendilek and Gulbeyaz (2011), who suggested a default pruning rate of 25 % and boosting trails of 10 to prune the tree in case of over-fitting.

The overall classification result for each of the three-channel SAR data was iteratively improved by performing several cycles of classification and validation procedures. After each classification step, the performance was evaluated on the basis of the testing samples and the training data base was then manually improved for the thematic classes and geographic regions with high classification errors.

However in the final classification maps, significant "salt and pepper" effects were evident in some cases and had a great impact on the level of accuracy. Therefore, a class-specific filtering approach was implemented before accuracy validation. Two filtering parameters were defined: the kernel size of the filter (filter window) and the filter scale. The filter scale represented the number of isolated points in the filter window. If the number of the center value in the filter window was less than or equal to the filter scale, this center value was replaced by the value with the maximum number in the filter window. In this study, the kernel size of the filter was set to 3×3; however, different filter scales were set for different classes. For example, the urban area was generally planted within homogenous and large regions and the filter scale for the urban class was therefore set to 3. Water bodies (especially in case of slim rivers and canals) required a filter scale of 1 to not impair the original structure of the water class after the filter process.

19.4.1.2 Results: Beijing Urban Area Classification and Expansion Analysis

Figure 19.3 is an urban spatial distribution map of Beijing for the period from 2006 to 2010 using proposed C5.0 rulesets classifier. To display the urban expansion, the vegetation and bare soil classes were combined as the "other" class. From the figure, it can be observed that Beijing expands from the inner city to the outskirts. According to urban area statistics, the proportion of the urban area of Beijing in 2006, 2007, 2008, 2009 and 2010 was 35.81 %, 37.16 %, 38.49 %, 40.45 % and 41.41 %, respectively. Due to the large-scale construction of venues for the Beijing 2008 Olympic Games, this result indicated that the urban area increased rapidly in 2006 and 2009.

Figure 19.4 shows changes for the megacity of Beijing from 2006 to 2010. Because the time frame from 2006 to 2010 is very short, the distribution of urban space in Beijing does not change much on the whole, especially in the downtown area, however, obvious signs of urban expansion are evident in the surrounding areas of the city. In particular, the urban area increases significantly in the southeast of Beijing, which is an effect of Beijing's efforts to develop the south of the city.

Based on the analysis of Beijing's urban distribution and the changes over the past 5 years, we obtained a general view of changes in urban development and the modes of development. First, the downtown area was moved out. In the face of population pressure in urban Beijing, the municipal government of Beijing adopted a policy on external population migration in the downtown area. For example, the populations in the Dongcheng and Xicheng Districts were successively migrated to the areas of Huilongguan, Daxing and Tongzhou (Fig. 19.4). Second, an axial expansion mode was adopted. So-called urban axial expansion means that urban land is expanded in a certain direction, creating a relatively narrow urban area. A recent prominent development mode in Beijing is its development toward the south and north along the medial axis (as shown in the blue box in Fig. 19.4), creating the medial-axis development corridor in Beijing. To apply for the list of world heritage of Beijing axis, Beijing has built an Asian Sports Village, Olympic venues and parks along the medial axis based on original characteristic architecture, which has enhanced Beijing's urban features by combining both ancient and modern

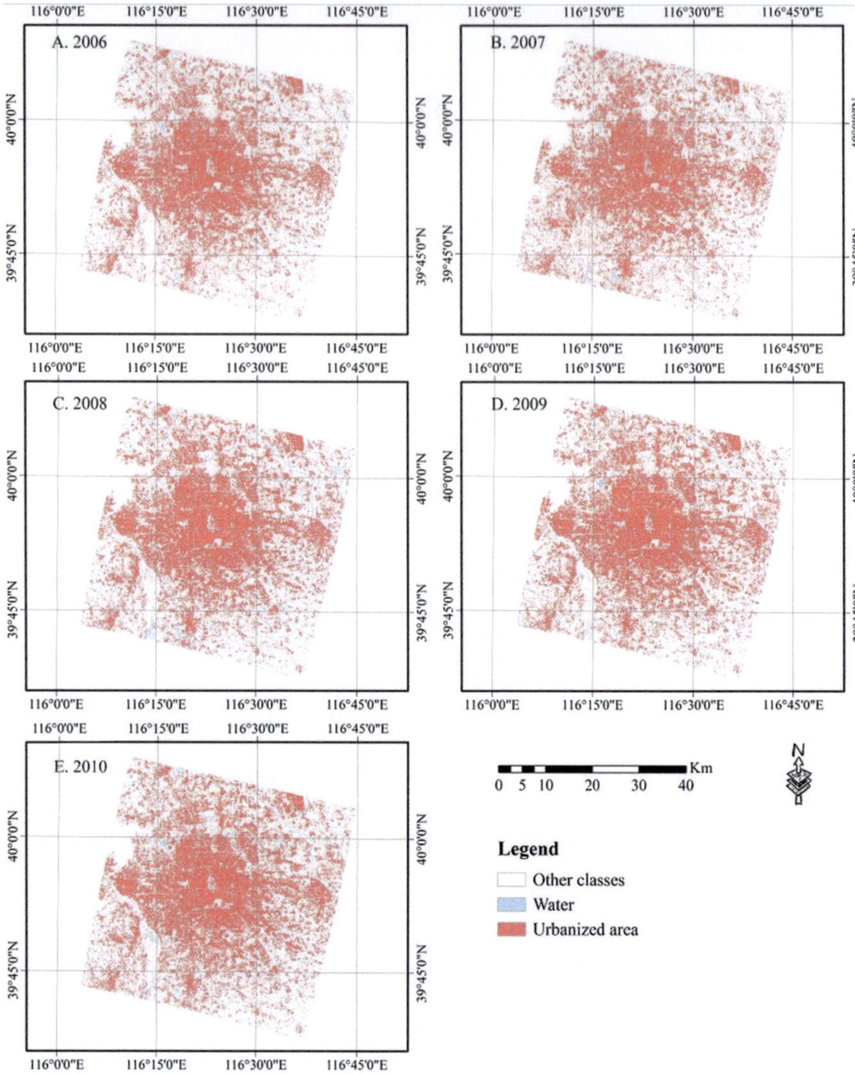

Fig. 19.3 Urban distribution maps derived from multi-temporal ASAR data

buildings. Third, a development mode in which the city expands and satellite towns are built was adopted. To relieve the pressure of population growth in Beijing, older urban areas have been reconstructed and people have been migrated; additionally, Beijing has gradually extended outwards, leading to the construction of satellite towns. Figures 19.3 and 19.4 show that Beijing has developed toward the surrounding areas year by year, mainly in the northern and eastern directions—the satellite towns of Daxing and Tongzhou were built in succession (see Fig. 19.5).

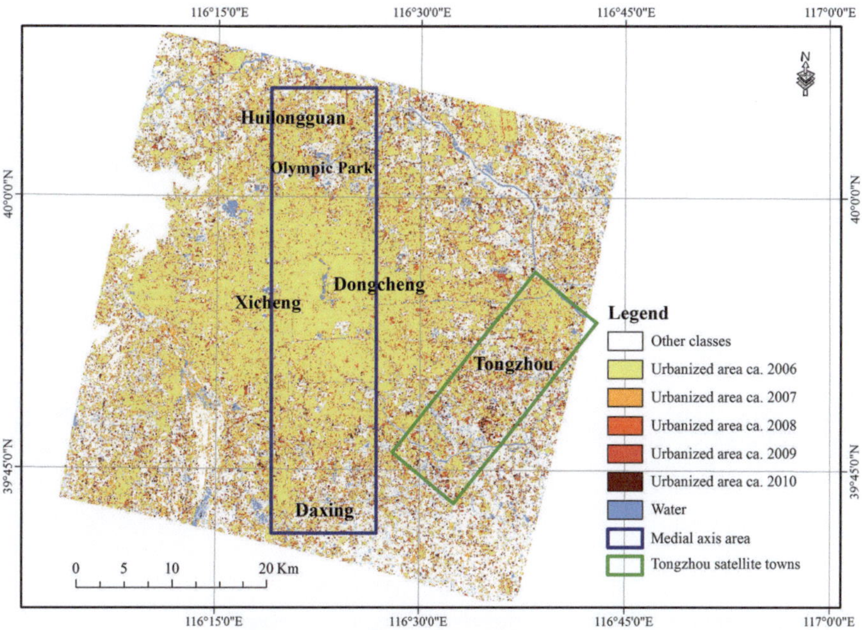

Fig. 19.4 Urban change detection map for the megacity of Beijing, China

These recent urban development changes in Beijing are mainly attributed to economic development, population growth and policy impact. There are three driving mechanisms of urban development. (1) Economic growth is a principle impetus of urban change. The regional gross domestic product (GDP) is an aggregative indicator that reflects the economic development of a region. The economy in Beijing has been developing rapidly since the reform and opening-up of China. Economic expansion and industrial optimization have led to an increased demand for space for non-agricultural industries and an increased gap between agricultural and non-agricultural land in terms of comparative benefits. The increase of non-agricultural practitioners in cities and the improvement of residential living conditions have also led to increased demands for land for construction, and the decreased percentage of agricultural employment has reduced the demand for agricultural land. Therefore, agricultural land, as represented by cultivated land, has changed its purpose and has become the incremental supply of land for construction. (2) Population growth has significantly stimulated urban land expansion and scaled economic development. Urban land expansion can facilitate population congregation within a certain territorial scope. The urban population increase has had the most direct impact on urbanization and urban expansion. The analysis and study of the urban changes in Beijing in 1975–2005 by Mu et al. (2007) show a very strong positive correlation between population growth, economic development and urban expansion, with correlation coefficients of 0.9913 and 0.9660, respectively. (3) Policy factors have also significantly boosted

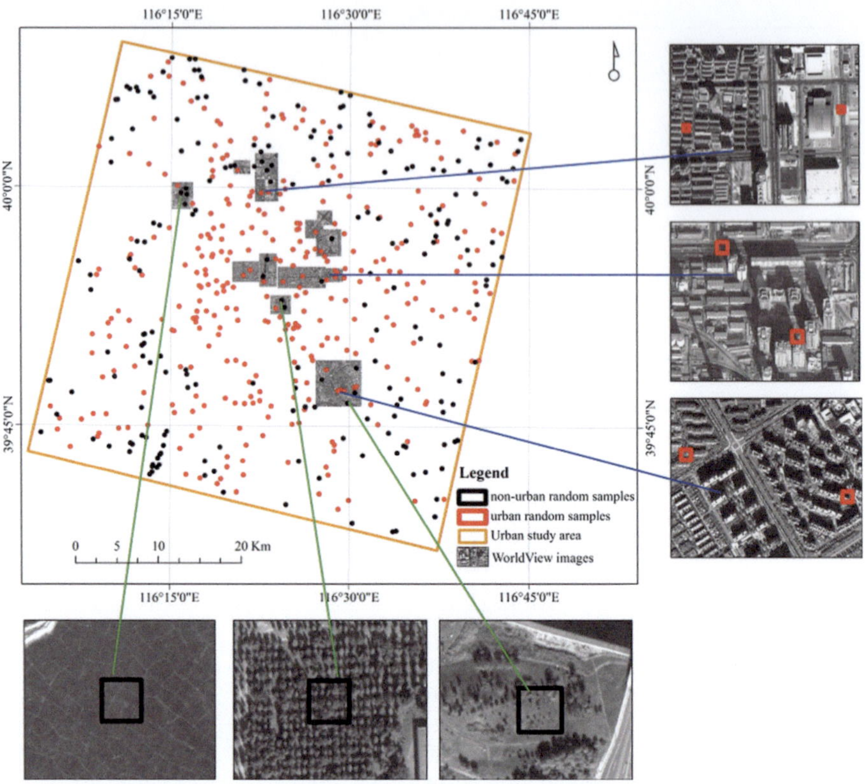

Fig. 19.5 Footprints of reference data and random validation samples for assessing the classification accuracy of the urban expansion product derived from ENVISAT/ASAR data

and affected land use changes in the area. On the one hand, hosting the Olympic Games had a significant impact on the urban infrastructure, road construction, etc. in Beijing. To host the Olympic Games, Beijing made a heavy capital investment in infrastructure, which significantly fueled economic growth and led to GDP growth. This has boosted the growth of the built-up urban areas in Beijing to some extent. On the other hand, Beijing's policies on inner city reconstruction, urban population migration, and the vigorous development of the south of Beijing city, have sped up Beijing's incremental development toward its surrounding areas. With the continued growth of population and economy in the future, it is foreseeable that Beijing will still expand, but the rate of expansion may be declining.

19.4.1.3 Accuracy Assessment

In addition to the cross-validation of the classification performance during the training process, the classification results were validated using an independent reference data set collected on the basis of Google Earth imagery and high spatial resolution WorldView. Frequently, in the past, Google Earth has been used as a reference for LULC classification validation because of the high geometric precision and the fine spatial resolution of Google Earth's imagery (Potere 2008; Benedek and Sziranyi 2009; Cohen et al. 2012; Dong et al. 2013; Leinenkugel et al. 2013). This study is focused on the extraction accuracy of the urban spatial distribution, therefore the water body, vegetation, and bare soil classes were combined as a non-urban class. In our research, the classification result for the urban and non-urban classes was validated using reference data collected from Google Earth and WorldView imagery via cross-validation. The validation sites (see Fig. 19.4) were distributed for urban and non-urban classes separately by stratified random sampling (Congalton and Green 1999) based on the proportions of the respective land cover classes. In the next step, these random validation points were converted into polygons and exported as a KML (Keyhole Markup Language) file. Sample polygons lying within the extent of high resolution Google Earth and WorldView imagery acquired between 2003 and 2013 were directly evaluated within the Google Earth and WorldView images. Finally, the classification accuracy was calculated using the commonly applied error matrix approach (Congalton 1991) by comparing the random validation points with the high-resolution WorldView and Google Earth imagery. The standard measures of classification accuracy, i.e., overall accuracy (OA) and over kappa (OK) coefficient, were derived from the matrix according to Foody (2002). Although Pontius and Millones (2011) proved that OK indices are flawed and have some limitations for purposes of classification accuracy assessment in remote sensing fields, OK indices have become a standard component of most every accuracy assessment and are considered a required component for most image analysis software packages that include accuracy assessment procedures (Congalton and Green 2009).

The assessment results are shown in Table 19.3—we see that the precision of our method for extracting an urban spatial distribution map based on SAR data has reached approximately 82 %. Our results show that the overall classification accuracies derived from five time steps of ASAR data were 84.4 %, 82.3 %, 82.4 %, 84.5 %, and 83.4 % in 2006, 2007, 2008, 2009, and 2010, respectively. The overall kappa (OK) was 68.7 %, 63.9 %, 64.1 %, 68.1 % and 66.5 % in 2006, 2007, 2008, 2009, and 2010, respectively. Analysis of the SAR imagery shows that the errors not only come from the shadow and layover of the SAR itself but also from different types of surface features having a similar scattering mechanism or back scattering value. In addition, artificial objects, such as bituminous pavement or runway demonstrate single scattering with a low back scattering coefficient and are prone to be confused with water bodies and bare earth. For example, high-rise buildings show volumetric scattering, causing confusion between buildings and forests due to the similar scattering mechanism.

Table 19.3 Accuracy assessment for the urban classification results

a. 2006				*b*. 2007			
Reference data				Reference data			
	Type	Urban	Non-urban		Type	Urban	Non-urban
Classified data	Urban	235	38	Classified data	Urban	300	73
	Non-urban	48	229		Non-urban	35	203
	OK (%)	68.7			OK (%)	63.9	
	OA (%)	84.4			OA (%)	82.3	
c. 2008				*d*. 2009			
Reference data				Reference data			
	Type	Urban	Non-urban		Type	Urban	Non-urban
Classified data	Urban	310	59	Classified data	Urban	327	65
	Non-urban	54	219		Non-urban	34	213
	OK (%)	64.1			OK (%)	68.1	
	OA (%)	82.4			OA (%)	84.5	

e. 2010			
Reference data			
	Type	Urban	Non-urban
Classified data	Urban	335	72
	Non-urban	46	259
	OK (%)	66.5	
	OA (%)	83.4	

19.4.2 Urban Subsidence Monitoring Using Multi-temporal ENVISAT/ SAR Imagery

Interferometric Synthetic Aperture Radar (InSAR) makes full use of the phase information of radar echoes. Initially, InSAR technology was mainly used to generate Digital Elevation Models (DEM) (Goldstein and Werner 1998). Later, it developed into Differential InSAR (DInSAR). With such technical features as high deformation measurement accuracy (cm level), high space resolution (tens of meters), highly automated data-processing and spatial surface observation (Gabriel et al. 1989), it has been gradually applied to surface deformation measurement. It has shown promise in the study of seismic deformation (Massonnet et al. 1993; Zebker et al. 1994), volcanic movement (Lu et al. 2000), earth surface subsidence (Galloway et al. 1998), landslides (Kimuar and Yamaguchi 2000), and glacier movement (Mark et al. 2005), etc. However, the accuracy and reliability of deformation measurements of DInSAR are affected by factors such as time decorrelation and atmospheric delay (Zebker et al. 1997; Rosen et al. 2000). To overcome the limitations of conventional DInSAR, Ferretti et al. (2001) proposed a method using differential interferometry phases to extract deformation only based on ground objects that have stable scattering properties, i.e., Permanent Scatterers (PS). This is called Permanent Scatterer Interferometry (PSInSAR). Following this, Small Baseline Subset InSAR (SBInSAR) based on correlated objects was proposed (Berardino et al. 2002; Lanari et al. 2004). Both of these approaches are collectively called Multi Temporal InSAR (MTInSAR) (Hooper 2008; Wauthier et al. 2013). These methods have been improved and perfected, enabling MTInSAR to achieve mm-level measurement accuracy to monitor micro-dynamic deformations of the earth's surface.

19.4.2.1 Fundamental Principle of MTInSAR Technology

MTInSAR technology primarily consists of PSInSAR and SBInSAR. Specifically, PSInSAR technology selects PS points of artificial buildings, rocks and other points with high coherence from a group of SAR time series images. Based on these PS points, which have stable scattering properties and are not greatly impacted by time and space decorrelation, it is possible to obtain reliable phase information, which can then be inverted for accurate surface deformation.

Considering that there are N differential interferograms, the phase of each PS pixel point may be expressed as follows:

$$\phi = \varphi_{def} + \varphi_{topo} + \varphi_{atm} + \varphi_{noise} \tag{19.1}$$

where φ_{def} is the deformation phase, φ_{topo} is the residual topographic phase, φ_{atm} is the atmospheric delay phase, and φ_{noise} are other noise phases. Because the atmospheric delay phase has a higher spatial correlation, the phase differentiation of the adjacent PS points may remove most of its impact. Considering the phase differentiation of adjacent PS points, the phase model for the deformation rate and residual topography can be built:

$$\Delta\varphi^m = \frac{4\pi}{\lambda} \cdot T^m \Delta v + \frac{4\pi}{\lambda R \sin\theta} \cdot B_\perp^m \Delta\varepsilon \tag{19.2}$$

where λ, R and θ are the wavelength of the incident wave, the slant distance between the sensor and ground object, and the incidence angle, respectively. $\Delta\varphi^m$ is the model phase difference of adjacent points in the differential interferogram m, Δv is the deformation rate difference between adjacent points, $\Delta\varepsilon$ is the elevation difference of the residual landform of two points, and T^m $B_\perp^m$ are the time and vertical baselines of the interferogram m. Defining the overall correlation coefficient (Ferretti et al. 2001):

$$\gamma = \left| \frac{1}{N} \sum_{m=1}^{N} \exp\left[j \cdot \Delta\varphi^m - j\left(k_v^m \cdot \Delta v + k_\varepsilon^m \cdot \Delta\varepsilon \right) \right] \right| \tag{19.3}$$

Taking this as the objective function, it is possible to solve for the unknown parameters Δv and $\Delta\varepsilon$ by maximizing from N interferograms. Then, the velocity and elevation parameters are integrated to obtain an estimation of the absolute velocity and vertical error on each PS point. Next, the known linear deformation phase and the vertical error phase are subtracted from the differential phase to obtain the components of the residual phase. By filtering in both space and time, it is then possible to remove the atmospheric delay signals, which have a high frequency in time and a low frequency in space and to obtain the information on residual nonlinear deformation. By now, all deformation phase components are solved.

The SBInSAR method may be used to obtain the low-resolution and large-scale deformation (Berardino et al. 2002; Lanari et al. 2004). It selects image pairs according to perpendicular baseline and temporal separation thresholds, and produces multiple interferogram subsets. Then, the Singular Value Decomposition (SVD) (Strang 1988; Golub and Van Loan 1996) is used to combine multiple small-baseline interferogram subsets to conduct a time series inversion for deformation. The SBInSAR method may effectively increase the coherence of the interferogram and solve the problem of over-large temporal sampling of SAR images.

19.4.2.2 Data Processing

(1) Generation of the interferogram
The open source software DORIS (Delft object-oriented radar interferometric software) (Kampes and Usai 1999) was used to generate the interferogram. The perpendicular baseline, the time interval, and the average Doppler frequency (see Fig. 19.6) among the SAR images were taken into consideration, and the April 2, 2008 SAR data were selected as the master image to maximize the coherence of the interferograms. For resampling, a 12-point raised cosine interpolation kernel was used. Then, the single master image was used to generate a total of 21 interferograms.

(2) Removal of the flat terrain and terrain phases
The accurate-orbit data provided by the ENVISAT satellite of Delft University of Technology (Netherlands) were used to remove the flat terrain phase in the interferogram (Scharroo and Visser 1998). The Band C 3″ DEM (Farr and Kobrick 2000; Rabus et al. 2003) generated by the United States' Shuttle Radar Topography Mission (SRTM) was used to remove the terrain phase.

(3) Time series processing
In this study, the StaMPS (Stanford Method for Permanent Scatterers) method was employed (Hooper et al. 2004, 2007). This method selected the PS points based on amplitude and phase stability. First, the amplitude of each pixel

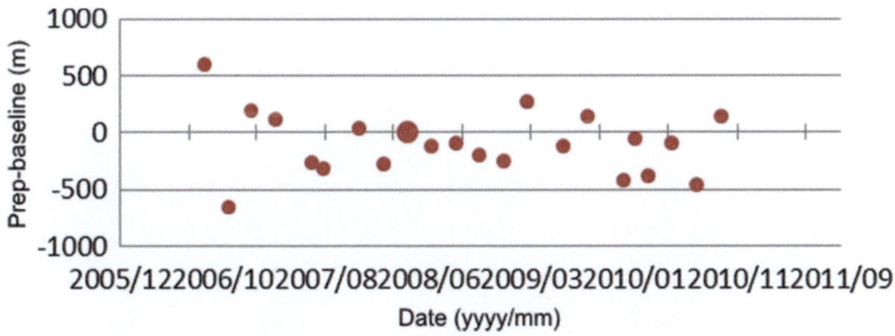

Fig. 19.6 Time baseline and perpendicular baseline of the intervention pair in the study area, where *red dots* represent the main image. The *large dot* represents the master image

was analyzed and the PS candidate points were selected. Then, a cycle analysis of the phase stability of the PS candidate points was conducted, and the PS points were refined. In the study area, the average PS point density reached 134/km^2 due to the presence of dense buildings. Finally, the errors caused by the DEM were estimated and removed. In this case, the phase difference between adjacent PS points was considered to be less than π so that correct unwrapping was possible. StaMPS used a 3D unwrapping method and considered the PS space and time factors. After unwrapping, different filters were used to extract the space-related errors and the surface deformation was obtained based on the reflection of the atmospheric delay and orbit errors in the master and slave images.

19.4.2.3 Results of Urban Subsidence Analysis

The StaMPS method was used to obtain the average speed (see Fig. 19.7) and time series deformation (see Fig. 19.8) in the study area from 2006 to 2010.

Analysis of Average Speed

To reduce the noise effect of a single PS point, the PS point mean value in the small area in Beijing where the original benchmark was located, including Yuyuantan, was selected as the reference, and the average speed diagram of the whole image was obtained (see Fig. 19.7). Figure 19.7 shows that the deformation in the downtown area of Beijing is small (approximately 1 mm/a) and is relatively stable during the research period. The subsidence areas are connected in blocks and are mainly distributed in a strip shape beyond the East 5th Ring Road and North 5th Ring Road of Beijing. Obviously, a subsidence funnel is formed at Guanzhuang outside the East 5th Ring Road and at Shahe outside the North 5th Ring Road (see Fig. 19.1), with a maximum yearly subsidence rate of more than 60 mm/a at Guanzhuang. A slight rise is observed in the northeast corner of the research area, i.e., the mountainous areas in the southeast of the Pinggu District. In addition, there

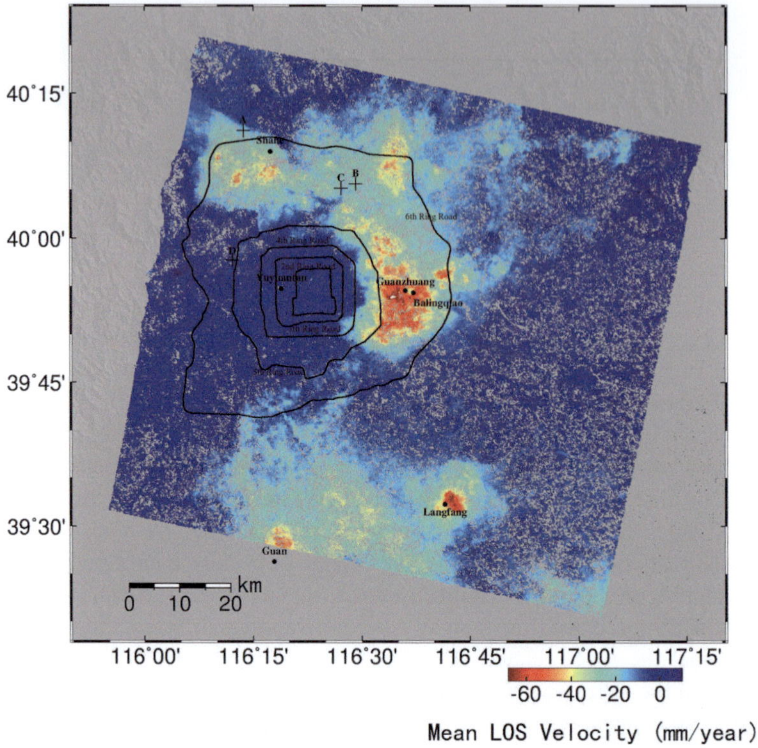

Fig. 19.7 Surface deformation speed (mm/a) in Beijing from 2006 to 2010. The background is the regional DEM. *Red* represents subsidence, and the *black cross stars* represent benchmark positions in Fig. 19.9

is also substantial subsidence in Langfang and Gu'an in the Hebei Province (see Fig. 19.1), which are adjacent to Beijing, with the maximum subsidence rate being more than 60 mm/a in Langfang City.

Subsidence is not obvious in the downtown area of Beijing, which is related to the stringent restrictions on groundwater mining. Because of the population explosion in Beijing, substantial expansion of urban areas during the research period and active social production activities in the areas outside the East 5th Ring Road and North 5th Ring Road, the surface subsidence has mainly been caused by excessive groundwater abstraction. In addition, surface subsidence in Beijing mainly develops in the middle and lower parts of the alluvial-proluvial fan plain, especially at the connection of the Wenyu River Alluvial-proluvial Fan, the Chaobai River Alluvial-proluvial Fan and the Yongding River Alluvial-proluvial Fan. This indicates that subsidence is also closely tied to geological features such as the hydrogeologic structure, especially the formation lithology and its structural features and that the area of substantial subsidence is in line with the thickness distribution of the compressible clay soil layer in the shallow layer (Jia et al. 2007).

Fig. 19.8 Time series displacement (rad) in the direction of the PS point line of sight in the study area from 2006–2010

Analysis of Time Series Deformation

The StaMPS method is based on the spatial correlation hypothesis of surface deformation, with no preset deformation model. Therefore, the long time-sequential dynamic deformation field can be generated (see Fig. 19.8). Subsidence was not prominent in the study area except in Shahe of Beijing from August 2006 to March 2007; instead, a rise was observed over a large area. This deformation pattern is likely related to environmental control measures taken by the government before the 2008 Olympic Games in Beijing, such as population migration and the shutdown of polluting enterprises such as the Shougang Group, which has substantially reduced groundwater mining. The subsidence layout developed from June 2007, which is even earlier, with subsidence increasing gradually. The total subsidence at Guanzhuang in the east of Beijing had exceeded 400 mm at the end of October 2010. The natural environment and infrastructure in Beijing changed substantially after the Olympic Games, and the population has grown rapidly since then. As stated above, the population explosion during the time period during which this research took place was mainly attributed to a surge of the external population. This has resulted in the rapid expansion of urban areas, which is mainly reflected by the former rural-urban continuum outside of the 5th Ring Road, which has increased groundwater utilization. In addition, a large-scale rise was observed in the study area in August 2008 and was restored in October. It is possibly related to the surface wave disturbance of the earthquake in Wenchuan County in Sichuan Province in May 2008 and requires further study based on continuous GPS and other data (Gu et al. 2009).

19.4.2.4 Validation

To validate the effectiveness and accuracy of the results, the time series deformation of MTInSAR and the leveling benchmark data were compared to confirm the reliability of the processing result of MTInSAR. In the study area, the land subsidence observation network of Beijing conducted 5 leveling observations during the summers of 2005, 2007, 2008, 2009 and 2010. Four groups of leveling benchmark data were selected (the cross stars in Fig. 19.7 represent benchmark positions, the cross stars in the northwest corner correspond to benchmark A, and the other cross stars correspond clockwise to the B, C, D points, respectively) and were projected in the direction of radar line of sight (RLOS). Deformation within the area was assumed to be related, and the average value of the time series displacement of the PS points in the 100 m * 100 m area with the benchmark as the center was obtained and compared with the benchmark (see Fig. 19.9). The mean MTInSAR time series displacements were approximately equal to the related leveling data, with a total mean error of -3.1 mm and a standard deviation of 7.5 mm. Obvious system bias can be observed in (a) and (c) in Fig. 19.9, which might be caused by the failure of the starting observation time at benchmarks to fully comply with MTInSAR. In addition, the MTInSAR result showed nonlinear

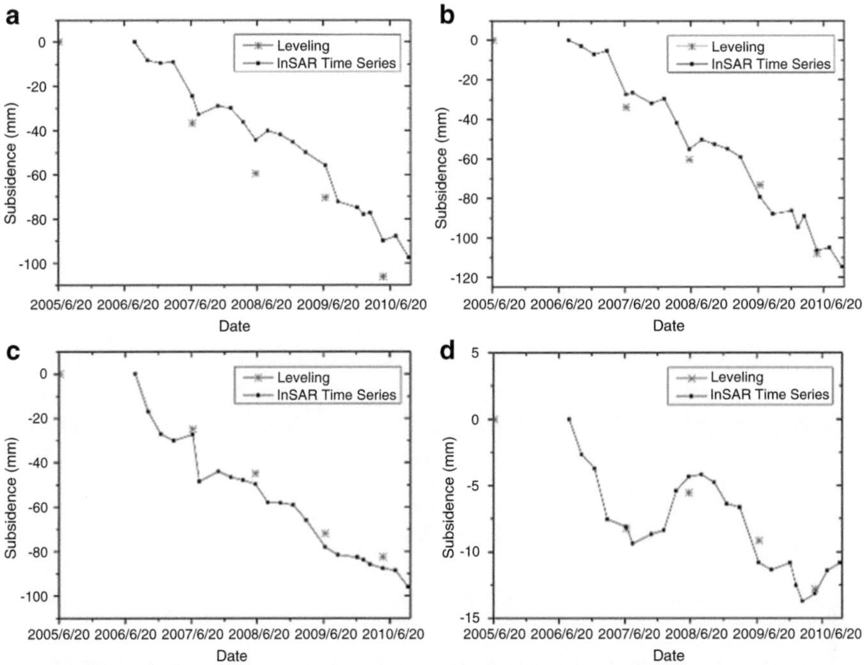

Fig. 19.9 Comparison between time series displacement of MTInSAR and benchmark data in the study area

deformation, but this was not obvious in the benchmark data, especially in (b) and (c), which might be due to the limited time for the benchmark sampling. Generally speaking, a good result for urban subsidence of Beijing city has been obtained and the validation shows the effectiveness and high accuracy of SAR time series monitoring.

19.5 Discussion and Conclusion

Based on ENVISAT/ASAR time series radar data, this chapter discusses and analyzes the applications for monitoring urban expansion and urban subsidence. The main conclusions are as follows:

(1) From ENVISAT/ASAR time series radar data, C5.0 rulesets classifier and Multi Temporal InSAR (MTInSAR) method, the urban expansion and urban subsidence of Beijing city has been effectively monitored with high precision. It shows the advantage of mapping the urban expansion and urban subsidence simultaneously, for which optical techniques do not work well.
(2) The changes of the urban space distribution of Beijing in 2006–2010 show that on the whole, the urban areas have not greatly changed because the time span

was short. Judging by the change trend, however, the main modes of expansion of Beijing include the axial development mode, which extends south and north along the medial axis, and the surrounding extension mode, which has strengthened the expansion of the Nancheng and Dongcheng Districts, with the building of satellite towns. The recent urban development changes in Beijing are mainly attributed to the following driving factors: economic development, population growth and policy impact.

(3) The features of the changes in urban subsidence in Beijing in 2006–2010 show that the deformation in the downtown area of Beijing is small and relatively stable. The subsidence areas are connected in blocks and are mainly arranged in a strip shape beyond the East 5th Ring Road and North 5th Ring Road of Beijing, with a maximum yearly subsidence rate of more than 70 mm/a. Ground surface subsidence in the study area was mainly caused by activities such as the excessive mining of groundwater and is also closely related to geological features such as the hydrogeologic structure, especially the formation lithology and structural features.

Although a good result for urban expansion and urban subsidence of Beijing city was obtained and validation showed the effectiveness and high accuracy of SAR for time series monitoring, some limitations for accurate monitoring were evident. (1) Due to the lack of long time series SAR datasets (only five years), our result cannot show the large urban area expansion of Beijing City, so the result cannot fully reveal the rule of urban expansion. Compared with optical remote sensing, this is an obvious disadvantage; for example, based on Landsat-TM datasets, the land surface change over 30 years can be monitored, so the acquisition of long time series SAR imagery is necessary for urban area monitoring in the future. (2) The intrinsic characteristics of radar layover and shadow effects also become severe challenges. (3) The coherence of the interferogram especially over a long time span is still a large problem for MTInSAR technology. To enhance the coherence, the Quasi-Permanent Scatterer technique and homogeneous filter should be gradually applied.

With the emergence of advanced SAR systems, the new generation of spaceborne SAR systems recently launched, such as RADARSAT-2, Sentinel-1, TerraSAR-X, and ALOS-2 (using C-, X-, and L-bands, respectively), have enhanced capabilities such as higher resolution (resolutions ranging from 3 to 10 m), dual/quad-polarization, more frequent revisits, and varying beam modes (scene swath and incidence angle). These enhanced capabilities will provide more useful time series information at spatial and temporal levels for accurately monitoring urban expansion and ground surface subsidence.

In the future, a joint spatial and temporal characteristic analysis combining both urban expansion and urban subsidence of Beijing will be conducted to achieve a deeper understanding of Beijing's urban development issues.

Acknowledgments The work was supported by grants from National Natural Science Foundation of China (Grant No. 41120114001 and 41201357), Innovation Foundation Project of CEODE,

CAS (Center for Earth Observation and Digital Earth, Chinese Academy of Sciences) (Grant No. Y2ZZ20101B), and "One-Three-Five" Strategic Planning of RADI (Institute of Remote Sensing and Digital Earth), CAS (Grant No. Y4SG0500CX).

References

Ban YF, Jacob A, Gamba P (2014) Spaceborne SAR data for global urban mapping at 30 m resolution using a robust urban extractor. ISPRS J Photogramm Remote Sens (in press). doi:10.1016/j.isprsjprs.2014.08.004

Benedek C, Sziranyi T (2009) Change detection in optical aerial images by a multilayer conditional mixed Markov model. IEEE Trans Geosci Remote Sens 47(10):3416–3430

Berardino P, Fornaro G, Lanari R, Sansosti E (2002) A new algorithm for surface deformation monitoring based on small baseline differential SAR interferograms. IEEE Trans Geosci Remote Sens 40:2375–2383

Brown de Colstoun EC, Walthall CL (2006) Improving global scale land cover classifications with multi-directional POLDER data and a decision tree classifier. Remote Sens Environ 100 (4):474–485

Cohen WB, Yang ZG, Kennedy R (2012) Detecting trends in forest disturbance and recovery using yearly Landsat time series: 2. TimeSync-Tools Calibr Valid 114(12):2911–2924

Congalton RG (1991) A review of assessing the accuracy of classifications of remotely sensed data. Remote Sens Environ 37(1):35–46

Congalton RG, Green K (1999) Assessing the accuracy of remotely sensed data: principles and practices. Lewis Publishers, Boca Raton/London/New York

Congalton R, Green K (2009) Assessing the accuracy of remotely sensed data: principles and practices, 2nd edn. CRC/Taylor & Francis, Boca Raton, p 183

Conradsen K, Nielsen AA, Schou J, Skriver H (2003) A teststatistic in the complex Wishart distribution and its application to change detection in polarimetric SAR data. IEEE Trans Geosci Remote Sens 41(1):4–19. doi:10.1109/TGRS.2002.808066

Dong J, Xiao X, Chen B, Torbick N, Jin C, Zhang G, Biradar C (2013) Mapping deciduous rubber plantations through integration of PALSAR and multi-temporal Landsat imagery. Remote Sens Environ 134:392–402

Evrendilek F, Gulbeyaz O (2011) Boosted decision tree classifications of land cover over Turkey integrating MODIS, climate and topographic data. Int J Remote Sens 32(12):3461–3483

Farr TG, Kobrick M (2000) Shuttle radar topography mission produces a wealth of data. Eos Trans Am Geophys Union 81:583–585

Ferretti A, Prati C, Rocca F (2000) Nonlinear subsidence rate estimation using permanent scatterers in differential SAR interferometry. IEEE Trans Geosci Remote Sens 39:8–20

Ferretti A, Prati C, Rocca F (2001) Permanent scatterers in SAR interferometry. IEEE Trans Geosci Remote Sens 39(l):8–19

Foody GM (2002) Status of land cover classification accuracy assessment. Remote Sens Environ 80:185–201

Freund Y, Schapire RE (1996) Experiments with a new boosting algorithm. Morgan Kaufman, San Francisco

Gabriel AK, Goldstein RM, Zebker HA (1989) Mapping small elevation changes over large areas: differential radar interferometry. J Geophys Res 94:9183–9191

Galloway DL, Hudnut KW, Ingebritsen SE, Phillips SP, Peltzer G, Rogez F, Rosen PA (1998) Detection of aquifer system compaction and land subsidence using interferometric synthetic aperture radar, Antelope Valley, Mojave Desert, California. Water Resour Res 34:2573–2585

Gamba P, Herold M (eds) (2009) Global mapping of human settlements: experiences, datasets, and prospects. CRC Press, Boca Raton

Gamba P, Aldrighi M, Stasolla M (2011) Robust extraction of urban area extents in HR and VHR SAR images. IEEE J Select Top Appl Earth Obs Remote Sens 4(1):27–34

Goldstein RM, Werner CL (1998) Radar interferogram filtering for geophysical applications. Geophys Res Lett 25(21):4035–4038

Golub GH, Van Loan CF (1996) Matrix computation [M]. Johns Hopkins University Press, Baltimore

Gu GH, Wang WX, Meng GJ (2009) Crustal movements and after the Wenchuan earthquake as detected by GPS observations. Geomatics Inform Sci Wuhan Univ 34(11):1336–1339

Guo HD (2001) Radar remote sensing applications in China. CRC/Taylor & Francis Books Ltd, Boca Raton, FL

Guo HD (2014) Scientific satellites for global change research, 2014. Science Press, Beijing

Guo HD, Li XW (2011) Technical characteristics and potential application of the new generation SAR for Earth observation (in Chinese). Chin Sci Bull (Chin Ver) 56:1155–1168

Guo HD, Li XW, Zhang L (2009) Study of detecting method with advanced airborne and spaceborne synthetic aperture radar data for collapsed urban buildings from the Wenchuan earthquake. J Appl Remote Sens 3(1) 031695:1–18. doi:10.1117/1.3153902

Guo HD, Wang XY, Li XW, Liu G, Zhang L, Yan SY (2010) Yushu earthquake synergic analysis using multimodal SAR datasets. Chin Sci Bull 55(31):3499–3503

Guo HD, Yang HN, Sun ZC, Li XW, Wang C (2014) Synergistic use of optical and PolSAR imagery for urban impervious surface estimation. Photogramm Eng Remote Sens 80 (1):91–102

Heliere F, Lin CC, Fois F, Davidson M, Thompson A, Bensi P (2009) BIOMASS: a p-band SAR earth explorer core mission candidate. In: Radar conference, IEEE, Pasadena, CA, USA, pp 1–6

Hooper A (2008) A multi-temporal InSAR method incorporating both persistent scatterer and small baseline approaches. Geophys Res Lett 35(16):L16302, 1–5. doi:10.1029/2008GL034654

Hooper A, Zebker H, Segall P, Kampes B (2004) A new method for measuring deformation on volcanoes and other non-urban areas using InSAR persistent scatterers. Geophys Res Lett 31:L23611

Hooper A, Segall P, Zebker H (2007) Persistent scatterer InSAR for crustal deformation analysis, with application to Volcán Alcedo. Galapagos J Geophys Res 112:B07407

Jia S, Wang HG, Luo Y (2007) The impacts of ground subsidence on urban construction in Beijing. City Geol 2(4):19–23 (in Chinese)

Kampes BM, Usai S (1999) Doris: the delft object-oriented radar interferometric software [C]. In: 2nd international symposium on operationalization of remote sensing, vol 16, Enschede, p 20

Kimuar H, Yamaguchi Y (2000) Detection of landslide ar.eas using satellite radar interferometry. Photogramm Eng Remote Sens 66(3):337–344

Lanari R, Mora O, Manunta M et al (2004) A small-baseline approach for investigating deformations on full-resolution differential SAR interferograms. IEEE Trans Geosci Remote Sens 42(7):1377–1386

Lee JS, Pottier E (2009) Polarimetric radar imaging from basics to applications. CRC Press/Taylor & Francis Group, London

Lee JS, Grunes MR, Grandi GD (1999) Polarimetric SAR speckle filtering and its implication for classification. IEEE Trans Geosci Remote Sens 37(5):2363–2373

Leinenkugel P, Kuenzer C, Oppelt N, Dech S (2013) Characterization of land surface phenology and land cover based on moderate resolution satellite data in cloud prone areas—a novel product for the Mekong Basin. Remote Sens Environ 136:180–198

Li XW, Pottier E, Guo HD, Ferro-Famil L (2010) Urban land cover classification with high-resolution polarimetric SAR interferometric data. Can J Remote Sens 36(3):236–247

Li XW, Zhang L, Guo HD, Sun ZC, Liang L (2012) New approaches of urban area change detection using multi-temporal RADARSAT-2 polarimetric SAR data. Can J Remote Sens 38(3):1–14

Lu Z, Mann D, Freymueller JT, Meyer DJ (2000) Synthetic aperture radar interferometry of Okmok volcano, Alaska: radar observations. J Geophys Res 105:10791–10806

Mark B, Seltzer G, Geoffrey O (2005) Evaluation of recent glacier recession in the Cordillera Blanca, Peru (AD 1962–1999): spatial distribution of mass loss and climatic forcing. Quat Sci Rev 24:2265–2280

Massonnet D, Rossi M, Carmona C, Adragna F, Peltzer G, Feigl K, Rabaute T (1993) The displacement field of the Landers earthquake mapped by radar interferometry. Nature 364:138–142

Millennium Ecosystem Assessment (2005) Ecosystems and human well-being: synthesis. Island Press, Washington, DC

Moreira A, Hajnsek I, Krieger G, Papathanassiou K, Eineder M, De Zan F, Younis M, Werner M (2009) TANDEM-L: monitoring the earth's dynamics with InSAR and Pol-InSAR. In: 4th international workshop on science and applications of SAR polarimetry and polarimetric interferometry, Frascati, Italy

Moser G, Serpico SB (2009) Unsupervised change detection from multichannel SAR data by Markovian data fusion. IEEE Trans Geosci Remote Sens 47(7):2114–2128. doi:10.1109/TGRS.2009.2012407

Mu FY, Zhang ZX, Chi YB, Liu B, Zhou QB, Wang CY, Tan WB (2007) Dynamic monitoring of built-up area in Beijing during 1973–2005 based on multi-original remote sensed images. J Remote Sens 11(2):257–268

Okada Y, Nakamura S, Iribe K, Tsuji M, Hariu K, Kankaku Y, Suzuki S, Shimada M (2014) PALSAR-2 initial mission check. IGARSS, Quebec, Canada, July, pp 3423–3414

Pesaresi M, Guo HD, Blaes X, Ehrlich D, Ferri S, Gueguen L, Halkia M, Kauffmann M, Kemper T, Lu LL, Marin-Herrera MA, Ouzounis GK, Scavazzon M, Soille P, Syrris V, Zanchetta L (2013) A global human settlement layer from optical HR/VHR RS data: concept and first results. IEEE J Select Top Appl Earth Obs Remote Sens 6(6):2102–2131

Pontius R, Millones M (2011) Death to Kappa: birth of quantity disagreement for accuracy assessment. Int J Remote Sens 32(4):407–4429

Potere D (2008) Horizontal positional accuracy of Google Earth's high-resolution imagery archive. Sensors 8:7973–7981

Quinlan JR (1993) C4.5: programs for machine learning. Morgan Kaufman Publishers, San Mateo

Rabus B, Eineder M, Roth A, Bamler R (2003) The shuttle radar topography mission- a new class of digital elevation models acquired by spaceborne radar. ISPRS J Photogramm Remote Sens 57:241–262

Rignot EJM, van Zyl JJ (1993) Change detection techniques for ERS-1 SAR data. IEEE Trans Geosci Remote Sens 31(4):896–906

Rosen PA, Hensley S, Joughin IR, Fuk KL, Madsen SN, Rodriguez E, Goldstein RM (2000) Synthetic aperture radar interferometry. Proc IEEE 88:333–382

Scharroo R, Visser P (1998) Precise orbit determination and gravity field improvement for the ERS satellites. J Geophys Res 103(C4):8113–8127

Schubert A, Small D, Meier E, Miranda N, Geudtner D (2014) Spaceborne SAR product geolocation accuracy: A SENTINEL-1 UPDATE. IGARSS, Quebec, Canada, July, pp 1465–1468

Séguin G (2010) Radarsat constellation preliminary design. In: 8th European conference on synthetic aperture radar. Offenbach, Berlin, pp 514–515

Shimada M (2009) Advance land-observation satellite (alos) and its follow-on satellite, ALOS-2. In: 4th International Workshop on Science and Applications of SAR Polarimetry and Polarimetric Interferometry, Frascati, Italy

Strang G (1988) Linear algebra and its applications, 3rd edn. Saunders, Philadelphia, p 5

Taubenböck H, Esch T, Felbier A, Wiesner M, Roth A, Dech S (2012) Monitoring urbanization in mega cities from space. Remote Sens Environ 117:162–176

Wauthier C, Cayol V, Poland M (2013) Nyamulagira's magma plumbing system inferred from 15 years of InSAR [J]. Geol Soc Lond, Spec Publ 380(1):39–65

Zebker HA, Rosen PA, Goldstein RM, Gabriel A, Werner CL (1994) On the derivation of co-seismic displacement fields using differential radar interferometry: the Landers earthquake. J Geophys Res 99:19617–19634

Zebker HA, Rosen PA, Hensley S (1997) Atmospheric effects in interferometric synthetic aperture radar surface deformation and topographic maps. J Geophys Res 102:7547–7563

Zhang L, Li XW, Guo HD (2010) Effect of aspect angle normalized PolSAR data on urban building detection. Can J Remote Sens 36(3):276–286

Chapter 20
SAR Time Series for the Analysis of Inundation Patterns in the Yellow River Delta, China

Claudia Kuenzer, Juliane Huth, Sandro Martinis, Linlin Lu, and Stefan Dech

Abstract Earth Observation using radar remote sensing is a valuable tool for the monitoring large scale inundation over time. This study performs a time series analysis using 18 ENVISAT/ASAR Wide Swath Mode data sets for the year 2008 and 13 TerraSAR-X Stripmap data sets for the year 2013/2014 to characterize inundation patterns in the Yellow River Delta, located in Shandong Province of China. Water surfaces are automatically derived using the software package WaMaPro, developed at the German Remote Sensing Data Center (DFD), of the German Aerospace Center (DLR), which allows an automatic classification using empirical thresholding. The temporal analysis allows the separation of different types of water bodies such as rivers, water storage basins, aquaculture, brine ponds, and agricultural fields based on inundation frequencies. This supports the understanding of the water dynamics in this highly variable study region. As ENVISAT data is not available anymore since April 2012, and as access to TerraSAR-X data is limited, Sentinel-1 data of the European Space Agency, ESA, are eagerly expected for the region. The good spatial resolution between 40 up to 5 m, as well as a dense temporal coverage, which allow to generate "true" SAR time series, and will help to lift annual analyses to the next level.

C. Kuenzer (✉) • J. Huth • S. Martinis
German Remote Sensing Data Center, DFD, Earth Observation Center, EOC,
German Aerospace Center, DLR, Oberpfaffenhofen, Germany
e-mail: claudia.kuenzer@dlr.de

L. Lu
Key Laboratory of Digital Earth Sciences, Institute of Remote Sensing and Digital Earth, RADI, Chinese Academy of Sciences, CAS, Beijing, China

S. Dech
German Remote Sensing Data Center, DFD, German Aerospace Center, DLR, Wessling, Germany

Institute for Geography and Geology, University of Wuerzburg, Wuerzburg, Germany

C. Kuenzer et al. (eds.), *Remote Sensing Time Series*, Remote Sensing and Digital Image Processing, DOI 10.1007/978-3-319-15967-6_20

20.1 Introduction: SAR Based Water Mapping

Although studies on optical time series analyses usually outnumber radar and synthetic aperture radar (SAR), based studies, SAR data has distinct advantages compared with optical data. Cloud cover and atmospheric influences, as well as limited illumination (polar night) do not impair data acquisition and quality, so that data collection is possible under all weather conditions and at all seasons. SAR based time series over the land surface have especially been exploited in the context of soil moisture retrieval (Wagner et al. 1999; Zhao et al. 2008; Kuenzer et al. 2009; Dorigo et al. 2012; Naemi et al. 2013; Wagner et al. 2013; Dostálová et al. 2014; Doubková et al. 2014) and biomass derivation (Santoro et al. 2002; Wagner et al. 2003; Askne and Santoro 2005; Cartus et al. 2011), as well as for Differential Interferometric SAR (DInSAR) applications to assess surface movements (Fielding et al. 1998; Amelung et al. 1999; Higgins et al. 2013).

However, the most common application for SAR based information derivation is the extraction of water surfaces from data of sensors such as ENVISAT ASAR (European Environmental Satellite, Advanced Synthetic Aperture Radar), Radarsat, ALOS Palsar (Advanced Land Observing Satellite - Phased Array type L-band Synthetic Aperture Radar), COSMO SkyMed, or TerraSAR-X. Although there are literally hundreds of studies, where authors investigated the potential of this data for flood mapping and inundation mapping based on single scenes or a few multitemporal data sets (Werle et al. 2001; Townsend 2001; Kasischke et al. 2003; Kiage et al. 2005; Lang et al. 2008; Martinez and Le Toan 2007; Hoque et al. 2010; Mason et al. 2010; Martinis and Twele 2010; Chaouch et al. 2011; Schumann et al. 2007; Kuenzer et al. 2013b; Martinis et al. 2014, 2015), only few studies have been published on the exploitation of extensive SAR time series for inundation analyses. Kuenzer et al (2013a) presented a time series analyses of overall 60 ENVISAT ASAR Wide Swath Mode (WSM) scenes at a spatial resolution of 150 m for the Mekong Delta in Vietnam, covering the years 2007–2011. This study, which has been the most comprehensive SAR based inundation mapping endeavour for this area contributed to a thorough understanding of the flood regime in the Mekong Delta. Greifeneder et al. (2014) employed nearly 700 ENVISAT ASAR WSM scenes for the entire Mekong Basin, covering the time span from 2007 to 2011, to generate a permanent water body product. In highly variable areas (rainy and dry seasons over the course of 1 year) a time series of at least one full year is needed to derive stable water surfaces in an accurate way.

In radar data water covered areas show very distinct backscattering characteristics, yielding no or an extremely low return signal. Therefore, water covered areas appear as very dark (black) surfaces in radar imagery and can easily be distinguished from other cover types. Water surfaces appear "smooth" to the incident radar waves, and thus specular reflection occurs (Zhou et al. 2000). This principle is also depicted in Fig. 20.1. Radar beams which hit dense canopy vegetation either experience surface backscattering or volume backscattering (Richards et al. 1987), and in special cases – e.g. if radar beams hit the trunks of trees (or e.g. buildings)

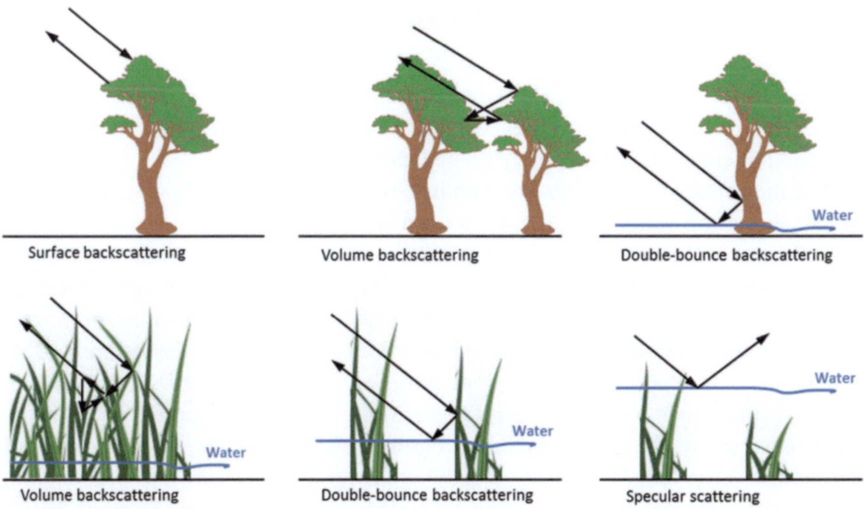

Fig. 20.1 Radar scattering processes over forests and wetlands, including specular scattering on undisturbed water surfaces (Modified based on Wang 2010)

double bounce backscattering occurs. On water surfaces, which are mostly covered by vegetation (mature rice fields, reeds etc.) volume backscattering is prevalent. However, on smooth, undisturbed water surfaces specular scattering occurs (see Fig. 20.1, lower right). The magnitude of these scattering processes also depends on the radar wavelength. C- and X-band radar data does usually not penetrate into tree canopies, while L-band data may allow for flood detection under canopies, however, usually only with low accuracy (Pierdicca et al. 2013; Hess et al. 1995, 2003).

It should be mentioned here that different terms for radar based water mapping exist. Strictly speaking, SAR sensors only receive a certain backscatter signal, which varies in intensity, and water surfaces usually yield a low (or no) return and therefore appear dark. However, different authors use different terms depending on the focus of the study. The term "water body mapping" is usually used, if the aim of the study is the derivation of a permanent water body map, or the delineation of water bodies in relatively unknown terrain. The term "flood mapping" usually refers to the delineation of water areas, which are normally not covered by water. "Flood mapping" is often undertaken during catastrophic flood events. The usage of the term at least indicates that people or infrastructure etc. are "at risk". The term "inundation" is commonly used in regions, where water coverage fluctuates (natural wetlands etc.), without harmful impacts on the human environment.

It is the goal of this study to analyse inundation dynamics in the Yellow River Delta of China. The delta is characterized by a highly dynamic coastal zone with intermittent wetlands, aquaculture, brine ponds, and partially irrigated agriculture. It shall be assessed how SAR time series can be of support to differentiate these.

20.2 Challenges with SAR Based Water Mapping

Challenges for SAR based water mapping or inundation mapping are manifold and can be grouped according to their origin into location related challenges, sensor related challenges, and data policy related challenges.

Location related challenges are circumstances in the area of observation, which hinder successful water mapping. Water can be mapped best if the water surface is smooth and undisturbed. High waves on water surfaces may lead to stronger backscatter signals, and therefore to an omission of these areas by a water detection algorithm. Furthermore, water surfaces which are covered by vegetation may yield higher return signals due to double bounce effects, or even surface scattering. This is a problem in reed covered lakes, mature wet rice fields (Kuenzer and Knauer 2013), or mangrove forests (Kuenzer et al. 2011), where the standing water below the plants can usually not be detected. Furthermore, sand dunes, streets, bare flat agricultural crop land, and radar shadow areas are easily mixed up with smooth open water areas and lead to an overestimation of the water extent (Martinis 2010).

Sensor related challenges include the wavelength of the sensor, incidence angles, and the possible polarization modes. As mentioned, C- and X-band SAR sensors have a lower penetration ability than L-band sensors and might therefore be less suited for the detection of water surfaces below canopies or vegetation in general (Hess et al. 1995, 2006). Furthermore, very low incidence angles also hamper water detection, which is why incidence angle correction is needed when analysing time series of data with varying incidence angles. Last but not least, several authors assessed impacts of polarization on water mapping results. It was found that horizontal, HH, polarized data is best suited for water detection and superior to cross polarized, HV, or vertically polarized, VV, data (Gstaiger et al. 2012; Henry et al. 2006).

However, the largest challenge for scientists aiming at time series analyses of SAR data for the assessment of water dynamics is data availability. Although a large number of radar and SAR sensors orbit our Earth, scientists can only access data of a few sensors free of charge. Most radar data is usually not accessible, or can only be ordered at high costs. Great exceptions have been the ENVISAT mission of the European Space Agency (ESA) with the ASAR sensor onboard. ASAR data has been made available free of charge in several spatial resolution modes from 2010 onwards. The data is available from 2002 to April 2012. ESA's Science Programme allowed access to all archived data and also enabled specific scheduling after submission of a project proposal for research applications. However, in archives still a lot of acquisitions were missing, and scheduling often conflicted with other data take requests (the three ASAR modes cannot be tasked in parallel). On average 10–30 ASAR WSM scenes could be collected for e.g. the Mekong Delta in the years 2007–2011, whereas – strictly according to overpass time and overlap coverage – about 40 WSM acquisitions should have been possible (Kuenzer et al. 2013a). The German Satellite TerraSAR-X provides SAR data in three different resolution modes from the year 2008 onwards. Scientists could receive data of all three modes free of charge for study areas of limited size. As TerraSAR-X data is also distributed commercially via Infoterra (now Airbus Defence and Space) to receive national, continental, or even global coverage was – and still is – not possible. Furthermore, even for restricted study areas many scenes can usually not be acquired due to competitive tasking for a commercial customer interested in other areas (TerraSAR-X does not acquire data continuously, but only when tasked). Similar to TerraSAR-X, data of the Japanese sensor ALOS Palsar 1 and 2 can be received for smaller areas upon request and the submission of scientific proposals. Due to this relatively restricted access to SAR data – especially since the failure of ENVISAT – very high hopes are now on the newly launched ESA Sentinel-1A satellite. Sentinel-1A (paired with the upcoming Sentinel-1B to be launched in 2016) is a constellation of C-band SAR satellites that deliver data in three different modes ranging from 40 m up to 5 m spatial resolution.

20.3 Study Area: The Yellow River Delta

The Yellow River Delta is located in Shandong Province in the East of China, where the Yellow River enters the Bohai Sea (see Fig. 20.2). It is the vulnerable end of a river whose total stream length exceeds 5,460 km. The delta itself covers an

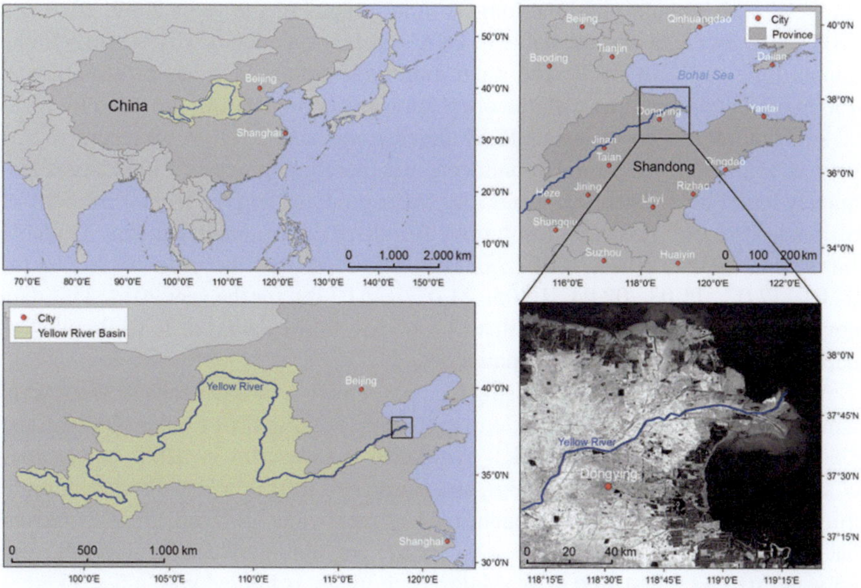

Fig. 20.2 Study area: the Yellow River Delta located in Shandong Province, China (Source: Kuenzer et al. 2014)

area of about 10,000 km^2 and hosts several larger towns; among them Dongying City – Dongying District's capital city – with over two million inhabitants.

Dongying District contains the river mouth and can be considered the delta district. It is characterized by large wetlands hosting a rich biodiversity on the one hand, as well as intensive agriculture, aquaculture, and oil fields, as well as scattered towns and villages on the other hand. Details on Yellow River Delta coastal processes, land use, oil production, and current in-situ research can be found in Kuenzer et al. (2014), and Ottinger et al. (2013). One major interest of stakeholders in the region is an improved understanding of inundation dynamics in the delta district. Dongying District is not only dissected by the Yellow River itself, but hosts a large variety of water bodies of different size and function. SAR time series analyses is one possible means to visualize the differences in inundation frequency and relate them to different types of land use.

20.4 Data and Methods: Employing WaMaPro

We performed the analyses of inundation dynamics for the Yellow River Delta, based on an annual time series of ENVISAT ASAR WSM data of 18 scenes, acquired in 2008. Dates of acquisition were 24.04., 27.04., 10.05., 14.05., 23.07., 07.08., 08.08., 10.08., 11.08., 26.08., 11.09., 15.09., 01.10., 16.10., 01.11., 04.11., 20.11., and 20.12.2008. Within the last years of ENVISAT acquisitions 2008 was the year, which could yield the longest time series. For the years 2009, 2010 and 2011 much less data was available in ESA's archives, which is why we present the annual time series of 2008. The radar amplitude data, which has a resolution of 150 m, was geo-corrected with the open source software BEAM, provided by ESA.

As communication with ENVISAT failed from early 2012 on, no recent ASAR data is available. For recent monitoring we therefore processed inundation frequency based on 13 thematic water / no water products derived from TerraSAR-X Stripmap data of 3 m spatial resolution. TerraSAR-X Stripmap data was available for 15.04., 29.05., 20.06., 12.07., 03.08., and 08.10. for the year 2013, and for 04.01., 17.02., 11.03., 02.04., 16.05., 21.07., and 03.09. for the year 2014. This data covers a relatively smaller area than the ASAR WSM data, but is well suited to depict local patterns with great spatial detail.

Water surfaces were derived using the open source software package WaMaPro, which has been developed at the German Remote Sensing Data Center (DFD) of the German Aerospace Center (DLR). WaMaPro is described in detail in Gstaiger et al. (2012), Kuenzer et al. (2013a), and Huth et al. (2015). WaMaPro uses simple thresholding and filtering approaches to extract water covered areas from the backscatter images. The output is a binary file containing values for water (1) and no-water (0). As a first step in WaMaPro each image is filtered with a 5×5 kernel standard convolution median filter to suppress the speckle, which is typical for radar imagery (P1). After this step two empirically chosen thresholds separate water

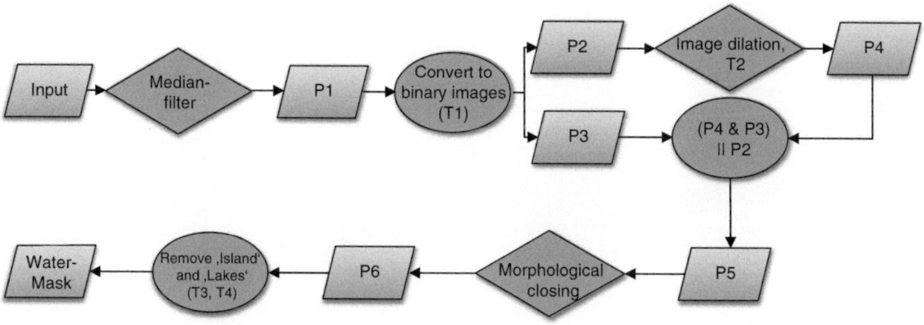

Fig. 20.3 Processing flow of DFD-DLRs software WaMaPro

from non-water pixels in a first approximation. The first threshold, T1, which has a lower value than the final water threshold, supports the definition of confident water pixels. Based on this intermediate product, a second threshold, T2, which has a higher value than the land threshold, supports the definition of confident land areas (P3). The two intermediate products (confident water and confident land) are the analyzed further. Buffer zones of two pixels, generated via dilatation, are applied to P2, resulting in product P4. The buffers define the transition zone from water to land, also represented by mixed pixels. The second threshold now enables the inclusion of the water pixels within this zone in the initial binary water mask. The temporary results P3 and P4 are now compared, and if coincidence occurs, the value (water or land, 0 or 1) is written to P5. Otherwise, the value from P2 is written to P5 (P4 & P3 || P2). In this way overestimated water pixels are excluded. Proceeding further, isolated pixels are removed via morphological image closing (P6) (see Fig. 20.3). The removal of so-called 'islands' and 'lakes' according to a defined maximum size (T3, T4) is mainly of relevance for higher resolution SAR data (e.g., TerraSAR-X data also processed in the course of the project), but does not affect ENVISAT ASAR derived results at 150 m resolution (Kuenzer et al. 2013a).

Once all water masks had been derived from the ASAR WSM and the TerraSAR-X data, the binary file for each of the sensors were added up, to yield products, depicting, how often a pixel has been flooded with respect to the number of observations available.

20.5 Results

Figure 20.4 depicts the results of inundation mapping based on the 18 ENVISAT ASAR WSM observations available for 2008. Areas west of the coastline, which have never been inundated (water covered) appear in white, whereas areas, which

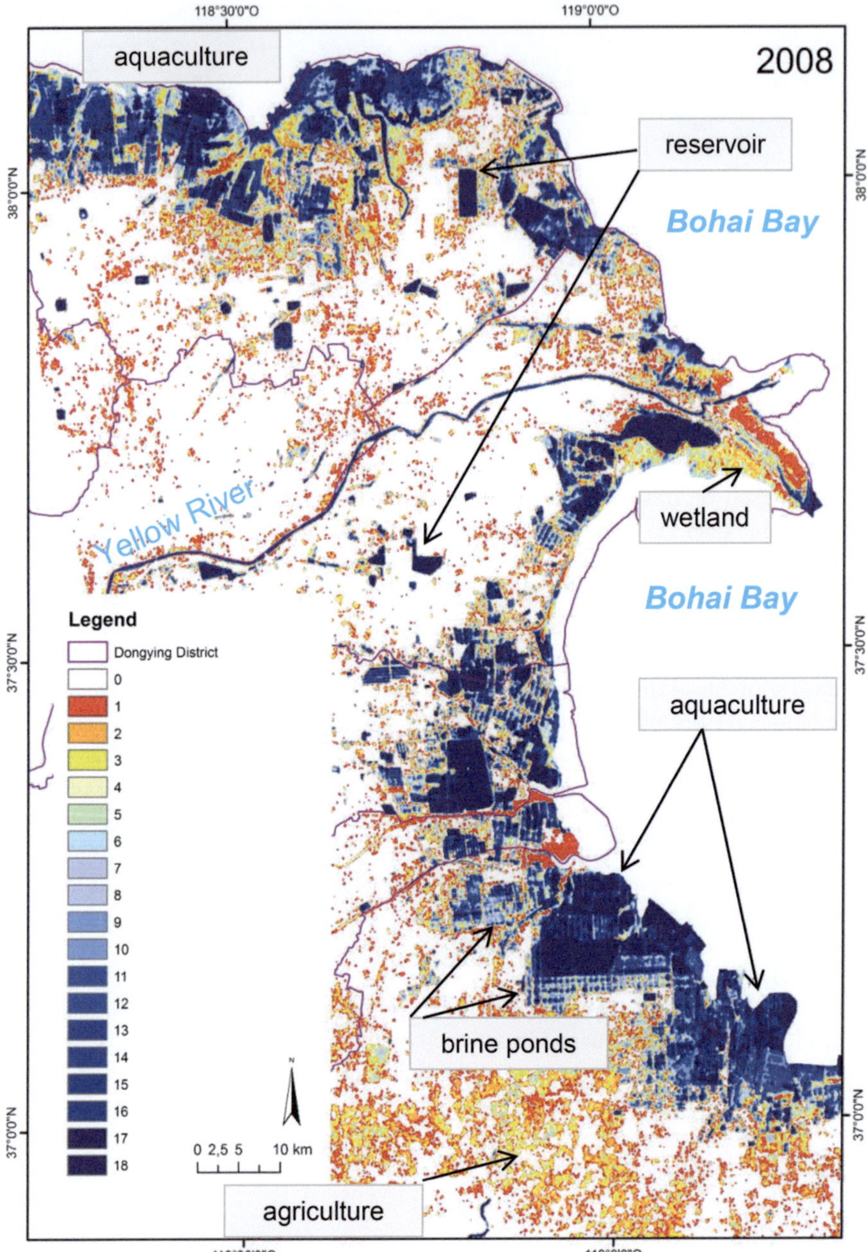

Fig. 20.4 Inundation frequency in the Yellow River Delta of China, as derived from 18 ENVISAT ASAR Wide Swath Mode data sets for the year 2008. Aquaculture, water reservoirs, brine ponds, and agriculture fields can be differentiated (Please note that Dongying District boundary outline (*purple*) includes mudflat areas of the ocean, which nowadays – after construction of a large sea dyke – do not belong to the land area anymore)

are always water covered during all 18 data takes appear in dark blue. Regions, which are rarely inundated, are presented in reddish, orange, and yellowish tones, whereas regions, which have been inundated for about half of the time, appear in light blue tones. The Yellow River dissects Dongying District and separates the northern part of the district from the southern part. North of the river, long-term inundated dark blue objects further inland are water storage reservoirs, whereas the frequently inundated areas along the northern coast of the delta are aquaculture areas. Reddish and yellowish regions in the central northern coastal zone are managed wetlands, which are rarely flooded. South of the Yellow River the river mouth area is characterized by rarely fully flooded wetland, and a permanent oval-shaped lagoon directly south of the river mouth. All frequently inundated (dark blueish) areas along the eastern coast of the Yellow River Delta are aquaculture areas, with scattered occurrences of brine ponds (salt farming), which occur in light blue tones, as they fall try during parts of the year. Reddish and yellowish region in the South are agricultural areas, which are rarely inundated (due to precipitation or irrigation). These findings were verified during field work in the study area. The SAR based time series allows for the differentiation of different land use types based on inundation frequency:

- River: always water covered
- Water storage basins: always water covered
- Aquaculture: always inundated or inundated for the largest part of the year (unless basins are cleaned or water is completely renewed), located at the coast
- Brine ponds: inundated for about half of the time, as basins need to fall dry so that salt can be harvested; located close to the coast
- Agricultural fields: rarely inundated and further away from the coast

Strictly speaking, the SAR time series should be combined with object oriented knowledge (size and shape of segments), and information on distance to the coast, to discern the above feature automatically. However, the inundation time series already allows for the differentiation of land use, which would not be possible based on optical data exclusively.

This is also demonstrated by Figs. 20.5 and 20.6. Both figures present a smaller area located directly south of the Yellow River mouth, where inundation patterns were mapped for 2013/2014 based on TerraSAR-X Stipmap data. Note that this region in 2013 can strongly differ from general patterns observed in 2008, as the Yellow River Delta is a highly dynamic environment, constantly reshaped by its inhabitants. Fig. 20.5 clearly lets us discern permanently inundated areas (river, canal, aquaculture), from less frequently inundated brine ponds (Fig. 20.6, zoom B), and from agriculture areas. In Fig. 20.5 (and 20.6) we can see that the area North of the Yellow River is characterized by agricultural fields, and also two dyked little oil fields exist (see upper zoom A in Fig. 20.6), whereas South of the river aquaculture and brine ponds prevail along the coast, whereas agricultural field characterize the hinterland.

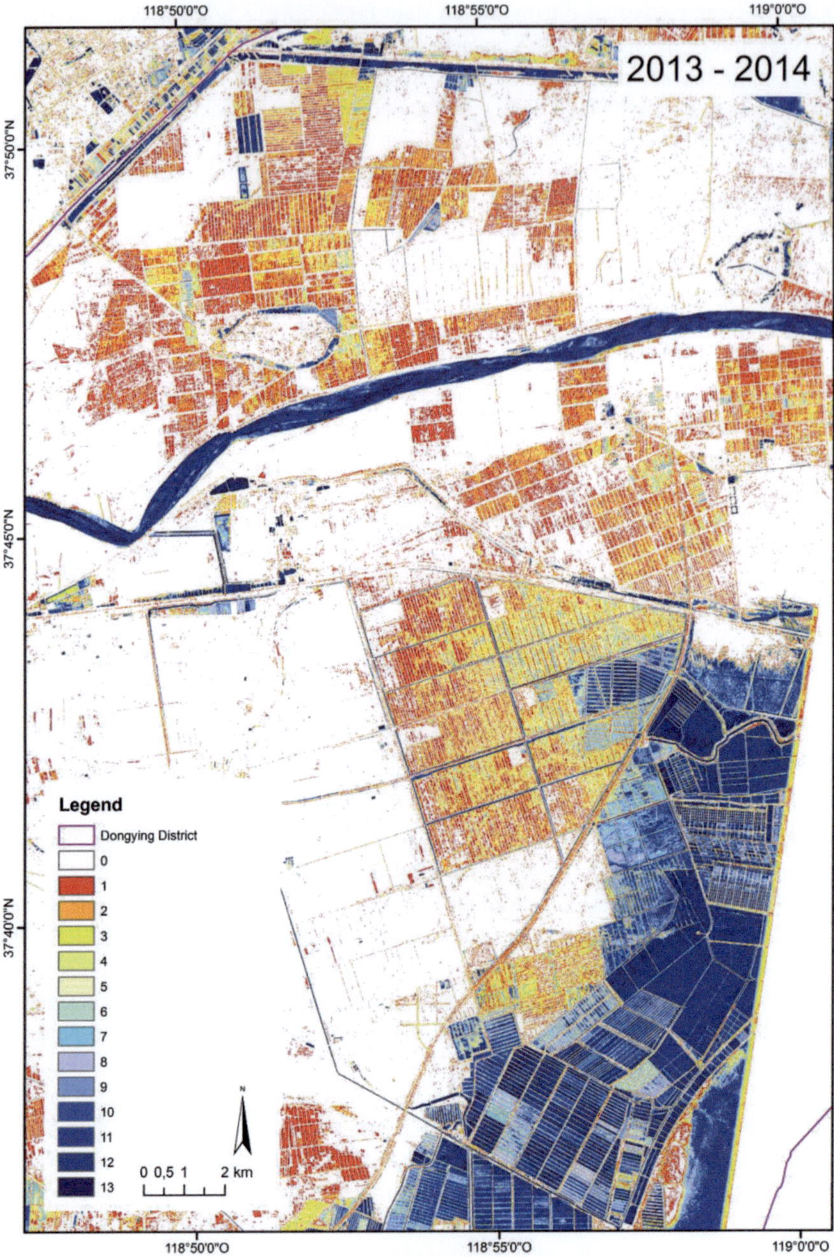

Fig. 20.5 Inundation frequency in the Yellow River Delta of China, as derived from 13 - TerraSAR-X Stripmap data sets for the year 2013/2014

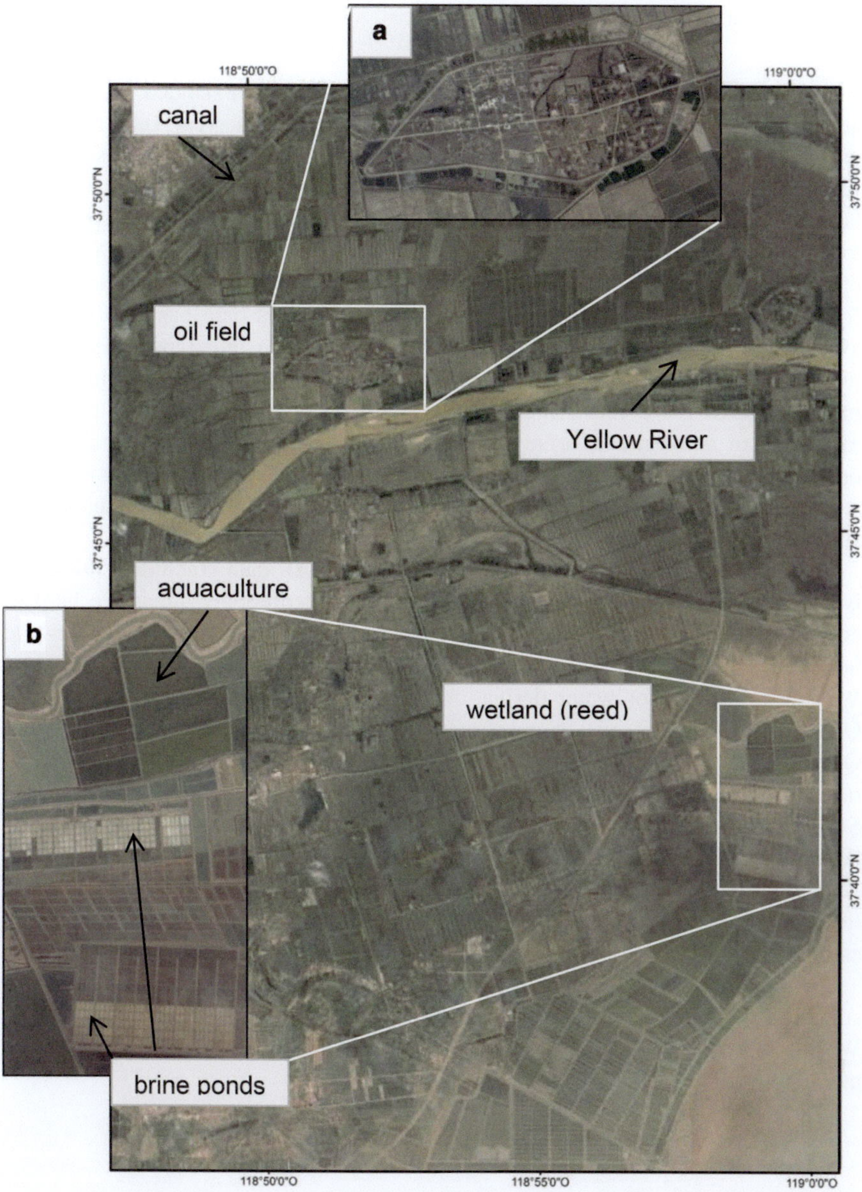

Fig. 20.6 Appearance of the area depicted in Fig. 20.5 in high resolution optical QuickBird data. Most water surface areas cannot be differentiated based on their optical appearance

20.6 Discussion

The brief study presented here could of course be well extended for analyses addressing inundation timing (first onset of inundation in the year, time of dry out etc.), and – if detailed land use maps for the respective years would exist – could furthermore directly related to such data products. However, for the year 2008 no land use map for the Yellow River Delta is available, and the delta is too dynamic as to use a product from another date. For validation of the TerraSAR-X derived product at 3 m resolution land use information derived from highest resolution data for the year 2013/2014 would be needed, however, this data has also not been available at the time of this study. However, frequent field campaigns into the area – including in October 2013 – allow us to define the functions of the individual inundated areas. We furthermore see a large potential for the fully automatic extraction of aquaculture, brine ponds, irrigated agriculture, and wetlands based on SAR derived inundation time series when combined with automatic image segmentation and analyses (size and shape of water polygons) and distance measures (e.g. to the coastline).

A weakness of most water mapping studies with SAR data - including this one – is a precise quantitative validation. Usually inundated/flooded areas are hard to access, and it is rare that scientists or stakeholder have the chance to map water-land boundaries exactly during the satellite overpass time. However, results of the tool WaMaPro utilized for water derivation from SAR data have been validated by Gstaiger et al. (2012) via comparison with highest resolution TerraSAR-X and QuickBird data as well as water-land boundary maps, resulting in detection accuracies of 95 %. However, as mentioned in the section on challenges with respect to SAR based water mapping, inundated areas can of course not be picked up if heavily covered by vegetation. As the Yellow River Delta contains larger wetlands near the river mouth, where shallow water is completely covered by reeds, it is likely that the inundated areas presented here are rather slightly under- than overestimated.

Furthermore, it should be mentioned that we consider 18 observations per year to still be a relatively low number of observations for "true" time series analyses. Weekly or even daily SAR observations at a resolution better than 100 m would be much preferred.

20.7 Conclusion

Time series of thematic "water / no water" information derived from SAR data of sensors such as ENVISAT ASAR WSM and TerraSAR-X Stripmap are a suitable means to characterize inundation patterns. For the Yellow River Delta of China they allow for the differentiation of various types of inundation. Aquaculture areas are usually water-covered throughout the whole year, whereas brine ponds (salt farming) fall dry for some parts of the year as salt needs to be harvested. Irrigated agricultural fields are only water covered during very few days of the year or after heavy rain events. Wetlands are typically characterized by alternating inundated and dry periods. These regions are most likely to be underestimated, as the Yellow River Delta contains many wetland areas, where reeds and other wetland plants cover shallow lagoons. In the Yellow River Delta large areas of aquaculture spread along the northern and eastern coast of the delta. Smaller brine pond fields can be found in surrounding of this use. The river mouth itself is characterized by a natural wetland environment, and the areas further inland of Dongying District are rarely inundated. Here only the Yellow River itself, a few canals, as well as some water reservoirs appear as constantly water covered areas.

As ENVISAT ASAR data is not available anymore since April 2012, and as TerraSAR-X data can usually only be received for smaller regions at random timely intervals, we currently centre our hopes on ESA's freely available Sentinel-1 data at resolutions higher than 40 m.

Acknowledgements This study was undertaken in the context of the DELIGHT project (www.delight.eoc.dlr.de) funded by the German Federal Ministry of Education and Research, BMBF. We are furthermore grateful for the constructive comments of two anonymous reviewers.

References

Amelung F, Galloway DL, Bell JW, Zebker HA, Laczniak RJ (1999) Sensing the ups and downs of Las Vegas—InSAR reveals structural control of land subsidence and aquifer-system deformation. Geology 27:483–486

Askne J, Santoro M (2005) Multitemporal repeat pass SAR interferometry of boreal forests. IEEE Trans Geosci Remote Sens 43(6):1219–1228

Cartus O, Santoro M, Schmullius C, Li Z (2011) Large area forest stem volume mapping in the boreal zone using synergy of ERS-1/2 tandem coherence and MODIS vegetation continuous fields. Remote Sens Environ 115:931–943

Chaouch N, Temimi M, Hagen S, Weishampel J, Medeiros S, Khanbilvardi R (2011) A synergetic use of satellite imagery from SAR and optical sensors to improve coastal flood mapping in the Gulf of Mexico. Hydrol Process 26:1617–1628

Dorigo W, deJeu R, Chung D, Parinussa R, Liu Y, Wagner W, Fernández-Prieto D (2012) Evaluating global trends (1988–2010) in harmonized multi-satellite surface soil moisture. Geophys Res Lett 39(L18405):1–7

Dostálová A, Doubková M, Sabel D, Bauer-Marschallinger B, Wagner W (2014) Seven years of Advanced Synthetic Aperture Radar (ASAR) Global Monitoring (GM) of surface soil moisture over Africa. Remote Sens 6(8):7683–7707

Doubková M, Dostálová A (née Hegyiová), van Dijk AIJM, Blöschl G, Wagner W, Fernández-Prieto D (2014) How do spatial scale, noise, and reference data affect empirical estimates of error in ASAR-derived 1 km resolution soil moisture? IEEE J Sel Top Appl Earth Obs Remote Sens 7(9):3880–3891. doi:10.1109/JSTARS.2014.2324657

Fielding EJ, Blom RG, Goldstein RM (1998) Rapid subsidence over oil fields measured by SAR interferometry. Geophys Res Lett 27:3215–3218

Greifeneder F, Wagner W, Sabel D, Naeimi V (2014) Suitability of SAR imagery for automatic flood mapping in the Lower Mekong basin. Int J Remote Sens 35(8):2857–2874

Gstaiger V, Gebhardt S, Huth J, Wehrmann T, Kuenzer C (2012) Multi-sensoral and automated derivation of inundated areas using TerraSAR-X and ENVISAT ASAR data. Int J Remote Sens 33(22):7291–7304

Henry JB, Chastanet P, Fellah K, Desnos YL (2006) Envisat multi-polarized ASAR data for flood mapping. Int J Remote Sens 27(10):1921–1929

Hess LL, Melack JM, Filoso S, Wang Y (1995) Delineationg of inundated area and vegetation along the Amazon floodplain with the SIR-C synthetic aperture radar. IEEE Trans Geosci Remote Sens 33(4):896–904

Hess LL, Melack JM, Novo EMLM, Barbosa CCF, Gastil M (2003) Dual-season mapping of wetland inundation and vegetation for the central Amazon basin. Remote Sens Environ 87:404–428

Higgins S, Overeem I, Tanaka A, Syvitski J (2013) Land subsidence at aquaculture facilities in the Yellow River Delta, China. Geophys Res Lett 40(15):3898–3902. doi:10.1002/grl.50758

Hoque R, Nakayama D, Matsuyama H, Matsumoto J (2010) Flood monitoring, mapping and assessing capabilities using RADARSAT remote sensing, GIS and ground data for Bangladesh. Nat Hazards 58:525–548

Huth J, Gebhardt S, Ahrens M, Hoffmann J, Kuenzer C (2015) Water surface derivation with WaMaPro: an easy to use radar data processing tool. Int J Remote Sens

Kasischke ES, Smith KB, Bourgeau-Chavez LL, Romanowicz EA, Brunzell S, Richardson CJ (2003) Effects of seasonal hydrologic patterns in south Florida wetlands on radar backscatter measured from ERS-2 SAR imagery. Remote Sens Environ 88(4):423–441

Kiage LM, Walker ND, Balasubramanian S, Babin A, Barras J (2005) Applications of Radarsat 1 synthetic aperture radar imagery to assess hurricane related flooding of coastal Louisiana. Int J Remote Sens 26(24):5359–5380

Kuenzer C, Knauer K (2013) Remote sensing of rice crop areas – a review. Int J Remote Sens 34 (6):2101–2139

Kuenzer C, Zhao D, Scipal K, Sabel D, Naeimi V, Bartalis Z, Hasenauer S, Mehl H, Dech S, Wagner W (2009) El Niño influences represented in ERS scatterometer derived soil moisture data. Appl Geogr 29(4):463–477

Kuenzer C, Bluemel A, Gebhardt S, Vo Quoc T, Dech S (2011) Remote sensing of mangrove ecosystems: a review. Remote Sens 3:878–928. doi:10.3390/rs3050878

Kuenzer C, Guo H, Leinenkugel L, Huth J, Li X, Dech S (2013a) Flood mapping and flood dynamics of the Mekong Delta: an ENVISAT-ASAR-WSM based Time Series Analyses. Remote Sens 5:687–715. doi:10.3390/rs5020687

Kuenzer C, Guo H, Schlegel I, Vo QT, Li X, Dech S (2013b) Scale and the capability of Envisat ASAR-WSM, TerraSAR-X Scansar, and TerraSAR-X stripmap data to assess urban flood situations: a case study in Can Tho Province of the Mekong Delta. Remote Sens 5:5122–5142. doi:10.3390/rs5105122

Kuenzer C, Ottinger M, Liu G, Sun B, Dech S (2014) Earth observation-based coastal zone monitoring of the Yellow River Delta: dynamics in China's second largest oil producing region over four decades. Appl Geogr 55:72–107

Lang MW, Kasischke ES, Prince SD, Pittman KW (2008) Assessment of C-band synthetic aperture radar data for mapping and monitoring Coastal Plain forested wetlands in the Mid-Atlantic Region, U.S.A. Remote Sens Environ 112(11):4120–4130

Martinez JM, Le Toan T (2007) Mapping of flood dynamics and spatial distribution of vegetation in the Amazon floodplain using multitemporal SAR data. Remote Sens Environ 108 (3):209–223

Martinis S (2010) Automatic near real-time flood detection in high resolution X-band synthetic aperture radar satellite data using context-based classification on irregular graphs. Dissertation, Ludwig-Maximilians-University Munich, Munich, Germany

Martinis S, Twele A (2010) A hierarchical spatio-temporal Markov model for improved flood mapping using multi-temporal X-band SAR data. Remote Sens 2(9):2240–2258

Martinis S, Twele A, Voigt S (2011) Unsupervised extraction of flood-induced backscatter changes in SAR data using Markov image modeling on irregular graphs. IEEE Trans Geosci Remote Sens 49(1):251–263

Martinis S, Twele A, Kersten J (2014) A fully automated TerraSAR-X based flood service. ISPRS J Photogramm Remote Sens. doi:10.1016/j.isprsjprs.2014.07.014 (in press)

Martinis S, Kuenzer C, Twele A (2015) Flood studies using synthetic aperture radar data. Remote Sens Water Resour Disasters and Urban Stud Taylor & Francis (submitted). http://www.crcpress.com/product/isbn/9781482217919

Mason DC, Speck R, Devereux B (2010) Flood detection in urban areas using TerraSAR-X. IEEE Trans Geosci Remote Sens 48(2):882–893

Naemi V, Leinenkugel P, Sabel D, Wagner W, Apel H, Kuenzer C (2013) Evaluation of soil moisture retrieval from the ERS and Metop scatterometers in the lower Mekong Basin. Remote Sens 5(4):1603–1623

Ottinger M, Kuenzer C, Liu G, Wang S, Dech S (2013) Monitoring land cover dynamics in the Yellow River Delta from 1995 to 2010 based on Landsat 5 TM. Appl Geogr 44:53–68

Pierdicca N, Pulvirenty L, Chini M, Guerriero L, Candela L (2013) Observing floods from space: Experience gained from COSMO-SkyMed observations. Acta Astronaut 3(80):122–133

Richards JA, Woodgate PW, Skidmore AK (1987) An explanation of enhanced radar backscattering from flooded forests. Int J Remote Sens 8(7):1093–1100

Santoro M, Askne J, Smith G, Fransson JES (2002) Stem volume retrieval in boreal forests from ERS-1/2 interferometry. Remote Sens Environ 81:19–35

Schumann G, Matgen P, Hoffman L, Hostache R, Pappenberger F, Pfister L (2007) Deriving distributed roughness values from satellite radar data for flood inundation modelling. J Hydrol 344:96–111

Townsend PA (2001) Mapping seasonal flooding in forested wetlands using multi-temporal radarsat SAR. Photogramm Eng Remote Sens 67(7):857–864

Wagner W, Lemoine G, Rott H (1999) A method for estimating soil moisture from ERS scatterometer and soil data. Remote Sens Environ 70:191–207

Wagner W, Luckman A, Vietmeier J, Tansey K, Balzter H, Schmullius C, Davidson M, Gaveau D, Gluck M, Le Toan T, Quegan S, Shvidenko A, Wiesmann A, Jiong Yu JJ (2003) Large-scale mapping of boreal forest in SIBERIA using ERS tandem coherence and JERS backscatter data. Remote Sens Environ 85(2):125–144

Wagner W, Hahn S, Kidd R, Melzer T, Bartalis Z, Hasenauer S, Figa-Saldanấ J, de Rosnay P, Jann A, Schneider S, Komma J, Kubu G, Brugger K, Aubrecht C, Züger C, Gangkofer U, Kienberger S, Brocca L, Wang Y, Blöschl G, Eitzinger J, Steinnocher K, Zeil P, Rubel F (2013) The ASCAT soil moisture product: a review of its specifications, validation results, and emerging applications. Meteorol Z 22(1):5–33

Wang Y (2010) Remote sensing of coastal environments. CRC Press/Taylor & Francis Group, Boca Raton, 457 pp

Werle D, Martin TC, Hasan K (2001) Flood and coastal zone monitoring in bangladesh with radarsat ScanSAR: technical experience and institutional challenges. John Hopkins APL Tech Dig 21(1):148–154

Zhao D, Kuenzer C, Fu C, Wagner W (2008) Evaluation of the ERS scatterometer derived soil water index to monitor water availability and precipitation distribution at three different scales in China. J Hydrometeorol 9:549–562

Zhou C, Luo J, Yang C, Li B, Wang S (2000) Flood monitoring using multi-temporal AVHRR and RADARSAT imagery. Photogramm Eng Remote Sens 66(5):633–638

MIX
Papier aus verantwortungsvollen Quellen
Paper from responsible sources
FSC® C105338

FSC
www.fsc.org

If you have any concerns about our products,
you can contact us on
ProductSafety@springernature.com

In case Publisher is established outside the EU,
the EU authorized representative is:
Springer Nature Customer Service Center GmbH
Europaplatz 3, 69115 Heidelberg, Germany

Printed by Libri Plureos GmbH
in Hamburg, Germany